Physics of Gravitating Systems   I

Equilibrium and Stability

A. M. Fridman
V. L. Polyachenko

Physics of
Gravitating Systems   I

Equilibrium and Stability

Translated by
A. B. Aries and Igor N. Poliakoff

With 85 Illustrations

Springer-Verlag
New York Berlin Heidelberg Tokyo

A. M. Fridman
V. L. Polyachenko
Astrosovet
Ul. Pyatnitskaya 48
109017 Moscow Sh-17
U.S.S.R.

Translators
A. B. Aries
4 Park Avenue
New York, NY 10016
U.S.A.

Igor N. Poliakoff
3 Linden Avenue
Spring Valley, NY 10977
U.S.A.

Library of Congress Cataloging in Publication Data
Fridman, A. M. (Alekseĭ Maksimovich)
  Physics of gravitating systems.
  Includes bibliography and index.
  Contents: 1. Equilibrium and stability
—2. Nonlinear collective processes. Astrophysical applications.
  1. Gravitation. 2. Equilibrium. 3. Astrophysics.
I. Poliachenko, V. L. (Valeriĭ L'vovich) II. Title.
QC178.F74 1984   523.01   83-20248

This is a revised and expanded English edition of: *Ravnovesie i ustoĭchivost' gravitiruĭushchikh sistem.* Moscow, Nauka, 1976.

Typeset by Composition House, Ltd., Salisbury, England.
Printed and bound by R. R. Donnelley & Sons, Harrisonburg, Virginia.

9 8 7 6 5 4 3 2 1

ISBN 978-3-642-87832-9   ISBN 978-3-642-87830-5 (eBook)
DOI 10.1007/978-3-642-87830-5

Dedication

This book, which has been written by physicists, originated from the ancient problems of astronomy which are related to the shapes and evolution of celestial bodies and their systems.

It has become a tradition among physicists to study individual fields of astrophysics. About 30 years ago this would have been a great event. At that time Professor D. A. Frank-Kamenetsky was one of the first Soviet physicists to begin to develop consistently the fundamental problems of astrophysics. Being a very erudite scientist he succeeded in influencing, by his passionate inquisitiveness, quite a number of young physicists who chose astrophysics as their speciality.

We dedicate this book to the memory of D. A. Frank-Kamenetsky—a prominent scientist and a man of pure soul.

Preface[1]

It would seem that any specialist in plasma physics studying a medium in which the interaction between particles is as distance-dependent as the interaction between stars and other gravitating masses would assert that the role of *collective effects* in the dynamics of gravitating systems must be decisive. However, among astronomers this point of view has been recognized only very recently. So, comparatively recently, serious consideration has been devoted to theories of galactic spiral structure in which the dominant role is played by the orbital properties of individual stars rather than collective effects. In this connection we would like to draw the reader's attention to a difference in the scientific traditions of plasma physicists and astronomers, whereby the former have explained the delay of the onset of controlled thermonuclear fusion by the "intrigues" of collective processes in the plasma, while many a generation of astronomers were calculating star motions, solar and lunar eclipses, and a number of other fine effects for many years ahead by making excellent use of only the laws of Newtonian mechanics. Therefore, for an astronomer, it is perhaps not easy to agree with the fact that the evolution of stellar systems is controlled mainly by collective effects, and the habitual methods of theoretical mechanics in astronomy must make way for the method of self-consistent fields.

---

[1] This extended preface to the Russian edition (1976) reflects the achievements in all the topics referred to therein over the last eight years. It especially concerns Part II, specifically the theory of multi-component systems and the problems of the nonlinear theory, the foundations of which were described in the Russian edition (for instance, the nonlinear theory of a gravitating disk, and the nonlinear waves and solitons traveling in such a disk).

Small oscillations of the medium can be considered as the simplest phenomenon in which collective effects are essential. The main purpose of this book is to treat systematically the theory of *small oscillations, equilibrium and stability* of gravitating systems. The theory to be presented below already has a variety of applications. Especially widespread recognition has been gained by its application to the problem of galactic spiral structure. Application of the results of the stability theory of gravitating disks with a central body to the problem of the law of planetary distances or of the rings of Saturn has gained a widespread reputation.

In this book the first priority is assigned to systems composed of a large number of gravitating masses *not colliding* with each other. Stellar systems of various sizes (for example, galaxies or globular star clusters), first of all, belong to such *collisionless* gravitating systems.

The foundations of the statistical mechanics of stellar systems (physical statistics of particles interacting in accordance with Newton's law) were laid down in the 1930s by Ambartsumian [128^{ad}]. One of the fundamental results in stellar dynamics (the proof of the "short-time scale" [129^{ad}]) had been derived while examining the evolution of stellar systems which makes use of the distribution function in phase space while depending on the fundamental integrals of motion. The approach of statistical mechanics of stellar systems is largely used in this book.

Theoretical investigations in stellar dynamics, particularly the problem of the stability of star systems, have risen to a completely new level upon penetration into this field of the research methods developed earlier in plasma physics. It is not out of place to mention here the pioneering investigations of collective effects in a gravitating medium carried out by Antonov [4], Lynden-Bell [281], Sweet [330], and others. They introduced the idea of instability as the spontaneous excitation of collective modes of oscillations of the medium, widely used in the theory of plasma instabilities. Thus, obsolete works devoted to the stability of orbits of individual particles were discarded from the theory of stability of gravitating systems, and the "one-particle" description of Newtonian mechanics was replaced by the statistical method for describing systems of many particles, which makes use of the distribution function concept that satisfies the kinetic equation. The *dispersion equation* method, usually applied in plasma theory, became the main method for studying stability. Also, many methods of solving the kinetic equation were taken from plasma theory—the method of "integration over trajectories," widely used in this book, may be cited as an example. Finally, in the theory of gravitational instability, different approximated research methods, such as the WKB method (or the method of local analysis), also became habitual.

The period of general passion for plasma methods and their somewhat one-sided application to gravitation was followed by a period of more attention to the possibilities of the theory, which have also revealed deep differences between plasma and the gravitating medium, are due to the difference in sign of interaction of electrically charged particles and gravitating masses and to the absence of screening of gravitational attraction.

We treat the question of the analogy between the plasma and gravitating media in the Introduction, where we also give the basic equations of the theory, and where the problem of gravitational (*Jeans*) instability of a *homogeneous* medium is discussed.

In Chapter I, which also has an introductory character, the stability theory of the simplest gravitating system—a homogeneous collisionless *flat layer*—is set forth. The methods used later can be readily illustrated, in particular, by this example.

Chapters II–IV deal with a study of equilibrium and stability of four "classical" figures of equilibrium: cylinder, sphere, ellipsoid, and disk.

In the first paragraphs of each chapter a review of results of the *equilibrium* theory of relevant systems is provided. It should be noted that the equilibrium theory of collisionless gravitating systems started to develop much earlier than stability theory. The works of Eddington [196, 197], Jeans [241, 242], Camm [179, 180], Freeman [202–204], and others played the main role in the formation of this theory. Primarily, works by Kuz'min [58–62], Weltmann [32], Bisnovatyi-Kogan and Zel'dovich [21–23] should be distinguished among the works of Soviet investigators.

Small perturbations of *cylindrical* configurations are considered in Chapter II. The investigation of stability of a collisionless cylinder with respect to arbitrary perturbations has been carried out by the authors and Mikhailovskii; also the beam instability in a gravitating medium was investigated in collaboration with Mikhailovskii [88]. "Flute"-type oscillations have been discussed by the authors (together with Shukhman) [112, 115] as well as by Antonov [14].

Chapter III is devoted to *spherically symmetric systems*. In this chapter we present the results which were obtained primarily by our group and by Antonov.

Chapter IV gives the analysis of stability of *ellipsoidal* gravitating systems, carried out by the authors together with Morozov and Shukhman.

The problem of the stability of *disk* systems (Chapter V) drew the attention of a large number of investigators.

As far as *applications* of the theory of stability of a gravitating disk to the problem of spiral structure formation of galaxies are concerned, the basic ideas and conclusions belong to Lin and Shu [269–272], Kalnajs [250, 289], Toomre [333, 334], and Lynden-Bell [289]. In spite of the undoubtedly interesting results achieved in these areas, we have arrived at the opinion that even the qualitative picture of the formation of spiral structure of galaxies depicted now is the only possible one.[2] For instance, in the process of the airal structure formation a distinct role can be played by *instabilities of the non-Jeans type*, which are discussed in Chapter VI of this book. These instabilities have been investigated by the authors together with Morozov,

---

[2] This opinion is shared by many authors (see, e.g., [11, 303, 304]). Reviews related to the problem of the spiral structure of galaxies also point out difficulties of the existing theory (e.g., [84]).

Fainstein, and Shukhman as well as by Bisnovatyi-Kogan and Mikhailovskii and Mark and Kulsrud.

Chapter VII is notable for the discussion of some questions of nonlinear theory. It presents the basic problems, which have now been solved (jointly with Mikhailovskii, Petviashvili, Frenkel, Churilov, and Shukhman). One section of this chapter is dedicated to the results of the "pancake" theory as derived by Zel'dovich and his co-workers.

Some astrophysical applications of the theory of equilibrium and stability of gravitating systems are treated in Part II.

We decided that it might be useful to have, for reference, tables of all the instabilities studied hitherto which can develop in various gravitating systems. There are two such tables in the book. The first one characterizes the Jeans instability of a multicomponent homogeneous medium. In commentaries to the table, among other things, a theorem is formulated of the Jeans instability in a multicomponent medium at rest. (The theorem of the number of instabilities for the general case of *moving* components is formulated in §1 of Chapter VI.) The second table characterizes all the non-Jeans instabilities and is placed in the last section (§12) of the Appendix. It contains: the geometry of the system considered, instability conditions, instability growth rates, typical frequencies of perturbed waves, and references to works in which this or that instability was first described. In commentaries to the table we describe briefly the physical mechanisms responsible for non-Jeans instabilities.

At the end of most of the chapters is a series of problems (and their solutions) which throw some light on details of mechanisms either operating in gravitating systems or somewhat removed from the basic class of questions.

It seemed reasonable to number the formulae separately within each section (and even in subsections), because references to formulae from other sections are comparatively rare.

Of course, we were unable to present the results of all authors on the subject of interest. When selecting any given work, we proceeded, first of all, from the reliability of results obtained in it. Therefore, we have omitted papers written on an almost purely "intuitive" level and those not supported by strict mathematical analysis. There is also considerable duplication of results. In such cases we selected, as a rule, the original work. It is possible, however, that taking into account the reservations made above, not all the papers which deserve to be mentioned have been cited in the book. We express our sincere apologies to the authors of those papers.

In order to read this book a general knowledge of the kinetic equation and of the basic concepts of potential theory is sufficient. Of course, for the reader familiar with the principles of plasma instability theory it will be easier to orientate himself in the subject of the book. Nevertheless, when writing the book, we have not supposed the reader to have any special knowledge of the physics of plasma instabilities (or of gravitational stability theory).

We acknowledge with gratitude our joint work with A. B. Mikhailovskii,

R. Z. Sagdeev, and Ya. B. Zel'dovich without whose support and advice this book would not have come into being.

The general plan of the book was discussed with A. B. Mikhailovskii, L. M. Ozernoy, and Ya. B. Zel'dovich, who made a number of suggestions which we tried to take into account in the course of our work on the manuscript.

We would especially like to thank our colleagues and friends I. G. Shukhman, A. G. Morozov, S. M. Churilov, and V. S. Synakh. Their valuable contribution is not restricted to providing the above-mentioned results but extends to the actual writing of some sections of Chapter VII. I. G. Shukhman together with A. G. Morozov wrote Section 2.2 and, together with S. M. Churilov, §4 and the supplement to it. Section 3 of this chapter is devoted to the results of the "pancake" theory obtained by Ya. B. Zel'dovich and his colleagues. This section was written at the authors' request by A. G. Doroshkevich. V. S. Synakh co-authored our first papers with the use of a computer. We wish to convey our sincere gratitude to all of them.

We are also grateful to G. I. Marchuk and A. G. Massevitch for their support during the preparation of the English edition.

We thank L. A. Chujanova for her great help in the design of the manuscript.

Moscow
April, 1984

A. FRIDMAN
V. POLYACHENKO

Contents (Volume I)

Introduction   1
§ 1. Basic Concepts and Equations of Theory   2
§ 2. Equilibrium States of Collisionless Gravitating Systems   6
§ 3. Small Oscillations and Stability   9
§ 4. Jeans Instability of a One-Component Uniform Medium   10
§ 5. Jeans Instability of a Multicomponent Uniform Medium   14
     5.1. Basic Theorem (on the Stability of a Multicomponent System with
         Components at Rest)   16
     5.2. Four Limiting Cases for a Two-Component Medium   17
     5.3. Table of Jeans Instabilities of a Uniform Two-Component Medium   18
     5.4. General Case of n Components   19
§ 6. Non Jeans Instabilities   19
§ 7. Qualitative Discussion of the Stability of Spherical, Cylindrical (and
     Disk-Shaped) Systems with Respect to Radial Perturbations   21

PART I

Theory

CHAPTER I
Equilibrium and Stability of a Nonrotating Flat Gravitating Layer   27
§ 1. Equilibrium States of a Collisionless Flat Layer   28
§ 2. Gravitational (Jeans) Instability of the Layer   31

§ 3. Anisotropic (Fire-Hose) Instability of a Collisionless Flat Layer 37
    3.1. Qualitative Considerations 37
    3.2. Derivation of the Dispersion Equation for Bending Perturbations
        of a Thin Layer 38
    3.3. Fire-Hose Instability of a Highly Anisotropic Flat Layer 40
    3.4. Analysis of the Dispersion Equation 41
    3.5. Additional Remarks 42
§ 4. Derivation of Integro-Differential Equations for Normal Modes of a
    Flat Gravitating Layer 43
§ 5. Symmetrical Perturbations of a Flat Layer with an Isotropic
    Distribution Function Near the Stability Boundary 50
§ 6. Perpendicular Oscillations of a Homogeneous Collisionless Layer 53
    6.1. Derivation of the Characteristic Equation for Eigenfrequencies 53
    6.2. Stability of the Model 63
    6.3. Permutational Modes 68
    6.4. Time-Independent Perturbations ($\omega = 0$) 69
Problems 69

CHAPTER II
Equilibrium and Stability of a Collisionless Cylinder 78
§ 1. Equilibrium Cylindrical Configurations 79
§ 2. Jeans Instability of a Cylinder with Finite Radius 83
    2.1. Dispersion Equation for Eigenfrequencies of Axial-Symmetrical
        Perturbations of a Cylinder with Circular Orbits of Particles 84
    2.2. Branches of Axial-Symmetrical Oscillations of a Rotating Cylinder
        with Maxwellian Distribution of Particles in Longitudinal Velocities 85
    2.3. Oscillative Branches of the Rotating Cylinder with a Jackson
        Distribution Function (in Longitudinal Velocities) 92
    2.4. Axial-Symmetrical Perturbations of Cylindrical Models of a More
        General Type 95
§ 3. Nonaxial Perturbations of a Collisionless Cylinder 97
    3.1. The Long-Wave Fire-Hose Instability 97
    3.2. Nonaxial Perturbations of a Cylinder with Circular Particle Orbits 100
§ 4. Stability of a Cylinder with Respect to Flute-like Perturbations 104
§ 5. Local Analysis of the Stability of Cylinders (Flute-like Perturbations) 109
    5.1. Dispersion Equation for Model (2), § 1 110
    5.2. Maxwellian Distribution Function 115
§ 6. Comparison with Oscillations of an Incompressible Cylinder 117
    6.1. Flute-like Perturbations ($k_z = 0$) 118
§ 7. Flute-like Oscillations of a Nonuniform Cylinder with Circular Orbits
    of Particles 119
Problems 125

CHAPTER III
Equilibrium and Stability of Collisionless Spherically Symmetrical
Systems 136
§ 1. Equilibrium Distribution Functions 138
§ 2. Stability of Systems with an Isotropic Particle Velocity Distribution 152
    2.1. The General Variational Principle for Gravitating Systems with the
        Isotropic Distribution of Particles in Velocities
        ($f_0 = f_0(E), f_0' = df_0/dE \leq 0$) 152

2.2. Sufficient Condition of Stability   155
2.3. Other Theorems about Stability. Stability with Respect to Nonradial
     Perturbations   158
2.4. Variational Principle for Radial Perturbations   160
2.5. Hydrodynamical Analogy   161
2.6. On the Stability of Systems with Distribution Functions That Do
     Not Satisfy the Condition $f_0'(E) \leq 0$ 163
§ 3. Stability of Systems of Gravitating Particles Moving On Circular
     Trajectories   164
3.1. Stability of a Uniform Sphere   164
3.2. Stability of a Homogeneous System of Particles with Nearly Circular
     Orbits   173
3.3. Stability of a Homogeneous Sphere with Finite Angular Momentum   174
3.4. Stability of Inhomogeneous Systems   179
§ 4. Stability of Systems of Gravitating Particles Moving in Elliptical Orbits   186
4.1. Stability of a Sphere with Arbitrary Elliptical Particle Orbits   186
4.2. Instability of a Rotating Freeman Sphere   190
§ 5. Stability of Systems with Radial Trajectories of Particles   193
5.1. Linear Stability Theory   193
5.2. Simulation of a Nonlinear Stage of Evolution   199
§ 6. Stability of Spherically Symmetrical Systems of General Form   207
6.1. Series of the Idlis Distribution Functions   209
6.2. First Series of Camm Distribution Functions (Generalized
     Polytropes)   219
6.3. Shuster's Model in the Phase Description   231
§ 7. Discussion of the Results   235
Problems   238

CHAPTER IV
Equilibrium and Stability of Collisionless Ellipsoidal Systems   246
§ 1. Equilibrium Distribution Functions   248
1.1. Freeman's Ellipsoidal Models   248
1.2. "Hot" Models of Collisionless Ellipsoids of Revolution   256
§ 2. Stability of a Three-Axial Ellipsoid and an Elliptical Disk   265
2.1. Stability of a Three-Axial Ellipsoid   265
2.2. Stability of Freeman Elliptical Disks   272
§ 3. Stability of Two-Axial Collisionless Ellipsoidal Systems   276
3.1. Stability of Freeman's Spheroids   276
3.2. Peebles–Ostriker Stability Criterion. Stability of Uniform
     Ellipsoids, "Hot" in the Plane of Rotation   284
3.3. The Fire-Hose Instability of Ellipsoidal Stellar Systems   290
3.4. Secular and Dynamical Instability. Characteristic Equation for
     Eigenfrequencies of Oscillations of Maclaurin Ellipsoids   294
Problems   296

CHAPTER V
Equilibrium and Stability of Flat Gravitating Systems   323
§ 1. Equilibrium States of Flat Gaseous and Collisionless Systems   327
1.1. Systems with Circular Particle Orbits   327
1.2. Plasma Systems with a Magnetic Field   334

1.3. Gaseous Systems   337
1.4. "Hot" Collisionless Systems   338
§ 2. Stability of a "Cold" Rotating Disk   343
2.1. Membrane Oscillations of the Disk   343
2.2. Oscillations in the Plane of the Disk   362
§ 3. Stability of a Plasma Disk with a Magnetic Field   380
3.1. Qualitative Derivation of the Stability Condition   380
3.2. Variational Principle   381
3.3. Short-Wave Approximation   385
3.4. Numerical Analysis of a Specific Model   387
§ 4. Stability of a "Hot" Rotating Disk   389
4.1. Oscillations in the Plane of the Disk   389
4.2. Bending Perturbations   401
4.3. Methods of the Stability Investigation of General Collisionless
      Disk Systems   404
4.4. Exact Spectra of Small Perturbations   412
4.5. Global Instabilities of Gaseous Disks. Comparison of Stability
      Properties of Gaseous and Stellar Disks   428
Problems   434

References   445

Additional References   459

Index   465

Contents (Volume II)

CHAPTER VI
Non-Jeans Instabilities of Gravitating Systems

CHAPTER VII
Problems of Nonlinear Theory

PART II

Astrophysical Applications

CHAPTER VIII
General Remarks

CHAPTER IX
Spherical Systems

CHAPTER X
Ellipsoidal Systems

CHAPTER XI
Disk-Like Systems. Spiral Structure

CHAPTER XII
Other Applications

Appendix

References
Additional References
Index

Introduction

The theory of gravitational stability has a history rich in events associated with the names of Laplace, Kovalevskaya, Maxwell, Lyapunov and Poincaré. These authors have created a fundamental and mathematically elegant theory of figures of equilibrium and stability of different geometrical shapes of *incompressible* fluid (sphere, ellipsoid, cylinder, ring, etc.). The results of their research have been summarized in a large number of reviews and monographs [15, 64, 77, 127]. Most recently, this work has been continued by Chandrasekhar and his co-workers, who have somewhat amplified and, above all, systematized the old results [148].

At first glance it seems paradoxical that up to the present time astronomers have made use only of a negligible portion of the enormous heritage of classical results of gravitational stability theory concerning deliberately stable figures of equilibrium approximating spherical shape. Mathematically, the most beautiful results (in particular, those about the critical figures of equilibrium of bi- and tri-axial ellipsoids which separate stable and unstable solutions) were not used in applications because of the absence of suitable astronomical objects. As a matter of fact, the applicability of the approximation of incompressible fluids is essentially limited to the study of equilibrium figures and oscillations of planets,[1] whose matter is in the solid phase, which is a rarity in the Universe. Most of the material is either in a very

---

[1] The program of the construction of the "bifurcation" theory of the planets with satellites (Poincaré) or that of double stars (Darwin's "division theory") which, for a long time were considered promising, proved after all, as is known, to be inconsistent [148].

1

rarefied plasma (or gaseous) state or collected into stars, clusters of which, e.g., galaxies, are, to a high degree of accuracy, *collisionless*. The well-known classical results obtained in incompressible fluid approximation already differ qualitatively from those obtained in the collisionless approximation, generally speaking, in the theory of equilibrium figures (even more so, when their stability is studied). Thus, for example, it is known that for an incompressible fluid spherically-symmetric configurations do not exist: a rotating sphere of incompressible fluid must inevitably evolve, thereby changing into a biaxial ellipsoid (and then into a triaxial ellipsoid if the initial angular momentum exceeds the critical one). In contrast, among collisionless systems stable rotating spherically symmetric systems [125] are known, indicating the principal absence of a relationship between the shape of the collisionless gravitating system and the angular momentum [49, 279].

Thus, the creation of the theory of gravitational stability of a *compressible* medium was connected with the need for a description of most of the objects in the Universe. A decisive step in this direction was taken, as is well known, by Jeans [242], who was the first to investigate the stability of homogeneous distributions of matter. Below we will turn to the Jeans problem and examine it in some detail but, first of all, let us systematize the subject.

§ 1 Basic Concepts and Equations of Theory

We shall be dealing with systems of a large number (N) of interacting particles which move in accordance with Newton's Second Law. For the description of such systems, different approximated models, statistical in nature, are employed. Instead of using exact positions and velocities of each particle at any time, these models employ a comparatively small number of quantities averaged in a specific way. The validity of either model depends on the character of the system under study. Such parameters as the number of particles, their density, the collision frequency of particles with each other, and so on are essential. Besides, the method of description also depends on characteristic times and spatial scales of those processes which we would like to study. Equations for any model must, under corresponding assumptions, be derived either directly from Newtonian equations of motion or from other models whose validity has been established earlier on the basis of the same Newtonian equations.

One of the most widely known models is *hydrodynamics* by use of which the motions of continuous media, fluid or gas, are described. In the hydrodynamical description the state of the system is characterized by the density $\rho(\mathbf{r}, t)$, pressure $P(\mathbf{r}, t)$, and velocity of motion of the fluid $\mathbf{v}(\mathbf{r}, t)$ at a given point. The basic hydrodynamical equation is the continuity equation

$$\frac{\partial \rho}{\partial t} + \mathrm{div}(\rho \mathbf{v}) = 0 \tag{1}$$

and the Euler equation

$$\frac{d\mathbf{v}}{dt} \equiv \frac{\partial \mathbf{v}}{\partial t} + (\mathbf{v}\nabla)\mathbf{v} = -\frac{1}{\rho}\frac{\partial P}{\partial \mathbf{r}} + \mathbf{F}, \tag{2}$$

where $\mathbf{F}$ is the force per unit mass, different from the pressure force, which we separated in the right-hand side of (2). For example, in magnetohydrodynamics, which describes the motion of a highly conducting fluid,

$$\mathbf{F} = \frac{1}{4\pi\rho}[\text{rot } \mathbf{H} \cdot \mathbf{H}], \tag{3}$$

and for particles, having charge e and moving in the electric and magnetic fields with strength $\mathbf{E}$ and $\mathbf{H}$,

$$\mathbf{F} = \frac{e}{m}\left(\mathbf{E} + \frac{1}{c}[\mathbf{v}\mathbf{H}]\right) \tag{4}$$

is the Lorentz force.

Most frequently $\mathbf{F}$ will imply the Newtonian gravity force; then

$$\mathbf{F} = -\text{grad } \Phi(\mathbf{r}, t), \tag{5}$$

where Φ is the gravitational potential.

The hydrodynamical description at $\mathbf{F} = 0$ is completed by adding to (1) and (2) a certain equation of state. But, if $\mathbf{F} \neq 0$, one has then to include the equations which describe the relationship of $\mathbf{F}$ with ρ and $\mathbf{v}$. In the plasma case these will be Maxwell's equations; in the case of gravitating medium, the Poisson equation

$$\Delta\Phi \equiv \frac{\partial^2\Phi}{\partial x^2} + \frac{\partial^2\Phi}{\partial y^2} + \frac{\partial^2\Phi}{\partial z^2} = 4\pi G\rho \tag{6}$$

(G = gravity constant).

The thermal motion of particles is most correctly taken into account by the methods developed in the kinetic theory. The kinetic method of description is also statistical in nature (systems, e.g., stellar, are regarded as statistical), but it is more general compared to hydrodynamics, so that hydrodynamical equations can be derived from kinetics.

The main quantity which is used in the statistical description of the behavior of systems of many particles in kinetics is the distribution function of particles $f(\mathbf{r}, \mathbf{v}, t)$. It has the meaning of mass (or number) of particles which are at an instant t in the unit element of volume of phase space in the vicinity of the point $(\mathbf{r}, \mathbf{v})$. Thus, according to this definition the mass dm (or the number dn) of particles in the range $\mathbf{r}, \mathbf{r} + d\mathbf{r}, \mathbf{v}, \mathbf{v} + d\mathbf{v}$ is equal to

$$dm(dn) = f(\mathbf{r}, \mathbf{v}, t)\, d\mathbf{r}\, d\mathbf{v}. \tag{7}$$

Objects of quite different nature may be "particles" in different cases. For instance, they may be stars, star clusters, galaxies and even galaxy clusters. The function f has been averaged over fluctuations which inevitably result from the thermal motion of the particles. Note that according to the definition the distribution function f must be nonnegative. We shall assume for simplicity that the masses of individual particles M are equivalent and equal unity: $M = 1$ (therefore we do not write M in the argument of the distribution function f).

The distribution function satisfies the Boltzmann kinetic equation

$$\frac{df}{dt} \equiv \frac{\partial f}{\partial t} + (\mathbf{v}\nabla_r)f + \mathbf{F}\frac{\partial f}{\partial \mathbf{v}} = C. \tag{8}$$

Here $\mathbf{F}$ denotes as before the force which acts upon the particle, d/dt is the total derivative along the (exact!) trajectory of the particle in the field of the force $\mathbf{F}$ ($d/dt = \partial/\partial t + \mathbf{v} \cdot \partial/\partial \mathbf{r} + d\mathbf{v}/dt \cdot \partial/\partial \mathbf{v}$, $d\mathbf{v}/dt = \mathbf{F}/M$, $M = 1$), and C is the so-called collisional term which defines the change of f due to collisions between particles.

The form of the collision term is dependent on the specific problem. In the collisionless approximation, which we shall mainly use below, it is assumed that $C = 0$. The *collisionless kinetic equation* obtained as a result is nothing other than the equation of continuity in phase space. This denotes the "incompressible" character of the flow of the "phase fluid"—the distribution function f remains unaltered:

$$\frac{df}{dt} = \frac{\partial f}{\partial t} + (\mathbf{v}\nabla_r)f + \mathbf{F}\frac{\partial f}{\partial \mathbf{v}} = 0 \tag{9}$$

along the particle path in phase space.

The density $\rho(r, t)$ is obtained by integration of the distribution function over the velocities:

$$\rho(\mathbf{r}, t) = \int f(\mathbf{r}, \mathbf{v}, t)\, d\mathbf{v}. \tag{10}$$

The kinetic equation (9), Poisson equation (6), and relation (10) form a complete set of equations which describes collisionless gravitational systems. This set is called the set of equations with a self-consistent field, or Vlasov's equations.

This set is quite analogous to the corresponding equations for a plasma. The interaction force between *electrically* charged particles (charges e_1 and e_2) is determined by Coulomb's Law

$$F = +\frac{e_1 e_2}{r^2}. \tag{11}$$

The Newtonian law of universal gravitation

$$F = -\frac{Gm_1 m_2}{r^2}, \tag{12}$$

and Coulomb's Law (11) coincide in all except the sign (and, of course, the relative force) of the interaction. The plus sign in (11) means that charges of like sign are repulsed and those of opposite signs are attracted. At the same time gravitational "charges" (i.e., masses m_1, m_2, ..., which are always "charges of the same sign"), according to (12), are attracted. Consequently, the Poisson equation for the electric potential ψ

$$\Delta\psi = -4\pi\rho, \tag{13}$$

differs from the Poisson gravitational equation (6) by the sign on the right-hand side.

The kinetic equation for a rarefied plasma medium is written in the form of (9) with force $\mathbf{F}$, which is given by expression (4). For potential motions of a one-component plasma in the absence of a magnetic field (when $\mathbf{E} = -\text{grad } \psi$) the kinetic equations in the two cases just examined coincide.

We remind the reader that the collisionless kinetic equation (9), describing the evolution of the single-particle distribution function $f = f(\mathbf{r}, \mathbf{v}, t)$, can be strictly derived from the exact kinetic equation (Liouville's equation) for the N-particle distribution function $\mathscr{F} = \mathscr{F}(\mathbf{r}_1, \mathbf{v}_1; \mathbf{r}_2, \mathbf{v}_2; \dots ; t)$ depending on the coordinates and velocities of all N particles of the system

$$\frac{\partial \mathscr{F}}{\partial t} + [\mathscr{F}, H] = 0, \tag{14}$$

where the Poisson bracket is

$$[\mathscr{F}, H] = \sum_\alpha \left(\frac{\partial \mathscr{F}}{\partial q_\alpha} \frac{\partial H}{\partial p_\alpha} - \frac{\partial \mathscr{F}}{\partial p_\alpha} \frac{\partial H}{\partial q_\alpha} \right),$$

H is the Hamiltonian of the system and q_α, p_α are the generalized coordinates and impulses. Equation (14) can also be given as

$$\frac{\partial \mathscr{F}}{\partial t} + \sum_\alpha v_\alpha \frac{\partial \mathscr{F}}{\partial x_\alpha} + \sum_\alpha F_\alpha \frac{\partial \mathscr{F}}{\partial v_\alpha} = 0. \tag{15}$$

Note that Liouville's theorem (15) is a rigorous result, ensuing from the Hamiltonian formulation of Newtonian mechanics. Equation (9) can be derived by integration of Eq. (15) over the phase coordinates of all particles, except one, on the assumption that F is represented as a product of the one-particle distribution function

$$\mathscr{F} = f(\mathbf{r}_1, \mathbf{v}_1, t) f(\mathbf{r}_2, \mathbf{v}_2, t) \cdots f(\mathbf{r}_N, \mathbf{v}_N, t). \tag{16}$$

Assumption (16), implying the statistical independence of phase distributions of different particles, is justified in the case when the interaction forces between individual pairs of particles ("collisions") are negligibly small as compared with the regular "smoothed" force $\mathbf{F}$. The latter may be due either to external influence or to the collective self-consistent action of all particles of the system. We shall mainly be interested in the latter case. The field is called self-consistent because it is produced by a particle distribution f which, in its turn, is determined by the same field.

As is well known [86, 138], for the plasma case the condition for neglecting collisions is expressed as the requirement that the number of particles in the sphere of Debye radius $r_d = \sqrt{T_e/4\pi n e^2}$ (n = density, T_e = electron temperature) is large, i.e.,

$$nr_d^3 \gg 1. \tag{17}$$

The Debye radius r_d determines the size of the unscreened interaction region: particles which are beyond the limits of the Debye sphere practically do not interact with those lying at its center. Therefore, the presence of r_d in criterion (17) is quite natural. In the gravitational case when, for example, a system, composed of N stars is considered in the volume with size R, it is merely required that N itself be large, $N \gg 1$, since the effect of screening is absent and the pair interaction must be, roughly speaking, N times (and more exactly $N/\ln N$ times) weaker than the collective interaction.

Note that in the particular case of a "cold" medium the kinetic description by means of Eqs. (9), (10) is equivalent to the hydrodynamical description [to Eqs. (1), (2) without pressure, $P = 0$]. The distribution function is then $f_0 \sim \delta[\mathbf{v} - \mathbf{v}_0(\mathbf{r})]$. The continuity equation (1) is derived from the kinetic equation (9) by integration over the velocities, and the Euler equation (2) by integration of the kinetic equation, multiplied by $\mathbf{v}$. Besides, such a result is fairly obvious even without a formalized derivation since there is no dispersion of thermal velocities in a cold medium.

The kinetic equation (9) is written in vector form. But when specified problems are solved, one has to write the kinetic equation (as well as the second basic equation of the theory—the Poisson equation) in some curvilinear coordinate system, and the choice is determined by the symmetry of the problem. In §1 of the Appendix we give the collisionless kinetic equations (and Poisson equation), written in different curvilinear coordinates as well as in the rotating reference system, which is convenient, for example, when studying the local properties of the stability of rotating systems.

Now consider separately the equations which describe equilibrium states and small oscillations of gravitating systems.

§2 Equilibrium States of Collisionless Gravitating Systems

These are described by the distribution function $f_0(\mathbf{r}, \mathbf{v})$, mass density $\rho_0(\mathbf{r})$ and by the related gravitational potential $\Phi_0(\mathbf{r})$.

Evidently they must satisfy the simplified system of Vlasov's equations (6), (9), (10), where $\partial/\partial t = 0$ is to be assumed:

$$\mathbf{v}\frac{\partial f_0}{\partial \mathbf{r}} - \frac{\partial \Phi_0}{\partial \mathbf{r}}\frac{\partial f_0}{\partial \mathbf{v}} = 0, \tag{18}$$

$$\Delta\Phi_0 = 4\pi G\rho_0, \tag{19}$$

$$\rho_0 = \int f_0 \, d\mathbf{v}. \tag{20}$$

Usually those systems which have some kind of spatial symmetry (e.g., flat or spherical) are considered. Allowance for symmetry simplifies, of course, to a great extent the general formulation of the problem contained in the set of equations (18)–(20). Note also that these equations describe the states stationary in an inertial system being at rest. In principle, configurations may exist to be stationary in a certain noninertial, e.g., rotating, coordinate system. We consider in detail such an example in Chapter IV — these are rotating triaxial ellipsoids. In the present case obviously there can be no stationarity in the inertial system since in it even the boundary of such an ellipsoid has to change with time. However, an equilibrium state exists in the coordinate system rotating with a definite angular velocity Ω. The system of functions f_0, Φ_0, ρ_0, which determines the steady state of this stellar ellipsoid, must satisfy Eqs. (19), (20) and the stationary kinetic equation in a rotating coordinate system (see §1 of the Appendix or Chapter IV).

Turn now to system (18)–(20). The kinetic equation (18), if considered as an equation for the distribution function f_0 by assuming the potential Φ_0 to be known, is a homogeneous differential equation in partial derivatives of the first order. The equations of particle motion under the action of the gravity force $(-\partial\Phi_0/\partial r)$ are the characteristics of this equation:

$$\frac{d\mathbf{r}}{dt} = \mathbf{v}, \qquad \frac{d\mathbf{v}}{dt} = -\frac{\partial\Phi_0}{\partial \mathbf{r}}. \tag{21}$$

Accordingly, a general solution for the kinetic equation (18) is an arbitrary function of the particle motion integrals in the field $\Phi_0(\mathbf{r})$.[2] There is also the natural physical requirement of unambiguity of the distribution function at any point of phase space of the system which, in turn, requires those integrals of motions which may serve as arguments of the distribution function to be single-valued. The equations of the characteristics (21) can (in Cartesian coordinates) be rewritten as

$$\frac{dx}{v_x} = \frac{dy}{v_y} = \frac{dz}{v_z} = \frac{dv_x}{-\partial\Phi_0/\partial x} = \frac{dv_y}{-\partial\Phi_0/\partial y} = \frac{dv_z}{-\partial\Phi_0/\partial z} = dt. \tag{22}$$

From (22) it is evident that there are all together six independent integrals, and, moreover, five of them do not contain the time t. Generally speaking, however, these integrals are single-valued in the local sense only: when continued analytically along the particle motion trajectory, their values may not coincide at different loops of the trajectory.

In general, only those integrals are single-valued whose existence is associated with the properties of uniformity of time (energy) or of homogeneity or isotropy of space (momenta and angular momenta, respectively). The rest of the integrals of motion are, as a rule, non-single-valued.

Thus, the particle energy in the field Φ_0 can always serve as one of the arguments of f_0:

$$E = v^2/2 + \Phi_0. \tag{23}$$

---

[2] This result is proved in the theory of equations of partial derivatives (see, e.g., [156]).

If the system we study possesses rotational symmetry around a certain axis z, one more single-valued integral of motion appears: the z-component of the particle angular momentum $L_z = rv_\varphi$ (in cylindrical coordinates). In the case of full spherical symmetry, the *vector* of angular momentum $\mathbf{L}$ is the integral, and, in particular, its value L [$L^2 = r^2(v_\theta^2 + v_\varphi^2)$ in spherical coordinates].

However, if the system is invariant with respect to translations along some direction (x), then the x-component of momentum p_x (or of velocity v_x) is the integral. Such systems must obviously be infinite. A plane layer, infinite and uniform in x, y (discussed in Chapter I), and a cylinder with an infinite generatrix (Chapter II) may serve as examples.

Systems having any additional symmetry properties may also have some additional single-valued integrals of motion. In this book, when studying equilibrium states of many simple systems, we shall often be faced with such a situation. In this connection we have to enumerate the integrals of motion — the distribution function arguments of the systems at issue — separately in each specific case.

With this we shall finish the general discussion of the kinetic equation (18). Let us only make one remark that will show that the existence of additional single-valued integrals of motion different from energy and angular momentum is not necessarily associated with some oversimplification or, perhaps, unrealistic assumption of the model systems mentioned. The remark concerns a classical example — the problem of the so-called *third integral* of motion. This problem arises in the determination of the minimum number of integrals of motion on which the distribution function of stars in the Galaxy may depend (see, e.g., [7]). The Galaxy is thought to possess axial symmetry with respect to the rotation axis (z). Consequently, one could attempt to construct models of the Galaxy in which the distribution function would depend merely on two obviously single-valued integrals of motion: the energy $E = (v_r^2 + v_\varphi^2 + v_z^2)/2 + \Phi_0(r, z)$ and the z-component of the angular momentum $L_z = rv_\varphi, f_0 = f_0(E, L_z)$. But v_r and v_z enter such a distribution function absolutely symmetrically, whence it follows, in particular, that the stellar velocity dispersion in r- and z-coordinates should coincide at any point. This, however, contradicts observations according to which the vertical component dispersion c_z of the velocity is less than the dispersion of the radial component c_r: $c_z < c_r$. Therefore (if one assumes the Galaxy to be stationary or near-stationary) there must exist at least one more, the third, integral, the dependence upon which would introduce the required asymmetry. It is the search for this integral that constitutes the problem not completely solved so far. It may, incidentally, be noted (see [7]) that in fact this third integral must, in the present case, also be the last one.

We shall now turn again to the system of equations (18)–(20). The potential Φ_0 which enters (18) and upon which, consequently, the distribution function f_0 is dependent, represents a self-consistent gravitational field: it is determined by the combined action of all the stars of the system. The need for a self-consistent solution makes the problem of finding the distribution function rather complicated so that a relatively small number of accurate solutions have been found hitherto. They are considered in the forthcoming chapters of this book.

§ 3 Small Oscillations and Stability

Consider now the general formulation of the problem concerning small oscillations of the system. Assume that at a certain instant the system which has been in an equilibrium state before, is driven out of this state. Then, at subsequent instants, it will be described by the functions f, Φ, ρ, which can be presented in the form

$$f = f_0 + f_1, \qquad \Phi = \Phi_0 + \Phi_1, \qquad \rho = \rho_0 + \rho_1. \qquad (24)$$

Here the quantities with index "0" correspond to the equilibrium unperturbed state, and those with index "1" characterize deviations from equilibrium. Let us call them disturbances. As a rule, we consider disturbances as small:

$$f_1 \ll f_0, \qquad \Phi_1 \ll \Phi_0, \qquad \rho_1 \ll \rho_0. \qquad (25)$$

Then, linearizing Eqs. (9), (10), (6), i.e., by substituting expansions (24) in them and neglecting small square terms, we obtain, on the one hand, the set of equations (18)–(20) which describe the initial steady state and, on the other hand, the set of equations for disturbances:

$$\frac{df_1}{dt} \equiv \frac{\partial f_1}{\partial t} + \mathbf{v}\,\frac{\partial f_1}{\partial \mathbf{r}} - \frac{\partial \Phi_0}{\partial \mathbf{r}}\frac{\partial f_1}{\partial \mathbf{v}} = \frac{\partial \Phi_1}{\partial \mathbf{r}}\frac{\partial f_0}{\partial \mathbf{v}}, \qquad (26)$$

$$\Delta\Phi_1 = 4\pi G\rho_1, \qquad (27)$$

$$\rho_1 = \int f_1\,d\mathbf{v}. \qquad (28)$$

In Eq. (26) df_1/dt now denotes the time derivative along the *unperturbed trajectory* (21).

The problem is simplified significantly if one seeks the solution for this system with time dependence of all disturbances of the type:

$$f_1(\mathbf{r}, t) = f_1(\mathbf{r})e^{-i\omega t}, \qquad \Phi_1(\mathbf{r}, t) = \Phi_1(\mathbf{r})e^{-i\omega t},$$
$$\rho_1(\mathbf{r}, t) = \rho_1(\mathbf{r})e^{-i\omega t}, \qquad \omega = \text{const.} \qquad (29)$$

Formally, the possibility of such a representation of the solution is associated with the fact that the coefficients of equations (26)–(28) are not time-dependent; that, in turn, is due to the stationarity of the initial equilibrium state. Upon substitution of (29) the operators $\partial/\partial t$ become numbers: $\partial/\partial t \rightarrow -i\omega$. Strict substantiation of such an approach is achieved by the Laplace method with the help of which the problem with the initial conditions is solved.

By solving the set of equations (26)–(28) with relevant boundary conditions, one finds the spectrum ω of frequencies at which the equations have non-trivial solutions. These frequencies are called *eigenfrequencies* and the solutions themselves *eigenfunctions*.

In general, the frequency ω turns out to be complex: $\omega = \omega_0 + i\gamma$, so that the time dependence of disturbances is f_1, Φ_1, $\rho_1 \sim e^{i\omega_0 t}e^{\gamma t}$. If $\gamma < 0$, the disturbances are damped out in time (and γ is called a damping *decrement*). At $\gamma = 0$ we deal with harmonic neutral oscillations. If, however, at least for one of the eigenoscillations of the system, the imaginary part of the frequency is positive ($\gamma > 0$), such a system is unstable: The corresponding disturbances increase exponentially in time and become formally infinite as $t \to \infty$. In this case γ is called an *increment* of instability.

To prove the *stability* of the system, apart from the absence of roots with $\gamma > 0$, one has to make sure of the absence of multiple real roots which may lead to power ($\sim t^n$) instabilities. Usually we restrict ourselves to a demonstration of the absence of *exponential* instability but leave the possibility of power instabilities open. A more thorough analysis is performed for a flat layer (Chapter I) and a cylinder (Chapter II).

§ 4 Jeans Instability of a One-Component Uniform Medium

It now appears logical to study, for the sake of illustration, small oscillations and stability of some simple gravitating system.

The simplest system imaginable is an infinite (in all directions) homogeneous medium. It was this system that was investigated first for stability in the Jeans classical work (see [242]).

Jeans assumed, as a consequence of the homogeneity and isotropy of a Newtonian world, uniformly filled with matter, that the gravitational force at any point is equal to zero and the system is in steady state.

Here it should be said that these suggestions of Jeans cannot be entirely correct because in reality the homogeneous infinite gravitating medium (in contrast to, e.g., the plasma case) ought to be nonstationary (see below). Therefore, strictly speaking, the Jeans problem must be solved on the *nonstationary* background, which, of course, somewhat alters the results. Nonetheless, the main qualitative conclusions drawn by Jeans prove to be justified also for an accurate consideration of stability of a contracting or expanding homogeneous world. The mathematical formulation of this problem is different on the whole, but the physics of instability studied by Jeans and now bearing his name remains unaltered. It involves a simple graphic meaning: gravitational contraction of rather large masses due to gravity.

This problem lies somewhat outside the main line of our book, devoted to the study of instabilities of *steady-state* systems. However, in this example it is most easy to introduce some concepts, important for further investigation (those of Jeans instability, critical wavelength, etc.). This simple example is also essential for explaining the basic difference between gravitating and plasma media. For these reasons we give below in some detail the solution for the Jeans problem in a traditional variant which belongs to the author himself. Then we shall mention some variations resulting from the non-stationary background.

The simplicity of the homogeneous case is due to the fact that the spatial dependence of perturbations may be chosen in the form of a plane wave: $f_1, \Phi_1, \rho_1 \sim e^{ikr}$ ($\mathbf{k}$ is a wave vector). Then the initial differential equations become a set of algebraic equations. By equating the determinant of this system to zero, we obtain the relation between the frequency of eigen-oscillations ω and the wave vector $\mathbf{k}$: $\omega = \omega(\mathbf{k})$—the *dispersion equation*.

For the description of perturbations of a "cold" medium, one can use the linearized hydrodynamical equations (1), (2) with pressure equal to zero:

$$\frac{\partial \mathbf{v}_1}{\partial t} = -\frac{\partial \Phi_1}{\partial \mathbf{r}}, \tag{30}$$

$$\frac{\partial \rho_1}{\partial t} + \rho_0 \, \mathrm{div} \, \mathbf{v}_1 = 0, \tag{31}$$

$$\Delta \Phi_1 = 4\pi G \rho_1. \tag{32}$$

Employing the above method,[3] the dispersion equation is readily obtained from this system:

$$\omega^2 = -\omega_0^2. \tag{33}$$

From Eq. (33) it follows that the considered gravitational system is unstable since one of the roots $\omega = +i\omega_0$ corresponds to the growth of perturbations. It is this conclusion that was reached by Jeans. Correspondingly, this main instability of a gravitating medium is termed the *Jeans instability*, and ω_0 is called the *Jeans frequency*.

We have already mentioned that the simplifications adopted by Jeans are not quite correct. This is most obvious in the case of cold matter. Indeed, an infinite system is logically considered as one resulting from the finite one with an unlimited increase in dimensions. But all such systems are non-stationary: they must contract due to gravity; moreover, the characteristic contraction time of a cold system is independent of its size. This is also evident from Eq. (33), which does not include the wavenumber k (or the wavelength $\lambda = 2\pi/k$). Therefore, a limiting system must also be considered as a nonstationary one as in the case of Newtonian cosmology [48].

Turn now, however, to the Jeans model of a *homogeneous* medium, but assume it now to be "hot." If a gaseous medium which possesses some finite pressure {which may be taken into account by adding to the right-hand

---

[3] In particular, in this case one may also proceed in the following way. Applying the operation div to Eq. (30) and expressing div $\mathbf{v}_1$ from (31), we get (taking into account that div grad $\Phi_1 = \Delta \Phi_1$ and making use of the Poisson equation) $\partial^2 \rho_1/\partial t^2 = \omega_0^2 \rho_1$, $\omega_0^2 = 4\pi G \rho_0$. Substituting $\rho_1 \sim e^{-i\omega t}$, we again have $\omega^2 = -\omega_0^2$.

side of Eq. (30) the term $[-(1/\rho_0)/\partial p_1/\partial \mathbf{r}]\}$ is considered, then for perturbations with the wave vector $\mathbf{k}$, it is easy to obtain the following dispersion equation instead of (33) (see, e.g., [48]):

$$\omega^2 = -\dot{\omega}_0^2 + k^2 D^2, \tag{34}$$

where D is the sound velocity in the medium $D^2 = \partial P_0/\partial \rho_0$. From this dispersion equation it follows that there is a definite critical wavelength λ_{cr} (which is also called the *Jeans wavelength*) which separates stable (at $\lambda < \lambda_{cr}$) and unstable ($\lambda > \lambda_{cr}$) perturbations. It is inferred from the condition $\omega^2 = 0$; therefore,

$$\lambda_{cr} = \frac{2\pi}{k_{cr}} = D\sqrt{\frac{\pi}{G\rho_0}}. \tag{35}$$

The dispersion equation for a cold medium (33) follows from (34) at $k \ll k_{cr}$, i.e. for the perturbations on a sufficiently large spatial scale. At the same time perturbations with a small wavelength propagate like sound waves:

$$\omega^2 \approx k^2 D^2. \tag{36}$$

Such behavior of perturbations is quite natural [48]. The gravitational potential of a bunch of a given density with a characteristic size λ is proportional to λ^2: $\Phi \sim Gm/\lambda \sim G\rho\lambda^3/\lambda \sim \lambda^2$ and the force is proportional to λ: $F \sim \partial\Phi/\partial r \sim \Phi/\lambda \sim \lambda$. On the other hand, the pressure gradient obviously generates a force inversely proportional to λ. Hence, for large wavelengths gravitational forces prevail (that leads to instability) and as $\lambda \to 0$ the pressure becomes dominant.

In a similar way the question of the effect of chaotic particle velocities on Jeans instability in a *collisionless* (stellar) system is also solved. But the character of this effect is evident beforehand, if only from the analogy between the dispersion of star velocities $\langle v^2 \rangle^{1/2}$ and the velocity of sound in a gaseous medium: Jeans instability occurring in the case of cold matter at all wavelengths must be suppressed by the dispersion of velocities on a small scale but remains in force on a large scale.

Thus, the critical Jeans wavelength similar to (35) must also arise here. Qualitative differences between the gaseous and collisionless systems reveal themselves for rather short wavelengths of perturbations (see, e.g., Problem 3 in Chapter III).

Here, however, one should make a similar remark to the case of a cold medium. Determining the size of a body at rest from the condition that at its every point the force of gravity is counterweighted by the force of pressure

$$\frac{1}{\rho}\frac{\partial P}{\partial r} = -\frac{GM(r)}{r^2} \tag{37}$$

(M is the mass within radius r), we get: $R \sim \lambda_{cr}$, i.e., this size is always of the order of the critical Jeans wavelength. Any gravitating systems whose size is significantly larger than λ_{cr} are nonstationary. The problem of the stability of a uniform gravitational medium should be, strictly speaking, investigated against a nonstationary background. In conventional hydrodynamics a correct analysis was carried out during an investigation of the stability of nonstationary Friedman models of the Universe (in the framework of the general theory of relativity) in the well-known work of Lifshitz [75], and in terms of the Newtonian theory such a problem was solved by Bonnor [168]. The main results turned out to be close to those obtained by Jeans. Perturbations with $\lambda > \lambda_{cr}$ are growing, though not exponentially ($\delta\rho/\rho \sim t^{2/3}$). The nonexponential (power) manner of perturbation growth is due to the fact that the undisturbed solution is time-dependent. A clear treatment of this question is contained in the book of Zel'dovich and Novikov [48]. The time dependence of the unperturbed density ρ_0 leads to a variability in Jeans frequency $\omega_0(t) \sim \sqrt{\rho_0(t)}$. In this case a logical generalization of Jeans law of growth of perturbation density $\rho_1(t) \sim e^{\omega_0 t}$, which would be valid at $\omega_0 = \text{const}$, is the following: $\rho_1(t) \sim \exp(\int \omega_0(t)\,dt)$. Such a modification of Jeans' formulae yields, as can be shown, a law of perturbation growth very similar to the exact solution by Lifshitz and Bonnor. Of course, one could not expect full coincidence. Indeed, the generalization of Jeans' law (WKB w.r.t. time) suggested above would be strictly valid only if the rate of perturbation growth were much greater than that of the change of the unperturbed quantities. This is, generally speaking, not the case. However, let us mention again that the approach adopted by Jeans in any case leads to the correct expression (35) for the critical wavelength λ_{cr} and to qualitatively accurate predictions: instability for perturbations with $\lambda > \lambda_{cr}$ and stability at $\lambda < \lambda_{cr}$. One should only bear in mind that the results thus attained are by no means applicable to stationary systems. Stationary gravitational systems have dimensions $R \sim \lambda_{cr}$; they must be essentially nonuniform and should be investigated in another way.

This point, which is essential in all of the subsequent exposition, will probably become clearer if one makes a comparison with the plasma medium. Langmuir oscillations of cold plasma obey, as is well known, the dispersion equation

$$\omega^2 = \omega_p^2, \tag{38}$$

where $\omega_p^2 = 4\pi n e^2/m$ is the square of the plasma frequency, n is the number of particles in a volume unit, and e and m are the mass and charge of an electron. The correspondence between Eqs. (38) and (33) is evident if one recalls the analogy between gravitational and electrostatic forces.

The relation $v_T/\omega_p = r_d$ in plasma determines the Debye radius of screening, outside of which the electrical potential is practically absent. As a rule, the Debye radius is much less than the other characteristic dimensions: the free path length, the experimental chamber dimensions, etc. This circumstance defines the quasineutrality of the plasma, as a result of which the natural state of the ionized gas is an equilibrium state in which the homogeneous plasma may be found.

The dimension of the equilibrium gravitating system is of the order of $\langle v^2 \rangle^{1/2}/\omega_0$ and to provide equilibrium one needs a particular pressure gradient for this system or a rotation with a corresponding angular velocity dependence on the radius.

Therefore, a steady-state gravitating system corresponds to the interior region of the Debye sphere in a plasma. In the case of stationary gravitating systems one often has to deal with complex boundary problems while in a plasma the model of a homogeneous infinitely extended background (quasineutral) already allows one to obtain many familiar effects.[4]

§ 5 Jeans Instability of a Multicomponent Uniform Medium [64a^{ad}]

If a spatially nonfinite two-component medium consists, for example, of cold and hot gas, then two Jeans scales appear: for a cold component λ_c and for a hot one λ_h. Let us look at how disturbances with different wavelengths behave in such a heterogeneous medium. Let us restrict ourselves to the case of one-dimensional disturbances propagating along the x-axis. Thus, all the disturbed values are dependent only on x; stationary parameters are assumed to be spatially uniform.

For the cold component of the medium the linearized equations have the form[5]

$$\frac{\partial v_c}{\partial t} = -\frac{\partial \Phi}{\partial x} - \frac{1}{\rho_{0c}} \frac{\partial P_c}{\partial x},$$

$$\frac{\partial \rho_c}{\partial t} + \frac{\partial}{\partial x}(\rho_{0c} v_c) = 0.$$

For the hot component we have the analogous set of equations

$$\frac{\partial v_h}{\partial t} = -\frac{\partial \Phi}{\partial x} - \frac{1}{\rho_{0h}} \frac{\partial P_h}{\partial x},$$

$$\frac{\partial \rho_h}{\partial t} + \frac{\partial}{\partial x}(\rho_{0h} v_h) = 0.$$

These sets of equations are connected by the Poisson equation

$$\frac{\partial^2 \Phi}{\partial x^2} = 4\pi G(\rho_c + \rho_h).$$

---

[4] Of course, under a local consideration (in the vicinity of a given point) of perturbations with scales small compared to a characteristic linear dimension R of the system, one may assume parameters of the stationary state equal to its values in a given point (WKB approximation).

[5] Here and below we omit the index "1" for disturbed values.

We assume disturbances to be adiabatic and the relation between pressure and density $P = P(\rho)$ for disturbed values to be linear:

$$P = c_s^2 \rho,$$

where $c_s^2 = \partial P_0/\partial \rho_0$ is the square of the sound velocity. Evidently, for each component of the medium we have its own equation of state.

For disturbances of the form $\sim e^{ikx - i\omega t}$, one can obtain

$$\rho_c = \rho_{0c} \frac{k^2 \Phi}{\omega^2 - k^2 c_{sc}^2},$$

$$\rho_h = \rho_{0h} \frac{k^2 \Phi}{\omega^2 - k^2 c_{sh}^2}.$$

Using the Poisson equation, we obtain the dispersion relation

$$\frac{\omega_{0c}^2}{\omega^2 - k^2 c_{sc}^2} + \frac{\omega_{0h}^2}{\omega^2 - k^2 c_{sh}^2} = -1,$$

where $\omega_{0c}^2 = 4\pi G \rho_{0c}$, $\omega_{0h}^2 = 4\pi G \rho_{0h}$.

Denoting the left side of this equation by $f(\omega^2)$, we represent schematically the function f in Fig. 1(a). Roots of the dispersion equation correspond to

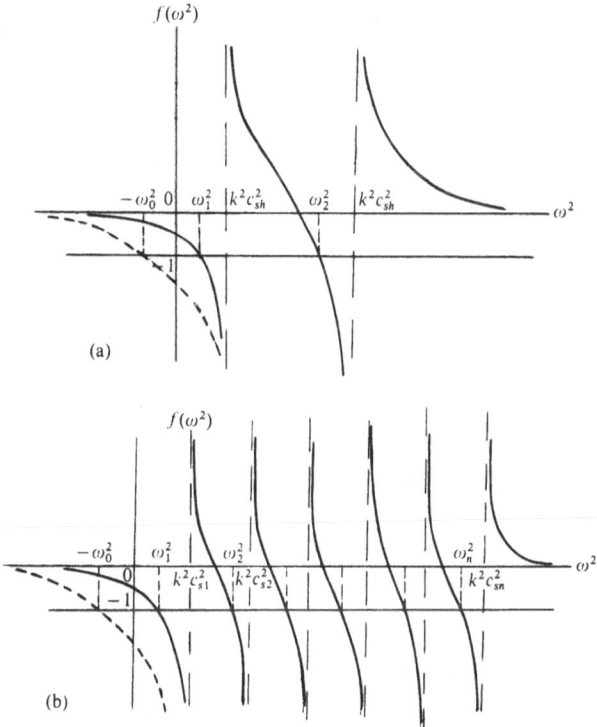

Figure 1. Jeans instability of a homogeneous gravitating medium in (a) a two-component system, and (b) an n-component system.

points of intersection of curves $f(\omega^2)$ with the straight line $f(\omega^2) = -1$. From Fig. 1(a) it is seen that the dispersion equation may have only one negative root $\omega^2 = -\omega_0^2$ when the left curve of $f(\omega^2)$ runs as the dotted line. But if the left curve $f(\omega^2)$ runs as the solid line, then both roots are positive: $\omega_{1,2}^2 > 0$.

The analysis of the roots of the dispersion equation for a two-component system may be easily generalized to the case of the n-component systems. For an arbitrary ith component of a gas its disturbed density equals

$$\rho_i = \rho_{0i} \frac{k^2 \Phi}{\omega^2 - k^2 c_{si}^2},$$

and so the dispersion equation for n-components has the form

$$\frac{\omega_{01}^2}{\omega^2 - k^2 c_{s1}^2} + \frac{\omega_{02}^2}{\omega^2 - k^2 c_{s2}^2} + \cdots + \frac{\omega_{0i}^2}{\omega^2 - k^2 c_{si}^2} + \cdots + \frac{\omega_{0n}^2}{\omega^2 - k^2 c_{sn}^2} = -1,$$

where $\omega_{0i}^2 = 4\pi G \rho_{0i}$.

Let us again denote the function on the left side of this equation by $f(\omega^2)$ which is schematically represented in Fig. 1(b). From Fig. 1(b) one can see that the dispersion equation may have only one negative root ($\omega^2 < 0$) for an n-component system as well as for a two-component one [this occurs in the case when the extreme left curve of $f(\omega^2)$ runs as the dotted line in Fig. 1(b)]. All the remaining roots are positive. Thus, one can formulate the following.

5.1 Basic Theorem (on the Stability of a Multicomponent System with Components at Rest) [64a[ad]]

In a heterogeneous system that consists of n uniform components, only one aperiodic instability may develop, provided that

$$\sum_{i=1}^{n} \frac{\omega_{0i}^2}{k^2 c_{si}^2} > 1.$$

The rest of the $n - 1$ collective modes are combined sound oscillations.

The question which we must answer now is: What wavelengths participate in the aperiodic instability of the heterogenous system? To answer this question, we shall restrict ourselves to the case of the two-component system, because it was shown above that, in the case of an arbitrary number of components, n, the dispersion relation may have only one negative root, $\omega^2 < 0$.

A solution of the dispersion equation for the two-component medium is

$$\omega^2 = \tfrac{1}{2}[-(\omega_h^2 + \omega_c^2) \pm \sqrt{(\omega_h^2 - \omega_c^2)^2 + 4\omega_{0h}^2 \omega_{0c}^2}],$$

where the notations $\omega_h^2 = \omega_{0h}^2 - k^2 c_{sh}^2$, $\omega_c^2 = \omega_{0c}^2 - k^2 c_{sc}^2$ are introduced.

5.2 Four Limiting Cases for a Two-Component Medium

5.2.1. Wavelength of a Disturbance is Less Than the Jeans Wavelengths in the Cold and Hot Components: $\lambda \ll \lambda_c,\ \lambda_h$. From the equivalent inequalities $k^2c_{sc}^2 \gg \omega_{0c}^2$, $k^2c_{sh}^2 \gg \omega_{0h}^2$, we obtain $\omega_c^2 \approx -k^2c_{sc}^2$, $\omega_h^2 \approx -k^2c_{sh}^2$ which, after substitution into the solutions of the dispersion equation, leads to two sound branches of oscillations,

$$\omega_1^2 \approx k^2c_{sc}^2, \qquad \omega_2^2 \approx k^2c_{sh}^2,$$

each of which is defined by the parameters of the hot and cold components of the medium, respectively.

5.2.2. The Wavelength of a Disturbance is Less Than the Jeans Wavelength in the Cold Component and Larger Than the Jeans Wavelength in the Hot Component: $\lambda_c \gg \lambda \gg \lambda_h$. The equivalent inequalities: $k^2c_{sc}^2 \gg \omega_{0c}^2$, $k^2c_{sh}^2 \ll \omega_{0h}^2$ may be written as $\omega_{0h}^2 \gg k^2c_{sh}^2 \gg k^2c_{sc}^2 \gg \omega_0^2$. Thus in this case the density of the hot component is much larger that the density of the cold component: $\omega_{0h}^2 \gg \omega_{0c}^2$.

From the system of inequalities we obtain $\omega_h^2 \approx \omega_{0h}^2$, $\omega_c^2 \approx -k^2c_{sc}^2$. Using these values of the squares of characteristic frequencies, we find the following solutions of the dispersion equation:

$$\omega^2 \approx \omega_{0h}^2,$$
$$\omega^2 \approx k^2c_{sc}^2.$$

The first of these solutions describes the Jeans instability, with an increment determined by the density of the hot component. The second solution describes sound oscillations in the heterogeneous medium, with a velocity coincident with the sound velocity in the cold component.

5.2.3. The Wavelength of a Disturbance is Larger Than the Jeans Wavelength in the Cold Component but Less Than the Jeans Wavelength in the Hot Component: $\lambda_h \gg \lambda \gg \lambda_c$. From the equivalent inequalities $k^2c_{sc}^2 \ll \omega_{0c}^2$, $k^2c_{sh}^2 \gg \omega_{0h}^2$, we find $\omega_h^2 = -k^2c_{sh}^2$, $\omega_c^2 = \omega_{0c}^2$. The solutions of the dispersion equation in this case will be the following:

$$\omega^2 = -\omega_{0c}^2, \qquad \omega^2 = k^2c_{sh}^2.$$

The first solution describes an aperiodical instability of the heterogeneous system, with an increment only equal to the Jeans increment of the cold component of the medium. The second solution describes the sound oscillations propagating in the heterogeneous system with a frequency which equals the sound frequency of oscillations in the hot component.

5.2.4. The Wavelength of a Disturbance is Larger Than the Jeans Wavelengths Both in the Cold and in the Hot Components: $\lambda \gg \lambda_c,\ \lambda_h$. From the equivalent system of the inequalities $k^2c_{sc}^2 \ll \omega_{0c}^2$, $k^2c_{sh}^2 \ll \omega_{0h}^2$, we find $\omega_h^2 \approx \omega_{0h}^2$,

$\omega_c^2 \approx \omega_{0c}^2$. The corresponding solutions of the dispersion equation are the following:

$$\omega^2 \approx -\omega_0^2,$$

$$\omega^2 \approx \frac{k^2 c_{sh}^2 \omega_{0c}^2 + k^2 c_{sc}^2 \omega_{0h}^2}{\omega_0^2},$$

where $\omega_0^2 \equiv \omega_{0h}^2 + \omega_{0c}^2$.

The first solution describes a Jeans instability of the heterogeneous system, with a square of the increment equal to the sum of the squares of the Jeans increments for the different components of the medium. The second solution describes combined sound oscillations.

For the sake of convenience we have summarized the results obtained above in the following table.

5.3 Table of Jeans Instabilities of a Uniform Two-Component Medium

Table I Jeans instabilities of a uniform two-component system.

Conditions on wavelengths	Equivalent inequalities	Ratio of densities	Two solutions of the dispersion equation	Comments
(1) $\lambda \ll \lambda_c, \lambda_h$	$k^2 c_{sc}^2 \gg \omega_{0c}^2$ $k^2 c_{sh}^2 \gg \omega_{0h}^2$	arbitrary	$\omega^2 = k^2 c_{sc}^2$ $\omega^2 = k^2 c_{sh}^2$	Two sound branches each of which is defined by parameters of hot and of cold, respectively
(2) $\lambda_c \gg \lambda \gg \lambda_h$	$k^2 c_{sc}^2 \gg \omega_{0c}^2$ $k^2 c_{sh}^2 \ll \omega_{0h}^2$	$\omega_{0c}^2/\omega_{0h}^2 \ll 1$	$\omega^2 = k^2 c_{sc}^2$ $\omega^2 = -\omega_{0h}^2$	Oscillations with the sound frequency of the cold component; instability determined by the parameters of the hot component
(3) $\lambda_h \gg \lambda \gg \lambda_c$	$k^2 c_{sc}^2 \ll \omega_{0c}^2$ $k^2 c_{sh}^2 \gg \omega_{0h}^2$	arbitrary	$\omega^2 = -\omega_{0c}^2$ $\omega^2 = k^2 c_{sh}^2$	Instability determined by density of the cold component; sound oscillations determined by the parameters of the hot component
(4) $\lambda \gg \lambda_c, \lambda_h$	$k^2 c_{sc}^2 \ll \omega_{0c}^2$ $k^2 c_{sh}^2 \ll \omega_{0h}^2$	arbitrary	$\omega^2 = -\omega_0^2$ $\omega^2 =$ $\dfrac{k^2 c_{sh}^2 \omega_{0c}^2 + k^2 c_{sc}^2 \omega_{0h}^2}{\omega_0^2}$ $\omega_0^2 = \omega_{0c}^2 + \omega_{0h}^2$	Jeans instability; combined sound oscillations

5.4 General Case of n Components

Evidently, it is easy to generalize these results to the case of an n-component medium. Summarizing, we shall bear in mind just this system.

Thus, (1) disturbances with a wavelength smaller than the minimum Jeans wavelength of the different components lead to the appearance of n different branches of sound oscillations, each of which is determined by the parameters of one subsystem; (2) disturbances with a wavelength larger than the Jeans wavelengths of m components lead to an instability (the square of the eigenfrequency turns out to be equal to the sum of the squares of the m (negative) Jeans frequencies, $\omega^2 = -\sum_{i=1}^{m} \omega_{0i}^2$) and also to the appearance of $n-1$ branches of the combined sound oscillations. For $m = n$ we obtain the last case (4) from the above.

The natural question is how in a heterogeneous system oscillations with a wavelength much larger than the largest among the Jeans wavelengths of the components may exist? The existence of oscillations with an arbitrary large wavelength in a heterogeneous system may be due to the fact that the oscillations in two different components are in opposite phases: minima in one of the components coincide with maxima in the other. Actually, from Fig. 1a, b, it is clear that, except for the extreme left-hand root which may correspond to an instability, any other ith root lies between $k^2 c_{si}^2$ and $k^2 c_{s(i+1)}^2$, i.e., $k^2 c_{si}^2 < \omega^2 < k^2 c_{s(i+1)}^2$. As follows from the expressions for perturbed densities, the greatest contribution for the ith root is made just by ρ_i and ρ_{i+2}. Since $\omega^2 - k^2 c_{si}^2 > 0$ and $\omega^2 - k^2 c_{s(i+1)}^2 < 0$, ρ_i and ρ_{i+2} are opposite in sign. The homogeneous system has lost this possibility; thus sufficiently large wavelengths always lead to a Jeans instability.

To avoid misunderstandings, we shall stress once more that the model of a homogeneous gravitating medium infinite in all *directions* has been considered above. To consider such systems as stationary is not quite correct. Even the model of a plane layer, i.e., of a system unbounded in two directions x, y, but bound in direction z, may be stationary. This is all the more valid for a cylindrical model (infinite only in one direction). The two models turn out to be *unstable* "according to Jeans" with respect to perturbations with wavelengths large enough (see Chapters I and II), but the increments of these instabilities vanish for infinitely large wavelengths $\lambda \to \infty$ ($k \to 0$). It is the latter circumstance that corresponds to the stationarity of the initial state of the system; growth of perturbations with $\lambda = \infty$ ($k = 0$) [as in the original Jeans model—see Eq. (34)] would correspond to non-steady-state.

§ 6 Non-Jeans Instabilities [88, 28, 114, 98–100, 23[ad]]

The models of a plane layer and of an infinite cylinder are convenient, in particular, for the investigation of some questions of the principle of the stability theory of gravitating systems. For instance, the possibility of spontaneous gravity oscillations in the gravitating medium by means of mechanisms other than the Jeansonian ones seems important. This problem is not so trivial as it may appear at first glance, since the equilibrium condition of the gravitating system rigidly links the gravity force at each point with other forces (pressure, centrifugal, etc.). For a long time any positive answer to this question was, one way or another, connected with the selection of a nonstationary model, with a violation of the equilibrium condition (for more detail see [20]). The problem implied that one should investigate possible instabilities of the gravitating medium under the *condition of the suppression of Jeans instability*; otherwise, it may, in particular, happen that the "newly discovered" instability is in fact a hidden Jeans instability. In [88], in the steady state model of a cylinder infinitely extended along the *z*-axis, the presence of *beam* instability in the gravitating medium was strictly proved, while in [28, 114], in the same model, the possibility of a *temperature-gradient* instability was shown. The increments (growth rates) of these instabilities do not exceed the Jeans one. For a specialist in plasma physics such a result is evident, since, translated into the plasma language, it means that $\gamma_{max} \sim \omega_{pe}$, where ω_{pe} is the plasma frequency.

Having investigated the beam and temperature gradient instabilities, there naturally arose the question of the possibility of the development of instabilities having increments much greater than Jeans increments in a gravitating gas. A positive answer to this question is given in [98–100] which demonstrates the possibility of building up such hydrodynamic instabilities as instability of tangential discontinuity or flute instability. For sufficiently short wavelengths (smaller than the Jeans) the gravitational effects are of no significance, and the above-mentioned instabilities develop in the same manner as in a normal (nongravitating) liquid. Note further that since the increments of these instabilities can be much larger than the Jeans ones, it is possible to investigate them quite strictly already in the model of an infinite (in all directions) medium: during the time of the growth of such perturbations the system practically does not change due to gravitational attraction. These instabilities were also investigated in simple *stationary* models (e.g., in the model of a cylinder). Chapter VI of this book is devoted to the non-Jeans instabilities enumerated above.

A qualitative description of the mechanisms of non-Jeans instabilities with a table of their basic characteristics is contained in §12 of the Appendix.

Approximately at the same time, the anisotropic (fire-hose) instability of collisionless gravitating systems in the simplest model of a nonrotating layer, uniform in the (x, y) plane, was discovered [263]. This instability often arises under conditions in a certain sense "additional" to the excitation conditions of the Jeans instability, for example, in a layer hot enough in the (x, y) plane. It is similar to the instability, known from elasticity theory, with respect to bending, arising in a metal rod if the latter is sufficiently strongly compressed on its sides. In gravitating systems the role of the compressing tension is played by the "pressure" (dispersion of peculiar velocities) of particles.

Thereafter the fire-hose instability was investigated in detail in [28[ad], 33[ad], 40[ad]]. Since this instability is one of the main instabilities in gravitational systems, which imposes strong restrictions on their possible equilibrium parameters, a corresponding theory is presented in detail (separately for systems with a different geometry) in Chapters I–V.

Another important instability developing in gravitational systems, rotating rapidly enough, is the instability of the so-called bar mode [93, 111, 252], i.e., the conversion of an initially axially symmetrical system into a "cucumber." This instability is actively discussed in connection with, for example, some promising versions of the theory of galactic spiral structure (see Chapters IV, V, VIII).

§ 7 Qualitative Discussion of the Stability of Spherical, Cylindrical (and Disk-Shaped) Systems with Respect to Radial Perturbations [19]

From the very beginning we have emphasized that the stability of the gravitating system as a whole (the collective effect) has, generally speaking, no bearing on the stability of the orbits of an individual particle. Only for some systems, arranged in a special way, the investigation of the stability of individual particles can simultaneously answer also the question of the stability of the whole system. We deal with radial oscillations of a gravitating ball (sphere) or cylinder in the case when the equilibrium is due to the rotation of masses which constitute them. The case of the *spherically symmetric* system with circular trajectories of all particles is illustrated in Fig. 2.

The stability of a sphere or a rotating cylinder to radial perturbations in which no crossings of the sheets occurs is evident since then each particle moves in a field of constant mass. For radial perturbations, the angular momentum of a particle is constant, and its stationary state corresponds to the minimum of energy (the Kepler problem) and therefore is stable. But

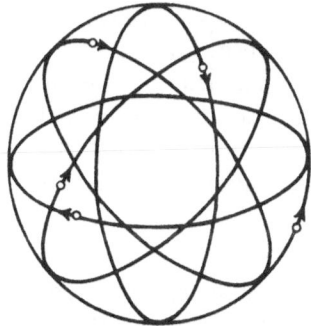

Figure 2. Einstein model of a spherical-symmetric cluster. All particles move in a circular trajectory about a common center of mass; directions of motion are indicated by arrows.

this simultaneously implies also the stability of the system as a whole (for perturbations of the type under consideration), since individual sheets oscillate independently of the others.

Let us now imagine that the system was subjected to nonradial perturbations which resulted in a change of the shape of the effective potential $U = U(r): U = U(r, \theta, \varphi)$. In this case the particle going off position r_0 to position r_1 $(\theta = \theta_0, \varphi = \varphi_0)$ can, for example, occupy the minimum point in the curve $U = U(r, \theta_0, \varphi_0)$. Thus, the stability of the system with respect to such perturbations is not evident a priori, and additional investigation is necessary (see §3, Chapter III).

Neither is the stability of a system evident with respect to radial perturbations with crossings of the sheets. Let us show [19] that crossing necessarily arises in a period of time, i.e. pairs of particles appear whose Euler coordinates coincide. If r_0 is the Lagrange radius of a particle, and r is its true Euler radius, then the presence of crossing means that $dr/dr_0 = 0$, i.e., particles of different sheets lie in the same radius. Let us specify a perturbation of the form $r = r_0 + \varepsilon z(r_0)$, $dr/dt = 0$, where $\varepsilon z(r_0)$ is a small perturbation.

Then the equations of motion determine the periodical trajectory of a particle

$$r = r_0 + \varepsilon z(r_0) \cos \Omega(r_0)t, \qquad \frac{dr}{dr_0} = 1 + \varepsilon z' \cos \Omega t - \varepsilon z \Omega' t \sin \Omega t,$$

where $\Omega(r_0)$ is the angular velocity of particle rotation.

Hence it is evident that, due to the dependence $\Omega(r_0)$, in the expression for dr/dr_0 there is a time dependent term. Therefore, for any initial small-perturbation crossing occurs. Exceptions are the homogeneous sphere and the homogeneous cylinder for which $\Omega(r_0) = $ const.

The effective potential energy U_{eff} of the sphere (see Fig. 2) equals

$$U_{\text{eff}} = \sum_i \left(-\frac{GM_i m_i}{r_i} + \frac{L_i^2}{2m_i r_i^2} \right), \tag{39}$$

where M_i is the mass inside $r = r_i$, m_i is the mass of the ith sphere sheet from the origin of the coordinate system, L_i^2 is the sum of squares of the angular momenta of particles of the ith sphere sheet. Note that the case is not excluded when $\mathbf{L}_i = \sum_k \mathbf{L}_{ik} = 0$, where $\mathbf{L}_{ik}$ is the angular momentum of the kth particle of the ith sheet, however, $L_i^2 = \sum L_{ik}^2 \neq 0$. Since the radial shifts preserve the angular momentum of each particle ($\mathbf{L}_{ik} = $ const), then for these shifts $L_i^2 = $ const. Let the effective potential energy of the system, as a result of radial perturbation, be equal to U'_{eff} which is obtained from formula (39) through the change $r_i \rightarrow r'_i$. The stability condition of the system is

$$\Delta U_{\text{eff}} = U'_{\text{eff}} - U_{\text{eff}} > 0. \tag{40}$$

In the case when sheet displacements $r_i \rightarrow r'_i$ do not result in their crossing, the effective potential energy of the ith sheet is equal to

$$U_{\text{eff}(i)} = -\frac{GM_i m_i}{r_i} + \frac{L_i^2}{2m_i r_i^2}. \tag{41}$$

The equilibrium condition of the ith sheet is

$$\frac{Gm_i M_i}{r_i^2} = \frac{L_i^2}{m_i r_i^3},$$

(42)

which is the condition of the minimum of the effective potential energy $(\sim \partial M_i/\partial r_i)$ is zero. In the case of sheet crossing, when the quantity of mass Any displacement of the ith sphere sheet toward an arbitrary point r_i' results in an energy increase (41). Since the above-stated is valid for any sheet, then for the system as a whole the equilibrium condition (40) is satisfied.

When determining the equilibrium conditions (42) from the minimum energy condition

$$\frac{\partial U_{\text{eff}}}{\partial r_i} = \frac{GM_i m_i}{r_i^2} - \frac{Gm_i}{r_i} \frac{\partial M_i}{\partial r_i} - \frac{L_i^2}{m_i r_i^3} = 0,$$

it was assumed that the second term of the left-hand side of the equality $(\sim \partial M_i/\partial r_i)$ is zero. In the case of sheet crossing, when the quantity of mass inside the ith sheet changes as a result of perturbation, the term $(Gm_i/r_i) \times (\partial M_i/\partial r_i) \neq 0$. To establish the influence of this term on stability, let us consider the systems composed of a central mass M and two layers (sheets) crossing as a result of perturbations. Then

$$U_{\text{eff}(1)} = -\frac{Gm_1 M}{r_1} + \frac{L_1^2}{2m_1 r_1^2},$$

(43)

$$U_{\text{eff}(2)} = -\frac{Gm_2(M + m_1)}{r_2} + \frac{L_2^2}{2m_2 r_2^2},$$

(44)

$$U'_{\text{eff}(1)} = -\frac{Gm_1(M + m_2)}{r_1'} + \frac{L_1^2}{2m_1 r_1'^2},$$

(45)

$$U'_{\text{eff}(2)} = -\frac{Gm_2 M}{r_2'} + \frac{L_2^2}{2m_2 r_2'^2}.$$

(46)

Expressions (43) and (44) define the energy of the sheets before their displacements. Let us assume that

$$r_1 < r_2, \qquad r_1' > r_2'$$

(47)

i.e., the "inner" sheet after radial perturbation becomes the "outer" one and vice versa. Determining ΔU_{eff} according to (40), we arrive at the following positive definite form:

$$\Delta U_{\text{eff}} = \frac{Gm_1 M(r_1' - r_1)^2}{2r_1 r_1'^2} + \frac{Gm_1 m_2(r_1' - r_2')}{r_1' r_2'} + \frac{Gm_2(M + m_1)(r_2' - r_2)^2}{2r_2 r_2'^2}.$$

(48)

The second term in (48) determines the change of the potential energy only on account of sheet permutation. According to (47), this term is always positive. Therefore, radial perturbations, resulting in crossing of spherical sheets, lead to an additional increase in the effective potential energy of the system.

A proof of the stability of a cylinder infinitely extended in z-direction with respect to radial perturbations can be made similarly to the above-considered sphere. In the case of sheet crossing there also arises in U_{eff} a positive term, causing additional stabilization of radial perturbations.

Thus it turns out that radial perturbations with crossings are "more stable" than those of the extension-compression type. The latter type of perturbation is stable in a sphere or cylinder because the equilibrium condition of a spherical or cylindrical sheet is the condition of the minimum of the effective potential energy of the sheet. This fact, in its turn, is the consequence of the fact that the field intensity at any point of the sphere or cylinder depends only on the interior mass with respect to the equipotential surface.

The stability of individual particle orbits is determined for the systems in question by very similar factors. If the particle is displaced from the equilibrium circular orbit, then, as can be readily shown (see, e.g., §1 of Chapter II), it will oscillate with an *epicyclic* frequency $\varkappa$ expressed by the formula

$$\varkappa^2 = 4\Omega^2 + r\frac{d\Omega^2}{dr}.$$

The frequency $\varkappa$ is an imaginary one (which corresponds to the instability of the orbit) if $\varkappa^2 < 0$. But this requires that the angular frequency Ω decay with the radius faster than $\sim 1/r^2$, at the same time diminishing at a maximum rate (in the Keplerian field) $\Omega \sim 1/r^{3/2}$. The least possible epicyclic frequency $\varkappa = \Omega$ does correspond to the latter case. For all other $\Phi_0(r)$ larger $\varkappa$ are obtained. The Keplerian field $\Phi_0(r)$ corresponds to the motion of particles without any sheet crossing, and consideration of crossing means deviation of $\Phi_0(r)$ from Kepler's law. Therefore, taking into account sheet crossings only "increases" the stability.

Recall that unlike a sphere (or a cylinder) the field strength in a gravitating *disk* is defined both by the interior and exterior mass. For this reason the trial body inside a gravitating sphere is in neutral equilibrium, while within the gravitating ring it is attracted to its nearest point. Radial perturbations easily laminate the gravitating disk (with circular particle orbits) into narrow rings [333] (see Chapter V).

PART I

THEORY

CHAPTER I

Equilibrium and Stability of a Nonrotating Flat Gravitating Layer

In the present chapter which essentially has an introductory purpose we shall restrict ourselves when investigating stability to the simplest model—that of a nonrotating flat layer. This model will permit us a rather detailed study of the properties of two important instabilities of gravitating systems: Jeans and anisotropic (see Introduction). More realistic models of flattened gravitating systems are also subject to these instabilities, provided that similar conditions hold. A somewhat more complicated rotating layer model will be considered later (see Chapter V).

The model of a plane-parallel mass distribution with volume density $\rho_0(z)$, depending only on one coordinate z (and, consequently, unbounded along x, y) was introduced by Oort [296] in order to study mass distribution in our Galaxy in the direction perpendicular to its equatorial plane ($z = 0$). The point is that the Galaxy is a very compressed disk-shaped system. Such a structure of the Galaxy seems to suggest that the rate of density change over the vertical (z) is much more than the rate of its change in the horizontal direction. Accordingly, as a first approximation when studying the nature of the vertical mass distribution in some specified place of the galactic disk, one can assume the flat sheet model, i.e., suggest that the density does not change at all along the plane (x, y) (preserving its values $\rho_0 = \rho_0(z)$ in a selected place).

Our Galaxy is classified as a spiral (S) galaxy, and the high degree of flatness is characteristic of all spiral galaxies. The same is valid for lens-shaped (SO) galaxies and for some other objects (see Chapter VIII). Thus, the model of a flat (generally rotating) sheet has a rather broad range of application as the simplest model for the description of localized equilibrium of all such systems in the vertical direction.

This model is also suitable for studying the behavior of *localized perturbations* of strongly flattened systems; and not only in the vertical (z) but also in the horizontal (x, y) directions. The latter is expecially interesting since it allows one to explain readily many facts which play an important role in the stability theory of *disk-shaped* systems.

In Section 1 a brief description of the equilibrium state of a collisionless flat sheet is given. In Section 2 we study Jeans perturbations in the long-wavelength limit, $k_\perp c \ll 1$ ($k_\perp$ is the component of the wave vector $\mathbf{k}$ in the plane (x, y), c is the layer's semithickness). Evidently, in this case the character of the density distribution along z-axis does not play any role. In the end of Section 2 we have computations of disturbances with arbitrary wavelengths but for the simplest layer model uniform in density. Section 3 is devoted to the investigation of the anisotropic (fire-hose) instability, and this is done again, as in Section 2, for the simplest cases: long-wavelength disturbances ($k_\perp c \ll 1$ which correspond to a "thin" layer) and for a uniform layer.

In Section 4 a derivation is presented of the integro-differential equations for symmetrical and antisymmetrical (with respect to the plane $z = 0$) normal modes of an arbitrary flat sheet (this derivation belongs to Mark [289a]). In the next section, (5), the equation describing symmetrical modes, is applied to the study of perturbations in the vicinity of the instability threshold ($k_\perp \simeq k_{cr}$). Finally, Section 6 is devoted to a detailed investigation into perturbations perpendicular to the plane (x, y) ($k_\perp = 0$) for a uniform-density layer.

We have included in the problems section the question of oscillations with $k_\perp c \gg 1$, as well as the question considered by Kalnajs [253] of the nonlinear evolution of perturbations of a homogeneous layer of the extension-compression type (with conservation of the homogeneous density), determination of the eigenfunctions of oscillations perpendicular to the plane (x, y), and, finally, the derivation of the dispersion equation for short-wavelength sheet oscillations.

§ 1 Equilibrium States of a Collisionless Flat Layer

We shall consider here stationary states of a flat nonrotating layer of collisionless particles. Possible equilibria and stability of analogous gaseous systems, being of great interest in connection with some variants of the theory of galaxy formation (e.g., in the "pancake" theory according to the terminology in [48]) and in the theory of planetary formation from a gas–dust protoplanetary cloud, will be considered in the appropriate sections of Part II, which is especially devoted to astronomical applications.

Take first of all the simplest sheet uniform in density (Fig. 3). If one is interested only in the motion of stars in the z-direction, then the system under consideration, in essence, is one-dimensional. Let the sheet density be

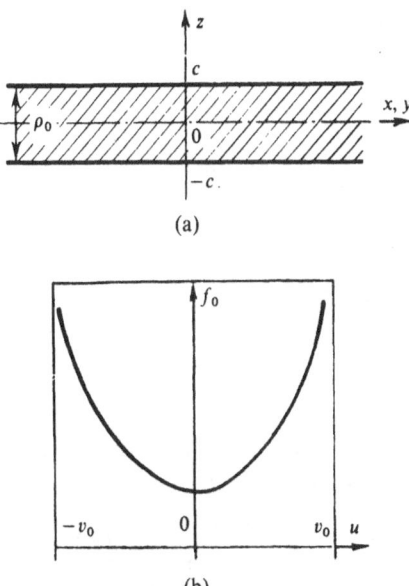

Figure 3. Plane gravitating layer: (a) general arrangement of a layer; ρ_0 is density; c is half-thickness; a system is homogeneous in the x- and y-directions; (b) equilibrium distribution function $f_0(v) \sim (v_0^2 - v^2)^{-1/2}$.

ρ_0 and the thickness $2c$. Then the gravitational potential $\Phi_0(z)$ inside the sheet is

$$\Phi_0 = \frac{\Omega_0^2 z^2}{2} + \text{const}, \tag{1}$$

where $\Omega_0^2 = 4\pi G \rho_0$. It is convenient to put $\Omega_0 = 1$, $c = 1$. The distribution function of the system in question must be dependent only on the energy of particle motion in the z-direction: $E = \frac{1}{2}v_z^2 + \frac{1}{2}z^2$. Indeed, the energy is the only time-independent integral for the case of one-dimensional motion which is evident, for example, from the system of characteristic equations

$$dt = \frac{dz}{v_z} = \frac{dv_z}{-\partial \Phi_0/\partial z}. \tag{2}$$

Thus, assuming $f_0 = f_0(E)$, we satisfy the kinetic equation

$$v_z \frac{\partial f_0}{\partial z} - z \frac{\partial f_0}{\partial v_z} = 0. \tag{3}$$

Now one should require coincidence of the density $\rho_0(z) = \int f_0(E_z) \, dv_z$ and the density $\rho_0(z)$ of the homogeneous layer, creating the potential (1). The function $\rho_0(z)$ can be written in the form

$$\rho_0(z) = \rho_0 \theta(1 - z^2), \tag{4}$$

where $\theta(x)$ is the single Heaviside function

$$\theta(x) = \begin{cases} 1, & x > 0, \\ 0, & x < 0. \end{cases} \tag{5}$$

Thus, the self-consistency condition is

$$\rho_0 \theta(1 - z^2) = \int f_0(E_z) \, dv_z. \tag{6}$$

By using $J = 1 - 2E_z$, Eq. (6) reduces to Abel's integral equation

$$\rho_0(x) = \rho_0 \theta(x) = \int_0^x \frac{f_0(J) \, dJ}{\sqrt{x - J}}, \tag{7}$$

where $x = 1 - z^2$. Abel's equation is solved (see, e.g., [132]) in a general form for the arbitrary function $\rho_0(x)$:

$$f_0(J) = \frac{1}{\pi} \frac{d}{dJ} \int_0^J \frac{\rho_0(x) \, dx}{\sqrt{J - x}}. \tag{8}$$

In this case $\rho_0(x) = \rho_0 \theta(x)$; we get [8]

$$f_0 = \frac{\rho}{\pi} \frac{\theta(1 - z^2 - v_z^2)}{\sqrt{1 - z^2 - v_z^2}} = \frac{\rho_0}{\pi} \begin{cases} J^{-1/2}, & J > 0, \\ 0, & J < 0. \end{cases} \tag{9}$$

This distribution function increases as the velocity increases from zero to $\sqrt{1 - z^2} \equiv v_0$ and thereafter breaks off sharply (Fig. 3(b)). It may be noted [8] that such a truncation is due to the need for a pronounced spatial boundary of the system and has no bearing on the escape velocity (which is infinite in the one-dimensional case).

The model under consideration is classified among systems with the quadratic potential, $\Phi_0(z) = z^2/2 + \text{const}$. Particles in such a potential execute harmonic oscillations in z at a frequency of Ω_0 (which we have assumed to be equal to unity). Other examples of such systems are treated in Chapters II–V.

In order to determine $f_0(E_z)$ in the case of a *nonhomogeneous* layer, it is easy to derive the equation similar to (7):

$$\frac{1}{\sqrt{2}} \rho_0(x) = \int_0^x \frac{f(J) \, dJ}{\sqrt{x - J}}, \tag{10}$$

where $x \equiv -\Phi_0(z)$, $J \equiv -E_z$. From this it follows that in order to determine the distribution function of a collisionless flat sheet with specified density $\rho_0(z)$, after determining the corresponding potential $\Phi_0(z)$, one has first of all to eliminate z from the equations

$$\rho_0(z) = \rho_0, \quad -\Phi_0(z) = x. \tag{11}$$

Using the function $\rho_0(x)$ thus obtained, the solution for Abel's equation (10) is solved by formula (8) [with substitution of $\rho_0(x) \to (1/\sqrt{2})\rho_0(x)$].

After that, one has only to verify the positiveness of the obtained distribution function over the whole range of the argument $J = -E_z$.

When studying oscillations of a layer in its plane, one should specify the dependence of the equilibrium distribution function on the velocities v_x and v_y. It can be arbitrary since v_x and v_y in the case in question are integrals of motion. Thus, the general form of the stationary distribution function for the gravitational layer is $f_0 = f(E_z, v_x, v_y)$.

Mark [289a] has found equilibrium solutions for isotropic velocity particle distributions of the following kind:

$$f_0 \sim (E_0 - E)^{n-3/2}\theta(E_0 - E), \tag{12}$$

where n is the constant parameter of the model and E denotes the total energy of the particle:

$$E = \frac{v_x^2 + v_y^2 + v_z^2}{2} + \Phi_0(z) = E_z + \frac{v_x^2 + v_y^2}{2}, \tag{13}$$

E_0 is the maximum possible energy (constant value).

A density corresponding to (12), as is readily calculated, is

$$\rho_0(z) = \text{const}[E_0 - \Phi_0(z)]^n. \tag{14}$$

To determine the potential $\Phi_0(z)$ [and then using formula (14) and the density $\rho_0(z)$], one should solve the nonlinear differential equation

$$\frac{d^2\Phi_0}{dz^2} = 4\pi G \,\text{const}[E_0 - \Phi_0(z)]^n. \tag{15}$$

The numerical solutions of the equation written for $n = 2-6$ in [289a] are presented as a table of dimensionless parameters, characterizing the models, and as graphs of the functions $[E_0 - \Phi_0(z)]/[E_0 - \Phi_0(0)]$, $\rho_0(z)/\rho_0(0)$. These functions decrease steadily when z changes from zero to $z = z_{max}$, where z_{max} is the first zero of the function $\rho_0(z)$, $\rho_0(z_{max}) = 0$.

Distribution functions similar to (12) ("polytropic") will also be discussed below (see the first sections of Chapters II and III), where they will be considered in more detail.

§ 2 Gravitational (Jeans) Instability of the Layer

Consider first of all the question of *long-wavelength* stability of a collisionless layer, i.e., we assume that the perturbation wavelength λ is large in comparison with the thickness c. (Short-wave oscillations ($k_\perp c \gg 1$) are briefly treated in Problem 1.)

As far as disturbances with arbitrary ("intermediate") wavelengths are concerned, they can be calculated by numerical methods only. The results of corresponding calculations are given at the end of the section.

Assume that when the layer is perturbed, no density, extension, or compression nodes of matter arise in the vertical direction. The dispersion equation for this type of perturbation (which we will refer to as "Jeans") can be derived in a simple way, by considering oscillations of an *infinitely thin* layer with surface density σ_0.

The simplicity of the infinitely thin layer model implies that in this limit the motion of matter in the plane of the layer (and this is particularly of interest to us) may occur independently of the motion perpendicular to the layer. Indeed, if the particle velocities at the initial moment (when the perturbation is applied) lie within the plane of the layer, then the motion will remain plane all the time because no forces in the perpendicular direction arise.

The perturbations interesting to us can be described by the one-dimensional kinetic equation

$$\frac{\partial f}{\partial t} + v_x \frac{\partial f}{\partial x} - \frac{\partial \Phi}{\partial x} \frac{\partial f}{\partial v_x} = 0. \tag{1}$$

Linearizing (1) and representing the perturbations in the form $\infty e^{ikx - i\omega t}$, we get

$$f_1 = - \frac{k \, \partial f / \partial v_2}{\omega - kv_2} \Phi_1; \tag{2}$$

hence we obtain the perturbation of the surface density:

$$\sigma_1 = -k\Phi_1 \int \frac{\partial f_0 / \partial v_x}{\omega - kv_x} dv_x. \tag{3}$$

If Im $\omega > 0$, then the integration in (3) may go along the real v_x-axis; however, the integral at Im $\omega \leq 0$ must be understood as an analytical continuation from the region Im $\omega > 0$. Thus, the integral over v_x in (3) must be calculated according to the well-known "Landau bypass rule" [65]: the contour of integration in the complex plane v_x bypasses the pole singularity $\omega = kv_x$ *from below*.

To obtain the dispersion relation, one has to use the Poisson equation, which in this case has the form:

$$\frac{\partial^2 \Phi_1(z)}{\partial z^2} - k^2 \Phi_1(z) = 4\pi G \sigma_1 \delta(z). \tag{4}$$

The solutions of the homogeneous equation, corresponding to (4) (the Laplace equation)

$$\Phi_+ \equiv \Phi_{1, z > 0} = c_1 e^{-kz}, \quad \Phi_- \equiv \Phi_{1, z < 0} = c_2 e^{kz} \quad (k > 0), \tag{5}$$

must be matched on the plane $z = 0$. The conditions of matching are

$$\Phi_+ = \Phi_- |_{z=0}; \quad \left(\frac{\partial \Phi_+}{\partial z} - \frac{\partial \Phi_-}{\partial z} \right)_{z=0} = 4\pi G \sigma_1. \tag{6}$$

By using them we get the following relation between the potential in the plane $z = 0$ and the surface density:

$$\Phi_1(z = 0) = - \frac{2\pi G \sigma_1}{k}. \tag{7}$$

A comparison of (7) and (3) leads to the sought dispersion equation

$$\varepsilon_0 \equiv 1 - \frac{2\pi G \sigma_0}{k} \int \frac{\partial f_0 / \partial v_x \, dv_x}{(\omega/k - v_x)} = 0. \tag{8}$$

In the case of a Maxwell velocity distribution in the (x, y) plane, we get from (8) (similarly to the plasma; see, e.g., [86])

$$\varepsilon_0 = 1 - \frac{2\pi G \sigma_0}{kT}\left[1 + i\sqrt{\pi}\,\frac{\omega}{kv_T}\,W\!\left(\frac{\omega}{kv_T}\right)\right] = 0, \tag{9}$$

where T is temperature and W is the Kramp function [134]

$$W(z) = e^{-z^2}\left(1 + \frac{2i}{\sqrt{\pi}}\int_0^z e^{x^2}\,dx\right). \tag{10}$$

As $T \to 0$ it follows from (9) that[1]

$$\omega^2 = -2\pi G \sigma_0 k, \tag{11}[1]$$

which means instability with the increment

$$\gamma = \sqrt{2\pi G \sigma_0 k}. \tag{12}$$

This is the Jeans instability of a flat layer.

The dispersion equation (11) was recently derived anew by Toomre [333]. He, in fact, studied perturbations of an nonhomogeneous rotating *disk*, but used a local approximation, for which both nonhomogeneity and rotation do not play any role. Therefore, the coincidence of the dispersion equation derived by him (in case of a cold disk) with equation (11) is not surprising. We shall return to this question in Chapter V.

By using the asymptotic representation of the Kramp function for large values of the argument [134]

$$W(z) \approx \frac{i}{\sqrt{\pi}z}\left(1 + \frac{1}{2z^2} + \frac{3}{4z^4} + \cdots\right), \qquad |z| \gg 1, \tag{13}$$

one can obtain the dispersion equation which takes into account the "thermal" correction, stabilizing the instability under investigation, Eq. (12):

$$\omega^2 \simeq -2\pi G \sigma_0 k + 3k^2 T. \tag{14}$$

Here it is assumed that the second term is small in comparison with the first one, $3k^2 T \ll 2\pi G \sigma_0 k$. To determine the stability boundary, let us consider dispersion equation (9) at $|\omega/kv_T| \ll 1$. Using the estimate $W(z)$ with small values of the argument [134]

$$W(z) \simeq 1 + \frac{2iz}{\sqrt{\pi}}, \qquad |z| \ll 1, \tag{15}$$

---

[1] The dispersion relation (11) for this limiting case (cold matter in the (x, y) plane) can be obtained in a simpler way by solving, instead of the kinetic equation, the hydrodynamical ones with zero pressure (similarly to the manner in the Introduction).

we have from (9)

$$\frac{\omega}{kv_T} \approx i\left(\alpha + \frac{2\alpha^2}{\sqrt{\pi}} + \cdots\right), \qquad \alpha = \frac{1}{\sqrt{\pi}}\frac{2\pi G\sigma_0 - kT}{2\pi G\sigma_0}, \qquad |\alpha| \ll 1. \quad (16)$$

When $\alpha > 0$, i.e., $2\pi G\sigma_0/T > k$, the perturbations grow, and when $\alpha < 0$ $(2\pi G\sigma_0/T < k)$ stability takes over. Thus, the critical wavenumber is

$$k_{cr} = \frac{2\pi G\sigma_0}{T}, \qquad (17)$$

and the relevant critical wavelength,

$$\lambda_{cr} = \frac{T}{G\sigma_0}. \qquad (18)$$

The increase in particle velocity (temperature) in the (x, y) plane shifts unstable perturbations toward the larger wavelength region, and the maximum achievable increment of instability [the latter seems to be of the order $\pi G\sigma_0/v_T$—see (14) and (17)] decreases.

It is interesting to compare oscillations of the *collisionless* layer considered above with those of the *incompressible* layer. These two limiting states (collisionless and incompressible) are opposites in a definite sense. More unexpected is the similarity between the two dispersion equations describing oscillations in either case.

Perturbations of an *incompressible* fluid layer have been studied by a number of authors (see reviews [13, 107]). Such perturbations require, evidently, a nonvanishing wave vector $\mathbf{k}$ in the (x, y) plane. They obey the equations

$$\text{div } \mathbf{v}_1 = 0, \qquad (19)$$

$$\frac{\partial \mathbf{v}_1}{\partial t} = -\text{grad }\frac{P_1}{\rho_0} - \text{grad } \Phi_1, \qquad (20)$$

$$\Delta\Phi_1 = 0. \qquad (21)$$

In addition, we must satisfy the condition that the total pressure P on the perturbed boundary be zero; it can be put in the form

$$(P_1 + \xi\nabla P_0)|_{z=c} = 0 \qquad (22)$$

(ξ is the vector of particle displacement), or, since

$$P_0 = \frac{\rho_0(1 - z^2)}{2}, \qquad (23)$$

then

$$(P_1 - \rho_0 \xi_z)|_{z=1} = 0, \qquad (24)$$

and in (23), (24) we put $c = 1$, $4\pi G\rho_0 = 1$. Equations (19)–(21) denote that both P_1 and Φ_1 must be harmonic functions:

$$\Delta P_1 = \Delta\Phi_1 = 0. \qquad (25)$$

Selecting solutions symmetric[2] with respect to the $z = 0$ plane, it is easy to find the dispersion equation

$$\omega^2 = -k \tanh k - \sinh k e^{-k}, \tag{26}$$

as well as the following relationships between the amplitudes of the different quantities characterizing perturbed motion:

$$\frac{v_x}{v_z} = i \frac{\cosh kz}{\sinh kz}, \qquad \frac{P_1}{\rho_0 \Phi_1} = -\frac{k}{e^k \cosh k}. \tag{27}$$

For perturbations having a wavelength greater than the layer thickness (i.e., at $k \ll 1$), we have from (26)

$$\omega^2 \approx -k. \tag{28}$$

These perturbations are unstable.[3]

In the opposite limit, $k \gg 1$, (26) yields the dispersion equation $\omega^2 \simeq +k$, which coincides with the dispersion equation describing deep-water waves in the external gravitational field g [67]: $\omega^2 = gk, g = \omega_0^2 c$. Such a coincidence is natural since the self-gravitation of a wave as $k \to \infty$ does not play any role.

The critical wavelength λ_{cr}, which separates the unstable (fairly long-wavelength) and stable (more short-wavelength) perturbations is obtained from (26) at $\omega^2 = 0$; it turns out to be $\lambda_{cr} \simeq 3.13\pi c$.

These are the main facts relevant to the stability of an incompressible layer.

We should like to call attention to the exact coincidence of dispersion equation (28) (if it is written in dimensional units) with dispersion equation (11) that describes long-wavelength disturbances of a "cold" collisionless layer. It should be pointed out right away that dispersion equations (28) and (11) physically describe quite different disturbances: in the case of (11) they describe the perturbation of surface density σ_1 which is created due to a change in localized density $\rho_0 \to \rho_0 + \rho_1$ while in the case of (28), that due to the curvature of the boundary with $\rho_1 = 0$.

Even dimensional analysis permits us to state that in both cases we should have $\omega^2 \infty - G\sigma_0 k$. Let us make clear the reasons for the *exact* agreement of the dispersion equations in the long-wavelength limit. In the case of symmetrical perturbations with $kc \ll 1$, the motions are concentrated in the plane of the layer,[4] and in both cases they can be described by the same quantities: $v_x(x \| \mathbf{k})$, Φ_1 and σ_1. However, these quantities satisfy the same

---

[2] Antisymmetric perturbations are stable (see, e.g., [209]).

[3] Fairly long-wavelength perturbations of any compressible layer infinite in the (x, y) plane are also unstable.

[4] For a cold collisionless layer this is quite evident, while for an incompressible layer this can be inferred from the above exact solution. In fact, from (27) it follows that at $kc \ll 1$ $|v_x/v_z| \approx 1/k_z \gg 1$.

equations. First, from (27) it follows that the pressure force in Euler's x-equation (20) may be neglected at $kc \ll 1$:

$$\left| \frac{\nabla_x P_1}{\rho_0} \middle/ \nabla_x \Phi_1 \right| = \left| \frac{P_1}{\rho_0 \Phi_1} \right| \sim kc \ll 1.$$

Therefore, it coincides with a corresponding equation for the cold collisionless layer

$$\frac{\partial v_x}{\partial t} = - \frac{\partial \Phi_1}{\partial x}.$$

Secondly, the perturbed surface density σ_1 generates the same potential Φ_1 independent of the means of producing σ_1 (due to the curvature of the boundary or to the localized density perturbation)

$$\Phi_1 = - \frac{2\pi G \sigma_1}{k}.$$

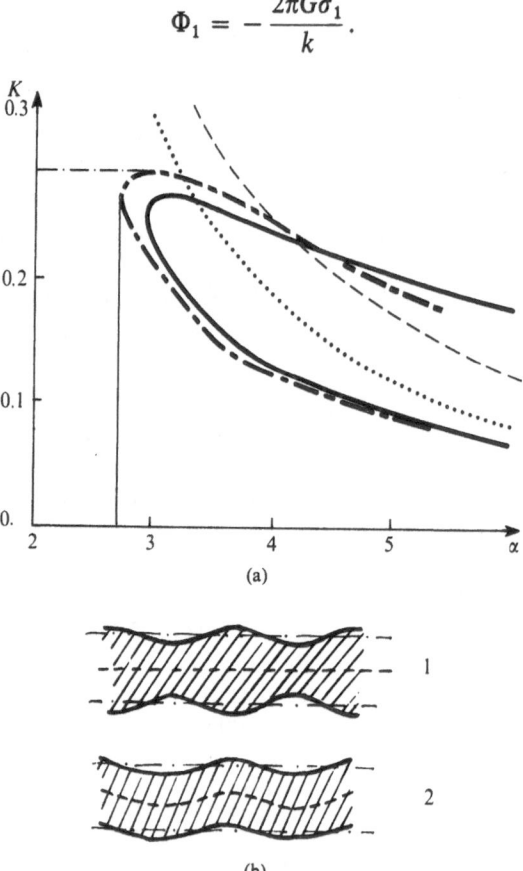

(a)

(b)

Figure 4. Jeans and fire-hose instabilities of a layer. (a) Curves corresponding to stability boundaries for perturbations of Jeans and fire-hose types. Solid line is the exact boundary of the fire-hose instability; dotted line is the exact boundary of the Jeans instability; dash-dotted curve is the approximate boundary for the fire-hose instability; pointed curve is the approximate boundary for the Jeans instability. (b) Schematic representation of the (1) Jeans and (2) fire-hose perturbations of the layer.

Finally, motion taking place in both cases predominantly in the plane of the layer must satisfy the continuity equation

$$\frac{\partial \sigma_1}{\partial t} + \sigma_0 \frac{\partial v_x}{\partial x} = 0.$$

In view of the above, the coincidence of dispersion equations in the two boundary cases under consideration is not surprising. A similar situation is encountered in Chapter V when we investigate the stability of a cold rotating disk.

The disturbances with arbitrary wavelengths ($\lambda \sim c$) must be calculated by means of a computer. The results of such a computation for the model of the uniform layer with Maxwellian distribution function in the (x, y) plane [and having form (9), §1 for motions along the z-axis]:

$$f_0 = \frac{\rho_0}{\pi} \frac{\theta(1 - z^2 - v_z^2)}{\sqrt{1 - z^2 - v_z^2}} \cdot \frac{e^{-v_x^2/2v_T^2}}{\sqrt{2\pi v_T^2}}, \tag{29}$$

are given in Fig. 4(a).

The dotted line in this figure is the stability boundary of the system described with respect to the Jeans perturbations. Naturally at large wavelengths, $kH \ll 1$ (accordingly $\alpha \gg 1$); these results agree with those considered above. On the other hand at the short-wavelength limit ($kH \gg 1$, $\alpha \to 0$), the curve of Fig. 4(a) is described by formula (12) of Problem 1 in the present chapter.

§ 3 Anisotropic (Fire-Hose) Instability of a Collisionless Flat Layer

3.1 Qualitative Considerations

Kulsrud et al. [263] (where the fire-hose instability was first investigated) give the following obvious interpretation of the conditions of appearance of this instability.

Imagine that in some region of a flat layer with linear dimension in x, equal to λ, a bending occurs (for example, upward with value h). The particle moving along the bending layer receives the action of two forces: the gravitational attraction F_{gr} and the centrifugal force F_c. These two forces act in opposite directions; moreover, the gravitation plays a stabilizing role, attempting to return the bent segment to its original position, while the centrifugal force promotes an increase in the bending. Hence follows the necessary condition of the instability: $F_c > F_{gr}$. We discuss below (in Section 4) the question of sufficiency of this condition [33^{ad}].

For perturbations with wavelength λ, much greater than the layer thickness c, $F_{gr} \approx 2\pi Gk\sigma_0 h$ (see Section 2.1, Chapter V). But since, on the other hand, $F_c \approx v_{Tx}^2/r$, where r is the curvature radius of the bending, $r \sim \lambda^2/h$, instability occurs if the condition

$$k^2 v_{Tx}^2 > 2\pi G \sigma_0 k \tag{1}$$

is satisfied. The latter can be written also in the following form [263]:

$$kc > \frac{v_{Tz}^2}{\alpha v_{Tx}^2}, \tag{2}$$

where α is a constant of the order of unity, given by the equality:

$$\alpha v_{Tz}^2 = 2\pi G \sigma_0. \tag{3}$$

The instability criterion is derived for sufficiently long-wave perturbations: $kc \ll 1$. As we shall see below, the disturbances with sufficiently small wavelength (of the order of the layer thickness) are stable. Thus the instability occurs at "intermediate" wavelengths and "supplements" from short wavelengths the unstable region with respect to Jeans disturbances.

3.2 Derivation of the Dispersion Equation for Bending Perturbations of a Thin Layer

Let us obtain the dispersion equation characterizing the fire-hose perturbations with wavelengths large compared to the layer thickness (that is, in the "thin layer" approximation).

We start from the particle motion equation along the z-axis:

$$\frac{d^2 z}{dt^2} = -\frac{\partial \Phi}{\partial z}. \tag{4}$$

Separate the particles which are situated near to the symmetry layer plane $z = 0$. The unperturbed vertical oscillations of these particles are approximately harmonic with the frequency $\omega = \omega_0 \equiv \sqrt{4\pi G \rho_0}$, where ρ_0 is the density at $z = 0$,

$$z = z_0 \cos \omega_0 t + \frac{v_0}{\omega_0} \sin \omega_0 t \equiv Z(z_0, v_0; t). \tag{5}$$

When the bending occurs, the disturbance ψ will be superimposed on the unperturbed motion (5); so now

$$z = Z + \psi. \tag{6}$$

Assuming the perturbations to be small, substitute (6) into (4), and, taking into account that

$$\ddot{Z} = -\omega_0^2 Z, \quad \frac{\partial \Phi}{\partial z} = \frac{\partial \Phi_0}{\partial z} + \frac{\partial \Phi_1}{\partial z} = \omega_0^2 z + \frac{\partial \Phi_1}{\partial z}, \tag{7}$$

(Φ_1 is the perturbation of the potential), we obtain [8] the following equation for the value of the displacement[5] ψ:

$$(\hat{D}^2 + \omega_0^2)\psi = -\frac{\partial\Phi_1}{\partial z}. \tag{8}$$

Here the operator $\hat{D}$ is the Lagrange derivative along the unperturbed particle trajectory; in the given case

$$\hat{D} = \frac{\partial}{\partial t} + v\frac{\partial}{\partial r} - \omega_0^2 z\frac{\partial}{\partial v_z}. \tag{9}$$

Let ψ now denote the Lagrange particle displacements at $z = 0$; for these particles $_z = 0$, $z = 0$, and so

$$\hat{D} = \frac{\partial}{\partial t} + v_\perp\frac{\partial}{\partial r_\perp}, \tag{10}$$

and in the case of a disturbance in the form of a plane wave $\infty\exp(-i\omega t + ikx)$

$$\hat{D} = -i(\omega - kv_x). \tag{11}$$

Due to the condition that the layer thickness be small $\omega_0 \gg |\omega|, |kv_{Tx}|$, and the solution of Eq. (8) may be written as

$$\psi = -\frac{1}{\omega_0^2}\left(1 - \frac{\hat{D}^2}{\omega_0^2}\right)\frac{\partial\Phi_1}{\partial z}. \tag{12}$$

Let us introduce now, analogously to the Jeans perturbations determined in the previous paragraph, the notion of the "fire-hose modes" of the thin layer as a lifting or a falling of each given layer segment as a whole. The forms of these disturbances are illustrated very clearly in Fig. 4(b) which we take from Ref. [263]. For these fire-hose modes the perturbation of the potential may be written in the form

$$\Phi_1 = z \cdot ae^{-i\omega t + ikx}, \qquad a = \text{const}; \tag{13}$$

so

$$\psi = -\frac{1}{\omega_0^2}\left[1 + \frac{(\omega - kv_x)^2}{\omega_0^2}\right]a. \tag{14}$$

Let us now average ψ over the distribution function $f_0(v_x)$ of the particles in the (x, y) plane:

$$\bar{\psi} = -\frac{1}{\omega_0^2}\left(1 + \frac{\omega^2 + k^2 v_T^2}{\omega_0^2}\right)a. \tag{15}$$

On the other hand, it is obvious that the Poisson equation in this case gives

$$a = (1 - kc)\omega_0^2\bar{\psi}. \tag{16}$$

---

[5] For a more detailed and strict derivation of Eq. (8) see §6 of this chapter.

Comparing (16) with the modified Eq. (15), we find the required dispersion equation [263]:

$$\omega^2 = 2\pi G \sigma_0 k - k^2 v_T^2. \tag{17}$$

This equation defines more precisely from the quantitative side the relations discussed already on the qualitative level above.

3.3 Fire-Hose Instability of a Highly Anisotropic Flat Layer [33[ad]]

It is clear from Eq. (17) that the collisionless layer must be unstable with respect to bendings if the particle velocity dispersion in the plane (x, y) is much larger than the velocity dispersion in the perpendicular (z) direction [as was assumed for the derivation of Eq. (17)]. It is convenient to introduce the notation $\alpha = v_{Tx}/v_{Tz}$ for the anisotropy of the velocity distribution of particles, where v_{Tx} and v_{Tz} are the velocity dispersions along the corresponding directions: moreover, we shall assume that $\alpha \gg 1$. According to (17) the instability occurs for perturbations with wavelengths $\lambda < \lambda_1$, $\lambda_1 \approx h\alpha^2$, where $2h$ is the layer thickness. In fact, as will be shown below [33[ad]], there is still one boundary of stability from the short-wavelength side, $\lambda_2 \approx h\alpha$: disturbances with shorter wavelengths are damping. For $\alpha \gg 1$ there is a wide region of instability: $\lambda_2 \lesssim \lambda \lesssim \lambda_1$; moreover, all these wavelengths are much larger than the layer thickness. The instability region narrows with the decrease of α and vanishes at some critical (maximum) anisotropy $\alpha_c \gtrsim 1$.

Below, the dispersion equation for long-wavelength ($\lambda \gg h$) perturbations of the flat layer is obtained. By means of this equation we shall determine the position of the unstable region (λ_1, λ_2) for $\alpha \gg 1$; then we shall trace the process of the gradual disappearance of this region with decreasing α and finally find the approximate value α_c.

Consider for simplicity the model of a collisionless uniform flat layer with the distribution function

$$f(z, v_z, v_x) = \frac{\rho_0}{\pi \omega_0 h} \left(1 - \frac{z^2}{h^2} - \frac{v_z^2}{\omega_0^2 h^2} \right)^{-1/2} \cdot F(v_x), \tag{18}$$

where ρ_0 is the density, $2h$ the layer thickness, $\omega_0 = \sqrt{4\pi G \rho_0}$ the frequency of particle oscillations along the z-axis, and F an arbitrary function of v_x. Let us use dimensionless variables, in which $h = \omega_0 = 1$. Assume the perturbations to be of long wavelength, i.e., $k_x h \ll 1$. However, the parameter $k v_{Tx}/\omega_0 \sim (kh)\alpha$ is not supposed to be small.

Let us consider the distribution function over v_x in the form of a superposition of flows;

$$F(v_x) = \int dv_0 F(v_0) \delta(v_x - v_0), \tag{19}$$

and then consider one particular flow with velocity v_0 and density $\delta\rho_0 = \rho_0 F(v_0)\, dv_0$. The perturbed distribution function of the flow (in the reference system where this flow is at rest) must have the form[6]

$$f = (1 - z^2 - v_z^2 - \chi)^{-1/2}\delta(v_x) + b(1 - z^2 - v_z^2)^{-1/2}\delta'(v_z); \qquad (20)$$

and for the small perturbations $\chi(t, x, z, v_z)$ and $b(t, x, z, v_z)$ one may obtain the following equations:

$$\frac{d\chi}{dt} = 2ikb \cdot (1 - z^2 - v_z^2) + 2v_z \frac{\partial\Phi_1}{\partial z}, \qquad (21)$$

$$\frac{db}{dt} = ik\Phi_1, \qquad (22)$$

where $d/dt = \partial/\partial t + v_z\,\partial/\partial z - z\,\partial/\partial v_z$, Φ_1 is the perturbation of the potential [$\infty \exp(ikx - i\omega t)$]. In the case of long waves ($kh \ll 1$) of interest to us the first term on the right-hand side of Eq. (21) is proportional to $(kh)^2$; so it may be omitted. It is easy to check that in this limit the localized density is not perturbed by bending oscillations: $\rho_1 = 0$; so we may put $\Phi_1 = z$. Then we find from (21)

$$\chi = -\frac{2z}{\omega^2 - 1}, \qquad (23)$$

where a factor, proportional to v_z, is omitted since it makes no contribution either to the boundary displacement or to the volume density. Thus the boundary displacement of the flow considered is determined by

$$\xi = -\frac{1}{2}\chi\Big|_{z=1} = \frac{1}{\omega^2 - 1}. \qquad (24)$$

Evaluating the perturbation of the potential corresponding to (24) and summing over all flows, we find (in the limit $kh \ll 1$)

$$\Phi_1 = -(1 - |k|)z \int \frac{F(v_0)\, dv_0}{(\omega - kv_0)^2 - 1}, \qquad (25)$$

where we took into account the Doppler shift of the frequency: $\omega \to \omega - kv_0$. Comparing with the intial $\Phi_1 = z$, we obtain the dispersion relation

$$1 + |k| + \int \frac{F(v_x)\, dv_x}{(\omega - kv_x)^2 - 1} = 0. \qquad (26)$$

3.4 Analysis of the Dispersion Equation

Assume, for example, a Maxwellian distribution $F = \exp(-v_x^2/v_{Tx}^2)/\sqrt{\pi}\,v_{Tx}$. The average thermal velocity along the z-axis is determined in the following way. The pressure along z is equal to $P_0 = \frac{1}{2}\rho_0\omega_0^2 h^2(1 - z^2/h^2)$, corresponding to a temperature $T_z = \frac{1}{2}\omega_0^2 h^2(1 - z^2/h^2)$. The temperature, averaged

---

[6] Similar δ expansions are considered in detail in the following chapters (in particular, see §§1, 3, Chapter III).

over the layer thickness, is $\overline{T}_z = \omega_0^2 h^2/3$. Finally, the average thermal velocity $v_{Tz} = (2\overline{T}_z)^{1/2} = \sqrt{\frac{2}{3}}\omega_0 h$ and the anisotropy $\alpha = \sqrt{\frac{3}{2}}v_{Tx}/\omega_0 h$.

Represent the dispersion equation in the form

$$\frac{1}{2}\frac{i\sqrt{\pi}}{|k|v_{Tx}}\left[W\!\left(\frac{\omega-1}{|k|v_{Tx}}\right) - W\!\left(\frac{\omega+1}{|k|v_{Tx}}\right)\right] = 1 + |k|, \qquad (27)$$

where $W(x)$ is the Kramp function. For $|\omega - 1|, |\omega + 1| \gg kv_{Tx}$ one may obtain from Eq. (27) the equation derived earlier (in dimensional variables):

$$\omega^2 \approx 4\pi G\rho_0 h|k| - \tfrac{1}{2}k^2 v_{Tx}^2. \qquad (28)$$

This equation describes the long-wavelength anisotropic (fire-hose) instability of a layer, which occurs for $kh > 3/\alpha^2$.

On the other hand, as follows from (27), the short-wavelength disturbances $(kv_{Tx}/\omega_0) \gg 1$ quickly decrease ("resolve") with a decrement $\gamma \sim kv_{Tx} \cdot [\ln(kv_{Tx}/\omega_0)]^{1/2}$.

It is clear that a second boundary of instability is in the region of $(kv_{Tx}/\omega_0) \backsim 1$ or $kh \sim 1/\alpha$ (in [263] an incorrect value $kh \sim 1$ was given for this boundary).

For an exact determination of the position of the instability region let us put $\omega = 0$ in (27). The resulting equation may be reduced to the form $(K \equiv |k|h)$

$$\frac{\sqrt{6}}{\alpha K}\exp\!\left(-\frac{3}{2\alpha^2 K^2}\right)\cdot\int_0^{\sqrt{3/2}/K\alpha} e^{t^2}\, dt = 1 + K. \qquad (29)$$

Equation (29) was solved numerically; the results are shown in Fig. 4(a). In particular, one can see from this figure that the instability region disappears at $\alpha = \alpha_c \approx 2.68$ (at the limit of the approximation used $kh \le 1$; in the given case $k_c h \approx 0.285$).

3.5 Additional Remarks

The computed exact stability boundary for bendings of model (18) is also shown in Fig. 4(a) (by the solid line). It is clear that the approximate analysis (dashed line) describes the instability well enough even at the limit of validity of the method used in §3.3 ($\alpha \sim 1$, $kh \le 1$).

In conclusion, we note that the qualitative study of the conditions under which the fire-hose instability occurs, performed by the authors of [263] (see §1), is slightly simplified compared to the real situation. A similar consideration reflects exactly the physics of the instability in a normal fire hose (where the returning force is produced by the elasticity of the walls), but it does not sufficiently reflect the specifics of collisionless systems we are interested in. In the case of the instability of a normal fire hose, its bending is conserved; fluid particles, coming back to their position in the bend, will increase the bending by a small amount; thus the perturbation is accumulated in an obvious way. But this is not obvious in the collisionless case. Here more accurate physical interpretation of the instability is analogous to the one given in §1, but it is necessary to consider the average motion of the layer as a whole and not the motion of the individual particles. In the limit when, during one oscillation along the z-axis, the particles are practically not displaced in x ($\lambda/v_{Tx} \gg 1\omega_0$), we obtain for average values an equation of the form

$$\frac{\partial^2 \xi_z}{\partial t^2} + \frac{v_{Tx}^2}{2}\frac{\partial^2 \xi_z}{\partial x^2} = F_{gr},$$

which admits the above-mentioned interpretation. The role of the fire-hose walls here plays the equilibrium gravitational field of the system. (Note that in the case of a plasma fire-hose instability a stationary magnetic field plays an analogous role.)

In the opposite limit, $\lambda/v_{Tx} \ll 1/\omega_0$, a particle during the time of one oscillation $1/\omega_0$ crosses many waves; the corresponding perturbations are damping.

The increasing influence of thermal motions in the (x, y) plane on vertical layer oscillations may also be described as the consequence of a "washing out" of the layer boundary—a decrease of the returning gravitational force due to the smallness of vertical displacements of particles with large horizontal velocities (see (24)).

If the flat system is immersed into an extensive "halo," the condition for the disappearance of the fire-hose instability permits us to obtain an estimate of its maximum mass.[7] Indeed, the equilibrium condition gives a relation between the gravitational force, the angular velocity of the centroid, and the average velocity of particular star motions in the equatorial plane of the system. The presence of a massive halo increases the gravitational force which must be compensated (for the same angular velocity) by an increase of particular velocities. This may cause an instability. The investigation of systems finite in the equatorial plane requires quantitative treatment (see below, Chapters II–V).

Note that apparently the fire-hose instability must also occur in plane gaseous systems in the presence of a sufficient anisotropy of the turbulent motions. In particular, it may be essential for the theory of galaxy formation.

§ 4 Derivation of Integro-Differential Equations for Normal Modes of a Flat Gravitating Layer

In the last two sections we have considered either long-wavelength ($\lambda \gg c$) disturbances of the layer, for which the structure of the latter in the vertical z direction plays no role, or arbitrary disturbances, but in the simplest model of a layer of uniform density. Below, the problem is studied in a more general

---

[7] An estimate of minimal mass of the "halo" may be obtained from the condition of stability with respect to barlike perturbations [301] (see Chapter IV, §3).

formulation: which equations describe arbitrary disturbances of a non-rotating collisionless layer of arbitrary structure?

First of all, solve the kinetic equation

$$\frac{df_1}{dt} \equiv \frac{\partial f_1}{\partial t} + \mathbf{v}\,\frac{\partial f_1}{\partial \mathbf{r}} - \frac{\partial \Phi_0}{\partial \mathbf{r}}\frac{\partial f_1}{\partial \mathbf{v}} = \frac{\partial \Phi_1}{\partial \mathbf{r}}\frac{\partial f_0}{\partial \mathbf{v}}, \tag{1}$$

(where in this case $\partial \Phi_0/\partial x = \partial \Phi_0/\partial y = 0, \partial \Phi_0/\partial z \neq 0$) by assuming the right-hand part of this equation

$$\frac{\partial \Phi_1}{\partial \mathbf{r}}\frac{\partial f_0}{\partial \mathbf{v}} \equiv Q$$

to be a known function.

Let us solve Eq. (1) by the method of "integration over trajectories" [138]. Recall (see Introduction) that d/dt is the derivative over the un-perturbed trajectory of a particle. Therefore, integrating (1) over t, one obtains the solution for this equation in the form

$$f_1(x, y, z, v_x, v_y, v_z, t) = \int_{-\infty}^{t} Q(x(t'),\, y(t'),\, z(t');\, v_x(t'),\, v_y(t'),\, v_z(t'),\, t')\, dt',$$

$$\tag{2}$$

where $x(t') \equiv x', \ldots, v_x(t') \equiv v'_x, \ldots$ are Cartesian coordinates and particle velocity components (moving on an unperturbed path) at the time t', if at the time t it is at a point of phase space with coordinates $(x, y, z; v_x, v_y, v_z)$. It is also assumed that $Q \to 0$ at $t' \to -\infty$ so that the integral on the right-hand side of Eq. (2) is convergent. With eigenoscillations $f_1 = \tilde{f}_1(x, y, z; v_x, v_y, v_z)e^{-i\omega t}, Q = \tilde{Q}(x, y, z; v_x, v_y, v_z)e^{-i\omega t}$, (2) can be reduced to the following form:

$$\tilde{f}_1(x, y, z; v_x, v_y, v_z) = \int_{-\infty}^{0} \tilde{Q}(x', y', z'; v'_x, v'_y, v'_z)e^{-i\omega t'}\, dt'; \tag{3}$$

moreover, now $x', \ldots, v'_x, \ldots$ denote the current (at time t') phase coordinates of a particle which traverses the point of phase space $(x, y, z; v_x, v_y, v_z)$ at time $t = 0$. Obviously the solution in the form (3) only has a meaning in the case when the eigenfrequency ω has a positive imaginary part: Im $\omega > 0$ (so that the convergence of the integral at $t' \to -\infty$ is guaranteed). For other ω, the solution of Eq. (1) has to be understood as an analytical continuation of (3) from the region Im $\omega > 0$.

Taking into account that in our case $f_0 = f_0(E_z, v_x, v_y)$, the right-hand side of (1) can be written in the form

$$\frac{\partial f_0}{\partial E_z}\left(\frac{d\Phi_1}{dt} - \frac{\partial \Phi_1}{\partial t}\right) + \left(\frac{\partial f_0}{\partial v_z} - v_x\frac{\partial f_0}{\partial E_z}\right)\frac{\partial \Phi_1}{\partial x}. \tag{4}$$

Owing to the homogeneity of the equilibrium state in the x and y directions, perturbations can be selected in the form proportional to

$$\exp[-i\omega t + i(k_x x + k_y y)],$$

where k_x and k_y are the corresponding components of the wave vector. Further, we shall assume that $k_y = 0$, $k_x = k$ (obviously this is possible without loss of generality).

Then it is easy to reduce the solution for initial equation (1) to the form (below we omit the tilde over the designations of the perturbed amplitudes):

$$f_1(z, v_z, v_x) = \Phi_1(z) \frac{\partial f_0}{\partial E_z}$$

$$+ i \left(\bar{\omega} \frac{\partial f_0}{\partial E_z} + k \frac{\partial f_0}{\partial v_x} \right) \int_{-\infty}^{0} \Phi_1(z') \exp[-i\omega t' + ik(x' - x)] \, dt',$$

$$(5)$$

where $\bar{\omega} \equiv \omega - kv_x$, and the integral is taken over the unperturbed particle trajectory which at $t' = 0$ passes through the point (z, x, v_z, v_x).

It is easy to change the variables z, v_z, v_x to the variables z, E_z, v_x. This is achieved [289a] by averaging (5) over the directions of the velocity v_z (for a given set of values z, E_z, v_x there are two orbits which differ in sign of v_z at time $t = 0$):

$$\bar{f}_1(z, E_z, v_x) = \Phi_1(z) \frac{\partial f_0}{\partial E_z} + \frac{1}{2} \sum_{\text{sgn} \, v_z} i \left(\bar{\omega} \frac{\partial f_0}{\partial E_z} + k \frac{\partial f_0}{\partial v_x} \right)$$

$$\times \int_{-\infty}^{0} \Phi_1(z') \exp[-i\omega t' + ik(x' - x)] \, dt'. \qquad (6)$$

The movement in x is evidently uniform,

$$x' = x + v_x t', \qquad v_x = v'_x, \qquad (7)$$

and the perpendicular (z) motion is determined from the law of energy conservation

$$\tfrac{1}{2}(v'_z)^2 + \Phi_0(z') = E_z = \tfrac{1}{2}v_z^2 + \Phi_0(z). \qquad (8)$$

The perturbed density ρ_1 is determined from $\bar{f}_1(z, E_z, v_x)$ by integration over the velocity

$$\rho_1 = \int \bar{f}_1 \, dv. \qquad (9)$$

If (6)–(9) are substituted in the right-hand side of the Poisson equation

$$\frac{d^2 \Phi_1}{dz^2} - k^2 \Phi_1 = 4\pi G \rho_1, \qquad (10)$$

we arrive at the integro-differential equation for the function $\Phi_1(z)$.

This equation can be simplified [289a]. Represent the solution for $\Phi_1(z)$ in the form of the sum of the even (φ) and odd (ψ) parts with respect to the z coordinate

$$\Phi_1(z) = \varphi(z) + \psi(z), \qquad (11)$$

where

$$\varphi(z) = \varphi(-z), \tag{12}$$

$$\psi(z) = -\psi(-z). \tag{13}$$

It can be readily shown that the equations for the functions $\varphi(z)$ and $\psi(z)$ are separable. Below, these separated equations will be written in an explicit form.

The integral over the infinite interval of t' in (6) can be reduced to a more convenient form. For example, in the case of even modes for an orbit, which at time $t' = 0$ arrives at point $z > 0$ with a positive velocity component $v_z > 0$, the above integral transforms in the following manner. Introduce a new time variable:

$$s = t' + \tau(z), \tag{14}$$

where $\tau(z)$ is the duration of particle motion from $z = 0$ to z (over the shortest path). Then

$$\int_{-\infty}^{0} \varphi[z'(t')] \exp[-i\omega t' + ik(x' - x)] \, dt'$$

$$= \int_{-\infty}^{\tau(z)} \varphi[z'(s)] \exp\{-i\bar{\omega}[s - \tau(z)]\} \, ds. \tag{15}$$

Further, we split the integral obtained in two: $\int_{-\infty}^{\tau(z)} = \int_{-\infty}^{0} + \int_{0}^{\tau(z)}$. The first one can be further split into:

$$\int_{-\infty}^{0} = \int_{-2\tau_0}^{0} + \int_{-4\tau_0}^{-2\tau_0} + \int_{-6\tau_0}^{-4\tau_0} + \cdots \tag{16}$$

[where $4\tau_0(E_z)$ is the period of particle oscillation in z]. It is readily shown, taking into account the symmetry of the mode $\varphi(z) = \varphi(-z)$, that on the right-hand side of (16) we have the sum of a geometrical progression with the denominator $q = \exp(2i\bar{\omega}\tau_0)$. Therefore, the integral of (15) is transformed to

$$\int_{0}^{\tau(z)} \varphi[z'(s)] \exp\{-i\bar{\omega}[s - \tau(z)]\} \, ds$$

$$+ [1 - \exp(2i\bar{\omega}\tau_0)]^{-1} \int_{-2\tau_0}^{0} \varphi[z'(s)] \exp\{-i\bar{\omega}[s - \tau(z)]\} \, ds, \tag{17}$$

then the last term in (17) can be reduced to the form

$$-\frac{\exp[i\bar{\omega}\tau(z)]}{i \sin[\bar{\omega}\tau_0(E_z)]} \int_{0}^{\tau_0(E_z)} \varphi[z'(s)] \cos[\bar{\omega}(\tau_0 - s)] \, ds. \tag{18}$$

In the derivation of (18) the symmetry of the function $\varphi(z)$ has been used again.

Similar expressions can also be written for the second orbit with a negative component v_z at the point z at $t = 0$. Summing up the contributions of the two orbits, we finally get the following integro-differential equation for even (symmetric) modes with respect to z:

$$\left[\frac{d^2}{dz^2} + w(z) - k^2\right]\varphi - \hat{L}_s\left[\left(\bar{\omega}\frac{\partial f_0}{\partial E_z} + k\frac{\partial f_0}{\partial v_x}\right)\varphi\right] = 0. \qquad (19)$$

In this equation

$$w(z) \equiv -4\pi G \int \frac{\partial f_0}{\partial E_z}\,dv, \qquad (20)$$

and $\hat{L}_s$ denotes the operator

$$\hat{L}_s f \equiv -4\pi G \int \left\{\hat{H}f + \frac{\cos[\bar{\omega}\tau(z)]}{\sin(\bar{\omega}\tau_0)}\hat{D}f\right\}dv, \qquad (21)$$

where, in turn, the operators $\hat{H}$ and $\hat{D}$ act according to the formulae

$$\hat{H}f \equiv \int_0^{\tau(z)} f(\xi)\sin[\bar{\omega}\tau(\xi, z)]\,d\tau(\xi),$$

$$\hat{D}f \equiv \int_0^{\tau_0} f(\xi)\cos[\bar{\omega}\tau(\xi, z_0)]\,d\tau(\xi), \qquad \tau(\xi, z) = \tau(z) - \tau(\xi), \qquad (22)$$

where $z_0 = z_0(E_z)$ is the amplitude of z-oscillations of a particle with energy E_z.

In a similar way the equation describing odd (antisymmetric) perturbations with respect to z [289a] are also obtained:

$$\left[\frac{d^2}{dz^2} + w(z) - k^2\right]\psi - \hat{L}_a\left[\left(\bar{\omega}\frac{\partial f_0}{\partial E_z} + k\frac{\partial f_0}{\partial v_x}\right)\psi\right] = 0, \qquad (23)$$

where the operator $\hat{L}_a$ is

$$\hat{L}_a \equiv -4\pi G \int \left\{\hat{H}f - \frac{\sin[\bar{\omega}\tau(z)]}{\cos(\bar{\omega}\tau_0)}\hat{D}f\right\}dv. \qquad (24)$$

Equations (19) and (23) are sufficient for the description of oscillations of flat layers with density $\rho_0(z)$, vanishing on the boundaries $z = \pm c$ smoothly, without any jump. Otherwise, one has also to take into account the perturbation of the potential due to the curvature of the boundaries.

To Eqs. (19) or (23) one should add the boundary conditions of matching with the exterior solution. Since the solution for the Laplace equation

$$\frac{d^2\Phi_1}{dz^2} - k^2\Phi_1 = 0, \qquad (25)$$

satisfying the vanishing condition at infinity is (assuming $k > 0$)

$$\Phi_1 = \begin{cases} e^{-kz}, & z > c, \\ e^{+kz}, & z < -c, \end{cases} \qquad (26)$$

the boundary conditions take the form

$$\frac{d\varphi}{dz} = \mp k\varphi \Biggr\}$$ (27a)

$$\left. \begin{array}{c} \\ \\ \end{array} \right\}, \quad \text{at } z = \pm c.$$

$$\frac{d\psi}{dz} = \mp k\psi \Biggr\}$$ (27b)

Equations (19) and (23) describe, in particular, the effects of the resonance interaction of particles with the gravitational potential wave. Thus, for a symmetrical wave the resonance corresponds to the zeros $\sin \bar{\omega}\tau_0$ of (21) in the denominator:

$$\bar{\omega} \equiv \omega - kv_x = \frac{m\pi}{\tau_0(E_z)} \qquad (m \text{ is integer}). \tag{28}$$

Consequently, those particles are resonant for which the wave frequency (taking into account the Doppler shift) is an integer fraction of the oscillation frequency $\omega_0 \equiv 2\pi/4\tau_0(E_z)$.

In the next section we shall use the equation for symmetric modes (19) for the investigation of the behavior of perturbations near the stability boundary $k \approx k_{cr}$ in the case of systems with isotropic velocity distribution functions.

To conclude the present section, we derive [289a], from Eqs. (23) and (27) describing the antisymmetric modes with respect to the vertical coordinate z, dispersion equation (17) of Section 3 for the bending oscillations of a "thin" layer ($c \ll \lambda$). Expand all the values characterizing the disturbance in powers of k ($\psi = \psi_0 + \psi_1 + \psi_2 + \cdots$). We assume, however, that the velocity dispersion of particles in the (x, y) plane is large, as compared to the velocity dispersion along the z-axis:

$$\frac{v_{Tx}^2}{v_{Tz}^2} \sim O(k). \tag{29}$$

In the lowest order of such a perturbation theory, the solution is known. It corresponds to the displacement of the layer as a whole along the vertical. Denoting the value of displacement by ξ obviously we have

$$\psi_0 = \Phi_0(z + \xi) - \Phi_0(z) \approx \xi \frac{d\Phi_0}{dz}, \tag{30}$$

where, as usual, $\Phi_0 = \Phi_0(z)$ means the unperturbed potential. This solution corresponds to the frequency $\omega = 0$ and, of course, satisfies Eq. (23) of the zero approximation:

$$\frac{d^2\psi_0}{dz^2} + w(z)\psi_0 = 0. \tag{31}$$

The equation for ψ_1 has the form

$$\frac{d^2\psi_1}{dz^2} + w(z)\psi_1 = \hat{L}_a\left[\left(\bar{\omega}\frac{\partial f_0}{\partial E_z} + k\frac{\partial f_0}{\partial v_x}\right)\psi_0\right]. \tag{32}$$

Derive the "condition of compatibility" of Eqs. (31) and (32)—it will just be the sought for dispersion equation. For this purpose, multiply the equation complexly conjugate to (31) by ψ_1, Eq. (32) by ψ_0^*, subtract one from the other, and integrate over z within the layer boundaries $(-c, c)$:

$$\psi_1 \frac{d\psi_0^*}{dz} - \psi_0^* \frac{d\psi_1}{dz} \bigg|_{-c}^{c} = - \int_{-c}^{c} dz\, \psi_0^* \hat{L}_a \left(\bar{\omega} \frac{\partial f_0}{\partial E_z} + k \frac{\partial f_0}{\partial v_x} \right) \psi_0. \tag{33}$$

Taking into account the boundary conditions

$$\frac{d\psi_1}{dz} \bigg|_{\pm c} = \mp k \psi_0 \bigg|_{\pm c}, \qquad \frac{d\psi_0}{dz} \bigg|_{c} = 0,\,^{8} \tag{34}$$

the left-hand side of (33) can be reduced to $2k|\psi_0(c)|^2$. Now transform the right-hand side of Eq. (33). The operator $\hat{L}_a$ involves, according to (24), the action of the operators $\hat{H}$ and $\hat{D}$. The operator $\hat{H}$ in this case yields $(\bar{\omega} \equiv \omega - kv_x)$:

$$\hat{H}f \approx \bar{\omega} \int_0^{\tau} f(\xi)[\tau - \tau(\xi)]\, d\tau(\xi),$$

$$f = \xi \left(\bar{\omega} \frac{\partial f_0}{\partial E_z} + k \frac{\partial f_0}{\partial v_x} \right) \cdot \frac{d\Phi_0}{dz}.$$

Since

$$\frac{d\Phi_0}{dz} = - \frac{dv}{dt},$$

then

$$\int_0^{\tau} \frac{d\Phi_0}{dz}\, d\tau = -v_z(\tau) + v_z(0), \qquad \int_0^{\tau} \frac{d\Phi_0}{dz} \tau\, d\tau = -\tau v + z,$$

$$\hat{H}f = \bar{\omega}\xi \left(\bar{\omega} \frac{\partial f_0}{\partial E_z} + k \frac{\partial f_0}{\partial v_x} \right) [\tau v_z(0) - z]. \tag{35}$$

The action of the operator $\hat{D}$ reduces to

$$\hat{D}f \simeq \xi \left(\bar{\omega} \frac{\partial f_0}{\partial E_z} + k \frac{\partial f_0}{\partial v_x} \right) v_z(0). \tag{36}$$

Thus

$$\hat{L}_a f \approx -4\pi G \xi \int dv \left[\bar{\omega} \left(\bar{\omega} \frac{\partial f_0}{\partial E_z} + k \frac{\partial f_0}{\partial v_x} \right) [\tau v_z(0) - z] \right.$$

$$\left. - \bar{\omega}\tau \left(\bar{\omega} \frac{\partial f_0}{\partial E_z} + k \frac{\partial f_0}{\partial v_x} \right) v_z(0) \right] = 4\pi G \xi z \int dv (\omega - kv_x)^2 \frac{\partial f_0}{\partial E_z}. \tag{37}$$

---

[8] The last equation (34) may also be considered as the condition of vanishing density $\rho_0(z)$ on the layer boundary:

$$\frac{d\psi_0}{dz} \bigg|_{\pm c} \sim \frac{d^2\Phi_0}{dz^2} \bigg|_{\pm c} = 4\pi G \rho_0 \bigg|_{\pm c} = 0.$$

Perform the remaining integration over z. Taking into account that $\partial f_0/\partial E_z = \partial f_0/\partial \Phi_0$, and

$$\int_{-c}^{c} dz \cdot z \cdot \frac{d\Phi_0}{dz} \frac{\partial f_0}{\partial E_z} = \int dz \cdot z \cdot \frac{df_0}{dz} = -\int_{-c}^{c} f_0 \, dz,$$

we obtain

$$\int dz \, \psi_0^* \hat{L}_a f = 4\pi G \xi^2 \int_{-c}^{c} dz \int dv(\omega - kv_x)^2 f_0. \tag{38}$$

Finally, we find that the "consistency condition" employed by us can actually be represented as the dispersion equation (17) of §3:

$$\omega^2 = 2\pi G \sigma_0 k - k^2 \overline{v_{Tx}^2}, \tag{39}$$

where

$$\overline{v_{Tx}^2} = \frac{1}{\sigma_0} \int_{-c}^{c} dz \int dv \, v_x^2 f_0. \tag{40}$$

The manner of averaging of v_x^2 (40) is quite natural.

§5 Symmetrical Perturbations of a Flat Layer with an Isotropic Distribution Function Near the Stability Boundary

The problem formulated in the section title was considered by Kulsrud and Mark [263, 289a]. In [263] the energetic principle was used, and in [289a] the equations for symmetrical normal modes derived in the previous section [see §4, (19)]. Below, we will follow mainly the latter.

Assume that $(k - k_{cr})$ is small and the following estimates are valid:

$$c_z = c_x \sim L/\tau \sim L(G\rho)^{1/2},$$

$$|\omega|\tau \sim \frac{|\omega|}{(G\rho)^{1/2}} \sim |k - k_{cr}|L \ll 1, \tag{1}$$

where c_x and c_z are the dispersions of the x and z components of the velocities and L is the characteristic thickness of the layer.

Then one can expand all the values over $(k - k_{cr})$:

$$\omega = \omega^{(1)} + \omega^{(2)} + \cdots, \qquad \varphi = \varphi^{(0)} + \varphi^{(1)} + \cdots, \tag{2}$$

where $\varphi^{(m)}$ and $\omega^{(m)}$ are of the order of $O[(k - k_{cr})^m]$.

In the lowest order of such a perturbation theory we obtain from Eq. (19) of the previous section

$$\left[\frac{d^2}{dz^2} + w(z) - k_{cr}^2 \right] \varphi^{(0)} = 0. \tag{3}$$

This equation must be solved together with the boundary conditions

$$\frac{d\varphi^{(0)}}{dz} \mp k_{\rm cr}\varphi^{(0)} = 0 \quad \text{at } z = \mp c \tag{4}$$

(for certainty, we assume that $k_{\rm cr} > 0$). Equation (3) is an equation for eigenvalues, which, as can be shown [263, 289a], determines subject to the boundary conditions (4) a unique critical wavenumber $k_{\rm cr}$ and a unique function $\varphi^{(0)}(z)$.

For equilibrium distribution functions (12) of §1 at $n = 2$–6 in [289a] the solutions of Eq. (4) are given in the form of plots. The functions $\varphi^{(0)}(z)$ are monotonously decreasing, with z increasing from 0 to c. In [289a], there is also a table of the relevant critical wavelength $\lambda_{\rm cr} = 2\pi/k_{\rm cr}$ in units of

$$\lambda_J \equiv \frac{c_z^2}{4\pi G\sigma_0}, \tag{5}$$

where σ_0 is the surface density of the layer. With a change of the model parameter n from 2 to 6, $\lambda_{\rm cr}/\lambda_J$ is monotonously decreasing from the value 34.4 to 28.9. At the same time, the relation c/λ_J increases in the same interval of the change of n from 10 to 18. Correspondingly, if $\lambda_{\rm cr}$ is measured in units of the full thickness of the layer $2c$, it turns out that, at $n = 2$, $\lambda_{\rm cr} = 1.72$, and, at $n = 6$, $\lambda_{\rm cr} = 0.80$. These ratios are near to unity.

In the following order of the perturbation theory we have

$$\left[\frac{d^2}{dz^2} + w(z) - k_{\rm cr}^2\right]\varphi^{(1)}(z) = 2k_{\rm cr}(k - k_{\rm cr})\varphi^{(0)} + \omega^{(1)}\hat{L}_s^{(1)}\varphi^{(0)}, \tag{6}$$

where (P is the principal value of the integral)

$$\hat{L}_s^{(1)}\varphi^{(0)} = 4\pi G \int (Q_1 + Q_2)\, dv_z, \tag{7}$$

$$Q_1 = P\int \frac{\partial f_0}{\partial E_z}\left\{\int_0^{\tau(z)} \varphi(\xi)\sin[k_{\rm cr}v_x\tau(\xi, z)]\, d\tau(\xi)\right.$$

$$\left. + \frac{\cos[k_{\rm cr}v_x\tau(z)]}{\sin(k_{\rm cr}v_x\tau_0)}\int_0^{\tau_0} \varphi(\xi)\cos[k_{\rm cr}v_x\tau(\xi, z_0)]\, d\tau(\xi)\right\} dv_x, \tag{8}$$

$$Q_2 = \frac{\pi i}{k_{\rm cr}}\sum_{m=-\infty}^{\infty}\left[\frac{\partial f_0}{\partial E_z}\bigg|_{v_x = m\pi/k_{\rm cr}\tau_0} R_m(z, E_z)\right], \tag{9}$$

$$R_m(z, E_z) \equiv \frac{1}{\tau_0(E_z)}\cos\left[m\pi\frac{\tau(z)}{\tau_0}\right]\int_0^{\tau_0} \varphi(\xi)\cos\left[\frac{m\pi}{\tau_0}\tau(\xi)\right] d\tau(\xi). \tag{10}$$

The boundary conditions on $\varphi^{(1)}$ are the following:

$$\frac{d\varphi^{(1)}}{dz} \pm k_{\rm cr}\varphi^{(1)} \pm (k - k_{\rm cr})\varphi^{(0)} = 0 \quad \text{at } z = \pm c. \tag{11}$$

Due to the parity of the function $f_0(E)$ over v_x, $Q_1 = 0$ and therefore there remain only contributions by resonant particles which are determined by the value Q_2 (the sum over m in the expression for Q_2 must, of course, involve only those resonant particles which are really present in the system).

The frequency $\omega^{(1)}$ is determined from the condition of the consistency of the solutions of Eqs. (3) and (6). This condition is obtained in a standard way. Multiplying the equation conjugate to (3) by $\varphi^{(1)}$ and Eq. (6) by $\varphi^{(0)*}$, subtracting, integrating over z from $-c$ to c, and using boundary conditions (4) and (11) for excluding derivatives at $z = \pm c$, we find

$$(k - k_{\mathrm{cr}})\left[|\varphi^{(0)}(c)|^2 + k_{\mathrm{cr}}\int_{-c}^{c} |\varphi^{(0)}|^2 \, dz\right]$$
$$= i\omega^{(1)}\frac{8\pi^2 G}{k_{\mathrm{cr}}}\sum_m \int \frac{\partial f_0}{\partial E_z}\bigg|_{v_y = m\pi/k_{\mathrm{cr}}\tau_0} \left|\int_0^{\tau_0} \varphi^{(0)}(\xi)\cos\left[\frac{m\pi\tau(\xi)}{\tau_0}\right] d\tau(\xi)\right|^2 \frac{dE_z}{\tau_0},$$

$$(12)$$

[where we have changed from integration over z, v_z to integration over E_z, $\tau(z)$].

Thus, it is seen that at $k < k_{\mathrm{cr}}$ the wave is an aperiodically increasing one, and at $k > k_{\mathrm{cr}}$ an aperiodically decreasing one, and, as we mentioned above, the growth (drop) of the wave occurs exclusively due to resonant particles.

The physical cause of such a situation is the following. Since at $k \approx k_{\mathrm{cr}}$ the wave is nearly stationary in time, most particles are able to cross (during the characteristic time of change of the potential) many wavelengths $\lambda \approx \lambda_{\mathrm{cr}} \sim c$ in the x-direction. Therefore, they almost do not interchange energy with the wave. Exceptions are resonant particles which receive coherent momenta from the wave.

Such a physical picture of instability at $k \approx k_{\mathrm{cr}}$ essentially differs from the picture of long wavelength Jeans instability (considered in §2) which has a purely hydrodynamical character and in whose development practically all the mass of the layer participates. We have seen that in the last case $\omega^2 \sim -k$, i.e., the characteristic time of perturbation growth $\tau \sim \sqrt{\lambda}$. With an increase in the wavelength, τ also increases but significantly slower than λ. Therefore, for rather large λ the greater portion of particles cannot cross (during time τ) a region of the order of λ, so that the particles remain in essence in the region of initial perturbation, where they form clusters due to gravity.

Let us also note a characteristic difference in the form of dispersion equations near k_{cr} for the gaseous ($\omega^2 \sim k - k_{\mathrm{cr}}$) and collisionless [$\omega \sim i(k_{\mathrm{cr}} - k)$] layers, associated directly with the different physics in the development of an instability in either case.

In [289a], also some generalizations of the results mentioned above were considered. Using Eq. (19) of the previous section, one may, for example, in a relatively simple way investigate the influence on the behavior of the perturbations of weak anisotropy distribution function f_0. It turns out that if the difference $[\partial f_0/\partial(\frac{1}{2}v_x^2) - \partial f_0/\partial E_z]$ is negative throughout (and diminishes rapidly in absolute value with increasing v_x^2), then $i\omega$ crosses zero at positive $(k - k_{cr})$, where k_{cr} is again determined from Eq. (3). This corresponds to our intuition: if the velocity dispersion (in the x-direction) diminishes, then we in reality expect that smaller wavelengths become unstable (see Fig. 4a).

§ 6 Perpendicular Oscillations of a Homogeneous Collisionless Layer

6.1 Derivation of the Characteristic Equation for Eigenfrequencies

We shall adopt the following presentation. First, we shall derive the characteristic equation determining the frequencies of eigenoscillations of a plane layer. In the simple example in question, the main ideas of some methods which can be used for the solution of this and other similar problems (see Chapters II–V) are presented. Thereafter, following mainly [8], we will make a more detailed study of oscillations of this model, and, in particular, we will provide a proof of its stability.

The difficulty of the problem consists in the fact that the corresponding distribution function (9), §1, tends to infinity at the boundary of the phase volume occupied by the system (see Fig. 3). Therefore, the conventional method of linearization of the kinetic equation by means of substitution of $f = f_0 + \varepsilon f_1$ ($\varepsilon f_1 \ll f_0$) in this case is, strictly speaking, unsuitable at least because it leads to diverging integrals in the calculation of the perturbed density $\rho_1 \sim \int (\partial f_0/\partial v)\, dv$. This is a formal reflection of the fact that here the change (due to a perturbation) of the boundary of the phase volume of the system necessarily has to be taken into account.

It should be noted that a need for the investigation of singular distribution functions, similar to (9), §1, or for example, δ-like, arises frequently in the study of the stabilities of the gravitating system models. We will encounter similar examples in the course of the book, and not only when solving some fundamental questions or such problems in which it is evident beforehand some "exoticity" of the distribution functions do not play any role, but also when considering the stabilities of quite realistic distributions when the spectra of singular functions of the type of (9), §1, serve as some "matrix", by which spectra of other distribution functions can be reconstructed (see §4, Chapter II, and §4.4 of Chapter V).

6.1.1. Method of Lagrange Shifts in Phase Space [8, 14]. Assume that the initial locations of all particles in phase space (z_0, v_0) statistically satisfy the distribution

$$f_0(z_0, v_0) = \frac{\rho_0}{\pi} \frac{1}{\sqrt{1 - z_0^2 - v_0^2}}. \tag{1}$$

The particles at time $t = 0$ occupy, according to (1), a phase volume confined by the circle

$$z_0^2 + v_0^2 = 1. \tag{2}$$

If there are no perturbations in the system, then the particle being at time $t = 0$ at the point of phase space with the coordinates (z_0, v_0) will at time t pass the point (z, v_z), and the new phase coordinates are expressed in terms of the old ones in the following way:

$$z = z_0 \cos t + v_0 \sin t, \qquad v_z = -z_0 \sin t + v_0 \cos t. \tag{3}$$

These formulae describe the natural *kinematic* evolution of the system, for which the form of the original distribution function (1) is conserved. When a small perturbation is imposed, the phase trajectories are slightly deformed:

$$z = z_0 \cos t + v_0 \sin t + \varepsilon\psi, \qquad v_z = -z_0 \sin t + v_0 \cos t + \varepsilon\varphi, \tag{4}$$

where two correction functions ψ and φ depending on t, z_0, and v_0 are introduced (ε is the small parameter of expansion, $\varepsilon \ll 1$). In reality, however, the knowledge of any one of the functions, for example ψ, is sufficient. The function φ is readily expressed in terms of ψ. For this purpose let us differentiate over time the first of Eqs. (4):

$$\dot{z} = v_z = -z_0 \sin t + v_0 \cos t + \varepsilon \frac{d\psi}{dt}. \tag{5}$$

Comparing (5) with the second of Eqs. (4), we find the relation:

$$\varphi = \frac{d\psi}{dt}. \tag{6}$$

The gravitational field strength, acting on the particle with the coordinate z, is related to ψ by

$$\ddot{z} = F = -\frac{\partial\Phi}{\partial z} = -z_0 \cos t - v_0 \sin t + \varepsilon \frac{d^2\psi}{dt^2}. \tag{7}$$

But on the other hand, we have

$$\frac{\partial\Phi}{\partial z} = \frac{\partial\Phi_0}{\partial z} + \varepsilon \frac{\partial\Phi_1}{\partial z} = z + \varepsilon \frac{\partial\Phi_1}{\partial z}, \tag{8}$$

so that (7) can be rewritten in the form

$$-z - \varepsilon \frac{\partial\Phi_1}{\partial z} = -(z_0 \cos t + v_0 \sin t) + \varepsilon \frac{\partial^2\psi}{\partial t^2}. \tag{9}$$

Hence taking (4) into account, we obtain the equation describing the evolution of the function ψ:

$$\frac{d^2\psi}{dt^2} + \psi = (\hat{D}^2 + 1)\psi = -\frac{\partial \Phi_1}{\partial z}. \qquad (10)$$

In this equation the operator $\hat{D} \equiv d/dt$ is the Stokes derivative in phase space (the time derivative along the unperturbed particle trajectory):

$$\hat{D} = \left(\frac{\partial}{\partial t}\right)_{z_0, v_0} = \frac{\partial}{\partial t} + v_z \frac{\partial}{\partial z} - z \frac{\partial}{\partial v_z}. \qquad (11)$$

If necessary, the right-hand side of (10), can be expressed in terms of ψ [8]. As a result, we get the integro-differential equation

$$\hat{D}^2\psi + \psi = \frac{1}{\pi} \int_{-\sqrt{1-z^2}}^{\sqrt{1-z^2}} \frac{\psi(z, v_z)\, dv_z}{\sqrt{1 - z^2 - v_z^2}}, \qquad (12)$$

which actually was investigated in [8].

However, it is easier to proceed in another way [14]. After determination from (10) the function $\psi(z, v_z)$, the perturbed density ρ_1 can be calculated by the formula

$$\rho_1 = -\rho_0 \frac{\partial \bar{\psi}}{\partial z}, \qquad (13)$$

where $\bar{\psi}$ means displacement averaged over velocities of particles, i.e.,

$$\bar{\psi} = \frac{1}{\pi} \int_{-\sqrt{1-z^2}}^{\sqrt{1-z^2}} \psi(z, v_z) \frac{dv_z}{\sqrt{1 - z^2 - v_z^2}}. \qquad (14)$$

Indeed, let us consider some "elementary flux" of particles, whose unperturbed velocities are within the interval v_z, $v_z + dv_z$. The unperturbed density corresponding to this flux is then

$$\delta\rho_0 = f_0(v_z)\, dv_z = \frac{\rho_0}{\pi} \frac{dv_z}{\sqrt{1 - z^2 - v_z^2}}. \qquad (15)$$

For a fixed flux, the continuity equation[9] evidently must be valid:

$$\frac{\partial \delta\rho_1}{\partial t} + \mathrm{div}(\delta\rho \cdot \mathbf{v}) = 0; \qquad (16)$$

or in Lagrange description:

$$\delta\rho_1 = \delta\rho - \delta\rho_0 = \delta\rho_0 \left[\frac{\partial(x_0, y_0, t_0)}{\partial(x, y, z)} - 1\right]. \qquad (17)$$

---

[9] Also the Euler equation (with pressure equal to zero). Each "elementary flux" can be described hydrodynamically since the velocities of all particles in it are fixed. Then the whole system should be considered as a (self-consistent) set of an infinite number of "elementary fluxes." Such an approach, which seems to be equivalent to kinetics, is called "multiflux hydrodynamics." It provides some visual evidence.

But the Jacobian of the transform $(x, y, z) \to (x_0, y_0, z_0)$ in the linear approximation is

$$\frac{\partial(x, y, z)}{\partial(x_0, y_0, z_0)} = 1 + \operatorname{div} \xi,$$

where ξ is the displacement vector of the particle flux $(r = r_0 + \xi)$; respectively, the Jacobian of the inverse transformation is

$$\frac{\partial(x_0, y_0, z_0)}{\partial(x, y, z)} = 1 - \operatorname{div} \xi; \tag{18}$$

therefore,

$$\delta\rho_1 = -\delta\rho_0 \operatorname{div} \xi. \tag{19}$$

The total perturbed density is determined from (19) by integration over the unperturbed velocities, which is equivalent to averaging the displacement ξ:

$$\rho_1 = -\rho_0 \operatorname{div} \bar{\xi}. \tag{20}$$

In the case in question of the one-dimensional movement (over z) and $\delta\rho_0$, corresponding to (15), formula (20) is evidently equivalent to (13).

It is convenient to introduce in (14) instead of v_z the new variable s, $v_z^2 = (1 - z^2)s^2$:

$$\bar{\psi} = \frac{1}{\pi} \int_{-1}^{1} \psi(z, s\sqrt{1 - z^2}) \frac{ds}{\sqrt{1 - s^2}}. \tag{21}$$

Equations (10), (13), and (21) together with the one-dimensional Poisson equation

$$\frac{d^2\Phi_1}{dz^2} = 4\pi G\rho_1 \tag{22}$$

form a complete system. The problem is to find all the solutions for this system of equations corresponding to eigenoscillations of the layer, i.e. having a time dependence $\sim e^{-i\omega t}$.

First let us consider only "true" eigenoscillations corresponding to collective oscillations of the system rather than to the motions of individual particles (the latter are considered separately below).

A characteristic feature of the system under investigation involves the possibility of selecting the perturbed potential Φ_1 in the form of finite polynomials in powers of z:

$$\Phi_1^{(n)} = a_0 z^n + a_1 z^{n-2} + \cdots \equiv \sum_{k=0}^{[n/2]} a_k(t) z^{n-2k}. \tag{23}$$

In subsequent chapters we will see that a similar property also holds for other homogeneous gravitating systems of ellipsoidal shape[10]: the perturba-

---

[10] A plane homogeneous layer, homogeneous spheres and cylinders, and a disk with a surface density $\sigma_0 \sim \sqrt{1 - r^2/R^2}$ can be considered as degenerate kinds of homogeneous ellipsoids.

tion of the potential $\Phi_1(x, y, z)$ can be represented in the form of finite polynomials in powers of Cartesian coordinates.

It is easy to check that in order to obtain the oscillation frequencies corresponding to (23), it is sufficient to retain during the calculations only the term of highest power z. The rest of the terms, though present in $\Phi_1^{(n)}$, prove to be insignificant for this purpose. This follows already from the very possibility of selecting the solution in the form of (23).

Omitting in (23) all the terms but the first, we put

$$\Phi_1^{(n)} \infty z^n, \tag{24}$$

where we now use the sign ∞ instead of the equality sign in (23). Accordingly, the averaging formula (21) is also written in the form

$$\bar{\psi} \infty \frac{1}{\pi} \int_{-1}^{1} \psi(z, izs) \frac{ds}{\sqrt{1 - s^2}}. \tag{25}$$

To determine the function ψ, corresponding to (24), one must solve the equation

$$(\hat{D}^2 + 1)\psi = -nz^{n-1}. \tag{26}$$

This can be done in different ways. Use, for example, such a method [14]: introduce instead of z, v_z new variables q, q_1: $q = z + iv_z$, $q_1 = z - iv_z$. In these variables the operator $\hat{D}$ takes the form

$$\hat{D} = -i\left(\omega + q\frac{\partial}{\partial q} - q_1\frac{\partial}{\partial q_1}\right).$$

The solution of the equation

$$\hat{D}g = Q(q, q_1), \tag{27}$$

with an arbitrary right-hand side $Q(q, q_1)$ is

$$g - \hat{D}^{-1}Q = i\int_0^\infty Q(qe^\tau, q_1e^{-\tau})e^{\omega\tau}\, d\tau. \tag{28}$$

We have to invert the operator $(\hat{D}^2 + 1)$, but

$$(\hat{D}^2 + 1)^{-1} = -\tfrac{1}{2}i[(\hat{D} - i)^{-1} - (\hat{D} + i)^{-1}]. \tag{29}$$

The operators $(\hat{D} \pm i)$ differ from $\hat{D}$ only in the substitution of $\omega \to (\omega \mp 1)$. Hence we get the required inversion formula

$$(\hat{D}^2 + 1)^{-1}Q = \int_0^\infty Q(qe^\tau, q_1e^{-\tau})\sinh \tau e^{\omega\tau}\, d\tau. \tag{30}$$

Thus, for ψ we will have

$$\psi = -n\int_0^\infty [z(qe^\tau, q_1e^{-\tau})]^{n-1}\sinh \tau e^{\omega\tau}\, d\tau. \tag{31}$$

Since $z = \frac{1}{2}(q + q_1)$, then under the integral in (31) z transforms to

$$\tfrac{1}{2}(qe^{\tau} + q_1 e^{-\tau}) = z \cosh \tau + iv_z \sinh \tau,$$

and we determine

$$\psi = -n \int_0^{\infty} (z \cosh \tau + iv_z \sinh \tau)^{n-1} \sinh \tau e^{\omega \tau}\, d\tau. \tag{32}$$

Averaging, according to (25), is performed by the formula

$$\bar{\psi} = -\frac{nz^{n-1}}{\pi} \int_{-1}^{1} \frac{ds}{\sqrt{1-s^2}} \int_0^{\infty} (\cosh \tau - s \sinh \tau)^{n-1} \sinh \tau e^{\omega \tau}\, d\tau. \tag{33}$$

Then calculating the density ρ_1, using formula (13) and using the Poisson equation (22), we arrive at the following characteristic equation:

$$\frac{1}{\pi} \int_{-1}^{1} \frac{ds}{\sqrt{1-s^2}} \int_0^{\infty} (\cosh \tau - s \sinh \tau)^{n-1} \sinh \tau e^{\omega \tau}\, d\tau = -1,$$

or, introducing $s = \cos \varphi$,

$$\frac{1}{\pi} \int_0^{\pi} d\varphi \int_0^{\infty} (\cosh \tau - \cos \varphi \sinh \tau)^{n-1} \sinh \tau e^{\omega \tau}\, d\tau = -1. \tag{34}$$

This equation, after some permutations, can be reduced to the form [8]

$$\sum_{k=-N}^{N} \frac{(N+k-1)!!(N-k-1)!!}{(N+k)!!(N-k)!!} \cdot \frac{1}{(\omega-k)^2 - 1} = -1, \tag{35}$$

where $N \equiv n - 1$. Let us postpone a detailed investigation of Eq. (35) until subsection 2. Note here only the simplest cases. At $N = 0$ $(n = 1)$ we get the trivial solution $\omega = 0$, $\psi = \text{const}$, corresponding to the displacement of the system as a whole. At $N = 1$ $(n = 2)$, there is a pair of complex conjugate solutions describing homogeneous expansion–compressions of the system:

$$\psi = (\pm i\sqrt{3}z - 2v_z)e^{\pm i\sqrt{3}t}. \tag{36}$$

These pulsations occur with a frequency of $\omega = \sqrt{3}$.

Let us now consider in the same example other methods for investigating stability.

Let us begin with a simple method which may be called *the method of variation of the phase volume of the system* [111, 113].

For an explicit accounting for the displacement of the boundary of the phase region of the system, we make use of a modification of the conventional perturbation theory. Modify the equilibrium distribution function f_0 in the following way:

$$f = \frac{\rho_0}{\pi} \frac{\theta(\varkappa^2 - v_z^2 - \varepsilon \chi)}{\sqrt{\varkappa^2 - v_z^2 - \varepsilon \chi}}, \qquad (\varkappa^2 \equiv 1 - z^2). \tag{37}$$

$\varepsilon\chi(z, v_z, t)$ is the perturbation, $\varepsilon \ll 1$. Such a form of the perturbation theory allows one automatically to take into account both the change of the boundary of the original phase volume of the system and local changes of the phase density.

Substitution of (37) into the kinetic equation yields the following equation for χ:

$$\frac{\partial\chi}{\partial t} + v_z\frac{\partial\chi}{\partial z} - z\frac{\partial\chi}{\partial v_z} = 2v_z\frac{\partial\Phi_1}{\partial z}. \tag{38}$$

It is evident that the left-hand side of (38) coincides with the conventional perturbation theory, but on the right-hand side, instead of $\partial f_0/\partial v_z$ we have $2v_z$. Φ_1 in Eq. (38) is a perturbed potential which is determined by the Poisson equation

$$\Delta\Phi_1 = 4\pi G\rho_1. \tag{39}$$

One has now to express the perturbed density ρ_1 in terms of the function χ. This can be achieved in the following way. The full density (unperturbed + perturbed) is

$$\rho = \frac{\rho_0}{\pi}\int_{-\varkappa}^{\varkappa}\frac{\theta(\varkappa^2 - v_z^2 - \varepsilon\chi)}{\sqrt{\varkappa^2 - v_z^2 - \varepsilon\chi}}\,dv_z. \tag{40}$$

Our aim now is to separate out in (40) the perturbed density ρ_1. Direct expansion of the subintegral expression in (40) over ε results in the already mentioned difficulties with diverging expressions. The proposed method of regularizations involves such a substitution of the integration variable in (40) that the subroot expression assumes (in new variables) the original form. For that purpose we introduce the expressions

$$\varkappa_1^2 = \varkappa^2 - \varepsilon\chi_0, \tag{41}$$

$$v_1^2 = v_z^2 + \varepsilon\chi_1, \tag{42}$$

where $\chi_0 = \chi(v = 0)$, $\chi_1 = \chi - \chi_0$, and transform the integral $(v_z \to v_1)$:

$$\rho = \frac{\rho_0}{\pi}\int_{-\varkappa_1}^{\varkappa_1}\frac{\theta(\varkappa_1^2 - v_1^2)}{\sqrt{\varkappa_1^2 - v_1^2}}\left(1 + \frac{\varepsilon\chi_1}{2v_1^2} - \frac{\varepsilon}{2v_1}\frac{\partial\chi_1}{\partial v_1}\right)dv_1. \tag{43}$$

Hence we obtain just the required rule of calculation of the perturbed density, using the known function χ:

$$\rho_1 = \varepsilon\frac{\rho_0}{2\pi}\int_{-\varkappa}^{\varkappa}\frac{\theta(\varkappa^2 - v_z^2)}{\sqrt{\varkappa^2 - v_z^2}}\left(\frac{\chi_1}{v_z^2} - \frac{1}{v_z}\frac{\partial\chi_1}{\partial v_z}\right)dv_z. \tag{44}$$

The first term (~ 1) in Eq. (43) yields the density ρ_0 in the layer. Its new boundaries are found from the condition

$$\varkappa_1^2 = \varkappa^2 - \varepsilon\chi_0(z) = 1 - z^2 - \varepsilon\chi_0(z) = 0, \tag{45}$$

(however, for the plane perturbations considered here the displacement of the boundary is negligible).

Thus, for the determination of the oscillation frequency spectrum and the corresponding eigenfunctions, one must solve the system of equations (38), (39), (44). The kinetic equation (38) is rather simple; it can be solved, for example, by the method of integration over trajectories (or by integration over the angle) [138].

We introduce the polar coordinates ρ, φ in phase space (z, v_z):

$$v_z = \rho \cos \varphi, \tag{46}$$

$$z = \rho \sin \varphi. \tag{47}$$

Then the kinetic equation (38) for eigenoscillations assumes the form

$$\frac{\partial \chi}{\partial \varphi} - i\omega\chi = 2n\rho^n \cos \varphi \, (\sin \varphi)^{n-1} + \cdots, \tag{48}$$

where the contribution by the leading power z in Φ_1 is singled out and the ellipsis denotes the contribution by the rest of the terms. The general solution of the nonhomogeneous differential equation (48) is constructed from the general solution of the homogeneous equation

$$\chi^{(0)} = C(\rho)e^{i\omega\varphi}, \tag{49}$$

and a partial solution of the inhomogeneous equation. The latter can be represented, for example, in such a form:

$$\chi^{(e)} = 2n\rho^n e^{i\omega\varphi} \int_0^\varphi e^{-i\omega x} \cos x \, (\sin x)^{n-1} \, dx. \tag{50}$$

Now one must impose the periodicity condition

$$\chi(\varphi) = \chi(\varphi + 2\pi), \tag{51}$$

which is the boundary condition for this problem. This leads, as is easily seen, to the following expression for the function $C(\rho)$:

$$C(\rho) = \frac{2n\rho^n e^{2\pi i\omega}}{1 - e^{2\pi i\omega}} \int_0^{2\pi} e^{-i\omega x} \cos x \, (\sin x)^{n-1} \, dx. \tag{52}$$

The solution for the kinetic equation, therefore, is

$$\chi = e^{i\omega\varphi} \left[\frac{e^{i\omega 2\pi}}{1 - e^{i\omega 2\pi}} \int_0^{2\pi} e^{-i\omega x} \cos x \, (\sin x)^{n-1} \, dx \right.$$

$$\left. + \int_0^\varphi e^{-i\omega x} \cos x \, (\sin x)^{n-1} \, dx \right] 2n\rho^n. \tag{53}$$

From the very beginning one could also have introduced the periodic solution for Eq. (48). For this purpose, one should represent the product of trigonometric functions on the right-hand side in the form of a sum of complex harmonics and use the fact that the solution for Eq. (48) for one harmonic

$$\frac{\partial \chi}{\partial \varphi} - i\omega\chi = e^{il\varphi} \qquad (l \text{ is integer}), \tag{54}$$

satisfying the required periodicity condition (51), is the following:

$$\chi_l = \frac{i}{\omega - l} e^{il\varphi}. \tag{55}$$

After calculation of the function χ, one can calculate, using formula (44), the perturbation of the density ρ_1 and, then, after solution of the Poisson equation, also the perturbed potential Φ_1, corresponding to this density, which again is obtained in the form of a polynomial, beginning with z^n. Equating the coefficients at z^n in the expression for Φ_1 obtained in such a way, and in the initial (24), we determine the required characteristic equation.

The calculations are readily completed and the general characteristic equation (35) is readily derived. A special way of solving the kinetic equation has, of course, no fundamental importance, though it can sometimes help to present the characteristic equation in an elegant way. The latter, however, has already been obtained [see (35)]; therefore, we will restrict ourselves in further calculations of the eigenfrequencies to only the simplest mode, corresponding to a homogeneous expansion–compression. This is quite sufficient for the illustration of the main ideas of various methods. Thus, assume $n = 2$; hence

$$\Phi_1 = z^2 + \text{const.} \tag{56}$$

It is clear that the required oscillations with constant layer density, $\rho_1 = \text{const} \neq \rho_1(z)$, correspond to this very mode. Instead of (48), in this case we have

$$\frac{\partial \chi}{\partial \varphi} - i\omega\chi = 4\rho^2 \cos \varphi \sin \varphi = 2\rho^2 \sin 2\varphi = \rho^2 \left(\frac{e^{2i\varphi} - e^{-2i\varphi}}{i} \right). \tag{57}$$

From this, according to (54), we get

$$\chi = \rho^2 \left(\frac{e^{2i\varphi}}{\omega - 2} - \frac{e^{-2i\varphi}}{\omega + 2} \right). \tag{58}$$

Now introduce again in (58) the variables (z, v_z):

$$\chi = \frac{1}{\omega - 2}(v_z + iz)^2 - \frac{1}{\omega + 2}(v_z - iz)^2 = (v_z^2 - z^2)\frac{4}{\omega^2 - 4} + \frac{4i\omega z v_z}{\omega^2 - 4}. \tag{59}$$

According to (41), (42), we separate out the function χ_1 (omitting the v_z-independent term χ_0):

$$\chi_1 = \frac{4}{\omega^2 - 4} v_z^2 + \frac{4i\omega z v_z}{\omega^2 - 4}. \tag{60}$$

The last term in (60) makes no contribution to ρ_1. We calculate ρ_1, substituting (60) into (44) and integrating:

$$\rho_1 = -\frac{2\rho_0}{\omega^2 - 4}. \tag{61}$$

Therefore the perturbed potential Φ_1 is

$$\Phi_1 = -\frac{z^2}{\omega^2 - 4}. \tag{62}$$

Comparing (62) and (56), we get

$$\frac{-1}{\omega^2 - 4} = 1, \qquad \omega^2 = 3 \tag{63}$$

in accordance with the previous method.

Of course, small oscillations of the simple system of the gravitating layer under consideration can also be investigated by other methods, including those closer to the traditional ones. In order to give a definite meaning to diverging expressions, it is necessary, strictly speaking, to introduce generalized functions. However, in this case one can proceed in the following way (similarly to [252]).

We will formally solve the kinetic equation

$$\frac{df_1}{dt} = \frac{\partial f_1}{\partial t} + \frac{\partial f_1}{\partial \varphi} = \frac{\partial \Phi_1}{\partial z} \frac{\partial f_0}{\partial v_z}, \tag{64}$$

linearized in the usual way. Let us assume, for the sake of simplicity, that $\Phi_1 \sim z^2$ again (we already know that the relevant oscillation frequency ω must be equal to $\sqrt{3}$). Substituting $\partial/\partial t \rightarrow -i\omega$, we get, instead of (64),

$$\frac{\partial f_1}{\partial \varphi} - i\omega f_1 = 2zv_z \frac{\partial f_0}{\rho \partial \rho} = 2\rho^2 \frac{\partial f_0}{\partial \rho^2} \sin 2\varphi. \tag{65}$$

We solve this equation by integration over the angle:

$$f_1 = \frac{\rho^2 f_0'}{\omega - 2} e^{2i\varphi} - \frac{\rho^2 f_0'}{\omega + 2} e^{-2i\varphi} = f_0' \left[(z^2 - v_z^2) \frac{-4}{\omega^2 - 4} + \frac{i4\omega z v_z}{\omega^2 - 4} \right]. \tag{66}$$

Now one has to calculate the integral

$$\rho_1 = \int f_1 \, dv_z \sim \frac{4}{\omega^2 - 4} \int v_z^2 \frac{\partial f_0}{\partial \rho^2} \, dv_z = \frac{2}{\omega^2 - 4} \int v_z \frac{\partial f_0}{\partial v_z} \, dv_z, \tag{67}$$

which, however, diverges. Evidently, the divergence is due to the singularity of f_0 at the boundary of the phase volume. We will smear slightly this singularity and regard f_0 as a limit of some sequence of "good" distribution functions $\{f_{0n}\}$, which have no singularity but are otherwise similar to f_0. Assume that this sequence of functions with continuous derivatives converges to f_0 at $n \rightarrow \infty$. Then one can perform the integration in (67) and take the limit $n \to \infty$. Integrating (67) by parts, we get

$$\frac{2}{\omega^2 - 4} \left\{ v_z f_{0n} \bigg|_{-\infty}^{\infty} - \int f_{0n} \, dv_z \right\}.$$

The function f_{0n} itself rather than the derivative stands now under the integral. Since f_0 is integrable, one can now take the limit $n \to \infty$ and then integrate. In this case the integral is simply equal to ρ_0. As for the term $v_z f_{0n} |_{-\infty}^{\infty}$, it is evidently equal to zero for any function f_{0n}. Thus, a rule of calculation of diverging integrals bearing a common character, i.e., applicable not only for the considered mode but also for other oscillation modes, is established. The expression obtained of the type of (67) must be formally integrated by parts and the off-integral term omitted.

Bring the calculation to an end. The perturbed density according to (67) is

$$\rho_1 = -\frac{2\rho_0}{\omega^2 - 4},$$

which is coincident with (61). Consequently, again $\omega^2 = 3$.

6.2 Stability of the Model

Turn now to the proof of the stability of the considered model of the plane self-gravitating layer. For that purpose, we will prove, first, that all the roots ω of the characteristic equation (35) are real and, second, that no two roots coincide. The former implies the absence of exponential instabilities growing $\sim e^{\gamma t}$. The latter (the absence of multiple roots) shows that there also are no power instabilities growing $\sim t^n$, $n > 0$.

6.2.1. Realness of the Eigenfrequencies. Let us first of all prove the following statement [14]. It turns out that in writing the dispersion equation in the form

$$F \equiv \sum_{m=0}^{N} \frac{\gamma_m}{(N - 2m + \omega)^2 - 1} + 1 = 0, \tag{68}$$

all the roots will be real if the following inequalities hold:

$$0 \le \gamma_m \le 1. \tag{69}$$

In fact, in this case, one can easily "localize" all the roots of Eq. (68). Consider now the set of values $\omega = -N - 2, -N, \ldots, N, N + 2$, which will be referred to as "intermediate." For each intermediate value of ω just one term of the sum in (68) is negative. It has a denominator (-1) so that due to (69) $F > 0$. But F changes sign at the point of the pole located in the middle of each of the $(N + 2)$ intervals between neighboring "intermediate" values of ω, and in the same interval F must change sign at least once. Therefore, each of these intervals must contain a root. Since, on the other hand, the power of Eq. (68) after being reduced to a common denominator also happens to be equal to $(N + 2)$, therefore all the roots are localized (and real).

In the example of the gravitating layer treated, according to (35)

$$\gamma_m = \frac{(2m - 1)!! \, (2N - 2m - 1)!!}{(2m)!! \, (2N - 2m)!!}.$$

Here the inequalities of (69) are evidently valid, since

$$\frac{(2m - 1)!!}{(2m)!!} < 1, \qquad \frac{(2N - 2m - 1)!!}{(2N - 2m)!!} < 1.$$

The same method is also applied in proving the realness of the roots of the characteristic equation describing the "flute" oscillations of a collisionless cylinder [14] (Chapter II, §4).

In the case treated here, all of the roots of the characteristic equation (35) can be "enumerated" also in another way [8]. Equation (35) reduces to the following equivalent multiplicative form:

$$-\frac{[\omega^2 - (N - 2)^2][\omega^2 - (N - 4)^2] \cdots \omega^2}{[\omega^2 - (N + 1)^2][\omega^2 - (N - 1)^2] \cdots [\omega^2 - 1]} = 1, \quad \text{at } N \text{ even,}$$

(70)

$$-\frac{[\omega^2 - (N - 2)^2][\omega^2 - (N - 4)^2] \cdots [\omega^2 - 1]}{[\omega^2 - (N + 1)^2][\omega^2 - (N - 1)^2] \cdots [\omega^2 - 4]} = 1, \quad \text{at } N \text{ odd.}$$

(71)

Prove the identity of (70), (71), and (35). The common denominator in (70) is

$$[\omega - (N + 1)][\omega - (N - 1)] \cdots [\omega + (N + 1)],$$

which coincides with the denominator in (35). The power of the numerator in the two cases is less than the power of the denominator; therefore, one has to test the coincidence of expansions in elementary fractions. The expansion coefficients in (70) can be calculated by the formula

$$a_{k+1} = \frac{P(\omega)}{2(k + 1)Q_1(\omega)}\bigg|_{\omega = k + 1},$$

where P is the numerator of the left-hand part of (70) and Q is the denominator divided by $[\omega^2 - (k + 1)^2]$. It can readily be tested that these coefficients coincide with the expansion coefficients in (35) which are determined in a quite elementary way. For odd N the verification is made using the same method. It should only be noted that in this case it seems at first glance that the common denominator is equal to

$$[\omega^2 - (N + 1)^2][\omega^2 - (N - 1)^2] \cdots [\omega^2 - 4]\omega,$$

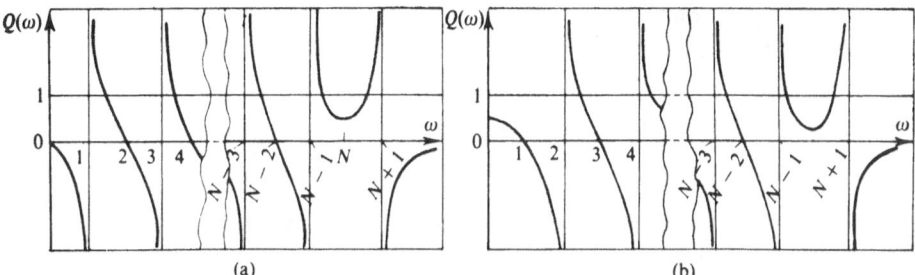

Figure 5. Graphic representation of a dispersion equation for natural frequencies of perpendicular oscillations of a homogeneous plane layer: (a) case of even N [Eq. (70)]; (b) case of odd N [Eq. (71)].

but in reality the elementary fractions with the denominator $1/\omega$ are omitted since they are present only in the terms of the sum with $k = \pm 1$,

$$\frac{1}{1 - (\omega \pm 1)^2} = \frac{1}{2}\left(\mp \frac{1}{\omega} + \frac{1}{2 \pm \omega}\right),$$

and cancel each other out.

The characteristic equation in the form of (70), (71) is quite readily investigated. Denote the left-hand part of (70) or (71) by $Q(\omega^2)$. An approximate plot of this function for the case of even N is presented in Fig. 5(a). In the range $0 < \omega < N$ the function $Q(\omega)$ changes sign whenever ω is equal to a natural number; moreover, Q passes then alternatively through zero or infinity of either sign. The roots of the characteristic equation correspond to the crossing points of $Q(\omega)$ and the horizontal straight line $Q = 1$. Their number is easily determined from the figure: there is one root in each interval $(1, 3), (3, 5), (5, 7), \ldots, (N - 3, N - 1)$ and two roots in the interval $(N - 1, N + 1)$; therefore, we have altogether $(N/2 + 1)$ roots.

A quite similar calculation for the odd N case (Fig. 5(b)) yields $([N/2] + 1)$ roots (here [ ] means an integer part). But this is all, as far as the roots of the characteristic equation (70) or (71) are concerned since the degree of this equation with respect to ω^2 after reduction to a common denominator is just $[N/2] + 1$. Therefore, the first part of the proof of stability is finished: it has been proved that all the roots are real.

6.2.2. Absence of Power Instabilities. Turn now to the more difficult second part of the problem—to the proof of the absence of multiple roots.

Let us explain first of all how in principle the presence of multiple roots can lead to power instabilities. If the power n of the polynomial expansion (23) Φ_1 is fixed, then one can obtain equations for time-dependent coefficients $a_k(t)$ of this expansion. Then the equation for the a_0 coefficient (with the highest power z^n) is homogeneous:

$$\hat{L}_n a_0(t) = 0. \tag{72}$$

When $a_0(t) \sim e^{-i\omega t}$ is substituted, it yields the characteristic equation (35) for frequencies of eigenoscillations corresponding to a given n. Assume that there are multiple frequencies among these. Let, for example, the root $\omega = \omega_0$ be a multiple of N_1. Then, as is known, the solutions of Eq. (72) corresponding to the frequency ω_0 are linear combinations of the following independent solutions:

$$e^{-i\omega_0 t}, te^{-i\omega_0 t}, \ldots, t^{N_1-1}e^{-i\omega_0 t}. \tag{73}$$

Therefore, in this case, a power instability is in principle possible.

The equations for the remaining coefficients in the expansion $\Phi_1^{(n)}$ are nonhomogeneous. The equation for the coefficient a_k can be written symbolically as

$$\hat{L}_{n-2k}a_k(t) = \Lambda(a_0(t), a_1(t), \ldots, a_{k-1}(t)), \tag{74}$$

where Λ is a linear function of the coefficients of higher degree of z in $\Phi_1^{(n)}(a_0, a_1, \ldots, a_{k-1})$ and $\hat{L}_{n-2k}$ is the linear differential operator, which at $\Lambda = 0$ and $a_k \sim e^{-i\omega t}$ yields eigenfrequencies of oscillations for the mode with the subscript $(n - 2k)$. Now assume that there is a common frequency $\omega = \omega_0$ among the oscillation modes (n) and $(n - 2k)$. This will lead to the *resonant* term $a_0(t) \sim e^{-i\omega_0 t}$ on the right-hand side of (74) which determines $a_k(t)$, and correspondingly to the solution for $a_k(t)$, growing as $te^{-i\omega_0 t}$.

As we have seen, there are no divisible roots among the solutions corresponding to one N, but some roots, corresponding to different N, could in principle coincide.

Further, it is convenient to recalculate all of the eigenfrequencies by introducing a two-index notation: ω_{jN}. The second index is N, and the former is that natural number opposite in parity to N which is the nearest to a given positive ω. At a given $N > 2$ the j index, as can easily be seen, for instance, from Fig. 5, assumes the values $1, 3, 5, \ldots, N + 1$ (if N is even) and similarly the values $2, 4, 6, \ldots, N + 1$ (if N is odd). Let us include also the value of $\omega = \sqrt{3}$ with $j = 2$ obtained for $N = 1$.

We will call a series a set of all values of ω_{jN} with the same j but with different N. It is evident that the second index of the terms of the series takes on the values $N = j - 1, j + 1, j + 3, j + 5, \ldots$.

Prove first the lack of coinciding roots within one and the same series j. For the first term we have

$$j - 1 < \omega_{j,j-1} < j, \tag{75}$$

and for the rest

$$j < \omega_{j,N} < j + 1 \qquad (N \geq j + 1). \tag{76}$$

Only the first series is an exception: for it, the term of the type of $\omega_{j,j-1}$ is omitted and it starts with the value $\omega_{1,2} = (\sqrt{15} - \sqrt{3})/2 = 1.071$. The inequalities (75) and (76) assure the lack of coincidence of the first term of the series with the subsequent ones.

Prove now another inequality, which "orders" the rest of the roots:

$$\omega_{j,N} > \omega_{j,N+2} \qquad (N = j + 1, \ldots). \tag{77}$$

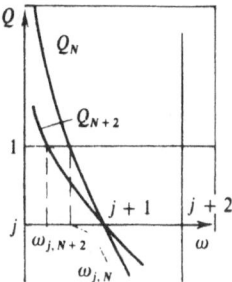

Figure 6. Diagram of the function $Q_N(\omega)$ and the function $Q_{N+2}(\omega)$ in sector $j < \omega < j + 2$.

When increasing N by a factor of 2, a factor less than unity in absolute value is added to the left-hand side of (70) or (71). But from $|Q_{N+2}(\omega_{jN})| < |Q_N(\omega_{jN})| = 1$ and $Q_N(j) = \infty$ (77)[11] follows (see Fig. 6). Inequalities (75), (76), and (77) show the lack of coinciding roots within the limits of one series.

One has to prove that the neighboring series themselves do not mutually overlap. As is evident from Fig. 5, in different (and only in neighboring!) series only roots of the kind of $\omega_{j+1,j}$ and $\omega_{j,j+1}$ can in principle coincide. Let us prove the validity of the following inequality:

$$\omega_{j+1,j} > \omega_{j,j+1} \qquad (j = 1, 2, \ldots). \tag{78}$$

By a rearrangement of the factors in (70) and (71), one can show that

$$\frac{Q_{N+1}(\omega + 1)}{Q_N(\omega)} = \frac{(\omega + 1)(\omega + N)}{\omega(\omega + N + 3)};$$

therefore at $\omega > N/2$ the inequality

$$|Q_{N+1}(\omega + 1)| < |Q_N(\omega)| \tag{79}$$

holds.

Let us apply this inequality to the comparison of the first terms of the series with $j = 2, 3, 4 \cdots$. In this case $N = j - 1$ and the interval $N < \omega < N + 1$ where $\omega > N/2$ is considered, so that inequality (79) is valid and yields

$$|Q_j(\omega_{j,j-1} + 1)| < |Q_{j-1}(\omega_{j,j-1})| = 1,$$

---

[11] The roots of the given series $\omega_{jN}(N \geq j + 1)$ are "contracted" with increasing N towards the boundary $\omega = j$, which is for them a point of condensation. This follows from the system of inequalities

$$\frac{Q_{N+2}(\omega)}{Q_N(\omega)} = \frac{N^2 - \omega^2}{(N + 3)^2 - \omega^2} < \left(\frac{N}{N + 2}\right)^2 \qquad (N \geq j + 1),$$

$$|Q_N(\omega)| < |Q_{j+1}(\omega)| \left(\frac{j + 1}{j + 3}\right)^2 \left(\frac{j + 3}{j + 5}\right)^2 \cdots \left(\frac{N - 2}{N}\right)^2 = \frac{|Q_{j+1}(\omega)|(j + 1)^2}{N^2}$$

so that $\lim_{N \to \infty} Q_N(\omega) = 0$; hence it follows that $\lim_{N \to \infty} \omega_{jN} = j$.

and, taking into account $Q_j(j + 1) = \infty$, it gives

$$\omega_{j+1,j} > \omega_{j,j-1} + 1 \qquad (j \geq 2). \tag{80}$$

If one, however, applies inequality (79) to $N = j + 1$, where $j = 1, 2, 3, \ldots$ in the interval $N - 1 < \omega < N$, we get

$$|Q_{j+2}(\omega_{j,j+1} + 1)| > |Q_{j+1}(\omega_{j,j+1})| = 1,$$

and because $Q_{j+2}(j + 1) = \infty$, then for the second terms of the series we have

$$\omega_{j+1,j+2} < \omega_{j,j+1} + 1 \qquad (j \geq 1). \tag{81}$$

The second series starts with the terms $\omega_{2,1} = \sqrt{3} = 1.732\ldots$; it is evident that $\omega_{2,1} > \omega_{1,2}$ (the value $\omega_{1,2} = \frac{1}{2}(\sqrt{15} - \sqrt{3}) = 1.07 \ldots$; has already been presented). Now, using (80) and (81), by induction we will obtain the required inequality (78). Prove that from $\omega_{j+1,j} > \omega_{j,j+1}$ it follows that $\omega_{j+2,j+1} > \omega_{j+1,j+2}$. In fact, using the inequalities (80), (78), and (81), we get

$$\omega_{j+2,j+1} > \omega_{j+1,j} + 1 > \omega_{j,j+1} + 1 > \omega_{j+1,j+2},$$

which was to be proved.

Thus the proof of the stability of "true" oscillations of a self-gravitating plane layer is brought to an end. It should only be noted that the derivation of the characteristic equation (70) or (71) fails at integral values of the frequency ω. It is clear, however, that the presence of these (real!) frequencies changes nothing in the problem of the system's stability. Such frequencies correspond to trivial "permutational" modes; they are discussed below, in subsection 3.

It is interesting that the system under consideration turned out to be stable (with respect to perturbations with $k_x = k_y = 0$) in spite of the "inverse population" of the energy levels: equilibrium distribution (1) is a growing function of particle energy $E \propto z^2 + v_z^2$.

Kulsrud and Mark [264] proved the stability of perpendicular oscillations of an arbitrary gravitating layer with a *decreasing* distribution function in $v_z(\partial f_0/\partial E_z < 0)$. The above result (together with numerical experiments [215]) shows that for stability the minimum on the distribution function $f_0(v_z)$ must be sufficiently deep and broad (see also [14a]).

6.3 Permutational Modes

Permutational modes describe permutations of particles without change of phase density ($f = \text{const}$). There are possible permutations both within the limits of one orbit and between different orbits. Since no perturbed gravitational field arises, we have

$$\hat{D}^2 \psi + \psi = 0. \tag{82}$$

For $\psi \sim e^{-i\omega t}$ we obtain

$$\frac{\partial \psi}{\partial t} = -i\omega\psi, \qquad \frac{\partial^2 \psi}{\partial \varphi^2} - 2i\omega \frac{\partial \psi}{\partial \varphi} - (\omega^2 - 1)\psi = 0.$$

The latter equation is solved in the general form

$$\psi = [A(\rho)e^{-i(\omega+1)\varphi} + B(\rho)e^{-i(\omega-1)\varphi}]e^{-i\omega t}. \tag{83}$$

The solution of (83) must be single-valued, which is guaranteed only at integer values of ω.

However, it is insufficient to satisfy condition (82). One has then to check the conservation of the original distribution function (1). This will lead to a relationship between the functions $A(\rho)$ and $B(\rho)$ [8].

6.4 Time-Independent Perturbations ($\omega = 0$)

These must be considered separately [8]. One can obtain two different solutions. One of them is simply a displacement of the system as a whole ($\psi = $ const). The second solution is the following:

$$\psi = b(\rho) \sin \varphi. \tag{84}$$

It has a clear geometrical meaning. Rotation of each orbit in the phase plane by a small angle $b(\rho)/\rho$ (different for different orbits) provides each individual particle with a displacement in the z-coordinate just equal to $b(\rho) \sin \varphi$. And the distribution function, evidently, does not change since the phase trajectories of all particles in this case are circumferences. This is one of the types of permutational solutions.

In a similar way [8] one can also treat perturbations with a linear time dependence:

$$\psi = \varphi_0 t. \tag{85}$$

One of them ($\psi = c_1 t$) represents a uniform movement of the system as a whole (a trivial case again). The other solution of the type of (85) is the following:

$$\psi = t b(\rho) \sin \varphi; \tag{86}$$

it corresponds to such a redistribution of masses for which the rotation periods change, but the system as a whole remains stationary.

Problems

1. Determine the long wavelength parts of oscillation branches of a plane gravitational layer homogeneous in density ($kc \ll 1$, $kv_{T\perp} \ll \omega$, $\omega_0 c \ll v_T$) as well as the short wavelength portion ($kc \gg 1$) of the Jeans branch (see beginning of §2).

Solution. The problem of the determination of the frequency spectrum of a "hot" layer in the (x, y) plane can be considered in principle as being solved when the dispersion equation for oscillations of a "cold" layer has been found. If the latter is written symbolically in the form ($F = \varepsilon_0 - 1$)

$$F(\omega, kc) = -1, \tag{1}$$

then obviously the general dispersion equation will have the form (perturbation $\sim e^{ikx}$);

$$\int f_0(v_x) F(\omega - kv_x, kc) \, dv_x = -1, \tag{2}$$

where $f_0(v_x)$ is the equilibrium distribution function [the singularities in this integral should be bypassed as usual, using the Landau rule (see §2)]. Accordingly, the thermal corrections to the frequency for long-wave ($kc \ll 1$) oscillations, satisfying the conditions $kv_T \ll \omega$, $\omega_0 c \ll v_T$, are obtained from the equation

$$\int f_0(v_x) F(\omega - kv_x, 0) \, dv_x = -1, \tag{3}$$

where

$$F(\omega, 0) = -1 \tag{4}$$

is in our symbolical presentation the dispersion equation for the frequencies of *perpendicular* oscillations.

All said above is, of course, also valid for any nonhomogeneous (in z) model, but in the case of a *homogeneous* layer we have explicit expressions for the function $F(\omega, 0)$ [the right-hand sides of Eqs. (70) and (71)]. The squares of eigenfrequencies of perpendicular oscillations of the homogeneous layer determined by Eqs. (70) and (71) can be considered as a limit (at $k \to 0$) for corresponding oscillation branches of $\omega_i^2(k)$. They can be classified (as we know from §§4 and 6) with respect to the evenness of the eigenfunctions [for example, of the perturbed potential $\Phi_1(z)e^{ikx}$] with respect to the middle plane of the layer, $z = 0$, and in the number of nodes n of the function $\Phi_1(z)$.

Let us determine the long-wavelength part of the oscillation branch which at $k \to 0$ turns into homogeneous expansions–compressions of the layer. In this case, according to Eq. (71) in §6 at $N = 1$, the function $F(\omega, 0) = 1/(\omega^2 - 4)$; therefore, Eq. (3) takes the form

$$\int f_0(v_x) \frac{dv_x}{(\omega - kv_x)^2 - 4} = -1. \tag{5}$$

For $\omega \gg kv_T$ this equation gives

$$\omega^2 \approx 3 - \tfrac{13}{2} k^2 v_T^2. \tag{6}$$

One can add to the number of even branches of oscillations of the layer [with the exception of those whose "terminals" are among the solutions of Eq. (71)] also the Jeans branch described in the beginning of §2. In the limit $k = 0$ it corresponds to the trivial solution $\Phi_1 = 0$. In §2, we have determined the Jeans branch at $kc \ll 1$: $\omega^2 = -2\pi G\sigma_0 k = -4\pi Gk\rho_0$. Determine now the short wavelength part of this branch of oscillations. For perturbations with $kc \gg 1$ one can neglect the curvature of the boundary (which decreases exponentially within the layer), and for the branch we are interested in one also can neglect the z-dependence of the perturbed potential. Then the solution of the linearized kinetic equation leads to the following expression for the density perturbation [similar to (3) in §2]:

$$\rho_1 = -k\Phi_1 \int \frac{\partial f/\partial v_x \, dv_x}{\omega - kv_x}. \tag{7}$$

Here the Poisson equation is

$$\Phi_1 = -\frac{4\pi G\rho_1}{k^2}, \tag{8}$$

so that the dispersion equation, ensuing from (7) and (8), has the form

$$\varepsilon_0 \equiv 1 - \frac{1}{k^2} \int \frac{\partial f_0/\partial v_x \, dv_x}{\omega/k - v_x} = 0. \tag{9}$$

For the Maxwell distribution f_0, Eq. (9) is equivalent to

$$\varepsilon_0 = 1 - \frac{1}{k^2 T}\left[1 + i\sqrt{\pi}\,\frac{\omega}{kv_T}\,W\!\left(\frac{\omega}{kv_T}\right)\right] = 0. \tag{10}$$

Therefore at $T \to 0$ we obtain as a natural result the usual Jeansonian instability of a cold infinite medium:

$$\omega^2 = -1. \tag{11}$$

This is apparently the strongest possible instability in the problem under consideration. We again obtain from (10) the condition of stabilization at $\omega = 0$:

$$k^2 T = 1. \tag{12}$$

At $kc \sim 1$ the criteria (12) and (17) of §2 yield, of course, consistent results for the critical temperature

$$T_{cr} \sim 4\pi G\rho_0 c^2. \tag{13}$$

Eliminating b from (21), (22), §3, we arrive at the equation for the function χ $(\sim e^{-i\omega t})$

$$\frac{d^2\chi}{dt^2} = -2k^2(1 - z^2 - v_z^2)\Phi_1 + 2\frac{d^2\Phi_1}{dt^2} + 2i\omega\frac{d\Phi_1}{dt}. \tag{14}$$

At $k = 0$, this equation reduces to (38), §6. The density perturbation and boundary displacement are still expressed by formulae (44) and (45) of §6, respectively. However, at $k \neq 0$ the polynomials in z no longer represent the perturbed potential $\Phi_1(z)$. The main difficulty of the solution of the problem is due to this fact. One can relatively simply determine long wavelength solutions in the form of a series in powers of kc. For example, the first correction to some frequency (ω_n) can be determined by representing the perturbed potential in the form

$$\Phi_1(z) = \alpha k^2 \Phi_1^{(n+2)}(z) + \beta\Phi_1^{(n)}(z) \qquad [\alpha, \beta = O(1)], \tag{15}$$

where $\Phi_1^{(n)}(z) = z^n + \cdots$ is the eigenfunction corresponding to the perpendicular oscillation with a frequency ω_n and $\Phi_1^{(n+2)}$ in the "neighboring" eigenfunction.

2. Test, by direct calculation (using the example of several simplest oscillations), that the eigenfunctions of the problem of small perpendicular oscillations of the layer are proportional to the differences of the "neighboring" Lagrange polynomials of identical parity:

$$\Phi_1^{(n)}(z) \propto P_n(z) - P_{n-2}(z), \tag{1}$$

where the even n correspond to symmetrical oscillations (with respect to the plane $z = 0$), and the odd n, to antisymmetrical oscillations. After this prove (1) for arbitrary n.

Solution. Let us restrict ourselves to even polynomials of power $n \leq 6$; i.e., assume that

$$\Phi_1 = dz^6 + az^4 + bz^2 + c. \tag{2}$$

By calculation of the mean displacement ψ, using formula (14), §6, we find

$$-\bar{\psi} = Az^5 + Bz^3 + Cz, \tag{3}$$

where

$$A = -\frac{3d}{64}\left[\frac{63}{4}Q(6) - 7Q(4) - \frac{5}{4}Q(2)\right], \tag{4}$$

$$B = \frac{15d}{32}\left[\frac{7}{4}Q(6) - 3Q(4) + \frac{3}{4}Q(2)\right] - \frac{a}{4}\left[\frac{5}{2}Q(4) - Q(2)\right], \tag{5}$$

$$C = -\frac{45d}{64}\left[\frac{1}{4}Q(6) - Q(4) + \frac{5}{4}Q(2)\right] + \frac{3a}{4}\left[\frac{1}{2}Q(4) - Q(2)\right] - \frac{b}{2}Q(2) \tag{6}$$

are introduced, and $Q(n) \equiv 1/(\omega - n) - 1/(\omega + n) = 2n/(\omega^2 - n^2)$. We calculate the perturbed density ρ_1 according to (13), §6:

$$\rho_1 = -\rho_0\frac{\partial\bar{\psi}}{\partial z} = 5A\rho_0 z^4 + 3B\rho_0 z^2 + C\rho_0. \tag{7}$$

Then we obtain the following equation for Φ_1:

$$\frac{d^2\Phi_1}{dz^2} = 5Az^4 + 3Bz^2 + C. \tag{8}$$

From this we find

$$\Phi_1 = \frac{A}{6}z^6 + \frac{B}{4}z^4 + \frac{C}{2}z^2 + D \qquad (D = \text{const}). \tag{9}$$

Comparing (9) and (2), we get the following system of equations:

$$\frac{A}{6} = d, \tag{10}$$

$$\frac{B}{4} = u, \tag{11}$$

$$\frac{C}{2} = b, \tag{12}$$

$$D = c. \tag{13}$$

First let us take $d \neq 0$. Then, from (10), we have the dispersion equation

$$-128 = \tfrac{63}{4}Q(6) - 7Q(4) - \tfrac{5}{4}Q(2), \tag{14}$$

which is easily reduced to the form of Eq. (35), §6, at $n = 6$. We shall rewrite Eq. (11) in the following way:

$$[4 + \tfrac{5}{8}Q(4) - \tfrac{1}{4}Q(2)]a = \tfrac{15}{32}[\tfrac{7}{2}Q(6) - 3Q(4) + \tfrac{3}{4}Q(2)]. \tag{15}$$

Taking $Q(6)$ from the dispersion equation (14) and substituting it into (15), we obtain for the right-hand side of (15)

$$-\tfrac{5}{3}[4 + \tfrac{9}{8}Q(4) - \tfrac{1}{4}Q(2)],$$

i.e., $a = -\tfrac{5}{3}$. With this value for a, Eq. (12) takes the form

$$[2 + \tfrac{1}{2}Q(2)]b = -\tfrac{5}{64}[\tfrac{9}{4}Q(6) - Q(4) - \tfrac{19}{4}Q(2)]. \tag{16}$$

But the combination $\tfrac{9}{4}Q(6) - Q(4)$ can be expressed using the dispersion equation (14); as a result the right-hand side of (16) reduces to

$$\tfrac{10}{7}[1 + \tfrac{1}{4}Q(2)],$$

so that $b = \tfrac{5}{7}$. Now, taking into account the fact that

$$P_6(z) = \tfrac{231}{16}z^6 - \tfrac{315}{16}z^4 + \tfrac{105}{16}z^2 - \tfrac{5}{16}, \qquad P_4(z) = \tfrac{35}{8}z^4 - \tfrac{15}{4}z^2 + \tfrac{3}{8},$$

one can easily show that

$$\Phi^{(6)} \infty z^6 - \tfrac{5}{3}z^4 + \tfrac{5}{7}z^2 + D \infty P_6(z) - P_4(z). \tag{17}$$

If we put $d = 0$, $a \neq 0$, then from (11) we get the dispersion equation corresponding to $n = 4$ in (35), §6, and from (12) we determine the coefficient $b = -\tfrac{6}{5}$. This again is in agreement with (1):

$$\Phi^{(4)} \infty z^4 - \tfrac{6}{5}z^2 + D \infty P_4(z) - P_2(z), \tag{18}$$

since

$$P_4(z) = \tfrac{35}{8}z^4 - \tfrac{15}{4}z^2 + \tfrac{3}{8}, \qquad P_2(z) = \tfrac{3}{2}z^2 - \tfrac{1}{2}.$$

Finally, in view of the arbitrariness of the constant D, one can always put:

$$\Phi^{(2)} \infty z^2 + D \infty P_2(z) - P_0(z) = \tfrac{3}{2}z^2 - \tfrac{3}{2}. \tag{19}$$

The case of antisymmetrical perturbations (n is odd) is treated in a similar way.

Now prove formula (1) for arbitrary n. In this case, it is more convenient to start from Eq. (26), §6:

$$(\hat{D}^2 + 1)\psi - -\frac{\partial \Phi_1}{\partial z} \tag{20}$$

and to prove, instead of (1), the equivalent relation

$$\Phi_1^{(N)}(z) = (1 - z^2)\frac{dP_N}{dz} \qquad (N = n - 1). \tag{21}$$

Then Eq. (20) reduces to

$$(\hat{D}^2 + 1)\psi = N(N + 1)P_N(z). \tag{22}$$

The operator D^{-1} acts on $P_N(z)$ according to the formula

$$\hat{D}^{-1}P_N(z) = \int_{-\infty}^{0} e^{-i\omega t}P_N(z \cos t + v_z \sin t)\, dt. \tag{23}$$

Respectively,

$$\psi = -N(N + 1)\int_{-\infty}^{0} dt\, e^{-i\omega t} \sin t P_N(z \cos t + v_z \sin t). \tag{24}$$

Averaging ψ over velocities and using the Poisson equation, we find

$$-P_N(z) = \int_{-\infty}^{0} \sin t e^{-i\omega t} \, dt \cdot \frac{1}{\pi} \int_{-\varkappa}^{\varkappa} P_N(z \cos t + v_z \sin t) \frac{dv_z}{\sqrt{\varkappa^2 - v_z^2}}, \qquad (25)$$

where $\varkappa \equiv \sqrt{1 - z^2}$. By means of change $v_z = \varkappa \sin \varphi$, transform the internal integral in (25); then we shall have

$$-P_N(z) = \int_{-\infty}^{0} dt e^{-i\omega t} \sin t \frac{1}{\pi} \int_{-\pi/2}^{\pi/2} P_N(z \cos t + \sqrt{1 - z^2} \sin t \sin \varphi) \, d\varphi. \qquad (26)$$

Now, of course, to calculate the integral over φ one has to make use of the familiar addition theorem [42]; in this case it yields

$$P_N(z \cos t + \sqrt{1 - z^2} \sin t \sin \varphi) = P_N(z)P_N(\cos t)$$

$$+ 2 \sum_{k=1}^{\infty} \frac{(N - k)!}{(N + k)!} P_N^k(z)P_N^k(\cos t) \cos\left(\frac{k\pi}{2} - k\varphi\right). \qquad (27)$$

But, as one can easily show, in integrating over φ, from the sum in (27) only the first term remains, which is proportional to $P_N(z)$. Thus, the proof of formula (1) is completed. Simultaneously, we obtain another fairly elegant form of notation for the dispersion equation for perpendicular layer oscillations:

$$1 + \int_{-\infty}^{0} dt e^{-i\omega t} \sin t \, P_N(\cos t) = 0. \qquad (28)$$

Note in conclusion that the first proof of formula (1) was given (in a different way) by Kalnajs [253].

3. Consider the nonlinear evolution of perturbations of a plane layer that conserve the spatial homogeneity of density (Kalnajs [253]).

Solution. To solve the problem, we shall use a method somewhat different from that used in the original work of Kalnajs (the idea of this method belongs to Antonov).

Represent the Lagrange coordinate of the particle at an arbitrary time $z = z(t)$ in the form of a linear combination of the original coordinate z_0 and the velocity $\dot{z}_0$:

$$z(t) = \alpha(t)z_0 + \beta(t)\dot{z}_0, \qquad (1)$$

where $\alpha(t)$ and $\beta(t)$ are unknown functions of time. Differentiating (1) in t, we shall determine the velocity $\dot{z}(t)$ of the particle at an arbitrary instant:

$$\dot{z}(t) = \dot{\alpha}z_0 + \dot{\beta}\dot{z}_0. \qquad (2)$$

Invert formulae (1), (2)

$$z_0 = \dot{\beta}z - \beta\dot{z}, \qquad (3)$$

$$\dot{z}_0 = -\dot{\alpha}z + \alpha\dot{z}, \qquad (4)$$

where it is taken into account that the Jacobian of the transformations (1) and (2) $(J = \alpha\dot{\beta} - \beta\dot{\alpha})$ is unity. Substituting (3) and (4) into formula (9), §1, for $f(z_0, \dot{z}_0)$, we get

$$f(z, \dot{z}) = f_0[z_0(z, \dot{z}), \dot{z}_0(z, \dot{z})] = \frac{\rho_0}{\pi} x^{-1/2}\theta(x), \qquad (5)$$

where

$$x \equiv 1 - (\dot{\alpha}^2 + \dot{\beta}^2)z^2 - (\alpha^2 + \beta^2)\dot{z}^2 + 2(\alpha\dot{\alpha} + \beta\dot{\beta})z\dot{z}. \qquad (6)$$

Let us calculate the density

$$\rho = \int f(z, \dot{z}) \, d\dot{z} = \frac{\rho_0}{\sqrt{\alpha^2 + \beta^2}}. \tag{7}$$

Here it is seen that the density ρ for perturbations of the type under consideration, (1) and (2), really remain constant within the layer with a semithickness

$$c = \sqrt{\alpha^2 + \beta^2}. \tag{8}$$

The last formula for c can be derived also from the condition for the maximum of the z-coordinate determined by expression (1) with the additional condition

$$z_0^2 + \dot{z}_0^2 = 1, \tag{9}$$

which is the law of energy conservation for particles that reach the boundary.

To obtain the "equations of motion," let us calculate the Lagrange function $L = T - U$, where T and U are the kinetic and potential energy of the layer, respectively, calculated for unit area.

The kinetic energy T is calculated directly

$$T = \tfrac{1}{2} \int \dot{z}^2 \, dm = \tfrac{1}{2} \iint (\dot{\alpha}^2 z_0^2 + \beta^2 \dot{z}_0^2 + 2\beta \dot{\alpha} z_0 \dot{z}_0) f_0(z_0, \dot{z}_0) \, dz_0 \, d\dot{z}_0$$

$$= \frac{\rho_0}{3} (\dot{\alpha}^2 + \beta^2). \tag{10}$$

In formula (10), f_0 is the equilibrium distribution function

$$f_0(z_0, \dot{z}_0) = \frac{\rho_0}{\pi} \frac{1}{\sqrt{1 - z_0^2 - \dot{z}_0^2}} \qquad (\omega_0^2 \equiv 4\pi G \rho_0 = 1, \, c_0 \doteq 1). \tag{11}$$

In the derivation of (10), the relations

$$\iint z_0^2 f(z_0, \dot{z}_0) \, dz_0 \, d\dot{z}_0 = \iint \dot{z}_0^2 f(z_0, \dot{z}_0) \, dz_0 \, d\dot{z}_0 = \frac{2\rho_0}{3},$$

$$\iint z_0 \dot{z}_0 f(z_0, \dot{z}_0) \, dz_0 \, d\dot{z}_0 = 0 \tag{12}$$

are employed.

In the determination of the potential energy in this case one has to be careful because of the infinite mass of the system considered. Thus, a conventional definition of the potential energy as the work which is required to construct a given layer out of matter originally at infinity is inapplicable here because the work calculated in this way is diverging. One can, however, alter the origin of the energy. It is convenient to adopt the potential energy of an infinitely thin layer (with the same surface density $\sigma_0 = 2\rho c$) as the origin. Then, the potential energy of the homogeneous layer with thickness c is, according to the definition, the deformation energy of an infinitely thin layer into a layer with thickness c. The potential energy thus determined is finite. It is readily calculated[12]:

$$U = \frac{1}{8\pi} \int \left[\left(\frac{\partial \Phi_0}{\partial z} \right)^2 - \left(\frac{\partial \Phi}{\partial z} \right)^2 \right] dz = \frac{8}{3} \pi G \rho^2 c^3 = \frac{2}{3} \rho_0 c. \tag{13}$$

---

[12] Note that (13) can be calculated as a special case of the general formula [180] for the potential energy of an *nonhomogeneous* layer with density $\rho(z)$ (determined by the method described above):

$$U = \int_{-\infty}^{\infty} z \frac{\partial \Phi}{\partial z} \rho(z) \, dz.$$

In (13), formulae

$$\frac{\partial \Phi_0}{\partial z} = 4\pi G \rho c = 2\pi G \sigma_0, \quad \text{for all } z, \tag{14}$$

$$\frac{\partial \Phi}{\partial z} = \begin{cases} 4\pi G \rho c, & z > c, \\ 4\pi G \rho z, & z < c, \end{cases} \tag{15}$$

are used. The index "0" corresponds to an infinitely thin layer.

Therefore, the Lagrange function can be written in the form

$$L = T - U = \frac{\rho_0}{3}(\dot{\alpha}^2 + \dot{\beta}^2 - 2\sqrt{\alpha^2 + \beta^2}). \tag{16}$$

It is convenient, instead of α, β, to introduce the polar coordinates c, φ,

$$c^2 = \alpha^2 + \beta^2, \qquad \alpha = c \cos \varphi, \qquad \beta = c \sin \varphi. \tag{17}$$

Then L can be rewritten as:

$$L = \frac{\rho_0}{3}(\dot{c}^2 + c^2 \dot{\varphi}^2 - 2c). \tag{18}$$

Constructing the equations of motion for the "generalized coordinates" $c, \varphi \equiv q_1, q_2$,

$$\frac{d}{dt}\frac{\partial L}{\partial \dot{q}_i} = \frac{\partial L}{\partial q_i} \qquad (i = 1, 2), \tag{19}$$

we find

$$\ddot{c} = c\dot{\varphi}^2 - 1, \tag{20}$$

$$c^2 \dot{\varphi} = \text{const.} \tag{21}$$

The constant in (21) is determined as $c_0^2 \dot{\varphi}_0$ (where the index "0" refers to the equilibrium state) and turns out to be unity. Then Eq. (20) can be rewritten in the form

$$\ddot{c} = \frac{1}{c^3} - 1. \tag{22}$$

Equation (22) coincides with the equation of a one-dimensional motion of the particle with the potential energy

$$W(c) = \frac{1}{2c^2} + c; \tag{23}$$

moreover, c plays the role of a coordinate. The plot of the $W(c)$ function is given in Fig. 7; from it, it is evident that the motion has an oscillative character. Equation (22) is readily integrated:

$$t - t_0 = \int^c \frac{dc}{\sqrt{2E - 1/c^2 - 2c}}, \tag{24}$$

where the "energy" is denoted by E:

$$E = \frac{1}{2}\dot{c}^2 + \frac{1}{2c^2} + c. \tag{25}$$

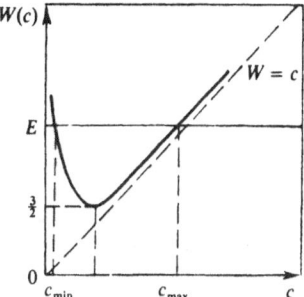

Figure 7. Diagram of the effective potential energy $W(c)$ for perpendicular oscillations of a layer, with preservation of uniformity of density.

It is evident that c oscillates between the "turning points" $c_{min}(E)$ and $c_{max}(E)$, which are determined by the equation

$$E = W = \frac{1}{2c^2} + c. \tag{26}$$

If we set

$$c = 1 + \varepsilon \quad (\varepsilon \ll 1), \tag{27}$$

then by linearizing Eq. (22) we obtain

$$\ddot{\varepsilon} + 3\varepsilon = 0, \tag{28}$$

i.e., the result familiar to us from linear theory: harmonic oscillations with a frequency $\omega = \sqrt{3}$.

The movement of the boundary is described by a simple law in another limiting case: at $E \to \infty$. Then, for most of the time $c \gg 1$, and in the radicand in formula (24) one can neglect the term $1/c^2$. As a result, the law of change of $c(t)$ similar to the law of free fall of a body in a constant gravitational field (with field strength $g = 1$) is obtained:

$$c(t) \approx c_0^{(1)} - \frac{(t - t_0)^2}{2} \quad (c_0^{(1)} \simeq E). \tag{29}$$

In [253], a table is given, which describes the boundaries c_{min}, c_{max}, periods and frequencies of oscillations of the type considered for different values of E.

In conclusion, let us note that Eq. (22) is a special case (for $\gamma = 3$) of the equation

$$\ddot{c} = c^{-\gamma} - 1 \tag{30}$$

that describes nonlinear automodel motions of the homogeneous gaseous layer with the adiabatic index γ.

Equilibrium and Stability of a Collisionless Cylinder

This chapter deals with cylindrical models, which may simulate stellar systems strongly elongated in one direction. We shall assume that there is no rotation around the x- and y-axes; at the same time, the cylinder as a whole and the particles can rotate about the z-axis. Elongated systems rotating with respect to the transverse axis (x or y) are treated in Chapter IV, which is devoted to ellipsoidal (two- or three-axial) systems. Such cylindrical systems seem to be of interest, first of all, as the simplest self-consistent models of gravitating systems for the investigation of some principal questions of the theory (see, for instance, §1, Chapter VI, where the possibility of the development of a beam instability in gravitating systems is shown).

Possible applications may be the so-called needle-shaped galaxies which are very extended along the rotation axis (for example, NGG 2685 [317]), or extended formations of the type of bridges and connections in some anomalous and disturbed galaxies [36]. However, at the present time little is known about the physical composition and structure of all these systems.[1]

We start with a brief review of equilibrium models of collisionless cylindrical configurations (§1). But the main aim of this chapter is a detailed discussion of possible instabilities of the cylinder (§§2, 3). Similarly to a collisionless layer investigated in the previous chapter, here there are

---

[1] An interesting hypothesis was also considered in [82^{ad}] namely that elliptical galaxies are perhaps elongated systems. However, in this case, an essential role must also be played by the limit dimensions of the real system as well as a possible rotation about the small axis.

also two main instabilities "supplementing" each other: Jeans (gravitational) and anisotropic (fire-hose).[2] The Jeans instability of a collisionless cylinder was considered in a large number of papers, regrettably far from being always correct, which has led to significant confusion. Clarity was introduced by paper [88], which is discussed in detail in §2. Instability occurs for all velocity dispersions for perturbations with sufficiently large wavelengths along the cylinder axis. However, with increasing velocity dispersion, the critical wavelength grows exponentially, i.e., practically, the system becomes stable (with respect to disturbances of the Jeans type).

The fire-hose instability of cylindrical models first investigated in [38^{ad}] is discussed in §3. It is also shown there that the simultaneous influence of both main types of instabilities leads to strong limitations on possible parameters of cylindrical systems. For instance, in sufficiently thin cylinders, the regions of Jeans and fire-hose instabilities are overlapping, so that such systems are absolutely unstable.

Perturbations independent of the longitudinal coordinate z, $k_z = 0$ ("flute") are, as a rule, stable. Their investigation constitutes the content of §4 and §5. In §4, an accurate investigation of eigenoscillations of a "hot" homogeneous cylinder, with the distribution function of partial kind, is made. In §5, an application of the method of the local dispersion equation for the analysis of small-scale perturbations of systems with nearly circular orbits is carried out (moreover, one is able to obtain a coherent picture of oscillation branches, which qualitatively remains correct also for perturbations on a larger scale). We emphasize here that the technique of investigation of flute perturbations in the case of a cylinder are largely coincident with the case of disturbances of disk systems which are of special interest in connection with the theory of spiral structure of galaxies (Chapter V); the difference is mainly due to the solution of the Poisson equation. In §6, a comparison of oscillations of a collisionless and incompressible cylinder, uniform in density, is made. Finally, in §7, flute oscillations of the inhomogeneous cylinder are considered.

Problems nos. 1–4 somewhat supplement the main text. The last two problems (nos. 5, 6) are devoted to the consideration of the nonlinear evolution of the simplest flute modes.

§ 1 Equilibrium Cylindrical Configurations

Since along the cylinder envelope no forces are present, the velocity distribution of particles over v_z can be arbitrary. Therefore, the dependence on v_z is omitted in the formulae of this paragraph.

---

[2] We do not consider here the possibility of such instabilities, like the two-beam instability which arises in the case of specially constructed ("beam") distribution functions of particles in longitudinal velocities. Such instabilities will be dealt with in Chapter VI.

Systems with circular orbits

$$f_0 = \rho_0(r)\delta(v_r)\delta(v_\varphi - v_0) \tag{1}$$

possess the simplest velocity distribution.

The dependence of the density $\rho_0(r)$ on the radius can be arbitrary. Equilibrium occurs due to an exact balance of gravity and centrifugal forces:

$$\frac{v_0^2}{r} = \frac{\partial \Phi_0}{\partial r}.$$

Here $\Phi_0(r)$ is the potential of the stationary gravitational field satisfying the Poisson equation

$$\frac{1}{r}\frac{d}{dr}\left(r\frac{d\Phi_0}{dr}\right) = 4\pi G\rho_0(r).$$

Further, we need first of all the simplest model of the homogeneous cylinder of density ρ_0 and radius R with a sharp boundary [20]:

$$\rho_0(r) = \rho_0\,\theta(R - r), \qquad \Phi_0(r) = \frac{\Omega_0^2 r^2}{2} + \text{const},$$

where $\Omega_0^2 \equiv 2\pi G\rho_0$, Ω_0 has the physical meaning of the rotation frequency of the particle in circular orbit (of arbitrary radius). If all particles revolve in circular orbits in one direction, then $v_{0\varphi} = \Omega_0 r$, i.e., the cylinder, as a whole, is revolving as a solid body at an angular velocity Ω_0.

One can construct also models of homogeneous cylinders with arbitrary *elliptical* orbits in the (x, y) plane [19]:

$$f_{0\gamma} = \frac{\rho_0}{\pi}\,\delta[(1 - \gamma^2)(1 - r^2) - v_x^2 - v_y^2] \qquad (|\gamma| \le 1). \tag{2}$$

The distribution function (2) is written in the reference system revolving at a velocity γ (in units of Ω_0); in this system, the cylinder as a whole is at rest. It can readily be tested that the distribution function (2) satisfies the required conditions: It is dependent on the integrals of motion of particles (E and L_z) and in the integration over v_x, v_y yields the density ρ_0 for $r \le 1$ and 0, if $r > 1$. At $\gamma = 0$, (2) describes a cylinder at rest, and for $|\gamma| = 1$, a cylinder with circular orbits of particles. For $|\gamma| \neq 1$, in the cylinder that is described by distribution function (2) there are particles with arbitrary elliptical orbits.

In the inertial reference system, distribution function (2) will be written as

$$f'_{0\gamma} = \frac{\rho_0}{\pi}\,\delta(1 - \gamma^2 - 2E + 2\gamma L_z). \tag{3}$$

For cylinders that are revolving as a solid body, the following "principle of superposition" is readily established: All functions of the form

$$f_0 = \int A(\gamma)f'_{0\gamma}\,d\gamma, \qquad \int_{-1}^{1} A(\gamma)\,d\gamma = 1, \tag{4, 5}$$

where $A(\gamma)$ is an arbitrary function of the parameter γ, are also possible distribution functions of homogeneous cylinders revolving as a solid body (with the density ρ_0). Formula (4) determines a rather wide set of models, including those with a quite realistic dependence on velocities of the equilibrium distribution.

It is essential that the oscillation spectra of the "composite" models of the type (4) are automatically determined using the known spectrum for the original model (3).

A similar principle of superposition also holds, as we shall see below, for disks, for which it was originally formulated in [21] (see §1, Chapter V).

One can also construct series of distribution functions of the type

$$f_0 \sim L_z^\beta [\Phi_0(a) - E]^\alpha, \tag{6}$$

$$f_0 \sim L_z^\beta [\Phi_0(a) - E + L_z^2/2a^2]^\alpha, \tag{7}$$

where α, β are constants and a is the radius. We will restrict ourselves to the consideration of the first of these series [22]. The density corresponding to it is

$$\rho_0 = a_1 r^\beta [\Phi_0(a) - \Phi_0(r)]^{\alpha + (\beta + 2)/2},$$

$$a_1 \equiv 2^{(\beta + 4)/2} \int_0^1 \varkappa^\beta \, d\varkappa \int_0^{\sqrt{1 - x^2}} (1 - x^2 - y^2)^\alpha \, dy.$$

Hence, in particular, it is evident that we shall have $\alpha > -1$, $\beta > -1$, since otherwise the last integral is divergent. The equation for the dimensionless potential

$$y = \Phi_1/\Phi_{1c}, \qquad r_0^{-2-\beta} = 4\pi G a_1 \Phi_{1c}^{\alpha + \beta/2}, \qquad \Phi_1 = \Phi_0(a) - \Phi_0(r)$$

is

$$xy'' + y' + x^{\beta + 1} y^{\alpha + (\beta + 2)/2} = 0 \qquad (x \equiv r/r_0). \tag{8}$$

At $\alpha + (\beta + 2)/2 = 1$, we get the solution in the form of the Bessel function:

$$y = J_0(2/(\beta + 2)x^{(\beta + 2)/2}).$$

To conclude this section, let us give a brief description of the so-called *epicyclic approximation*, which is in common use in investigations of systems with orbits close to circular ones. In case of cylinders, we shall deal with the motions of particles in the (x, y) plane.

In the polar coordinates r, φ, the equations of motion are written as

$$\ddot{r} - r\dot{\varphi}^2 = -\frac{\partial \Phi_0}{\partial r}, \qquad \frac{d}{dt}(r^2 \dot{\varphi}) = -\frac{\partial \Phi_0}{\partial \varphi}, \tag{9, 10}$$

For the axial-symmetrical potential Φ_0, Eq. (10) gives a law of conservation of angular momentum of a particle:

$$r^2 \dot{\varphi} = r v_\varphi = L_z = \text{const.} \tag{11}$$

Then Eq. (9) reduces to

$$\ddot{r} = -\frac{\partial \Phi_0}{\partial r} + \frac{L_z^2}{r^3} = -\frac{\partial W}{\partial r}, \tag{12}$$

where the effective potential energy is

$$W = \Phi_0 + \frac{L_z^2}{2r^2}. \tag{13}$$

The full energy E is

$$E = \frac{1}{2}\left(\dot{r}^2 + \frac{L_z^2}{r^2}\right) + \Phi_0 = \text{const.} \tag{14}$$

Assume now that the movement of the particle is nearly circular:

$$r(t) = r_0 + r_1(t) \qquad (r_1 \ll r_0), \tag{15}$$

where $r_0 \neq r_0(t)$ is the radius of the equilibrium circular orbit corresponding to the same E and L_z. Substituting (15) into Eq. (12), we find

$$\frac{L_z^2}{r_0^3} = \frac{\partial \Phi_0}{\partial r}\bigg|_{r=r_0}, \tag{16}$$

$$\ddot{r}_1 + \varkappa^2 r_1 = 0, \tag{17}$$

$$\varkappa = \left[\frac{\partial^2 \Phi_0}{\partial r^2}\bigg|_{r=r_0} + \frac{3L_z^2}{r_0^4}\right]^{1/2}. \tag{18}$$

Equation (16) yields the radius of the circular orbit r_0, and Eq. (17) describes the deviation of the real radius $r = r_0 + r_1$ from r_0. The nature of these deviations is determined by the value $\varkappa$ which is called the *epicyclic frequency*. If $\varkappa^2 > 0$, then the particle oscillates with respect to r_0 with frequency $\varkappa$:

$$r_1 = a \sin(\varkappa t + \alpha). \tag{19}$$

The amplitude of these oscillations a (called the *radius of the epicycle*) is obviously connected with the amplitude of oscillations v_{r0} of the velocity $\dot{r}_1 : a = v_{r0}/\varkappa$. The mean value of v_{r0} is of the order of magnitude of the velocity dispersion c_r; hence we get the estimate

$$a \sim \frac{c_r}{\varkappa}. \tag{20}$$

Introducing the notation $\Omega_0(r_0)$ for the local angular velocity of rotation of a particle in circular orbit with the angular momentum L_z

$$L_z = r_0^2 \Omega_0(r_0), \tag{21}$$

we rewrite the condition in (16) as

$$\Omega_0^2 r_0 = \frac{\partial \Phi_0}{\partial r_0}. \tag{22}$$

By determining $\partial^2\Phi_0/\partial r_0^2$ from this and substituting into definition (18) we find the following expression for the epicyclic frequency:

$$\varkappa = 2\Omega_0\left(1 + \frac{r\Omega_0'}{2\Omega_0}\right)^{1/2}, \qquad \Omega_0' \equiv \frac{d\Omega_0}{dr_0}. \tag{23}$$

This formula is generally used for the calculation of $\varkappa$.

The case $\varkappa^2 < 0$ means, obviously, an *instability* of the circular orbit (in the linear approximation).

§ 2 Jeans Instability of a Cylinder with Finite Radius [88, 90]

For the systems to be studied in this chapter, the coordinate-time dependence of small disturbances may obviously be chosen in the form

$$f_1(r)\exp(-i\omega t + ik_z z + im\varphi),$$

where r, φ, z are cylindrical coordinates. First of all, we shall be interested in "Jeans" and "fire-hose" disturbances analogous to those depicted in Fig. 4(b) for a flat layer. Here they must be approximately like those in Fig. 8. From the figure it is seen that these perturbations differ, in particular, in the type of symmetry; to the "Jeans" ones corresponds $m = 0$, while to bending ("fire-hose") ones, $m = 1$. The former conserve the symmetry axis of the system, while the latter bend it. In this section axial-symmetrical perturbations are considered: in Sections 2.1–2.3, using the simplest model of a uniform cylinder with circular orbits of particles in the (x, y) plane, and in Section 2.4, for models of a more general kind. Non-axial-symmetrical disturbances ($m \neq 0$) and, in the first place, "fire-hose" modes with $m = 1$ are discussed in §3.

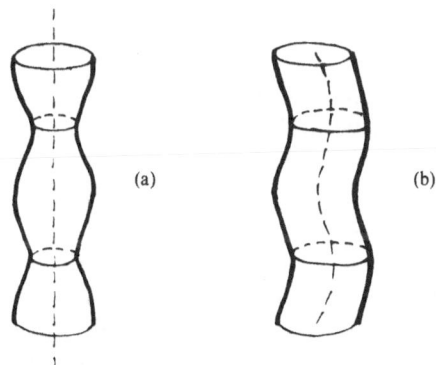

Figure 8. General appearance of the (a) Jeans and (b) fire-hose perturbations of a cylinder.

2.1 Dispersion Equation for Eigenfrequencies of Axial-Symmetrical Perturbations of a Cylinder with Circular Orbits of Particles

When the potential Φ deviates from its equilibrium value, each group of particles having originally the velocity $\mathbf{v}_0 = \mathbf{e}_\varphi v_{0\varphi} + \mathbf{e}_z v_{0z}$ will move at a speed $\mathbf{v}$, satisfying the equation $d\mathbf{v}/dt = -\nabla\Phi$. If the mass density of this group of particles originally equaled $\rho_0(v_{0z})$, then at $\Phi \neq \Phi_0$, these particles will be characterized by the density $\rho(v_{0z})$ such that $\partial\rho(v_{0z})/\partial t + \mathrm{div}(\rho\mathbf{v}) = 0$.

We linearize the equations of motion and continuity by denoting small deviations by the index "1." Recall that we have restricted ourselves so far to the analysis of axial-symmetrical perturbations $\partial/\partial\varphi = 0$. As a result, we determine the perturbation of the density $\rho_1(v_{0z})$ of a group of particles with the unperturbed velocity v_{0z}:

$$\rho_1(v_{0z}) = -\rho_0(v_{0z})\left(\frac{\Delta_r\Phi_1}{\omega'^2 - 4\Omega_0^2} - \frac{k_z^2}{\omega'^2}\Phi_1\right) - \frac{\partial\rho_0}{\partial r}\frac{\partial\Phi_1/\partial r}{\omega'^2 - 4\Omega_0^2},$$

where $\omega' \equiv \omega - k_z v_{0z}$, $\Delta_r = (1/r)(\partial/\partial r)(r\,\partial/\partial r)$ is the radial part of the Laplacian. Integrating $\rho_1(v_{0z})$ over all v_{0z}, we get the full value of density perturbation:

$$\rho_1(r, t) = \int_{-\infty}^{\infty} \rho_1(r, t, v_{0z})\, dv_{0z}.$$

Substituting $\rho_1(r, t)$ into the Poisson equation, we arrive at the following equation for Φ_1:

$$\frac{1}{r}\frac{\partial}{\partial r}\left(r\varepsilon_\perp\frac{\partial\Phi_1}{\partial r}\right) - k_z^2\varepsilon_\parallel\Phi_1 = 0, \tag{1}$$

where

$$\varepsilon_\perp = 1 + \omega_0^2\int_{-\infty}^{\infty}\frac{f_0(v_z)\,dv_z}{\omega'^2 - 4\Omega_0^2},$$

$$\varepsilon_\parallel = 1 + \omega_0^2\int_{-\infty}^{\infty}\frac{f_0(v_z)\,dv_z}{\omega'^2}. \tag{2}$$

Equation (1) is written in the form characteristic for a problem of plasma oscillations in a magnetic field. But if we substitute now in (2) the square of the Jeans frequency for the negative square of the plasma frequency, $\omega_0^2 \rightarrow -\omega_p^0 \equiv -4\pi n_0 e^2/m$, and the doubled frequency of rotation of the gravitating particles by the cyclotronic frequency of charged particles (with a charge e and mass m) in the magnetic field B_0, $2\Omega_0 \rightarrow \omega_B \equiv eB_0/mc$, then the expressions $\varepsilon_\perp$, $\varepsilon_\parallel$ turn out to coincide with the transversal and longitudinal components of the tensor of permittivity of an electron plasma. In such a transition, Eq. (1) coincides with the differential equation for the electric potential of electrostatic (potential) oscillations of a plasma cylinder of homogeneous density located in a homogeneous magnetic field $B_0 = B_{0z}$.

The derivation of the dispersion equation using (1) is similar to the plasma case. Inside the cylinder, where the stationary density is homogeneous, the values $\varepsilon_\parallel$ and $\varepsilon_\perp$ are independent of the radius, so that (1) reduces to a Bessel equation and has the solution

$$\Phi_1^{(1)} = c_1 J_0(k_\perp r) \qquad (r < R), \tag{3}$$

where $k_\perp$ is defined by the relation

$$k_\perp^2 = -k_z^2 \frac{\varepsilon_\parallel^{(0)}}{\varepsilon_\perp^{(0)}}, \tag{4}$$

and the superscript (0) means that $\varepsilon_\parallel$ and $\varepsilon_\perp$ refer to the interior region of the cylinder. Outside the cylinder (in a vacuum), the Macdonald function

$$\Phi_1^{(2)} = c_2 K_0(k_z r) \qquad (r > R) \tag{5}$$

is the solution of (1). Here and further we assume $k_z > 0$. At the boundary of the cylinder, the solutions of (3) and (5) are linked by two relations. The one is the continuity condition of the potential

$$\Phi_1^{(1)}(R) = \Phi_1^{(2)}(R). \tag{6}$$

The other is obtained by integration of (1) along the transition layer $(R - \delta, R + \delta)$ with a further trend $\delta \to 0$. It has the form

$$\varepsilon_\perp^{(0)} \left(\frac{\partial \Phi_1^{(1)}}{\partial r} \right)_{r=R} = \left(\frac{\partial \Phi_1^{(2)}}{\partial r} \right)_{r=R} \tag{7}$$

and is similar to the electrodynamical condition of continuity of the normal component of the vector of electrical induction. By means of (3)–(7), we get the dispersion equation

$$\varepsilon_\perp^{(0)} \frac{k_\perp J_0'(k_\perp R)}{J_0(k_\perp R)} = \frac{k_z K_0'(k_z R)}{K_0(k_z R)}. \tag{8}$$

2.2 Branches of Axial-Symmetrical Oscillations of a Rotating Cylinder with Maxwellian Distribution of Particles in Longitudinal Velocities

In the case of a Maxwell distribution function

$$f_0(v_z) = (\pi v_T^2)^{-1/2} \exp(-v_z^2/v_T^2) \tag{9}$$

the expressions for $\varepsilon_\perp^{(0)}$ and $\varepsilon_\parallel^{(0)}$ have the form (the superscript (0) will be omitted in the following):

$$\varepsilon_\perp = 1 + \frac{i\sqrt{\pi}}{4} \frac{\omega_0^2}{\Omega_0 |k_z| v_T} \left[W\left(\frac{\omega + 2\Omega_0}{|k_z| v_T} \right) - W\left(\frac{\omega - 2\Omega_0}{|k_z| v_T} \right) \right], \tag{10}$$

$$\varepsilon_\parallel = 1 - 2 \frac{\omega_0^2}{k_z^2 v_T^2} \left[1 + i\sqrt{\pi} \frac{\omega}{|k_z| v_T} W\left(\frac{\omega}{|k_z| v_T} \right) \right], \tag{11}$$

where W is the Kramp function.

Equation (8) takes a more simple form in the limiting cases of perturbations, long wavelengths and short wavelengths in z, i.e., at $k_z R \ll 1$ and $k_z R \gg 1$. Let us consider them separately.

2.2.1. Long-Wavelength ($k_z R \ll 1$), Large-Scale ($k_\perp R \ll 1$) Perturbations.[3]

In the case of $k_z R \ll 1$, $K_0(k_z R) \sim \ln(1/k_z R)$, so that from (8) it follows

$$\varepsilon_\perp \frac{R k_\perp J_0'(k_\perp R)}{J_0(k_\perp R)} = -\frac{1}{\ln(1/k_z R)}. \tag{12}$$

Due to the smallness of the right-hand side of this equation, the numerator of the left-hand side must be close to zero. This is possible, in particular, at $k_\perp R \ll 1$, and then, taking into account (4), the dispersion equation reduces to the form

$$1 + \frac{k_z^2 R^2}{2} \ln\left(\frac{1}{k_z R}\right) \varepsilon_\parallel = 0. \tag{13}$$

Neglecting terms of the order of $k_z^2 R^2$ in comparison with unity and using (11), we reduce it to the form

$$1 - \frac{R^2 \omega_0^2}{v_T^2} \beta \left[1 + i\sqrt{\pi} \frac{\omega}{|k_z| v_T} W\left(\frac{\omega}{|k_z| v_T}\right)\right] = 0, \tag{14}$$

where $\beta = \ln(1/k_z R)$.

In considering the limiting cases of large and small $\omega/|k_z| v_T$, one can show that Eq. (14) has no roots $\omega(k_z)$, corresponding to weakly damped oscillations. In fact, at $|\omega| \gg |k_z| v_T$ from (14) we have

$$\omega^2 = -\beta k_z^2 v_0^2, \tag{15}$$

where $v_0 \equiv v_{0\varphi}(R)$ is the linear velocity of the boundary particles. This solution describes aperiodically increasing or aperiodically decreasing perturbations:

$$\text{Re } \omega = 0, \qquad \text{Im } \omega = \pm \beta^{1/2} |k_z| v_0. \tag{16}$$

The solution with Im $\omega > 0$ corresponds to the Jeans instability. It is valid (the condition $\omega \gg k_z v_T$), if the thermal scatter is not very great:

$$v_T^2 \ll \beta^2 v_0^2. \tag{17}$$

With increasing thermal scatter, the instability increment decreases. In fact, in the reverse limiting case, at $|\omega| \ll k_z v_T$, from (14) ensues the expression for the frequency

$$\omega = -\frac{i}{\sqrt{\pi}} |k_z| v_T \left(\frac{1}{2\beta} \frac{v_T^2}{v_0^2} - 1\right), \tag{18}$$

---

[3] In subsection 6 of §1 in Chapter VI, we call these surface perturbations.

which is valid (the condition $|\omega| \ll |k_z| v_T$) if

$$\left| \frac{1}{2\beta} \frac{v_T^2}{v_0^2} - 1 \right| \ll 1.$$

With a fairly large thermal scatter, $v_T^2 = 2\beta v_0^2$, the increment of the above instability vanishes. At still larger v_T, as is seen from (18), perturbations are damped aperiodically.

Thus, the Jeans instability of large-scale perturbations with $k_z R \ll 1$ is suppressed if the longitudinal thermal scatter is large enough:

$$v_T^2 > 2\beta v_0^2. \tag{19}$$

From the definition of β it follows that the last condition cannot be met for all k_z at any large but finite value of v_T. Perturbations for which

$$k_z R < \exp\left(-\frac{v_T^2}{2v_0^2} \right) \tag{20}$$

are unstable.

This means that an infinitely long cylinder is unstable for any large but finite thermal scatter of particles in longitudinal velocities.

Using condition (20) and the expressions (16) and (18), we arrive at an estimate of the increment

$$\gamma \simeq \varepsilon e^{-\varepsilon^2} \Omega_0, \tag{21}$$

where $\varepsilon = v_T^2 / v_0^2 \gg 1$. It is evident that with a large ε the increment of Jeans instability is exponentially small.

The growth of large-scale perturbations may be viewed as the result of an increase in the effective Jeans frequency in the case of small k_z and $k_\perp$. Actually, as is evident from (16) or (18), the role of the characteristic frequency of the collective motions is played by the quantity $\sqrt{\beta}\omega_0$, and not simply by ω_0', as is the case in small-scale perturbations (see below, subsection 2.2.2). The effect of increasing the frequency of oscillations in perturbations having small wavenumbers is well known in the theory of plasma oscillations [135].

2.2.2. Long-Wave ($k_z R \ll 1$), Small-Scale ($k_\perp R \gtrsim 1$) Perturbations (Jeans Branches). Equation (12) for the perturbations with $k_\perp R \gtrsim 1$ is approximately satisfied if

$$k_\perp R = k_{\perp n} R \equiv \lambda_n^{(1)}, \tag{22}$$

where $\lambda_n^{(1)}$ is the nonzero root of the equation

$$-J_0'(\lambda_n^{(1)}) = J_1(\lambda_n^{(1)}) = 0. \tag{23}$$

Evaluating λ_n and $k_{\perp n}$ and using (4), the dispersion equation may be represented in this case as

$$k_{\perp n}^2 \varepsilon_\perp + k_z^2 \varepsilon_{\parallel} = 0. \tag{24}$$

For $\varepsilon_\perp$ and $\varepsilon_\parallel$ from (10) and (11) this equation denotes that

$$\frac{k_\perp^2}{k_z^2} = \frac{1 - 2\dfrac{\omega_0^2}{k_z^2 v_T^2}\left[1 + i\sqrt{\pi}\,\dfrac{\omega}{k_z v_T}W\!\left(\dfrac{\omega}{k_z v_T}\right)\right]}{1 + \dfrac{i\sqrt{\pi}}{4}\dfrac{\omega_0^2}{k_z^2 v_T^2}\left[W\!\left(\dfrac{\omega + 2\Omega_0}{k_z v_T}\right) - W\!\left(\dfrac{\omega - 2\Omega_0}{k_z v_T}\right)\right]}.$$ (25)

The case $k_z R \ll 1$, $k_\perp R \gtrsim 1$ under consideration corresponds to the inequality $k_\perp \gg k_z$. As one can see from Eq. (25), this inequality may be satisfied if $\Omega_0 \gg k_z v_T$.

We show that in this case there are solutions of two types: (1) $\omega \ll \Omega_0$; (2) $\omega^2 \approx 2\Omega_0^2$. Indeed for $\omega \gg k_z v_T$ (in the opposite limit the proof is trivial: $\omega \ll k_z v_\perp \ll \Omega_0$) the "permittivity" reduces to the hydrodynamic form

$$\varepsilon_\parallel = 1 + \frac{\omega_0^2}{\omega^2} = \frac{\omega^2 + 2\Omega_0^2}{\omega^2},$$ (26)

$$\varepsilon_\perp = 1 + \frac{\omega_0^2}{\omega^2 - 4\Omega_0^2} = \frac{\omega^2 + \omega_0^2 - 4\Omega_0^2}{\omega^2 - 4\Omega_0^2},$$ (27)

and Eq. (25) may be transformed to

$$\frac{k_z^2}{k_\perp^2} = \frac{\omega^2 + \omega_0^2 - 4\Omega_0^2}{4\Omega_0^2 - \omega^2} \cdot \frac{\omega^2}{\omega^2 + 2\Omega_0^2} \ll 1.$$ (28)

The inequality (28) may be satisfied in two cases: (1) $\omega^2 \ll \Omega_0^2$ (Jeans branches); (2) $\omega^2 \approx -\omega_0^2 + 4\Omega_0^2$ (rotational branches). As shown in §2.1 the kinetic equation for the rotating gravitating cylinder by means of the substitution $\omega_0^2 \to -\omega_p^2$, $2\Omega_0 \to \omega_B$ (ω_p = the plasma frequency, ω_B = the cyclotron frequency) reduces to the kinetic equation for charged particles in a magnetic field. Making such a substitution in case (2) we obtain the square of the frequency in the form $\omega^2 = \omega_p^2 + \omega_B^2$, which determines the frequency of the rotational branch of oscillations in a plasma. We shall come back to rotating branches in subsection 2.2.3; here let us continue the investigation of the Jeans branches. These branches exist for $\omega \gg k_z v_T$ as well as for the opposite case. Here $\omega \ll \Omega_0$ (see above), and Eq. (25) may be reduced to

$$1 - 4\frac{\omega_0^2}{k_\perp^2 v_T^2}\left[1 + i\sqrt{\pi}\,\frac{\omega}{k_z v_T}W\!\left(\frac{\omega}{k_z v_T}\right)\right] = 0.$$ (29)

In the case $\omega \gg k_z v_T$ we obtain from Eq. (29) the aperiodic growth of perturbations

$$\omega^2 = -2\frac{k_z^2}{k_\perp^2}\omega_0^2$$ (30)

under the condition

$$v_T^2 \ll 4v_0^2.$$ (31)

In the opposite case, $\omega \ll k_z v_T$, we have the following expression for the frequency:

$$\omega = \frac{i}{\sqrt{\pi}}\left(1 - 4\frac{\omega_0^2}{k_\perp^2 v_T^2}\right)k_z v_T, \tag{32}$$

which is valid if (the condition $\omega \ll k_z v_T$):

$$\left|1 - 4\frac{\omega_0^2}{k_\perp^2 v_T^2}\right| \ll 1.$$

These perturbations are aperiodically damping with a decrement (32) under the condition

$$v_T^2 \gtrsim 4\omega_0^2/(k_\perp^2)_{\min}. \tag{33}$$

Since $(k_\perp)_{\min} \simeq 3.8/R$, the condition (33) will be satisfied if

$$\frac{v_T}{v_0} \gtrsim \frac{2\sqrt{2}}{\lambda_n^{(1)}} \approx 0.74. \tag{34}$$

With this criterion only large-scale perturbations (Jeans type) considered in subsection 2.2.1 remain unstable.

2.2.3. Long-Wave ($k_z R \ll 1$), Small Scale ($k_\perp R \gtrsim 1$) Perturbations (Rotational Branches). In the previous subsection we called the branch $\omega^2 \simeq 2\Omega_0^2$ rotational by analogy with plasma branches. For such a frequency expressions for the "permittivity" $\varepsilon_\parallel$ [(26)] and $\varepsilon_\perp$ [(27)] can be further simplified:

$$\varepsilon_\parallel \approx 2, \qquad \varepsilon_\perp \approx -\frac{\omega^2 - 2\Omega_0^2}{2\Omega_0^2}. \tag{35}$$

Substituting the expressions (35) into Eq. (4), we find

$$\frac{\omega^2 - 2\Omega_0^2}{4\Omega_0^2} \approx \frac{k_z^2}{k_\perp^2}. \tag{36}$$

Equation (12) becomes

$$\frac{2\Omega_0^2 - \omega^2}{2\Omega_0^2}\frac{k_\perp R J_1(k_\perp R)}{J_0(k_\perp R)} \approx \frac{1}{\ln(1/k_z R)},$$

or, after the substitution of (36),

$$-\frac{2}{k_\perp R}\frac{J_1(k_\perp R)}{J_0(k_\perp R)} \approx \frac{1}{k_z^2 R^2 \ln(1/k_z R)}.$$

This equation at $k_\perp R \gtrsim 1$ and $k_z R \ll 1$ is approximately satisfied, if

$$k_\perp R \simeq k_{\perp n} R \equiv \lambda_n^{(0)}, \tag{37}$$

where $\lambda_n^{(0)}$ is the root of the equation:

$$J_0(\lambda_n^{(0)}) = 0. \tag{38}$$

Taking into account the imaginary terms of Eq. (25), which are exponentially decreasing for $|\omega \pm 2\Omega_0|, |\omega| \gg k_z v_T$, we find the damping increment

$$\text{Im } \omega = -\sqrt{\pi} \frac{\omega_0^2}{|k_z| v_T} \left(\frac{\chi_1}{4\sqrt{2}} + \frac{\omega_0^2}{k^2 v_T^2} \chi_2 \right). \tag{39}$$

Here χ_1, χ_2 are positive and are determined by the relations

$$\chi_1 = \exp\left[-\left(0.6 \frac{\Omega_0}{k_z v_T} \right)^2 \right] - \exp\left[-\left(3.4 \frac{\Omega_0}{k_z v_T} \right)^2 \right],$$

$$\chi_2 = \exp\left[-\frac{2\Omega_0^2}{k_z^2 v_T^2} \right].$$

The term on the right-hand side of (39), proportional to χ_1, is the damping decrement of rotational oscillations caused by the resonance interaction of rotating particles with waves (resonance of the type of $\omega = k_z v_z \pm 2\Omega_0$). This is the analogue of cyclotron damping of plasma oscillations in a magnetic field. The term on the right-hand side of (39), proportional to χ_2, is the damping increment of rotational oscillations due to Cherenkov interaction with resonant particles (resonance of the type $\omega = k_z v_z$).

From (39) it is seen that the ratio Im ω/Re ω decreases exponentially with decreasing $|k_z|$. Therefore, at fairly small $|k_z|$ these oscillations may be considered as undamped.

2.2.4. Short-Wavelength $(k_z R \ll 1)$ Perturbations.

Since at $k_z R \gg 1$ $K_0(k_z R) \approx \sqrt{\pi/2}(k_z R)^{-1/2} e^{-k_z R}$ Eq. (8) reduces to the form

$$\varepsilon_\perp^{(0)} \frac{k_\perp R J_1(k_\perp R)}{J_0(k_\perp R)} \simeq k_z R. \tag{40}$$

As $k_z R \gg 1$, as in the opposite case $k_z R \ll 1$, considered above, the dispersion equation (40) describes perturbations of two types: Jeans and rotational. Our aim is to determine the short-wave "tails" of these branches.

On the right-hand side of Eq. (40), we have a large quantity. Therefore, the dispersion equation is approximately satisfied if $k_\perp$ satisfies the condition (37), and λ_n is determined from Eq. (38). The dispersion equation in this case also is reduced to the form of (24), however, with somewhat different values of $k_{\perp n}$. In the perturbation analysis of a Maxwellian medium (in longitudinal velocities), one may proceed from Eq. (25) by assuming here, however, that $k_z \gg k_\perp$.

For a cold cylinder, Eqs. (40) and (4) are

$$\frac{(\omega^2 - 2\Omega_0^2)}{(\omega^2 - 4\Omega_0^2)} \frac{k_\perp R J_1(k_\perp R)}{J_0(k_\perp R)} \approx k_z R, \tag{41}$$

$$\frac{k_\perp^2 R^2}{k_z^2 R^2} = -\frac{\omega^2 + 2\Omega_0^2}{\omega^2} \frac{\omega^2 - 4\Omega_0^2}{\omega^2 - 2\Omega_0^2}. \tag{42}$$

We seek solutions with $k_z R \gg 1$, $k_\perp \ll k_z$. According to (42), the latter inequality is satisfied in one of the two following cases: (1) $\omega^2 \approx -2\Omega_0^2 = -\omega_0^2$; (2) $\omega^2 \approx 4\Omega_0^2$. The former corresponds, evidently, to perturbations of the Jeans type, and the latter to perturbations of the rotational type (at $v_T \to 0$).

It is easy to show that, for Jeans disturbances, Eq. (41) is roughly satisfied [under condition (42)], if $k_\perp$ satisfies the conditions in (37) and (38).

Similarly to the short-wave $k_z R \gg 1$, rotational perturbations, the transversal wavenumber $k_\perp$ must be taken from the discrete series $k_{\perp n}$, as determined by Eqs. (22) and (23).

One can also investigate the hot medium ($v_T \neq 0$) and show that in this case also, the conditions (37), (38) and (22), (23) are asymptotically satisfied (for the Jeans and rotational branches, respectively).

At rather large k_z ($k_z v_T > \omega_0$ and $k_z \gg k_\perp$), all of the solutions of Eq. (8) correspond to strongly damped disturbances. Consider, for instance, the Jeans branches. In this limit $\varepsilon_\perp \approx 1$, $J_0(k_\perp R) \approx 0$, and

$$\varepsilon_\parallel \simeq 0. \tag{43}$$

This is the dispersion equation for the case under consideration. It follows from physical considerations, that it coincides with the dispersion equation in the Jeans problem of the stability of the homogeneous gravitating background. On the other hand, (43) differs from the analogous plasma equation only in the substitution of $-\omega_0^2 \leftrightarrow \omega_{pe}^2$. Therefore, a further conclusion can be drawn from the paper of Landau [65] concerning oscillations of the electron plasma. For that purpose, let us turn, in the dispersion equation $\varepsilon_\parallel = 0$, to the integral form

$$\frac{\omega_0^2}{k_z} \int \frac{\partial f_0/\partial v_z}{\omega - k_z v_z}\, dv_z = 1. \tag{44}$$

For the case $k_z v_T \gg \omega_0$, in the calculation of the integral in (44), it is sufficient [65] to restrict oneself to residue at the point $v_z - \omega/k_z$, so that Eq. (44) will be reduced to

$$\sqrt{2\pi}\, \frac{i\omega}{\omega_0(k_z a)^3} \exp\left(-\frac{\omega^2}{2\omega_0^2(k_z a)^2}\right) = 1 \qquad \left(a \equiv \frac{v_T}{\omega_0}\right). \tag{45}$$

This equation differs from the analogous plasma equation in the substitution of $+1$ for -1 on the right-hand side. Assuming $\mathrm{Im}\,\omega \equiv \gamma \gg \mathrm{Re}\,\omega \equiv \omega_1$, we take the module in both parts of (45)

$$\xi e^{\xi^2/2} = \frac{1}{\sqrt{2\pi}}(k_z a)^2, \tag{46}$$

where

$$\xi \equiv \frac{-\gamma}{\omega_0 k_z a}. \tag{47}$$

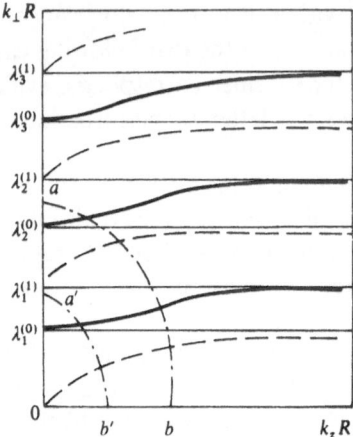

Figure 9. Relationship between wave numbers k_z and $k_\perp$ for axisymmetric perturbations of a cylinder having circular particle orbits [26]: solid lines—rotating oscillation branches; broken lines—Jeansonian branches; dot-and-dash curves separate the unstable (under the curves) and stable (over the curves) segments of Jeansonian branches; curve $a'b'$—the same as ab, but with different values ΩR and v_{Tz} (or Δ—for Jackson distribution function).

On the other hand, the phase factor of the left-hand side of (45) is (in the same approximation)

$$\exp\left(-\frac{i\gamma\omega_1}{\omega_0^2 a^2 k_z^2}\right). \tag{48}$$

Since a positive value is on the right-hand side of (45), this factor must be $+1$. Hence it follows that $\omega_1 \approx 0$. The damping decrement is determined from (47)

$$\gamma = -\omega_0 k_z a\xi = -k_z v_T \xi, \tag{49}$$

whereas ξ is defined by Eq. (46) (mainly $\xi \sim \ln^{1/2} k_z a$). We obtain the expected result: an aperiodical damping of such perturbations. The physical meaning of this damping involves the running-away of the particles from the region of perturbation due to their thermal movement. This has no bearing on the usual Landau damping (due to resonant wave-particle interaction)—as well as the damping of the corresponding waves in the plasma.

The results described in this section and concerning the dependence of $k_\perp(k_z)$ are summed up in Fig. 9 [26].

2.3 Oscillative Branches of the Rotating Cylinder with a Jackson Distribution Function (in Longitudinal Velocities) [26, 88]

$$f_0(v_z) = \frac{\Delta}{\pi} \frac{1}{v_z^2 + \Delta^2}, \tag{50}$$

where Δ is the analog of thermal scatter of velocities.

This case is convenient in that (unlike, for instance, the Maxwellian medium) all integrals over v_z, arising in the derivation of the dispersion equation, are taken in their ultimate form in elementary functions.[4]

If we introduce the dimensionless values

$$z \equiv \left(\frac{\omega}{\Omega_0} + i\frac{k_z\Delta}{\Omega_0}\right)^2, \qquad x \equiv \frac{k_z}{k_\perp}, \qquad \alpha = k_z R \qquad \left(k_\perp R = \frac{\alpha}{x}\right), \qquad (51)$$

then relation (41) will be written as

$$1 + \frac{2}{z-4} = -x^2\left(1 + \frac{2}{z}\right), \qquad (52)$$

and the condition of linking the derivatives of the potential in (7) takes the form

$$\varepsilon_\perp = 1 + \frac{2}{z-4} = x\frac{K_1(\alpha)\,J_0(\alpha/x)}{K_0(\alpha)\,J_1(\alpha/x)}, \qquad (53)$$

Solving (52) with respect to z and substituting it into (53), we have

$$z = 1 \pm \sqrt{1 + \frac{8x^2}{1+x^2}}, \qquad (54)$$

$$\frac{z-2}{z-4} = \frac{1 \mp \sqrt{1 + 8x^2/(1+x^2)}}{3 \mp \sqrt{1 + 8x^2/(1+x^2)}} = x\frac{K_1(\alpha)\,J_0(\alpha/x)}{K_0(\alpha)\,J_1(\alpha/x)}. \qquad (55)$$

The latter equation determines for each $\alpha \equiv k_z R$ an infinite discrete set $x \equiv k_z/k_\perp$ (see Fig. 9). Of course, we had the same also in the case of the Maxwellian distribution function considered earlier. We shall see that the *main* properties of solutions in these two cases coincide. However, the presence in the equations for the Maxwellian distribution function of the probability integrals from the complex argument (Kramp function rather complicates the analysis).

The solution put down above [(54), (55)] contains only the elementary and simplest cylindrical ($K_{0,1}, J_{0,1}$) functions. In addition, there are singled out initially the rotational and Jeans branches of oscillations [correspondingly, the upper and lower signs in (54), (55); see also below]. An analysis of these equations is readily made.

Consider, first of all, long-wave perturbations on the z-axis ($k_z R \equiv \alpha \ll 1$). In this case, for two types of oscillations in (55) we have

$$\frac{1 - \sqrt{1 + 8x^2/(1+x^2)}}{3 - \sqrt{1 + 8x^2/(1+x^2)}} \approx x\frac{J_0(\alpha/x)}{J_1(\alpha/x)}\frac{1}{\alpha\ln(2/\alpha)}, \qquad (56)$$

$$\frac{1 + \sqrt{1 + 8x^2/(1+x^2)}}{3 + \sqrt{1 + 8x^2/(1+x^2)}} \simeq x\frac{J_0(\alpha/x)}{J_1(\alpha/x)}\frac{1}{\alpha\ln(2/\alpha)}. \qquad (57)$$

---

[4] Note the case of the distribution function in the form of a "step", $f_0 = \rho_0/2v_0 = \text{const}$ for $v_z \in [-v_0, v_0]$, is simpler.

If $\alpha/x \equiv k_\perp R \to 0$, then a solution is available only for the (Jeans) branch (57)

$$x^2 \approx \frac{\alpha^2}{4} \ln\left(\frac{2}{\alpha}\right), \qquad z \approx -\alpha^2 \ln\left(\frac{2}{\alpha}\right), \tag{58}$$

or, in dimensional units,

$$\omega \approx -ik_z\left[\Delta \pm \Omega R \ln^{1/2}\left(\frac{2}{k_z R}\right)\right]. \tag{59}$$

This equation is similar to (15), (18) for the Maxwellian distribution. Similar conclusions may be also drawn from (15), (18), and (59). For any thermal dispersion, sufficiently long waves,

$$\ln\left(\frac{2}{k_z R}\right) > \frac{\Delta^2}{\Omega^2 R^2}, \qquad k_z < \frac{2}{R}e^{-\Delta^2/\Omega^2 R^2} \tag{60}$$

remain unstable. At $\Delta \gg v_0$, the critical wavelength is exponentially great, while the instability increment is correspondingly small.

In addition to the solution of (58) for $\alpha/x \to 0$, there are two infinite sets of solutions with $\alpha/x \to$ const [as $\alpha \to 0$], one for each type of waves (56) and (57). In the case of perturbations of the rotational type (56)

$$x \approx \frac{\alpha}{\lambda_n^{(0)}}, \qquad \frac{\alpha}{x} > \lambda_n^{(0)}, \qquad z \approx 2, \qquad \omega \simeq \Omega\sqrt{2}, \tag{61}$$

where $\lambda_n^{(0)}$ are the solutions of the equation $J_0(\lambda_n^{(0)}) = 0$. These oscillations are always stable. In the case of the Jeans branches (57),

$$x \approx \frac{\alpha}{\lambda_n^{(1)}}, \qquad \frac{\alpha}{x} > \lambda_n^{(1)}, \qquad z \approx -4\frac{\alpha^2}{\lambda_n^{(1)2}}, \qquad \omega \approx i\left(k_z\Delta \pm 2\frac{k_z v_0}{\lambda_n^{(1)}}\right), \tag{62}$$

where $\lambda_n^{(1)}$ denote non-zero solutions for the equation $J_1(\lambda_n^{(1)}) = 0$. These oscillations are stable for sufficiently large thermal scatter:

$$\Delta > \frac{2v_0}{\lambda_1^{(1)}} \qquad (\lambda_1^{(1)} \simeq 3.8). \tag{63}$$

Now let α and x be arbitrary values. The left-hand side of Eqs. (55) does not change its sign when x $(0 < x < \infty)$ changes; therefore α also changes in accordance with x, so that on each branch of oscillations $J_0(\alpha/x)$ and $J_1(\alpha/x)$ do not change sign.

Thus, $\alpha/x \equiv k_\perp R$ when α changes in $0 \le \alpha \le \infty$ is within the intervals

$$\lambda_n^{(0)} \le \frac{\alpha}{x} \le \lambda_n^{(1)} \qquad (n \ge 1), \qquad \text{for rotational branches,}$$

$$2\left(\ln\frac{2}{\alpha}\right)^{-1/2} \le \frac{\alpha}{x} \le \lambda_1^{(0)}, \qquad \text{for the Jeans branch (58),} \tag{64}$$

$$\lambda_n^{(1)} \le \frac{\alpha}{x} \le \lambda_{n+1}^{(0)} \qquad (n \ge 1), \qquad \text{for the rest of the Jeans branches.}$$

The region of perturbation stability (of Jeans type) is determined by the condition

$$\frac{k_z^2 \Delta^2}{\Omega_0^2} > \sqrt{1 + \frac{8x^2}{1 + x^2}} - 1, \tag{65}$$

or

$$\left(\frac{\alpha}{x}\right)^2 > \alpha^2 \left\{ \frac{8}{[1 + (\Delta/v_0)^2 \alpha^2]^2 - 1} - 1 \right\}$$

The curve ab (see Fig. 9), separating the regions of stability and instability, crosses the ordinate at $\alpha/x = 2v_0/\Delta \equiv a$, and the abscissa at the point $\alpha = \sqrt{2}v_0/\Delta \equiv b$. As is seen from Fig. 9 at $2v_0/\Delta < \lambda_n^{(1)}$ there are n unstable branches; as $2v_0/\Delta < \lambda_1^{(1)}$, there remains one unstable branch (58).

2.4 Axial-Symmetrical Perturbations of Cylindrical Models of a More General Type

Let us now depart from the assumption of uniformity of the cylinder and from the circular character of the orbits of all particles in the (x, y) plane. It is clear, first of all, that the structure of the cylinder in (x, y) must not play any role for perturbations with "longitudinal" wavelengths $\lambda = 2\pi/k_z$, with values greater than the transversal size of the cylinder R (i.e., for $k_z R \ll 1$). Considering such perturbations, we arrive at the model of a "thin" cylinder—"filament" (compare with the beginning of Section 2 of Chapter I). It is possible here to examine generally the Jeans instability (the fire-hose instability in this simplest model will be treated in Section 3).

For oscillations along the axis of the "filament" displacement of particles, Z satisfies the equation

$$\hat{D}^2 Z = -\frac{\partial \Phi_1}{\partial z}, \tag{66}$$

which is similar to (10) in §6 of Chapter I; the square of the frequency of oscillations ($\omega_0^2 = 1$ in the case of a flat sheet) does not enter Eq. (66) for the simple reason that along the z-axis the system in question is infinite, so that in the unperturbed state the particles do not oscillate along z.

For the particles on the axis of a "thin" system, we obtain from (66), where $\Phi_1 = \exp(ik_z z)$:

$$Z = \frac{ik_z}{(\omega - k_z v_z)^2}, \tag{67}$$

so that the mean displacement

$$\bar{Z} \equiv \chi = \int \frac{ik_z f_0}{(\omega - k_z v_z)^2} \, dv_z. \tag{68}$$

The relevant density perturbation is

$$\rho_1 = -\rho_0 \frac{\partial \chi}{\partial z} = \rho_0 k_z^2 \int \frac{f_0 \, dv_z}{(\omega - k_z v_z)^2}. \qquad (69)$$

The Poisson equation in this case yields

$$\Phi_< = A I_0(k_z r) \approx A \qquad (k_z r \ll 1). \qquad (70)$$

The constant A is determined from the conditions of matching the inner and vacuum solutions:

$$A = \frac{4\pi G \rho_1 K_0'(k_z)}{W_0 k_z^2},$$

where the system radius is taken for the sake of brevity to be equal to 1: $R = 1$, I_0, K_0 are the relevant cylindrical functions, $W_0 = K_0'(k_z) I_0(k_z) - I_0'(k_z) K_0(k_z) = 1/k_z$ is the Wronskian. In the approximation under consideration of a thin gravitating "filament" (C is the Euler constant)

$$A \approx \frac{4\pi G \rho_1}{k_z^2} \left(1 + \frac{k_z^2}{2} \ln \frac{C k_z}{2} - \frac{k_z^2}{4} \right).$$

Thus, the coupling between Φ_1 and ρ_1 in the case of thin systems following from the Poisson equation is

$$\Phi_< = 4\pi G \rho_1 \left(\frac{1}{2} \ln \frac{C k_z}{2} - \frac{1}{4} \right). \qquad (71)$$

Comparing with (69), we obtain the dispersion equation:

$$\int_{-\infty}^{\infty} \frac{f_0 \, dv_z}{(\omega - k_z v_z)^2} = \left[k_z^2 \left(\ln \frac{C k_z}{2} - \frac{1}{2} \right) \right]^{-1}. \qquad (72)$$

In particular, for $|\omega| \gg k_z v_T$ from (72), we find the expression (15) already known to us, for the increment of the long-wavelength Jeans instability of a cylinder "cold" in z:

$$\gamma^2 \approx k_z^2 |\ln k_z|. \qquad (73)$$

For a Maxwellian distribution function the right-hand side of (72) will be written in the form

$$-\frac{2}{k_z^2 v_T^2} \left[1 + i\sqrt{\pi} \frac{\omega}{|k_z| v_T} W\left(\frac{\omega}{|k_z| v_T} \right) \right].$$

The stability boundary corresponds to $\omega = 0$, i.e. instability occurs at

$$\frac{v_T^2}{2} \equiv v_{Tz}^2 < \left| \ln\left(\frac{C k_z}{2} \right) \right| + \frac{1}{2}. \qquad (74)$$

To conclude this section, let us note that the problem of perturbations of a cylinder with arbitrary wavelengths must, generally, be solved by numerical methods. In Fig. 10, the dotted line represents the stability

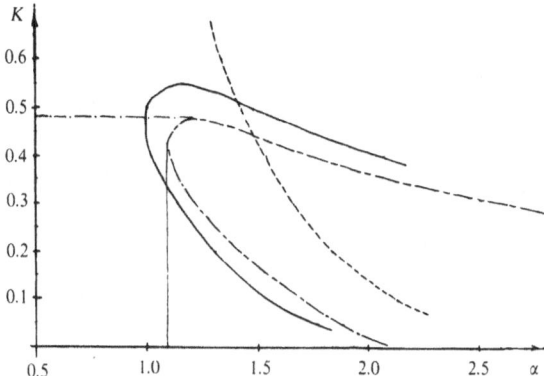

Figure 10. Stability boundaries of the cylinder for perturbations of the Jeans and fire-hose types. Solid line is the exact boundary of the fire-hose instability; dotted line is the exact boundary of the Jeans instability; dash-dotted line is the approximate boundary of the fire-hose instability.

boundary obtained with the computer[5] for Jeans perturbations of a cylinder uniform in density with the distribution function in the (x, y) plane [Eq. (2) of §1] at $\gamma = 0$. Thus, this cylinder does not rotate; the equilibrium is provided by the pressure of particles; the distribution in longitudinal velocities v_z was assumed to be Maxwellian.

§ 3 Nonaxial Perturbations of a Collisionless Cylinder[90, 38^{ad}]

In the first papers dealing with the bending perturbations of a cylinder, a conclusion is drawn about their instability. Moreover, the stable solution is assumed to be qualitatively evident on the assumption that with weak "lateral" perturbations the cylinder is similar to a taut spring in terms of stability. A correct analogy is, however, the opposite: In reality, the cylinder is like a compressed rather than a taut spring: chaotic velocities of stars are similar to pressure rather than tension. Therefore, it would be natural to expect that it must be subjected to a fire-hose (anisotropic) instability.

3.1 The Long-Wave Fire-Hose Instability

To begin with, we shall derive a dispersion equation [38^{ad}] in the approximation of a very much elongated, or, what is the same, a very "thin" system (compare again the beginning of §2, Chapter I). This approximation leads to simplifications, such as in the determination of the system's response to an imposed perturbation of the gravitational potential Φ_1 (from the kinetic equation) as well as in solving the Poisson equation, which must connect this response to the field Φ_1.

It is evident that for "thin" systems it is sufficient to calculate only displacements of particles, which in the unperturbed state are on the axis of the system

---

[5] Calculations performed by S. M. Churilov on our request.

$(z = v_z = 0)$. With bending perturbations, all particles are displaced (at $k_\perp R < 1, k_z R < 1$) in an identical way. This simply yields the solution of the kinetic equation, which then becomes one-dimensional.

With such perturbations, the density ρ remains unchanged; thus perturbations of the potential are due only to the curvature of the boundary.

By assuming the frequency of transversal oscillations of particles in the unperturbed potential Φ_0 to be unity, write, instead of the kinetic equation, the equation for transversal (along the x-axis) displacements of particles X (compare with Eq. (10), §6, Chapter I):

$$(\hat{D}^2 + 1)X = -\frac{\partial \Phi_1}{\partial x}, \tag{1}$$

where the operator $\hat{D}$ being, in the general case,

$$\hat{D} = -i\omega + v\frac{\partial}{\partial r} - \frac{\partial \Phi_0}{\partial r}\frac{\partial}{\partial v} \qquad (\omega = \text{the frequency}),$$

of the particles on the axis ($v_x = v_y = 0, x = y = 0$), becomes

$$\hat{D} = -i\omega + v_z\frac{\partial}{\partial z}. \tag{2}$$

Assuming

$$\Phi_1 = -(x + iy)e^{ik_z z}, \tag{3}$$

we get from (1)

$$X = -\frac{1}{(\omega - k_z v_z)^2 - 1}. \tag{4}$$

The mean displacement of particles on the axis is

$$\psi \equiv \overline{X} = \int_{-\infty}^{\infty} Xf_0(v_z)\, dv_z, \tag{5}$$

where $f_0(v_z)$ is the distribution function of particles (normalized to unity).

Calculate now the gravitational potential corresponding to this displacement ψ (perturbation of the density $\rho_1 = 0$). For the potentials outside the "filament" $\Phi_>$ and inside it $\Phi_<$ we obtain by the standard method

$$\Phi_< = AI_1(k_z r)e^{i\varphi + ik_z z}, \tag{6}$$

$$\Phi_> = BK_1(k_z r)e^{i\varphi + ik_z z}, \tag{7}$$

where I_1 and K_1 are Bessel and McDonald functions. The potential in this case is produced by a simple sheet with a surface density $\sigma_0 = \rho_0\psi$, and situated on the surface of the unperturbed cylinder ($R = 1$). Therefore, from the linking conditions of the solutions of (6) and (7) we obtain

$$A = \frac{2\psi K_1(k_z)}{k_z W_1},$$

where $W_1 \equiv I_1(k_z)K'_1(k_z) - I'_1(k_z)K_1(k_z) = -1/k_z$; thus

$$A = -2\psi K_1(k_z). \tag{8}$$

Since $k_z < 1$, we have, by taking a corresponding representation of the $K_1(k_z)$ function [42] (C is the Euler constant),

$$A = -2\left(\frac{1}{k_z} + \frac{k_z}{2}\ln\frac{Ck_z}{2} - \frac{k_z}{4}\right)\psi. \tag{9}$$

This relation, which may be rewritten in the form

$$\Phi_< = -\left(1 + \frac{k_z^2}{2}\ln\frac{Ck_z}{2} - \frac{k_z^2}{4}\right)re^{i\varphi}\psi, \tag{10}$$

yields, in the case of elongated systems, the required coupling by the $\Phi_<$ potential and by the ψ displacement. In our case (10) plays the same role as the coupling $\Phi_< = -2\pi G\sigma_1/k$, employed in §2 of Chapter I, between the potential and the surface density of strongly flattened systems. For Jeans perturbations of elongated systems a similar role is played by formula (71) of §2. Substituting ψ from (5) and (4) and comparing with (3), we get the dispersion equation

$$\int_{-\infty}^{\infty} \frac{f_0(v_z)\,dv_z}{(\omega - k_z v_z)^2 - 1} = -1 + \frac{k_z^2}{2}\ln\frac{Ck_z}{2} - \frac{k_z^2}{4}. \tag{11}$$

In the simplest case $(\omega - k_z v_T)^2 \ll 1$, whence it follows that

$$\omega^2 = k_z^2\left(\frac{1}{2}\left|\ln\frac{Ck_z}{2}\right| - v_{zT}^2 - \frac{1}{4}\right), \tag{12}$$

so that with thermal velocities sufficiently high (more precisely, with sufficiently high anisotropy of the velocity distribution of particles)

$$v_{zT}^2 > \frac{1}{2}\left|\ln\frac{Ck_z}{2}\right| + \frac{1}{4}, \tag{13}$$

the fire-hose (anisotropic) instability occurs. Equation (11) also remains valid when $\omega \sim 1$ (there must be only $k_z < 1$). In this case, for the Maxwellian distribution function, Eq. (11) was solved numerically; in Fig. 10, the dash–dot line divides the regions of stable and unstable solutions. It is seen that the instability vanishes at a sufficiently small scale of the perturbation ($k_z R < 0.48$) or at sufficiently small anisotropy of the velocity distribution [$\alpha \equiv (\sqrt{3}/2)v_{Tz} < 1.09$].

Of course, numerically the last two results could prove to be not very accurate, since they occur on the boundary of applicability of the approximation adopted. However, the results of computations[6] of the boundary of the stability for fire-hose perturbations of a uniform nonrotating cylinder with the distribution function in the (x, y) plane [Eq. (2), §1] (at $\gamma = 0$)

---

[6] Performed by S. M. Churilov at our request.

coincide with the simple approximation obtained above, not only qualitatively but also quantitatively. This is evident from the comparison of the
solid and dash–dot lines in Fig. 10 (the former corresponds to the numerical
calculation without any approximations).

For "thin" systems (the limit $k_z R \ll 1$), as one can see from comparison
of formula (13) with formula (74) of the previous section, the unstable
regions for Jeans and fire-hose branches intersect; thus sufficiently elongated
systems are universally unstable (with respect to at least one of the instabilities considered). In any case Fig. 10 shows clearly enough that the stability
condition imposes a very strong restriction on the parameters of elongated
systems.

3.2 Nonaxial Perturbations of a Cylinder with Circular Particle Orbits

The derivation of the dispersion equation for arbitrary perturbations of the
homogeneous cylinder is made similarly to the case $m = 0$. First of all, the
inhomogeneous $\rho_0 = \rho_0(r)$ and arbitrarily rotating $[\Omega_0 = \Omega_0(r)]$ cylinder
is considered. Using the method described above, one can obtain the following differential equation for the perturbed potential $(\sim e^{im\varphi})$ [90]:

$$\frac{1}{r}\frac{d}{dr}\left(\varepsilon_\perp r \frac{d\Phi_1}{dr}\right) - \frac{m^2}{r^2}\varepsilon_\perp \Phi_1 - k_z^2 \varepsilon_\parallel \Phi_1 - \frac{2m}{r}\Phi_1$$

$$\times \int \frac{1}{\omega'}\frac{d}{dr}\left(\frac{\omega_0^2 \Omega_0 f_0}{\Delta}\right) dv_z = 0, \tag{14}$$

where

$$\omega' \equiv \omega - m\Omega_0 - k_z v_z, \qquad \Delta \equiv \omega'^2 - \varkappa^2,$$

$$\varkappa^2 \equiv 4\Omega_0^2\left(1 + \frac{r\, d\Omega_0/dr}{2\Omega_0}\right), \tag{15}$$

$$\Omega_0^2(r) = \frac{1}{r^2}\int_0^r \omega_0^2 r\, dr = \frac{2GM(r)}{r^2}, \qquad \omega_0^2 = 4\pi G\rho_0(r),$$

$M(r)$ is the mass inside the radius r; $\varepsilon_\perp$ and $\varepsilon_\parallel$ are determined again by the
formulae (2), §2; f_0 is assumed to be normalized to unity; $\int f_0\, dv_z = 1$.
Note further a possible dependence of the longitudinal temperature v_{Tz}
on the radius [v_{Tz} enters $f_0(v_z)$ as a parameter]. Equation (14) for the inhomogeneous cylinder will be considered below, in §7 (as well as in Chapter
VI in studying gradient instabilities). Here, however, we shall turn to the
homogeneous cylinder $\rho_0 = \text{const}$ and $v_{\varphi 0} = \Omega_0 r$, $\Omega_0 = \text{const}$. Equation
(14) is used for obtaining the required boundary condition (the substitute
for (7), §2). It can be determined by integration of equation (14) over the

transition layer with an infinitesimal thickness δ:

$$\frac{\partial \Phi_1^{(2)}}{\partial r}\bigg|_{r=R} = \varepsilon_\perp \frac{\partial \Phi_1^{(1)}}{\partial r}\bigg|_{r=R} - \frac{2m}{R}\Phi_1(R)\omega_0^2\Omega_0 I, \qquad (16)$$

where $\Phi_1^{(1)}$ denotes the internal $(r < R)$ and $\Phi_1^{(2)}$ the external $(r > R)$ solution,

$$I \equiv \int_{-\infty}^{\infty} \frac{f_0(v_z)\,dv_z}{\omega'\Delta}. \qquad (17)$$

The perturbed potential inside the homogeneous cylinder is proportional to the Bessel function (of order m)

$$\Phi_1^{(1)} = AJ_m(k_\perp R) \qquad (r < R, A = \text{const}), \qquad (18)$$

where again $k_\perp$ is determined by the relation (4), §2:

$$k_\perp^2 = -\frac{k_z^2 \varepsilon_\parallel}{\varepsilon_\perp}. \qquad (19)$$

At $m \neq 0$, the transversal wave number $k_\perp$ is, generally speaking, a complex number (see below). In vacuum, Φ_1 is the MacDonald function

$$\Phi_1^{(2)} = BK_m(k_z r) \qquad (r > R, B = \text{const}). \qquad (20)$$

The boundary condition for the continuity of the potential in (6), §2,

$$\Phi_1^{(1)}(R) = \Phi_1^{(2)}(R), \qquad (21)$$

remains constant.

Using (18) and (20), instead of (16) and (21), we get

$$B\frac{\partial K_m}{\partial r} = \varepsilon_\perp^{(0)} A \frac{\partial J_m}{\partial r} - \frac{2m\Omega_0}{R}AJ_m\omega_0^2 I, \qquad (22)$$

$$BK_m = AJ_m. \qquad (23)$$

The condition of the existence of the nontrivial solution for the coefficients A, B is [90]

$$k_z R \frac{K_m'(k_z R)}{K_m(k_z R)} = k_\perp R\varepsilon_\perp^{(0)} \frac{J_m'(k_\perp R)}{J_m(k_\perp R)} - 2m\Omega_0\omega_0^2 I. \qquad (24)$$

This equation, along with relation (19), in principle, solves the problem. Consider the limiting cases.

3.2.1. Long-Wave ($k_z R \ll 1$), Large-Scale ($k_\perp R \ll 1$) Perturbations. Put first of all $m = 1$. Using the well-known asymptotic formulae for cylindrical functions (further we assume for brevity $R = 1$),

$$\frac{k_z K_1'(k_z)}{K_1(k_z)} \simeq -1 + k_z^2 \ln\frac{Ck_z}{2}, \qquad \frac{k_\perp J_1'(k_\perp)}{J_1(k_\perp)} \simeq 1 - \frac{k_\perp^2}{4}, \qquad (25)$$

and putting in formulae (10), (11) §2, for $\varepsilon_\parallel$ and $\varepsilon_\perp$, $\omega = 0$,

$$\varepsilon_\perp \simeq \tfrac{1}{3}(1 - \tfrac{7}{9}k_z^2 v_T^2), \qquad \varepsilon_\parallel \simeq 3(1 + k_z^2 v_T^2), \qquad I \simeq \tfrac{1}{3}(1 + \tfrac{5}{9}k_z^2 v_T^2), \qquad (26)$$

we obtain the following equation, determining the stability boundary of a system [38^{ad}]:

$$2v_{Tz}^2 = v_{Tc}^2 \simeq |\ln(Ck_z/2)| + \tfrac{3}{4}. \tag{27}$$

For $k_z R \ll 1$ this equation coincides practically with Eq. (13), as was to be expected. From (24) and (19) one may easily obtain also the dispersion equation for $k_z \ll 1$, $k_\perp \ll 1$, $\omega \ll 1$ [38^{ad}]:

$$\omega^2 \simeq \tfrac{1}{2}k_z^2[-\ln(Ck_z/2) - v_T^2 + \tfrac{3}{4}], \tag{28}$$

where it was assumed that $v_T^2 > 1$. Equation (28) for $v_T^2 > v_{Tc}^2$ describes the fire-hose instability of a cylinder with circular particle orbits.

Note that in the region $\alpha \equiv v_T^2/\Omega_0^2 R^2 \sim 1$ and $k_z R \sim 1$ the dispersion equation (24), (19) lead to results essentially different from those given in Fig. 10. Briefly, one can say that the unstable region of a cold cylinder is much wider than in Fig. 10. Of course, this is due to the full degeneration of the system (rotation along exactly circular orbits). This same reason also explains the instability of a cold cylinder for all $m > 1$. The corresponding dispersion equation which may be obtained from (24) and (19) for $m > 1$ and $k_z \ll 1$, $k_\perp \ll 1$, is

$$\omega^2 \simeq \frac{k_z^2}{2}\left(\frac{2m-1}{m(m^2-1)} - v_T^2\right). \tag{29}$$

Equation (29) for $v_T^2 > v_{Tc}^2 \equiv (2m-1)/m(m^2-1)$ describes an instability. It is natural to interpret this instability (especially for large m) as the fire-hose instability at the edge of the cylinder. Obviously flutes arising there and rotating practically together with the cylinder ($\omega \simeq 1 - 1/m$) are unstable with respect to bending along the z-direction because the cylinder is cold. From the physics of the fire-hose instability it is clear that the introduction of a velocity dispersion in the equatorial plane of the cylinder will make it stable at least for modes with large enough m.

3.2.2. Long-Wave ($k_z R \ll 1$), Small-Scale ($k_\perp R \gtrsim 1$) Perturbations.

Suppose initially that the velocity dispersion along the z-axis is absent. Then $\varepsilon_\perp, \varepsilon_\|$, and I arc given by the formulae ($\omega_* \equiv \omega - m\Omega_0$):

$$\varepsilon_\| = 1 + \omega_0^2/\omega_*^2, \qquad \varepsilon_\perp = 1 + \omega_0^2/(\omega_*^2 - 4\Omega_0^2), \qquad I = 1/\omega_*(\omega_*^2 - 4\Omega_0^2). \tag{30}$$

The dispersion relation $\omega(k_z)$ is given by Eqs. (24) and (19) which, in the case of a cold cylinder, are reduced to the following:

$$\frac{k_z R K_m'(k_z R)}{K_m(k_z R)} = \frac{k_\perp R J_m'(k_\perp R)}{J_m(k_\perp R)}\frac{\omega^{*2} - 2\Omega_0^2}{\omega^{*2} - 4\Omega_0^2} - \frac{2m\Omega_0\,\omega_0^2}{\omega^*(\omega^{*2} - 4\Omega_0^2)}, \tag{31}$$

$$\frac{k_\perp^2}{k_z^2} = -\frac{\omega^{*2} + 2\Omega_0^2}{\omega^{*2}}\cdot\frac{\omega^{*2} - 4\Omega_0^2}{\omega^{*2} - 2\Omega_0^2}. \tag{32}$$

According to the assumption $k_z \ll k_\perp$ from the latter equation it follows that two types of solutions are possible: those with $\omega^* \ll \Omega_0$ and those with $\omega^{*2} \simeq 2\Omega_0^2$.

For the first type of solution, from (31) and (32) we get

$$\frac{k_\perp R J'_m(k_\perp R)}{J_m(k_\perp R)} \simeq \pm \frac{mik_\perp}{k_z}, \tag{33}$$

$$\omega^* \simeq \pm 2i\Omega_0 \frac{k_z}{k_\perp}. \tag{34}$$

The right-hand side of (33) contains a large module; therefore, we have

$$k_\perp R \approx k_{\perp n}^{(m)} R \equiv \lambda_n^{(m)}, \tag{35}$$

and $\lambda_n^{(m)}$ are the roots of the Bessel function:

$$J_m(\lambda_n^{(m)}) = 0. \tag{36}$$

The approximate solution of Eq. (33) is

$$k_\perp \approx \lambda_n^{(m)}/R \mp ik_z/m. \tag{37}$$

One of the solutions of (34) corresponds, obviously, to a Jeans instability with the increment

$$\gamma_n^{(m)} \simeq \frac{2k_z R}{\lambda_n^{(m)}} \Omega_0. \tag{38}$$

For the second type of solution, with $\omega^2 \approx 2\Omega_0^2$, we find in a similar way:

$$\omega^2 \approx \left(2 + \frac{4k_z^2}{k_\perp^2}\right)\Omega_0^2, \tag{39}$$

$$k_\perp R \approx \lambda_n^{(m)} + \frac{2\lambda_n^{(m)}}{m(1 \pm \sqrt{2})} \frac{k_z^2}{k_\perp^2}. \tag{40}$$

These solutions are stable.

The question of the stabilization of the Jeans instability mentioned above (taking into account the velocity dispersion along the cylinder axis) is considered in the same way as in subsection 2.2.2. For perturbations with $k_z v_T \ll \Omega_0$, the dispersion equation reduces to the form of (26), but with a somewhat different set of discrete $k_\perp$. Perturbations die out if the criterion

$$v_T > 2\sqrt{2}v_0/\lambda_n^{(m)}, \tag{41}$$

similar to (34) §2, is satisfied. At $m = 1$, the criteria in (41) and (34), §2, coincide, and at $m > 1$, according to (41), a smaller velocity dispersion v_T (the minimum $\lambda_n^{(m)}$ is obtained at $n = 1$, $m = 1$) is required for stability.

3.2.3. Short-Wave ($k_z R \gg 1$) Perturbations.

At $k_z R \gg 1$, Eq. (31) is reduced to

$$\frac{\omega^{*2} - 2\Omega_0^2}{\omega^{*2} - 4\Omega_0^2} \frac{k_\perp R J'_m(k_\perp R)}{J_m(k_\perp R)} \approx -k_z R + \frac{4m\Omega_0^3}{\omega^*(\omega^{*2} - 4\Omega_0^2)}. \tag{42}$$

In this case again, due to $k_z \gg k_\perp$, we have two possibilities of satisfying Eq. (32): (1) $\omega^2 \approx -2\Omega_0^2 = -\omega_0^2$, (2) $\omega^2 \approx 4\Omega_0^2$. It is evident that the first solution corresponds to the short-wave part of the Jeans perturbation and the second to rotational perturbations. Their long-wave asymptotics were discussed in the previous subsection.

Just as above, we obtain:

(1) for the Jeans type of perturbations

$$\omega \approx \pm i\Omega_0\sqrt{2} = \pm i\omega_0, \tag{43}$$

$$k_\perp R \approx \lambda_n^{(m)} - \tfrac{2}{3}\lambda_n^{(m)}/k_z R; \tag{44}$$

(2) for rotational perturbations

$$\omega \approx \pm(4 - \tfrac{4}{3}k_\perp^2/k_z^2)^{1/2}\Omega_0, \tag{45}$$

$$k_\perp R \approx \mu_n^{\pm(m)}, \tag{46}$$

where $\mu_n^{\pm(m)}$ are the roots of the equations

$$\frac{\mu^\pm J_m'(\mu^\pm)}{J_m(\mu^\pm)} = \pm m. \tag{47}$$

The stabilization condition of the Jeans instability [(43)] is obviously the following:

$$k_z v_T \gtrsim \omega_0 \tag{48}$$

(perturbations with $k_z v_T \gg \omega_0$ die out rapidly; see the end of subsection 2.2.4). This condition is satisfied if the stability criterion for long-wave perturbations [(34), §2] has been satisfied.

§ 4 Stability of a Cylinder with Respect to Flute-like Perturbations

In this section, the stability of the hot model for the cylinder (2), §1 with respect to flute perturbations ($k_z = 0$) is investigated [14, 112, 115]. The dispersion equation can be derived by one of the methods described in the previous chapter.

We shall illustrate the application of the variation method of the phase volume for a given system in Problem 2, where the partial perturbation spectrum (keeping the density constant) is derived. To derive the general dispersion equation, let us use the method of Lagrange displacements in phase space.

Let the Cartesian displacement components be denoted by $X(x, y, v_x, v_y, t)$ and $Y(x, y, v_x, v_y, t)$. The equations describing the evolution of X and Y are of the same form as in the one-dimensional case [8, 14]:

$$(\hat{D}^2 + 1)X = -\frac{\partial \Phi_1}{\partial x}, \qquad (\hat{D}^2 + 1)Y = -\frac{\partial \Phi_1}{\partial y}. \tag{1}$$

The operator $\hat{D}$ is again the Stokes derivative along the unperturbed trajectory:

$$\hat{D} = \frac{d}{dt} = \frac{\partial}{\partial t} + v_x \frac{\partial}{\partial x} + v_y \frac{\partial}{\partial y} - x \frac{\partial}{\partial v_x} - y \frac{\partial}{\partial v_y}. \tag{2}$$

The general pattern of calculations also remains mainly the same. By imposing the polynomial potential $\Phi_1(x, y)$ (with indefinite coefficients) we determine X and Y in solving Eq. (1). Then we average X and Y over velocities at any point of (x, y). For X, this is accomplished by using the formula

$$\bar{X}(x, y) = \frac{1}{2\pi} \int_0^{2\pi} X(x, y, -\gamma y + \rho \cos \varphi, \gamma x + \rho \sin \varphi) \, d\varphi, \tag{3}$$

where ρ denotes the module of the peculiar velocity of the particle:

$$\rho = \sqrt{(1 - \gamma^2)(1 - x^2 - y^2)}, \tag{4}$$

and the angle φ gives its direction. The formula for the averaging of Y is similar to (3).

Then, one has to calculate the density perturbation (see §6, Chapter I)

$$\rho_1 = -\rho_0 \left(\frac{\partial \bar{X}}{\partial x} + \frac{\partial \bar{Y}}{\partial y} \right), \tag{5}$$

and the radial displacement of the boundary

$$\delta R = \left(\frac{x}{R} \bar{X} + \frac{y}{R} \bar{Y} \right)_{r=R}. \tag{6}$$

Finally, calculating the perturbation potential Φ_1 using ρ_1 and δR and comparing it with the original expression, we shall determine the dispersion equation and the coefficients not determined earlier in the polynomial representation of Φ_1.

Let us specify a polynomial potential. In polar coordinates one can always write

$$\Phi_1^{(n, m)} = \left(\sum_{l=0}^{n} a_l r^{2(n-l)} \right) r^m e^{im\varphi}. \tag{7}$$

Each eigenfunction of (7) is identified by two indices: m = the azimuthal number which characterizes the angular dependence of the potential, and n determines the power of the polynomial over r ($n > 0$).

For the calculation of $\bar{X}$ and $\bar{Y}$ it is convenient to use the following relation, which is easily proved:

$$\frac{1}{2\pi} \int_0^{2\pi} f(x + w \cos \varphi, y + w \sin \varphi) \, d\varphi = \sum_{n=0}^{\infty} \frac{1}{(n!)^2} \left(\frac{w}{2} \right)^{2n} \Delta^n f(x, y), \tag{8}$$

where $\Delta \equiv \partial^2/\partial x^2 + \partial^2/\partial y^2$ is the Laplacian operator. For the potential of the form (7) one can, after rather cumbersome calculations involving relation (8), derive equations for the coefficients a_l:

$$a_l(n - l)(n - l + m) + \int_{-\infty}^{0} dt\, e^{-i\omega t} \sin t \sum_{k=0}^{n} (-1)^k \frac{\sin^{2k} t}{(k!)^2} (1 - \gamma^2)^k$$

$$\times (\cos t + i\gamma \sin t)^m \sum_{s=0}^{l} (-1)^{l-s} a_s (\cos^2 t + \gamma^2 \sin^2 t)^{n-s-k-1}$$

$$\times \frac{(n - s + m)!(n - s)!}{(n - s + m - k)!(n - s - k)!(l - s)!(k - l + s)!}$$

$$\times [2 \cos t(n + m - k - s)(n - l) - (\cos t + i\gamma \sin t)m(k + s - l)] = 0. \tag{9}$$

In particular, at $l = 0$ we have the dispersion equation[7] $(n \geq 1)$

$$1 + \frac{1}{n(n + m)} \int_{-\infty}^{0} dt\, e^{-i\omega t} \sin t \sum_{k=0}^{n} (-1)^k \frac{\sin^{2k} t}{(k!)^2} (1 - \gamma^2)^k$$

$$\times (\cos t + i\gamma \sin t)^m \cdot (\cos^2 t + \gamma^2 \sin^2 t)^{n-k-1} \frac{n!(n + m)!}{(n + m - k)!(n - k)!}$$

$$\times [2(n + m - k)n \cos t - km(\cos t + i\gamma \sin t)] = 0. \tag{10}$$

We shall not reproduce here the details of all calculations giving (9). In essence, they are elementary. Let us make only some remarks.

1. The integrals of the type as in Eq. (9), arise in a natural way if the "equations of motion" (1) are solved by the method of "integration over trajectories" [138]. In §4, Chapter I [see formulae (1)–(3)] we have already considered the application of this method for equations like

$$\hat{D}g(x, y, v_x, v_y, t) \equiv \frac{d}{dt} g(x, y, v_x, v_y, t) = Q(x, y, v_x, v_y, t). \tag{11}$$

The solution of more complex equations (1) is obtained after the familiar solution of equations like (11) by the method described in §6, Chapter I.

2. The potential Φ_1 of the surface layer due to the displacement of the boundary (6) is (where $\delta R = \text{const } e^{im\varphi}$):

$$\Phi_1 = \text{const} \frac{2\pi GR}{m} \left(\frac{r}{R}\right)^m e^{im\varphi}, \qquad r < R, \quad m \neq 0;$$

$$\Phi_1 = 0, \qquad m = 0. \tag{12}$$

In comparing (12) with (7), it is at once understood that, practically, δR must be calculated only for the "surface modes" for which $\rho_1 \equiv 0$ (see Problem 2). From Eq. (9), at $l = 1, 2, \ldots, n$, one determines success-

---

[7] If one is interested only in the derivation of the dispersion equation, only the "leading" term of expansion (7) may be taken into account (omitting the rest) $\Phi_1 \sim r^{2n+m} e^{im\varphi}$.

fully the a_l coefficients of the polynomial Φ_n^m. It can be represented in the form

$$\Phi^{(n,m)} = \left(\sum_{p=0}^{n} (-1)^p b_p r^{2(n-p)} \right) r^m e^{im\varphi}, \tag{13}$$

where the positive b_p coefficients are determined from the recurrent relation

$$b_p = b_{p-1} \cdot \frac{(n+m+1-p)(n+1-p)}{p(2n+m-p)}. \tag{14}$$

Using (14), let us express the common term of the b_p series through the first coefficient (which obviously is arbitrary):

$$b_k = b_0 \cdot \frac{n!m!(n+m+k-1)!}{(n-k)!k!(m+k)!(n+m-1)!}. \tag{15}$$

For instance, at $n = 2$, $m = 0$ from (13) and (15) we get $\Phi^{2,0} = r^4 - \frac{4}{3}r^2 +$ const, and $n = 1$, $m = 2$ belongs to the potential $\Phi^{1,2} = (r^2 - 1)r^2 e^{2i\varphi}$.

It is easy to show that the dispersion equation (10) has the structure of Eq. (68), §6, Chapter I:

$$\sum_{k=0}^{N} \frac{\gamma_k}{(N-2k+\omega)^2 - 1} = -1 \qquad (N \equiv 2n + m + 1). \tag{16}$$

However, the expressions for the γ_k coefficients turn out to be very cumbersome (see [14]). Correspondingly cumbersome is the proof of the stability of the system which reduces to the proof of the inequalities (see §6, Chapter I):

$$0 < \gamma_k \leq 1. \tag{17}$$

The general proof, using a number of complicated methods, is given in [14]. We shall give (in Problem 3) only a part of this proof that establishes the stability of "fairly small scale" modes.

Also stable are all composed models (4), §1. For them, instead of (16), we shall obviously have

$$\sum_{k=0}^{N} \frac{\overline{\gamma_k}}{(N-2k+\omega)^2 - 1} = -1,$$

$$\overline{\gamma_k} = \int_{-1}^{1} \gamma_k(\gamma) A(\gamma)\, d\gamma. \tag{18}$$

This equation retains its previous form (16). Also the inequalities similar to (17) remain valid:

$$0 \leq \overline{\gamma_k} \leq 1. \tag{19}$$

This gives the stability of any composed model.

Note that the dispersion equation itself is given in [14] in the form somewhat different from (10):

$$\frac{1}{2\pi} \int_0^\infty d\tau \int_0^{2\pi} (\psi^{n+m}\psi_1^{n-1} + \psi^{m+n-1}\psi_1^n) \sinh \tau e^{-\omega\tau}\, d\varphi = -1 \qquad (n \geq 1), \tag{20}$$

where

$$\psi = \cosh \tau - \gamma \sinh \tau - \sqrt{1 - \gamma^2}\, e^{i\varphi} \sinh \tau, \qquad (21)$$

$$\psi_1 = \cosh \tau + \gamma \sinh \tau - \sqrt{1 - \gamma^2}\, e^{-i\varphi} \sinh \tau. \qquad (22)$$

It is obtained in a natural way if Eqs. (1) are solved by the method which we have used in the previous chapter (§6). Let us note the following simple form of the dispersion equation for radial oscillations of a cylinder ($m = 0$):

$$1 + \frac{1}{n} \sum_{k=1}^{n} (-1)^k \frac{4k^2 a_k}{\omega^2 - 4k^2} = 0, \qquad (23)$$

where

$$a_k \equiv \sum_{s=k}^{n} (-1)^s \left(\frac{1 - \gamma^2}{4}\right)^{s-1} \frac{(n + s - 1)!(2s - 1)!}{(n - s)![(s - 1)!]^2 (s - k)!(s + k)!}. \qquad (24)$$

Let us consider separately the flutelike ($k_z = 0$) oscillations of the *cold* cylinder [20]. Corresponding frequencies can be obtained either from the general dispersion equation (10) by the limiting transition at $|\gamma| \to 1$, or directly by solving the hydrodynamical equations of motion in the x, y plane, which in this case are very simple. By any method it is easy to demonstrate that the eigenfrequencies are equal to

$$\omega_{1,2} = m \pm \sqrt{2}, \qquad \omega_3 = m, \qquad n \geq 1, \qquad (25)$$

$$\omega_1 = \omega_2 = m - 1, \qquad n = 0. \qquad (26)$$

Pay attention to the fact that the eigenfrequencies are independent of the radial wave number n in (25): at a fixed type of angular symmetry (m for the cylinder) the radial dependence of the eigenfunction may be arbitrary. The same degeneration is characteristic for homogeneous spheres with *circular orbits* of particles (see §3, Chapter III, where this question is treated in more detail).

The frequency in (26) corresponding to "surface" oscillations ($\rho_1 = 0$) of the cold cylinder is multiple. Therefore, one could expect the presence of a power instability. In Problem 5, the proof of the presence of a power instability (an increase in $\sim t$) is given for partial oscillations of the surface type, at which the original circular cylinder is deformed into a cylinder of elliptical cross section [9].

Let us consider also the possibility of power instabilities connected with the presence of multiple roots [14]. In this respect, the cylindrical model differs from that considered in §6, Chapter I, concerning the homogeneous layer, which turned out to be completely stable. For example, we have just found the multiple roots in (26) at $\gamma = 1$, corresponding to the same mode. Resonance is also possible [the effect of the higher terms in the polynomial expansion of the potential $\Phi_1(x, y)$ on the terms of lower orders—see §6, Chapter I]. Resonance occurs if the eigenfrequencies for the modes (n, m) and $(n - k, m)$ coincide at some $k > 0$. The simplest case of the resonance occurs at $n = 1$, $m = 2$ (Problem 4). According to Antonov's suggestion [14], the set of resonant values of γ is countable and dense everywhere within the interval $(-1, 1)$. Therefore, the model of the homogeneous cylinder in this section is stable but excluding a set of values of the parameter γ when there is a power $(\sim t)$ instability.

In addition to the above "true" oscillations of the cylinder, there is a large number of various trivial solutions (permutational and "dusty," according to the terminology adopted in [14]). We do not dwell on these solutions since they do not affect the stability of the system (integer frequencies correspond to them).

§ 5 Local Analysis of the Stability of Cylinders (Flute-like Perturbations)

For fairly short wavelength perturbations ($n \gg 1$), the characteristic equation (10) derived in the previous section becomes very cumbersome. On the other hand, at $n \gg 1$, the eigenfunctions [for example, the perturbed potential $\Phi_1(r)$, see formula (13), §4] are rapidly oscillating. Therefore, here in principle one may use the local method by assuming that

$$\Phi_1 \sim e^{ikr + im\varphi} \tag{1}$$

and $kr \gg m$. However, for an arbitrary value of the parameter γ, the calculations still remain cumbersome. They become simple if $1 - \gamma^2 \ll 1$, i.e., if the motion of particles is fairly close to circular. At the same time, the information thus obtained is rather full; in particular, qualitatively it holds also for longer wave oscillations (or for hotter cylinders). It is also important that in the approximation of short wavelengths one can with identical success investigate the oscillations not only of homogeneous but also of inhomogeneous $\rho_0 = \rho_0(r)$ and differentially rotating $[\Omega_0 = \Omega_0(r)]$ systems. An analytical consideration of large-scale perturbations in such systems is practically impossible.

The derivation of the dispersion equation can, as usual, be divided into two stages. The former involves the solution of the linearized kinetic equation, with a given perturbation of the potential Φ_1. As a result, we determine the perturbed distribution function, and by integrating it over velocities, the perturbation of the density ρ_1. Note also that ρ_1 thus calculated is often called the "response" of the system (to the specified perturbation of the gravity field).

In the second stage, we impose the condition of self-consistency requiring that according to the Poisson equation the same Φ_1 correspond to the "response" of the system ρ_1 to the given field Φ_1 (as calculated in the first stage of the derivation from the *kinetic* equation), i.e., that this field is produced by the perturbation of the density ρ_1.

For short-wave perturbations $kr \gg 1$, the Poisson equation reduces to the simple relation

$$\rho_1 = -\frac{k^2}{4\pi G} \Phi_1. \tag{2}$$

5.1 Dispersion Equation for Model (2), § 1

In the first stage, the main features of using the short-wave approximation are the same for cylindrical and disk systems. Therefore, the derivation that follows below is largely a simple repetition of Lin and Shu's calculations [271] in the derivation of the dispersion equation for short-wave perturbations of the disk (see Section 4.1, Chapter V).

Let us avail ourselves of the method of variation of the phase volume (Chapter I, §6). The calculations are quite similar to the case of the flat layer. The full density (unperturbed + perturbed) is

$$\rho = \frac{\rho_0}{\pi} \int_0^{2\pi} \int_0^{\varkappa} \delta[\varkappa^2 - v^2 - \varepsilon\chi(r, \varphi, v_r, v_\varphi)]v \, dv \, d\alpha, \tag{3}$$

where

$$\varkappa^2 = (1 - \gamma^2)(1 - r^2), \qquad \tan \alpha = \frac{v_\varphi}{v_r}, \qquad v^2 = v_r^2 + v_\varphi^2. \tag{4}$$

Hence we can derive the formula for the calculation of the perturbed density,

$$\rho_1 = \frac{\rho_0}{2\pi} \int_0^{2\pi} d\alpha \int_0^{\varkappa} dv \, \frac{\partial \chi}{\partial v} \, \delta(\varkappa^2 - v^2), \tag{5}$$

and the equation of the perturbed boundary,

$$\varkappa^2 - \varepsilon\chi_0 = 0, \tag{6}$$

where

$$\chi_0 = \chi(v = 0), \qquad \chi_1 = \chi - \chi_0. \tag{7}$$

The equation for the χ function has the form

$$\frac{d\chi}{dt} = 2v_r \frac{\partial\Phi_1}{\partial r} + \frac{2v_\varphi}{r} \frac{\partial\Phi_1}{\partial\varphi} = 2\left(\frac{d\Phi_1}{dt} - \frac{\partial\Phi_1}{\partial t}\right), \tag{8}$$

where d/dt is the full derivative along the unperturbed trajectory of the particle in the rotating reference system [in which the original distribution function has the form of (2), §1].

The general equation (8) for the χ function in the case of short-wave oscillations of a fairly cold cylinder reduces to the following:

$$-i(v - k'v_r)\chi + \left(v_r \frac{\partial\chi}{\partial v_\varphi} - v_\varphi \frac{\partial\chi}{\partial v_r}\right) = 2ik'v_z \tilde{\Phi}_1, \tag{9}$$

where $v = \omega/2$, $k' = k/2$, $\tilde{\Phi}_1$ is the amplitude of the perturbed potential.

Equation (9) can be solved by the method of integration over the angle [138]. For this purpose, we introduce the polar coordinates v, s in velocity space:

$$v_r = v \cos s, \qquad v_\varphi = v \sin s. \tag{10}$$

Then Eq. (10) becomes

$$-\frac{\partial \chi}{\partial s} + i(v - k'v \cos s)\chi = -2ik'\tilde{\Phi}_1 v \cos s. \tag{11}$$

The differential equation of this form occurs in the physical plasma kinetics [138]. Its solution can be represented in various ways. The most convenient seems to be the form of the solution obtained by Lin and Shu [271]:

$$\chi = 2\tilde{\Phi}_1 \cdot Q, \tag{12}$$

where

$$Q = 1 - \frac{v\pi}{2\pi \sin v\pi} \int_{-\pi}^{\pi} \exp\{i[vs - k'v_r \sin s - k'v_\varphi(1 + \cos s)]\} ds. \tag{13}$$

To obtain (12), let us write the general solution of Eq. (11):

$$-\chi/2\tilde{\Phi}_1 = e^{iF(s, v)}\left\{C(v) - ik' \int_0^s e^{-iF(s', v)}v \cos s' ds'\right\}, \tag{14}$$

where $F(s, v) \equiv vs - k'v \sin s$.

The "constant" of integration $C(v)$ can be found from the periodicity condition

$$\chi(s) = \chi(s + 2\pi). \tag{15}$$

The integral in $\chi(s + 2\pi)$ appearing due to the use of (15) should be written as

$$\int_0^{s+2\pi} = \int_0^s + \int_s^{s+2\pi}.$$

Making in the last integral the substitution $s' = s + s''$, one can obtain for the bracketed expression in Eq. (14):

$$C - ik' \int_0^s = e^{-ik'v \sin s}(-ik') \frac{e^{2\pi iv}}{1 - e^{2\pi iv}} \int_0^{2\pi} e^{-ivs' + ik'v \sin(s + s')} v \cos(s + s') ds'.$$

Simple transformations of the right-hand side of this formula (partial integration, substituting $\pi - s' = s''$) reduce the solution to the form (12).

Now one has to determine the perturbed density ρ_1 using formula (3), and one can perform the integration over velocities and reduce the "response" ρ_1 to the form

$$\rho_1 = \rho_0 \Phi_1 \frac{k'}{\pi} \frac{v\pi}{\sin v\pi} \int_{-\pi}^{\pi} ds\, e^{ivs} \cos \frac{s}{2} \cdot J_1\left[-2\kappa k' \cos \frac{s}{2}\right]. \tag{16}$$

Taking into account the Poisson equation (2), we then arrive at the dispersion equation[8]

$$\frac{k_1}{G\rho_0} = \frac{v\pi}{\sin v\pi} \int_{-\pi}^{\pi} ds\, e^{ivs} \cos \frac{s}{2} J_1\left(2k_1 \cos \frac{s}{2}\right), \tag{17}$$

[8] Equation (17) may be easily generalized also for the case of disturbances with $k_z \neq 0$.

where $k_1 = k'\varkappa = k/2\sqrt{(1 - \gamma^2)(1 - r^2)}$. The integral in (17) is again expressed by the Bessel function [42] and we ultimately get

$$2k_1 = \frac{v\pi}{\sin v\pi} [J_{v+1}(k_1)J_{-v}(k_1) + J_v(k_1)J_{1-v}(k_1)]. \tag{18}$$

The dispersion equation (18) determines the function

$$v = v(k_1), \qquad v = \frac{\omega}{2}, \qquad k_1 \equiv \frac{k}{2}\sqrt{(1 - \gamma^2)(1 - r^2)}. \tag{19}$$

Note that the value $2k_1$ can be written as $k(\overline{v^2})^{1/2}/2\Omega_0$, where $(\overline{v^2})^{1/2} = \sqrt{(1 - \gamma^2)(1 - r^2)}$ is the mean square velocity: It is equal (in order of magnitude) to the ratio of the mean size of the epicycle $(v^2)^{1/2}/2\Omega_0$ and the perturbation wavelength $1/k$. The dimensionless frequency v is defined as the ratio of the frequency ω (measured in the rotating reference system) and the epicyclic frequency $\varkappa = 2\Omega_0$.

According to the assumption $k \gg 1$, $(1 - \gamma^2) \ll 1$, k_1 can be both small and large. To construct a qualitative picture of the oscillation branches corresponding to the dispersion equation (18) let us consider the limiting cases.

5.1.1. $k_1 \ll 1$. In the limit $k_1 = 0$, we must obtain the eigenfrequencies of oscillations of the cold cylinder $v = \pm 1/\sqrt{2}$ [formula (25), §4]. Perform the limiting transition in (18). Since at $k_1 \ll 1$

$$J_v(k_1) = \left(\frac{k_1}{2}\right)^v \frac{1}{\Gamma(1 + v)}, \tag{20}$$

the bracket on the right-hand side of Eq. (18) is

$$[\ ] \simeq \frac{k_1}{2}\left[\frac{1}{\Gamma(1 - v)\Gamma(2 + v)} + \frac{1}{\Gamma(1 + v)\Gamma(2 - v)}\right]. \tag{21}$$

Using the relations

$$\Gamma(\alpha + 1) = \alpha\Gamma(\alpha), \qquad \Gamma(1 - \alpha)\Gamma(\alpha) = \pi/\sin \pi\alpha,$$

transform (21) to the form

$$[\ ] \simeq \frac{k_1}{2}\left[\frac{\sin v\pi}{\pi v(v + 1)} - \frac{\sin v\pi}{\pi v(v - 1)}\right]. \tag{22}$$

Substituting (22) into (18), we get

$$2 = -1/(v^2 - 1), \qquad \text{i.e.,} \qquad v^2 = 1/2, \qquad \omega^2 = 2.$$

If we expand the Bessel functions in the dispersion equation (18) up to the term $\sim k_1^4$, we find

$$2 \simeq -\frac{1}{v^2 - 1} - \frac{3k_1^2}{2(v^2 - 1)(v^2 - 4)} + \frac{3k_1^4}{8(v^2 - 1)(v^2 - 9)}. \tag{23}$$

Hence one can first determine the behavior of the branch $v^2 \simeq \frac{1}{2}$ for small k_1:

$$v^2 \approx \tfrac{1}{2} + \tfrac{3}{14}k_1^2. \tag{24}$$

Furthermore, there are the solutions for Eq. (23) near the "resonance" frequencies ± 2 and ± 3:

$$v^2 \approx 4 - \tfrac{3}{14}k_1^2, \tag{25}$$

$$v^2 \approx 9 - \tfrac{3}{136}k_1^4. \tag{26}$$

From expansions up to terms of the order of $k_1^{2(n-1)}$ similar to (23), one can find the behavior of the branch $v^2 \simeq n^2$ near the resonance frequencies $\pm n$:

$$v^2 \simeq n^2 - \alpha k_1^{2(n-1)} \qquad (\alpha > 0). \tag{27}$$

5.1.2. $k_1 \gg 1$. In this limit, by using the asymptotic representation of the Bessel function,

$$J_v(k_1) \simeq \sqrt{\frac{2}{\pi k_1}} \cos\!\left(k_1 - \frac{\pi v}{2} - \frac{\pi}{4}\right), \tag{28}$$

we readily find

$$v^2 \simeq n^2 \left[1 + (-1)^{n+1}\frac{2 \cos 2k_1}{\pi k_1^2}\right] \qquad (n = 1, 2, \ldots). \tag{29}$$

Now one can represent the full qualitative picture of the behavior of flute-like ($k_z = 0$) branches of oscillations of the homogeneous collisionless cylinder (Fig. 11).

Let us consider the problem of how the approximate picture is related to the accurate solution investigated in §4. We have seen there that at any fixed finite index of the n mode (to which corresponds an effective radial wave number k) and at fixed m, there is a finite number of eigenfrequencies ω_i. With increasing n, this number increases and tends to infinity at $n \to \infty$.

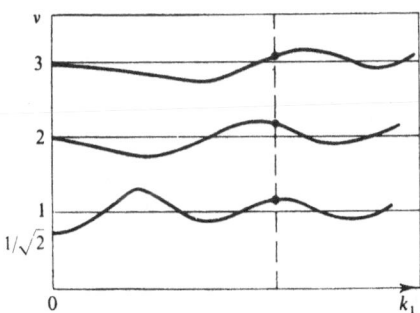

Figure 11. Branches of flute oscillations $v = v(k_1)$ of the cylinder with distribution function (2), §1, with $(1 - \gamma^2) \ll 1$.

In the limit $n \to \infty$ ($k_1 \to \infty$), all frequencies, as follows from (29), tend to the resonant $v \to \pm n$ ($\omega \to \pm 2n$). Such a result is quite logical. From the Poisson equation (2) it is evident that at $k \to \infty$, for oscillations with $\rho_1 \simeq$ const, $\Phi_1 \to 0$, i.e. the self-gravity of the wave does not affect the oscillations of particles near the arbitrary point r. Then only oscillations of infinitely thin layers (or individual particles) remain in the common gravity field which may occur on the epicyclic frequency $\varkappa$ (or its harmonics). For the homogeneous cylinder, the epicyclic frequency $\varkappa = 2\Omega_0$ is independent of the radius and the described oscillations of layers, essentially not related mutually by the interaction, may occur with formally consistent phases. It is, however, evident that in the case of inhomogeneous systems when $\varkappa = 2\Omega_0(1 + r\Omega_0'/2\Omega_0)^{1/2} = \varkappa(r)$ only *singular* oscillations are eigen [at $k \to \infty$ at which any layer with its epicyclic frequency $\varkappa = \varkappa(r_0)$ oscillates, and all the remaining are in the unperturbed state].

The convergence of the oscillation frequencies at $n \to \infty$ towards resonant ones can also be inferred from the accurate expressions for the coefficients γ_k in the dispersion equation (18), §4. It turns out that all $\gamma_k \to 0$. Accordingly, at larger n [and, to be more precise, at larger $k_1 \sim n\sqrt{(1 - \gamma^2)}$] the values of individual terms of the sum in Eq. (18), §4, differ essentially from zero only in the neighborhood of special (integer) points towards which the roots contract:

$$\omega \to 2(n - k) \qquad (k = 0, 1, \ldots, 2n + m - 1) \qquad (30)$$

(here ω is the frequency in the rotating frame of reference).

One can easily understand the oscillating character of the dependence $v(k_1)$ (with crossings of curves $v = n$ in Fig. 11) on some fixed branch of oscillations. Take, for example, the simplest mode ($m = 0, n = 2$). Obviously, it does not apply to the number of the short-wave modes, but the essence of the matter does not change in spite of this fact. The dispersion equation in this case is the following:

$$1 + \frac{2 - 3\varepsilon}{\omega^2 - 4} + \frac{3\varepsilon}{\omega^2 - 16} \qquad (\varepsilon \equiv 1 - \gamma^2). \qquad (31)$$

Here, the mode index n is fixed, to which corresponds the fixed wave number k; however, $k_1 \sim k\sqrt{(1 - \gamma^2)}$ changes with γ, and the manner of the dependence $\omega^2(\gamma) \sim \omega^2(k_1)$ may be traced.

From (31), it is evident that in this case there are two branches of the $\omega^2(\gamma) \sim \omega^2(k_1)$ function. They are presented in Fig. 12. On one of them, ω^2 increases monotonically from $\omega^2 = 2$ at $\varepsilon = 0$ to $\omega^2 = 9 - \sqrt{13} \simeq 5.4$ at $\varepsilon = 1$, intersecting the straight line $\omega = 2$ ($v = 1$) at $\varepsilon = \frac{2}{3}$. The second branch decreases monotonically from $\omega^2 = 16$ ($v = 2$) at $\varepsilon = 0$ to $\omega^2 = 9 + \sqrt{13} \simeq 12.6$ at $\varepsilon = 1$. The behavior of the two branches is in qualitative agreement with the original portions of the lower branches in Figure 11 (although we have now considered one of the modes with largest scale).

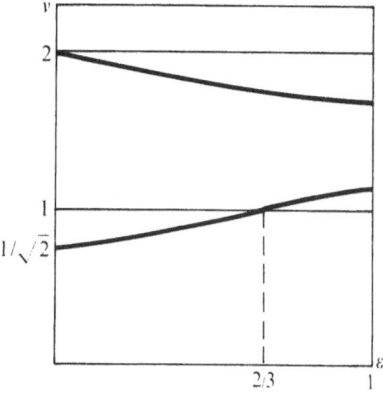

Figure 12. Relation $v = v(\varepsilon)$ for the simplest oscillation mode ($n = 2$, $m = 0$).

5.2 Maxwellian Distribution Function

The dispersion equation describing short-wave oscillations of a rather cold cylinder can also be obtained for other distributions of random velocities. Of special interest for applications is the Maxwellian distribution

$$f_0 = \frac{\rho_0}{\pi v_T^2} \exp\left(-\frac{v^2}{v_T^2}\right) \qquad (v^2 = v_r^2 + v_\varphi^2) \tag{32}$$

Of course, strictly speaking, the velocity distribution in a real system (for example, of stars in a galaxy) cannot be Maxwellian. In particular, it must have a cutoff at escape velocity of the system. However, at a "low temperatures" (for sufficiently cold systems) there are few particles with high velocities in (32), so that they can be neglected.

As mentioned above, a similar problem concerning the short-wave oscillations of disk systems with orbits close to circular, was solved by Lin and Shu [271] (Section 4.1, Chapter V), and they considered inhomogeneous (and differentially rotating) systems. The latter, in particular, does not increase the complexity of the calculations (which is natural for the local approximation). The equilibrium distribution function considered in [271] is the so-called Schwarzschield distribution

$$f_0 = \frac{\rho_0}{2\pi c_r c_\varphi} \exp\left(-\frac{v_r^2}{2c_r^2} - \frac{v_\varphi^2}{2c_\varphi^2}\right), \tag{33}$$

and it was assumed that the radial and azimuthal velocity dispersions are linked by the (Lindblad) relation $c_r = (2\Omega_0/\varkappa)c_\varphi$ where $\varkappa = 2\Omega_0(1 + r\Omega_0'/2\Omega_0)^{1/2}$ is the epicyclic frequency. The distribution in (33) becomes Maxwellian (32) in the case of the system rotating as a solid body ($\Omega_0' = 0$).

Thus, we may use Lin and Shu's expression for density perturbations with the Schwarzschield distribution function:

$$\frac{\rho_1}{\rho_0} = - \frac{k^2 \Phi_1}{\varkappa^2 (1 - v^2)} \mathscr{F}_v(x), \qquad (34)$$

where $v = (\omega - m\Omega_0)/\varkappa$, $x = k^2 c_r^2/\varkappa^2$, and $\mathscr{F}_v(x)$ is the so-called "reduction factor,"

$$\mathscr{F}_v(x) = \frac{1 - v^2}{x} \left[1 - \frac{v}{2 \sin v\pi} \int_{-\pi}^{\pi} e^{-x(1 + \cos s)} \cos vs \, ds \right]. \qquad (35)$$

The derivation of formula (34), because of its importance for the theory of density waves in galaxies, is given in Section 4.1, Chapter V [though it does not contain anything fundamentally new in comparison with the derivation of formula (16)]. Also we give there the plot of the $\mathscr{F}_v(x)$ function (Fig. 42). It is unity at $x = 0$, i.e., for a cold system $c_r = 0$, and monotonically decreases with increasing x, i.e., with an increase of the mean velocity dispersion.

$\mathscr{F}_v(x)$ becomes small at $x \gtrsim 1$. Thus, this function takes into account the decrease ("reduction") of the effect of the gravity field on the particles having radii of epicycles comparable or larger than the perturbation wavelength.

The Poisson equation (2) and relation (34) give the dispersion equation

$$\frac{\varkappa^2}{4\pi G \rho_0} (1 - v^2) = \mathscr{F}_v(x). \qquad (36)$$

In particular, for a disk rotating as a solid body, (36) can be represented in the form similar to (18):

$$2x = 1 - \frac{v\pi}{2\pi \sin v\pi} \int_{-\pi}^{\pi} e^{-x(1 + \cos s)} \cos vs \, ds. \qquad (37)$$

It is clear that the value of x in this case plays the same role as the value of k_1 in the previous item. Considering also the limiting cases $x \ll 1$ and $x \gg 1$, one can arrive at the picture of oscillation branches depicted in Fig. 13.

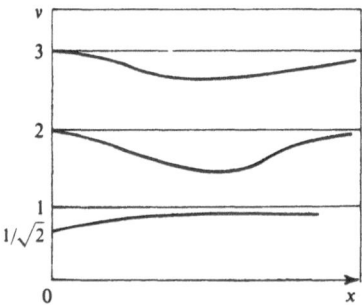

Figure 13. Branches of flute oscillations $v = v(x)$ of a homogeneous cylinder with a Maxwellian distribution function with respect to thermal velocities of particles in the plane of rotation.

In particular, at $x \to \infty$, the frequencies tend to the resonance $\nu = n$ according to

$$\nu \approx n\left(1 - \frac{1}{2\sqrt{2\pi}\, x^{3/2}}\right),\tag{38}$$

while the behavior of $\nu(x)$ at small x for the first three branches in Fig. 13 is

$$\nu_1^2 \simeq \tfrac{1}{2} + \tfrac{3}{7}x,\tag{39}$$

$$\nu_2^2 \simeq 4 - \tfrac{3}{7}x,\tag{40}$$

$$\nu_3^2 \simeq 9 - \tfrac{3}{34}x^2.\tag{41}$$

The picture presented in Fig. 13 resembles very much the cyclotron branches of plasma oscillations at $k_z = 0$ [86].

§ 6 Comparison with Oscillations of an Incompressible Cylinder

Let us briefly mention some results of investigations of the gravitational stability of the incompressible fluid cylinder.

The dispersion equation describing small perturbations of a *nonrotating* liquid cylinder is [13, 186] (disturbances $\sim e^{ikz + im\varphi}$)

$$\frac{\omega^2}{4\pi G\rho_0} = -\frac{kRI_m'(kR)}{I_m(kR)}[K_m(kR)I_m(kR) - \tfrac{1}{2}],\tag{1}$$

where I_m is the Bessel function of the imaginary argument and K_m is the Macdonald function. In particular for radial perturbations ($m = 0$)

$$\frac{\omega^2}{4\pi G\rho_0} = -\frac{kRI_0'(kR)}{I_0(kR)}[K_0(kR)I_0(kR) - \tfrac{1}{2}].\tag{2}$$

From (2), it can be inferred that at $k < k_{cr} \equiv 1.0668/R$, i.e., on fairly long waves (in comparison with the radius), the cylinder is unstable. The maximum increment occurs at $k_z R \simeq 0.580$. For axially asymmetrical perturbations ($m \neq 0$) Eq. (1) yields only stable roots [from the theory of cylindrical functions it is known that $K_m(x)I_m(x) < \tfrac{1}{2}$ for all $m \neq 0$].

Compare the depicted picture of oscillations of the incompressible cylinder with the corresponding results for the collisionless cylinder. Since we have considered *rotating* cylinders, we must compare them with large-scale ($k_\perp R \ll 1$) perturbations of the Jeans branch (the lowest in Fig. 9); for them the rotation obviously does not play any significant role.

We have seen above that the collisionless cylinder with a fairly high velocity dispersion of the particles along the rotational axis z is practically stable[9]: The instability increment is exponentially small. The collisionless

---

[9] Of course, this is only true if the distribution function of the v_z particle has no "beam" character (see Chapter VI).

cylinders, cold along the z axis, are, of course, unstable, and the instability increment is of the same order as that for incompressible cylinders ($\sim \omega_0$).

However, the physical causes of the instability in either case are quite different. The instability of the incompressible cylinder is the reflection of the natural trend of a liquid gravitational mass towards the division into nearly spherical clusters. For this system, it turns out energetically profitable to bend the boundary. As shown by Chandrasekhar [186], a change in the gravitational potential energy ΔW, as the cylinder is deformed, is negative just for those modes, for which the cylinder is unstable according to Eq. (2). It turns out that

$$\Delta W = \text{const} \cdot \varepsilon^2 [K_m(k_z R)I_m(k_z R) - \tfrac{1}{2}], \tag{3}$$

where ε is the amplitude of deformation of the boundary (at $r = R$), from which follows the identity of the conditions $\omega^2 < 0$ and $\Delta W < 0$ (the energy principle).

As far as the collisionless cylinder is concerned, the gravitational energy is mainly released during compression that changes the local matter density (at $k_z R \ll 1$). It can be shown that for the modes considered, the contribution to the perturbation of the potential from the deformation of the boundary ξ_r, is less than the contribution of the local density ρ_1 roughly by a factor $\beta = \ln(1/k_z R)$.

In the case of collisionless cylinders at $k_z \neq 0$, there are generally no modes of "*incompressible*" type. For example, for a cylinder cold in the plane of rotation, their presence would mean (at $m = 0$), according to §2,

$$1 + \frac{\omega_0^2}{\omega^2 - 4\Omega_0^2} = 1 + \frac{\omega_0^2}{\omega^2} = 0, \tag{4}$$

i.e., a condition which cannot be met.

6.1 Flute-like Perturbations ($k_z = 0$)

With respect to perturbations with $k_z = 0$, incompressible cylinders are stable at $\Omega_0^2 < 2\pi G\rho_0$.[10] Cylinders rotating faster are unstable. The physical reason for the instability of the incompressible cylinder involves a decrease in the effective "force of gravity" on its surface with an increase in the rotational velocity, and this effect becomes stronger than the stabilization due to the conservation of angular momentum of the particles around the z-axis [13].

The oscillation frequency in the frame of reference rotating together with the cylinder is given by

$$(\omega + \Omega_0)^2 = (m - 1)(2\pi G\rho_0 - \Omega_0^2). \tag{5}$$

[10] By the way, at $\Omega_0^2 > 2\pi G\rho_0$, the equilibrium of the cylinder can be realized only on account of *negative* pressure.

In particular, in the limiting case $\Omega_0^2 = 2\pi G\rho_0$ (i.e., for the cold cylinder, the equilibrium pressure is zero) from (5) we get the universal frequency

$$\omega = -\Omega_0 \quad \text{(for all } m\text{)}; \tag{6}$$

it coincides with the frequency for similar oscillations of the collisionless cylinder (§4). This is not unexpected, if one takes into account the similarity of the equilibrium states in the two cases of interest and the fact that perturbations of this type are of "incompressible" character ($\rho_1 = 0$) also for the collisionless cylinder.

Of course, the frequency spectrum of oscillations of the collisionless cylinder is much more complex than that of the incompressible (generally any liquid) cylinder. It is composed of an infinite number of branches, which implies the presence of additional degrees of freedom connected with the distribution of particles in velocity space. These branches can in principle be restored by starting from oscillation frequencies of the hot cylinder in the rotation plane ($k_z = 0$) and by extending them onto the region $k_z \neq 0$. The characteristic equation (10), §4, contains the "ends" (limiting frequencies at $k_z \to 0$) for all branches but one (Jeans), which corresponds to $k_z \to 0$, $\omega \to 0$.

Noteworthy is a fairly close analogy between the oscillation spectra of the collisionless cylinder and the model considered earlier for the homogeneous thin layer. In a similar way, in particular, the thermal ($\sim k_z^2 v_T^2$) corrections to the frequencies of flute-like oscillations of the cylinder (see Problem 1, Chapter I) are determined.

§ 7 Flute-like Oscillations of a Nonuniform Cylinder with Circular Orbits of Particles

Oscillations of the inhomogeneous cylinder having circular orbits of particles (in the plane of rotation x, y) obey Eq. (14), §3. Below, we shall be interested in flute-like perturbations ($k_z = 0$), for which the equation

$$\frac{1}{r}\frac{d}{dr}\left(r\varepsilon_\perp \frac{d\Phi_1}{dr}\right) - \varepsilon_\perp \frac{m^2}{r^2}\Phi_1 - \frac{2mI}{r}\Phi_1 = 0, \tag{1}$$

$$\varepsilon_\perp = 1 + \frac{\omega_0^2}{\omega_*^2 - \varkappa^2}, \qquad I = \frac{1}{\omega_*}\frac{d}{dr}\left(\frac{\omega_0^2 \Omega_0}{\omega_*^2 - \varkappa^2}\right), \tag{2}$$

$$\omega_0^2 \equiv 4\pi G\rho_0(r), \qquad \omega_* = \omega - m\Omega(r), \qquad \varkappa^2 = 4\Omega^2 + r(\Omega^2)', \tag{3}$$

is obtained.

If one considers a cylinder limited in the (x, y) plane by radius R, then Eq. (1) must further satisfy the boundary conditions

$$\varepsilon_\perp \left.\frac{d\Phi_{1<}}{dr}\right|_R - \left.\frac{2m\omega_0^2 \Omega \Phi_{1<}}{R\omega(\omega_*^2 - \varkappa^2)}\right|_R = \left.\frac{d\Phi_{1>}}{dr}\right|_R, \tag{4}$$

$$\Phi_{1<}|_R = \Phi_{1>}|_R, \tag{5}$$

which connect the internal $\Phi_{1<}$ and the external (vacuum) $\Phi_{1>}$ potentials. The latter is evidently the solution of the Laplace equation

$$\Phi_{1>}(r) = \text{const} \cdot r^{-m}, \tag{6}$$

where the constant is defined by the value of the "multipole momentum," produced by the redistribution of the mass of the cylinder due to the perturbation (as well as by the curvature of the boundary). If the density $\rho_0(r)$ vanishes smoothly (without a jump) at $r \to R$, then (4) reduces simply to the requirement of continuity of the derivative of the potential on the boundary.

In the limiting case of a uniform cylinder, we get, instead of (1)–(5),

$$\varepsilon_\perp^{(0)} \Delta \Phi_1 = 0, \qquad \varepsilon_\perp^{(0)} = 1 + \frac{2\Omega_0}{(\omega - m\Omega_0)^2 - 4\Omega_0^2}, \tag{7}$$

$$\varepsilon_\perp^{(0)} \frac{d\Phi_{1<}}{dr}\bigg|_R - \frac{4m}{R(\omega - m\Omega_0)} \frac{\Omega_0^3 \Phi_{1<}}{(\omega - m\Omega_0)^2 - 4\Omega_0^2}\bigg|_R = \frac{d\Phi_{1>}}{dr}\bigg|_R, \tag{8}$$

$$\Phi_{1<}|_R = \Phi_{1>}|_R. \tag{9}$$

There are, as we are aware, solutions for Eq. (7) of two types, which can be called "internal" and "surface" solutions. For the former

$$\Delta \Phi_1 \neq 0, \qquad \varepsilon_\perp^{(0)} = 0;$$

therefore

$$\omega_*^2 \equiv (\omega - m\Omega_0)^2 = 2\Omega_0^2. \tag{10}$$

The boundary conditions in (8) and (9) in this case can be satisfied by requiring that the "multipole momentum" produced by the perturbation is zero. For that purpose, it is sufficient, for example, that the condition

$$\int r^m \rho_1(r)\, dr = \int r^m \Delta \Phi_{1<}\, dr = 0, \qquad \frac{d\Phi_{1<}}{dr}\bigg|_R = \Phi_{1<}|_R = 0 \tag{11}$$

is satisfied. Then

$$\Phi_{1>}(r) = 0, \tag{12}$$

and the boundary conditions in (8) and (9) are satisfied in a trivial way. For "surface" perturbations

$$\Delta \Phi_{1<} \equiv \Delta \Phi_{1>} = 0, \tag{13}$$

$$\Phi_{1<} = ar^m, \qquad \Phi_{1>} = br^{-m}, \tag{14}$$

while the conditions in (8) and (9) yield

$$\frac{2(\omega + \Omega_0)^2}{\omega(\omega + 2\Omega_0)} = 0, \tag{15}$$

i.e., the familiar result, the multiple root $\omega = -\Omega_0$.

In connection with the solutions for the homogeneous cylinder let us make two remarks:

1. The choice of the radial dependence of eigenfunctions for the first type of perturbations ("internal"), as we saw, is nearly arbitrary. In particular, they can be localized in infinitesimally thin regions Δr near an arbitrary $r = r_0$.

At the same time, the "surface" solutions are localized (near the surface $r = R$) only for harmonics with sufficiently large azimuthal numbers m. The "surface" modes with $m \sim 1$ are always large-scale: the perturbation covers the entire system.

2. For radial perturbations ($m = 0$) only solutions of the first type are possible. Consider now sequentially the radial and nonradial oscillations of the *inhomogeneous* cylinder.

Radial oscillations are described by a very simple equation:

$$\frac{1}{r}\frac{d}{dr}\left(r\varepsilon_\perp \frac{d\Phi_1}{dr}\right) = 0, \qquad \varepsilon_\perp = \frac{\omega^2 - \varkappa_1^2}{\omega^2 - \varkappa^2}, \qquad \varkappa_1^2 \equiv \varkappa^2 - \omega_0^2 = 2\Omega_0^2(r). \tag{16}$$

The solutions for this equation are

$$\frac{d\Phi_1}{dr} = \text{const} \cdot \delta[\omega^2 - 2\Omega^2(r)]. \tag{17}$$

They are nontrivial when the frequencies are in the interval

$$\sqrt{2}\Omega_{\min} < |\omega| < \sqrt{2}\Omega_{\max}, \tag{18}$$

where $\Omega_{\min}$ and $\Omega_{\max}$ are the minimum and maximum angular velocities of the particles in the differentially rotating cylinder, respectively.

For instance, assume $\omega = \sqrt{2}\Omega(r_0)$; then the gravity force

$$\frac{d\Phi_1}{dr} = \text{const} \cdot \delta(r - r_0), \tag{19}$$

while the perturbed density is

$$\rho_1 \sim C_1 \cdot \delta'(r - r_0). \tag{20}$$

This solution evidently describes oscillations of an infinitely thin layer $r = r_0$, while all the remaining layers of the cylinder are at rest (or, in other words, the amplitude of their oscillations is zero). The choice of r_0 inside the cylinder is arbitrary, the described type of solutions may correspond to any specific r_0, i.e., to oscillations of the corresponding layer with its "own" frequency $\sqrt{2}\Omega_0$.

Such solutions form a complete set: Directly from Eq. (16) it can be readily proved that there are no reasonable *nonsingular* solutions, which would correspond to oscillations of the whole system as a whole (rather than individual layers).

Nonradial oscillations must satisfy an essentially more complex equation (1) and boundary conditions (4) and (5).

In this case, as we will see further, there are two types of solutions: "global," which embrace the entire system, and "local," which are concentrated near an arbitrary point $r = r_0$. "Global" perturbations within the limit of the uniform cylinder become "surface" modes, which do not perturb the local density ρ_1. These oscillations have no analogues in the case of radial oscillations considered above.

"Local" perturbations at $\rho_0 \to$ const change into "internal" modes. Here the existence of "local" perturbations is not obvious since perturbations at $m \neq 0$ in some selected layer affects the rest of the layers (unlike the case of radial oscillations). From this it follows that, in any event, such localized perturbations, if any, must be singular, in order that one may neglect the perturbation of all the mass of the cylinder, apart from that concentrated in a selected thin layer. This condition is really satisfied, although the manner of singularity at $m \neq 0$ does differ from the δ-like one (as at $m = 0$).

Let us turn now to the direct analysis of Eq. (1). Note above all that the value $\varepsilon_\perp$ can be represented as $\varepsilon_\perp = (\omega_*^2 - 2\Omega^2)/(\omega_*^2 - \varkappa^2)$ since from the equilibrium condition it follows that $\varkappa^2 = \omega_0^2 + 2\Omega_0^2$. Performing the differentiation in Eq. (1) and multiplying it by $\omega_*(\omega_*^2 - 2\Omega^2)(\omega_*^2 - \varkappa^2)$, rewrite this equation in the following way:

$$\omega_*(\omega_*^2 - 2\Omega^2)(\omega_*^2 - \varkappa^2)\Phi_1''$$

$$+ \left[\omega_*(\omega_*^2 - 2\Omega^2)(\omega_*^2 - \varkappa^2)\frac{1}{r} - \omega_*(\omega_*^2 - \varkappa^2)(2m\Omega'\omega_* + 4\Omega\Omega') \right.$$

$$\left. + \omega_*(\omega_*^2 - 2\Omega^2)(2\varkappa\varkappa' + 2m\Omega'\omega_*) \right]\Phi_1'$$

$$- \left[\omega_*(\omega_*^2 - 2\Omega^2)(\omega_*^2 - \varkappa^2)\frac{m^2}{r^2} + \frac{2m}{r}(\omega_*^2 - \varkappa^2)(\omega_0^2\Omega)' \right.$$

$$\left. + \frac{2m}{r}\omega_0^2\Omega(2\varkappa\varkappa' + 2m\Omega'\omega_*) \right]\Phi_1 = 0. \tag{21}$$

The original equation presented in such a form does not contain any frequency-independent denominators. The local dispersion relation corresponding to (21) is obtained by equating to zero the coefficient of the highest derivative.

We have, according to (21), three possibilities:

(1) $\qquad \omega_* = 0, \qquad$ i.e., $\quad \omega = m\Omega(r),$ $\qquad\qquad$ (22)

(2) $\qquad \omega_*^2 = 2\Omega^2, \qquad$ i.e., $\quad \omega = m\Omega(r) \pm \sqrt{2}\Omega(r),$ $\qquad$ (23)

(3) $\qquad \omega_*^2 = \varkappa^2, \qquad$ i.e., $\quad \omega = m\Omega(r) \pm \varkappa(r).$ $\qquad$ (24)

Obviously, the last of these possibilities corresponds to such oscillations, at which no density-potential perturbations occur, so that individual particles oscillate independently of the others (see §1). It would also be formally incorrect to study such oscillations on the basis of Eq. (21), for in the derivation of this equation we had, in expressing the velocities v_{r1} and $v_{\varphi 1}$ by the potential Φ_1, to divide by $(\omega_*^2 - \varkappa^2)$.

The other two possibilities correspond to *collective* oscillations of the system. Equations (22) and (23) define three regions of the continuous spectrum of oscillations of the system:

$$\omega = m\Omega(r), \qquad\qquad m\Omega_{min} < \omega < m\Omega_{max}, \qquad (25)$$

$$\omega = (m + \sqrt{2})\Omega(r), \quad (m + \sqrt{2})\Omega_{min} < \omega < (m + \sqrt{2})\Omega_{max}, \quad (26)$$

$$\omega = (m - \sqrt{2})\Omega(r), \quad (m - \sqrt{2})\Omega_{min} < \omega < (m - \sqrt{2})\Omega_{max}, \quad (27)$$

from which the first one exists only in the case of nonaxial-symmetrical oscillations ($m \neq 0$).

Such a number of frequencies is in agreement with an elementarily determined number of degrees of freedom for oscillations of each individual layer. Really, generally speaking, (for $m \neq 0$), the movement is fully determined if (at the initial time) three independent amplitudes—of velocities v_{r1} and $v_{\varphi 1}$ and of density ρ_1 (or of potential Φ_1; ρ_1 and Φ_1 are connected by the Poisson equation) are specified. For axially-symmetrical perturbations, the number of degrees of freedom decreases by unity due to conservation of particle angular momentum by such perturbations.

Investigating in a conventional manner the behavior of the solutions near $r = r_0$ we define

$$\left.\begin{array}{l} \Phi_1^{(1)} = c_1(r - r_0), \\ \Phi_1^{(2)} = c_2 + c_3(r - r_0)\ln|r - r_0|, \end{array}\right\} \quad \begin{array}{l} \text{in the case of (25),} \\ \omega = m\Omega(r_0), \end{array} \quad (28)$$

$$\left.\begin{array}{l} \Phi_1^{(1)} = c_1', \\ \Phi_1^{(2)} = c_2' \ln|r - r_0|, \end{array}\right\} \quad \begin{array}{l} \text{in the case of (26), (27),} \\ \omega = (m \pm \sqrt{2})\Omega(r_0) \end{array} \quad (29)$$

where c_1, c_2, c_3, c_1', and c_2' denote constants. The eigenfunctions that correspond to the frequencies of the continuous spectrum (25)–(27) turn out, as was to be expected, to be singular.[11]

In all cases, where one can suspect the existence of a continuous spectrum of frequencies (which is usually connected with a singularity of the eigenfunctions, as is the case in the problem considered), it is more natural to suggest another approach consisting of solving the problem of the time evolution of smooth initial perturbations [184]. Let us consider perturbations located near some fixed point r_0, and first of all let

$$\omega - m\Omega(r_0) = \sqrt{2}\Omega(r_0). \qquad (30)$$

The system under consideration is rather simple, and the formulated problem can be solved in different ways [128, 129, 184]. However, a particular simple technique was suggested in [79] for the solution of similar problems in plasma instability theory. Expand all the quantities that enter the coefficients of Eq. (21) in small $x = (r - r_0)$ and restricted everywhere only by

---

[11] The problem of the existence of nonsingular solutions for Eq. (1) of the "internal" type (belonging to a discrete spectrum) remains open.

the principal terms. Then we get

$$[\omega - (m + \sqrt{2})\Omega(r_0) + ax]\Phi'' + b\Phi' + c\Phi = 0, \tag{31}$$

where

$$a = -(m + \sqrt{2})\Omega'(r_0),$$

$$b = -\left.\frac{2m\Omega'\omega_* + 4\Omega\Omega'}{\omega_* + \sqrt{2}\Omega}\right|_{r=r_0} = (m + \sqrt{2})\Omega'(r_0) = -a,$$

$$c = \left.\frac{m\omega_0^2}{2r_0\Omega}\left[\frac{(\omega_0^2\Omega)'}{\omega_0^2\Omega} - \frac{(x^2)' + m\sqrt{2}(\Omega^2)'}{2\Omega^2 + r(\Omega^2)'}\right]\right|_{r=r_0},$$

$\Phi' = \partial\Phi/\partial x$, $\Phi'' = \partial^2\Phi/\partial x^2$. Substituting $\omega \rightarrow i\partial/\partial t$, $\Phi_1 \rightarrow \chi e^{-i(m+\sqrt{2})\Omega(r_0)t}$, we arrive at the equation

$$i\frac{\partial\chi''}{\partial t} + ax\chi'' + b\chi' + c\chi = 0. \tag{32}$$

Solve this equation for the initial perturbation which, in the narrow region $\Delta x = \Delta r$, near the point $r = r_0$ ($x = 0$) (where it is sufficient to restrict oneself in the expansions of the unperturbed values to the first terms) can be represented in the form

$$\chi(t = 0) = e^{ik_0 x}, \tag{33}$$

and let us assume that $k_0\Delta r \gg 1$.

The solution may be sought in the following form:

$$\chi(r, t) = e^{ik(t)x}\psi(t), \tag{34}$$

where $k(t)$ and $\psi(t)$ are two unknown functions. Substituting (34) into (32), we readily find

$$\chi(x, t) = \left(1 + \frac{at}{k_0}\right)^{(b/a - 2)}\exp\left[i(k_0 + at)x + \frac{ic}{a}\left(\frac{1}{k_0 + at} - \frac{1}{k_0}\right)\right]. \tag{35}$$

Here $b/a = -1$; therefore, for example, the asymptotic (at larger t) solution can, with due regard for the above substitution $\Phi_1 \rightarrow \chi$ and the definition a, be presented in the form

$$\Phi_1(r, t) = \text{const} \cdot \frac{\exp[-i\omega(r)t]}{t^3},$$

$$\omega(r) = m\Omega(r) + \sqrt{2}\Omega(r). \tag{36}$$

This formula also illustrates convincingly the very existence of the continuous spectrum of frequencies $\omega_* = \sqrt{2}\Omega(r)$.

It is easy to obtain the solution for Eq. (32) also for other types of initial perturbations (such as the Gaussian packet of the form $\exp[\sim x^2/(\Delta r)^2]$ [79]. We shall not write, however, the corresponding solutions since the qualitative picture of the development of the perturbation is clear from the simplest solution given above.

In a similar way one can determine also solutions for $\omega_* = 0\,[\omega = m\Omega(r)]$ located near the point $r = r_0$. It is quite evident that in this case the problem is again reduced to the equation of the type (32), but with $b = 0$ and certain other coefficients a, c. Accordingly, on the asymptotic $(t \to \infty)$ we then have

$$\Phi(r, t) = \frac{\text{const}}{t^2}\exp[-i\omega(r)t], \tag{37}$$

$$\omega(r) = m\Omega(r),$$

i.e., these perturbations die somewhat more slowly than those treated earlier.

"Global" oscillations of the inhomogeneous cylinder in an explicit form are readily determined in two cases: for a nearly homogeneous cylinder [when $|\rho(R) - \rho(0)|/\rho(r) \ll 1$] and for "surface" perturbations with $m \gg 1$ (which are concentrated near $r = R$).

Problems

1. Find the approximate dispersion equation and determine the stability boundary for large-scale $(k_\perp R \ll 1)$ axial-symmetrical perturbations of an arbitrary collisionless cylinder, inhomogeneous in the (x, y) plane (Antonov and Nezhinskii [10]).

Solution. According to formula (3), §2, the surfaces of constant perturbed density $\rho_1 = \text{const}$ (as well as equipotential surfaces $\Phi_1 = \text{const}$) where $k_\perp R \ll 1$ become planes parallel to the plane (x, y). It may be suggested that, also in the general case of the cylinder inhomogeneous (and hot) in x, y, there will be such perturbations. In order to consider them, it is natural to adopt the following model. The particles with identical values of coordinate z and velocity v_z, should conventionally be grouped into disks. Let us assume these disks to be "rigid," i.e., assume that also in the perturbed state they migrate as a whole. The thermal velocities of particles in the (x, y) plane in this model are taken into account indirectly by the fact that the distribution of the surface density, coincident with the distribution of the volume density $\rho_0(r)$ in the initial stationary model, is ascribed to the disk. Assume the disks to remain always oriented parallel to the (x, y) plane and to pass freely through each other.

The distribution function $f(z, v_z)$ of "rigid" disks satisfies the *one-dimensional* kinetic equation

$$\frac{\partial f}{\partial t} + v_z\frac{\partial f}{\partial z} - \frac{\partial \Phi}{\partial z}\frac{\partial f}{\partial v_z} = 0. \tag{1}$$

Let f be normalized by unity:

$$\int f\,dv_z = 1. \tag{2}$$

The linearized kinetic equation can be written in the form $(v_z \equiv v)$

$$\frac{\partial f_1}{\partial t} + v\frac{\partial f_1}{\partial z} + \varphi\frac{\partial f_0}{\partial v} = 0, \tag{3}$$

$$\varphi = -\frac{\partial \Phi_1}{\partial z} = \frac{1}{M}\int\limits_{-\infty}^{\infty}\!\!\int F(z_1 - z)f_1(z_1, v_1, t)\,dz_1\,dv_1; \tag{4}$$

$F(z_1 - z)$ is the force, by which the disk with the coordinate z_1 attracts the other disk (having the coordinate z); M is the disk mass.

Substituting $f_1(z, v, t) = \tilde{f}(v)e^{-i\omega t + ikz}$ and introducing a new integration variable $\xi = z_1 - z$, we shall obtain

$$-i\tilde{f}(v)(\omega - kv) + \frac{1}{M}\frac{\partial f_0}{\partial v}\int_{-\infty}^{\infty} F(\xi)e^{ik\xi}\, d\xi \int_{-\infty}^{\infty} \tilde{f}(v_1)\, dv_1 = 0. \tag{5}$$

This gives the amplitude of the perturbed distribution function $\tilde{f}(v)$:

$$\tilde{f}(v) = \frac{2}{M}\frac{\partial f_0/\partial v}{(\omega - kv)}\int_0^{\infty} F(\xi)\sin k\xi\, d\xi \int_{-\infty}^{\infty} \tilde{f}(v_1)\, dv_1, \tag{6}$$

which takes into account that $F(\xi)$ is odd.

Integrating (6) over v and eliminating $\int_{-\infty}^{\infty}\tilde{f}(v)\, dv$ in the equation thus obtained, we arrive at the dispersion equation

$$1 - \frac{2}{M}\int_{-\infty}^{\infty}\frac{\partial f_0/\partial v}{\omega - kv}\, dv \int_0^{\infty} F(\xi)\sin k\xi\, d\xi = 0. \tag{7}$$

The stability boundary is defined by the condition $\omega = 0$, which yields the equation

$$1 + \frac{2}{M}\int_{-\infty}^{\infty}\frac{\partial f_0/\partial v}{kv}\, dv \int_0^{\infty} F(\xi)\sin k\xi\, d\xi = 0. \tag{8}$$

Apply the Maxwellian distribution function in longitudinal velocities

$$f_0 = \frac{1}{\sqrt{\pi}v_T}\exp\left(-\frac{v^2}{v_T^2}\right). \tag{9}$$

Then (8) is reduced to the following form:

$$\frac{v_T^2}{4} = \frac{1}{kM}\int_0^{\infty} F(\xi)\sin k\xi\, d\xi. \tag{10}$$

This equation defines the critical velocity dispersion v_{Tcr} for each value of the wave number k: $v_{Tcr} = v_{Tcr}(k)$ or, vice versa, the critical wave number k_{cr} for a given dispersion: $k_{cr} = k_{cr}(v_T)$. Let us calculate $F(\xi)$ for fixed density $\rho_0 = \rho_0(r)$:

$$F(\xi) = G\int_{-R}^{R}\int_{-\sqrt{R^2-x^2}}^{\sqrt{R^2-x^2}}\int_{-R}^{R}\int_{-\sqrt{R^2-x_1^2}}^{\sqrt{R^2-x_1^2}}\frac{\xi\rho_0(\sqrt{x^2 + y^2})\rho_0(\sqrt{x_1^2 + y_1^2})\, dx\, dx_1\, dy\, dy_1}{[(x - x_1)^2 + (y - y_1)^2 + \xi^2]^{3/2}}. \tag{11}$$

Substituting (11) into (10) (by changing the order of integration and performing the integration over ξ), we have

$$v_{Tcr}^2 = \frac{4G}{M}\iiiint K_0(k\sqrt{(x - x_1)^2 + (y - y_1)^2})\rho_0(\sqrt{x^2 + y^2})$$
$$\times \rho_0(\sqrt{x_1^2 + y_1^2})\, dx\, dx_1\, dy\, dy_1, \tag{12}$$

where $K_0(z)$ is the MacDonald function. Taking into account that $K_0(z)$ is a positive and monotonically decreasing function, we infer that the expression on the right-hand side of (12) is positive and diminishes monotonically with increasing k. At $k \to 0$, $v_{Tcr}^2 \to \infty$; therefore, $K_0(\alpha k) \to \infty$ at $k \to 0$. On the other hand, $v_{Tcr}^2 \to 0$ at $k \to \infty$, because $K_0(\alpha k) \to 0$ at $k \to \infty$. This evidence is sufficient for the qualitative representation of $v_{Tcr}^2(k)$ (Fig. 14).

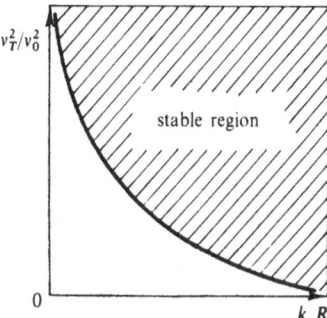

Figure 14. Qualitative representation of the boundary separating stability regions (above the curve) and instability regions (below the curve) for arbitrary collisionless cylinders [10].

Consider in more detail the question regarding the long-wave instability of the cylinder. Assuming that in (12) $k \to 0$, $v_T^2 \to \infty$, we get

$$v_T^2 = -\frac{2G}{M} \iiiint \{\tfrac{1}{2} \ln[(x - x_1)^2 + (y - y_1)^2] + \ln \tfrac{1}{2}k + C\} \rho_0(\sqrt{x^2 + y^2})$$

$$\times \rho_0(\sqrt{x_1^2 + y_1^2})\, dx\, dx_1\, dy\, dy_1,$$

or

$$v_T^2 = -\frac{2}{M} \iint \rho_0(r)\Phi_0\, dx\, dy - GM(C + \ln \tfrac{1}{2}k), \tag{13}$$

where C is the Euler constant and Φ_0 is the logarithmic potential corresponding to the disk of density $\rho_0(r)$. Introducing the effective radius R_e, according to

$$\iint \rho_0 \Phi_0\, dx\, dy = 2GM^2 \ln R_c, \tag{14}$$

we have from (13)

$$k \sim \frac{2}{R_c} e^{-v_T^2/4GM - C}, \tag{15}$$

or

$$\lambda \sim \pi R_e\, e^{v_T^2/4GM + C}. \tag{16}$$

Besides, if we introduce σ_1^2, using the formula

$$\sigma_1^2 = \frac{\iiiint (v_x^2 + v_y^2) f_0\, dx\, dy\, dv_x\, dv_y}{M}, \tag{17}$$

then, as can be shown, formula (16) will take the form

$$\lambda \sim \pi R_e\, e^{v_T^2/4\sigma_1^2 + C}. \tag{18}$$

For a nonrotating disk, σ_1^2 is obviously the mean velocity dispersion of particles in the (x, y) plane. In the case of a cold rotating disk, σ_1^2 is simply related to the mean rotation velocity $v_{0\varphi}$.

In the special case of a uniform cylinder $\rho_0(r) = \rho_0 = \text{const}$ (12) is reduced, as shown in [10], to the following equation:

$$v_T^2 = \frac{8\pi^2 G \rho_0^2 R^2}{kM} [1 - 2I_1(kR)K_1(kR)] \qquad (R_e = Re^{-1/4}). \qquad (19)$$

Comparing (19) with the results of an accurate consideration of the stability of this model in §2, one may make sure that the approximate method stated above really leads to *qualitatively* correct inferences at $k_\perp R \ll 1$, $k_z R \lesssim 1$ (the beginning of the lower Jeans branch in Fig. 9).

In [10], the stability boundaries have also been found for some other density distributions in the cylinder: $\rho = \rho_0 e^{-r^2/a^2}$, $\rho = \rho_0(a^2 + r^2)^{-3/2}$.

In conclusion note that, by using this method, also large-scale (in the direction z, Fig. 3, $k_z c \ll 1$) perturbations of a flat layer can be treated.

2. Using the method of phase volume variation (Chapter I, §6), derive the dispersion equation for surface oscillations of the flute type ($\rho_1 \equiv 0$, $k_z = 0$) of a uniform cylinder with the distribution function in (2), §1.

Solution. The perturbation theory necessary for the derivation of the dispersion equation is given at the beginning of §5. In the general case, the perturbed potential within the cylinder (§4)

$$\Phi_1^{(n, m)} \sim r^{2n+m} e^{im\varphi}. \qquad (1)$$

The solution for the kinetic equation $d\chi/dt = 2\, d\Phi_1/dt + 2i\omega\Phi_1$ by the method of "integration over trajectories" will be written in the form (χ, $\Phi_1 \sim e^{i\omega t}$):

$$\chi(r, v_r, v_\varphi) e^{im\varphi} \sim 2e^{im\varphi}\left(r^{2n+m} + i\omega \int_{-\infty}^{0} r'^{2n+m} e^{im\varphi' - i\omega t'}\, dt'\right), \qquad (2)$$

where the functions $r'(t')$ and $r'(t')e^{i\varphi'(t')}$ (the trajectories of the particle in a system rotating with a velocity γ) are given by the formulae

$$r'^2 = v^2 \sin^2 t' + \gamma^2 r^2 + r^2(1 - \gamma^2)\cos^2 t' + rv_r \sin 2t' + 2\gamma r v_\varphi \sin^2 t', \qquad (3)$$

$$r'e^{i\varphi'} = \tfrac{1}{2}\{e^{-i(1+\gamma)t'}[r(1-\gamma) - (v_\varphi - iv_r)] + e^{i(1-\gamma)t'}[r(1+\gamma) + (v_\varphi - iv_r)]\}. \qquad (4)$$

The integral in (2) is convenient to calculate, by representing the subintegral expression as

$$(r'^2)^n (r'e^{i\varphi'})^m. \qquad (5)$$

For the "surface" perturbations which are to be dealt with further, the potential

$$\Phi_1^{(0, m)} \sim r^m e^{im\varphi}. \qquad (6)$$

The calculations in this case are particularly simple, resulting in a compact dispersion equation

$$2m = -\frac{1}{1-\gamma^2}\left[\frac{\omega}{2^{m-1}} \sum_{k=0}^{m} C_m^k \frac{(1-\gamma)^k(1+\gamma)^{m-k}}{\omega - m(1-\gamma) + 2k} - 2\right], \qquad (7)$$

where C_m^k is the number of combinations. This equation contains only stable solutions. At $\gamma \to 1$, it yields

$$(\omega + 1)^2 = 0. \qquad (8)$$

The power instability [9] corresponding to the multiple root in (8) is due to the fact that the orbits are assumed to be circular (and $k_z = 0$). Any small thermal scatter stabilizes this instability. Consider the case $m = 2$. From (7), we have [112]:

$$\omega^3 + 6\gamma\omega^2 + (12\gamma^2 - 3)\omega + 2\gamma(4\gamma^2 - 2) = 0. \tag{9}$$

All the roots of this equation are different. The condition of realness for all the roots of the cubic equation yields $\gamma^2 \leq 1$, which is always satisfied. Equation (8) ensues also in the hydrodynamical model of an incompressible cylinder, where the pressure is assumed to be zero. Although in [9] only the case $m = 2$ is investigated, the situation is similar also for larger m.

3. Using directly the dispersion equation (20), §4, prove the stability with respect to small-scale perturbations of flute type ($k_z = 0$) [14].

Solution. The realness of the eigenfrequencies for perturbations that satisfy the criterion

$$L \equiv (n + m - 1)(n - 1)(1 - \gamma^2) \geq 2 \tag{1}$$

can be proved in the following way [14]. The integrals over φ on the left-hand side of Eq. (20), §4, are expanded in powers of e^τ into series of the form

$$\frac{1}{2\pi} \int_0^{2\pi} \psi^{n+m}\psi_1^{n-1}\, d\varphi = \sum_{k=0}^{N} \beta_k e^{(2m-N)\tau}, \tag{2}$$

$$\frac{1}{2\pi} \int_0^{2\pi} \psi^{n+m+1}\psi_1^n\, d\varphi = \sum_{k=0}^{N} \beta_k' e^{(2m-N)\tau}. \tag{3}$$

In these expansions τ may be assumed to be an arbitrary complex number. The dispersion equation with the help of (2) and (3) is reduced to the form of (18), §4, with

$$\gamma_k = \beta_k + \beta_k'. \tag{4}$$

As we know, to prove the realness of the roots, it is sufficient to establish the validity of the inequality

$$\gamma_k \leq 1. \tag{5}$$

We shall prove the stronger inequalities

$$\beta_k \leq \tfrac{1}{2}, \qquad \beta_k' \leq \tfrac{1}{2}. \tag{6}$$

For this purpose, we shall need only merely imaginary values of $\tau = it$. Denote the left-hand sides of (2) and (3) correspondingly by $F(t)$ and $F'(t)$. Eliminating ψ_1, using the identity

$$\psi_1 = \frac{1 - \sqrt{1 - \gamma^2}\, e^{-i\varphi} \sinh \tau\psi}{\cosh \tau - \gamma \sinh \tau}, \tag{7}$$

we expand then the subintegral expressions in powers of $e^{i\varphi}$ and calculate the ensuing integrals. Thus we find

$$\frac{F(t)}{(\cos t - i\gamma \sin t)^{m+1}} = \sum_{k=0}^{n-1} \frac{(-1)^k(n + m + k)!(n - 1)!}{(k!)^2(n - 1 - k)!(n + m)!} [(1 - \gamma^2)\sin^2 t]^k, \tag{8}$$

$$\frac{F'(t)}{(\cos t - i\gamma \sin t)^{m-1}} = \sum_{k=0}^{n} \frac{(-1)^k(n + m - 1 + k)!n!}{(k!)^2(n - k)!(n + m - 1)!} [(1 - \gamma^2)\sin^2 t]^k. \tag{9}$$

Calculate now $F(t)$ at $t = t_1 \equiv \arcsin \sqrt{2/L}$ ($L \geq 2$ according to the assumption):

$$\frac{F(t)}{(\cos t - i\gamma \sin t)^{m+1}} = -\frac{m+3}{(n+m+1)(n-1)}$$

$$+ \sum_{k=3}^{n-1} \frac{(-1)^k (n+m+k)!(n-1)!}{(k!)^2 (n-1-k)!(n+m)!} \left[\frac{2}{(n+m+1)(n-1)} \right]^k .$$

$$(10)$$

The terms of the sum $\sum_{k=3}^{n-1}$ in (10) are grouped into pairs: the negative terms are followed by a positive (or a zero at the end of the sum). The module of their ratio is less than unity:

$$\frac{|a_{k+1}|}{|a_k|} = \frac{2(n+m+1)(n-k-1)}{(k+1)^2(n+m+1)(n-1)} \leq \frac{(n+m+k+1)(n-k-1)}{(n+m+1)(n-1)} \leq 1$$

$$(k = 3, 4, \ldots). \quad (11)$$

Therefore $\sum_{k=3}^{n-1}$ in (10) is negative (the first term in $\sum_{k=3}^{n-1}$ is negative). Therefore, also the entire right-hand side of (8) at $t = t_1$ is negative. On the other hand, at $t = 0$, it is unity. Because of continuity, for some intermediate $t = t_0$ we should have $F(t_0) = 0$.

Hence it follows that $\beta_k \leq \frac{1}{2}$. Take equation (2) at $t = t_0$. Multiplying it by $\exp[i(N - 2v)t_0]$ with some integer v and separating out on both sides the real parts, we find

$$\mathrm{Re}\{F(t_0)\exp[i(N - 2v)t_0]\} = \sum_{k=0}^{N} \beta_k \cos 2(k - v)t_0 \geq \beta_v - \sum_{k \neq v} \beta_k = 2\beta_v - 1. \quad (12)$$

Since $F(t_0) = 0$, it yields $\beta_v \leq \frac{1}{2}$.

The proof of the inequality $\beta_k' \leq \frac{1}{2}$ is performed in a quite similar way if one starts with the substitution

$$t = \arcsin \sqrt{\frac{2}{(n+m)(1 - \gamma^2)n}}$$

[which is acceptable due to the inequality $(n + m)n > (n + m + 1)(n - 1)$ identical to $n > 1$].

4. Investigate the possibility of a resonance influence of the higher terms in the polynomial expansion of the potential on the lower terms for the flute-like mode with the indices $n = 1, m = 2$ (§4) (Antonov [14]).

Solution. Such an influence exists (and leads to a power increase of the perturbation). This follows from the fact of the intersection of the branches lying in the interval $1 < \omega < 3$ for the modes ($n = 1, m = 2$) and ($n = 0, m = 2$). The existence of the intersection can be proved in the following way. Write down the accurate dispersion equations for these modes:

$$\frac{(2 + \gamma)(1 - \gamma)^2}{4[(\omega - 3)^2 - 1]} + \frac{(2 + 3\gamma + 3\gamma^2)(1 - \gamma)}{4[(\omega - 1)^2 - 1]}$$

$$+ \frac{(2 - 3\gamma + 3\gamma^2)(1 + \gamma)}{4[(\omega + 1)^2 - 1]} + \frac{(2 - \gamma)(1 + \gamma)^2}{4[(\omega + 3)^2 - 1]} + 1 = 0$$

for ($n = 1, m = 2$) and

$$\frac{(1 - \gamma)}{2[(\omega - 1)^2 - 1]} + \frac{(1 + \gamma)}{2[(\omega + 1)^2 - 1]} + 1 = 0$$

for $(n = 0, m = 2)$. Consider now γ close to $1: \gamma = 1 - \delta, \delta \ll 1$. Then, using the above equations, we readily find

$$\text{for } (n = 1, m = 2): \quad \omega \equiv \omega_1 \simeq 2 - (\tfrac{6}{7})\delta,$$
$$\text{for } (n = 0, m = 2): \quad \omega \equiv \omega_2 \simeq 2 - (\tfrac{2}{5})\delta,$$

i.e., near $\gamma = 1$, $\omega_1 < \omega_2$. However, at $\gamma \to 0$, we have the inverse inequality: $\omega_1 > \omega_2$, since $\omega_1 \to 2$ and $\omega_2 \to \sqrt{3}$. Therefore, at some point on the segment $0 \leq \gamma \leq 1$ the branches must intersect.

5. Investigate the nonlinear stability of radial pulsations of the collisionless cylinder (3), §1, retaining density homogeneity (Antonov and Nuritdinov [12]).

Solution. A similar problem was considered by us in the previous chapter (Problem 3). As done there, let us describe the cylinder deformation by using the linear transform

$$\mathbf{r} = \alpha(t)\mathbf{r}_0 + \beta(t)\mathbf{v}_0, \qquad \mathbf{v} = \dot{\alpha}(t)\mathbf{r}_0 + \dot{\beta}(t)\mathbf{v}_0, \tag{1}$$

coupling the initial $\mathbf{r}_0, \mathbf{v}_0$ and current $\mathbf{r}, \mathbf{v}$ Lagrange coordinates and velocities of particles. The initial distribution function can be represented as

$$f_0 = \frac{\rho_0}{\pi} \delta\{1 - \gamma^2 - r_0^2 - v_0^2 + 2\gamma[\mathbf{r}_0 \cdot \mathbf{v}_0]\}. \tag{2}$$

Inverting the transformation formulae in (1) and substituting into (2), we have

$$f = \frac{\rho_0}{\pi} \delta\{1 - \gamma^2 - (\dot{\alpha}^2 + \dot{\beta}^2)r^2 - (\alpha^2 + \beta^2)v^2 + 2(\alpha\dot{\alpha} + \beta\dot{\beta})\mathbf{rv} + 2\gamma[\mathbf{rv}]\}, \tag{3}$$

where it is taken into account that the Jacobian of the transform (1) is 1:

$$\alpha\dot{\beta} - \dot{\alpha}\beta = 1. \tag{4}$$

If (3) is integrated over the velocity $\mathbf{v}$, this yields the density depending only on time (but not on spatial coordinates):

$$\rho = \frac{\rho_0}{\alpha^2 + \beta^2}. \tag{5}$$

Thus, transform (1) really retains the property of density homogeneity. Now one can write the equation of motion of an individual particle in the variable potential

$$\Phi = \frac{1}{2} \frac{\rho}{\rho_0} r^2, \tag{6}$$

which corresponds to the density in (5):

$$\frac{d^2\mathbf{r}}{dt^2} = -\frac{\mathbf{r}}{\alpha^2 + \beta^2}. \tag{7}$$

Taking (1) into account, we obtain the following equation for $\alpha(t)$ and $\beta(t)$:

$$\ddot{\alpha} = -\frac{\alpha}{\alpha^2 + \beta^2}, \qquad \ddot{\beta} = -\frac{\beta}{\alpha^2 + \beta^2}. \tag{8}$$

They formally agree with the equation of motion of the particle in the central-symmetrical potential $U = \ln(\alpha^2 + \beta^2)/2$; moreover, α and β play the role of Cartesian coordinates ($\alpha \sim x, \beta \sim y$) of the particle in the plane of motion. In this case, one of

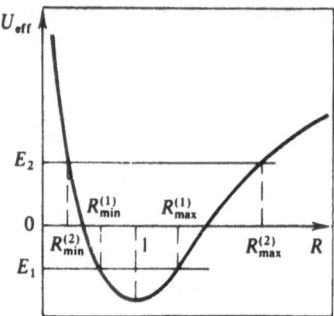

Figure 15. Effective potential energy $U_{\mathrm{eff}}(R)$ for radial oscillations of a cylinder, in the course of which spatial density uniformity is retained.

course may pass from α, β to the corresponding polar coordinates R, φ:

$$R(t) = \sqrt{\alpha^2 + \beta^2}, \qquad \tan \varphi(t) = \frac{\beta}{\alpha}. \tag{9}$$

It can be easily understood that $R(t)$ is a variable external radius of the cylinder $R^2 \sim \rho^{-1}$. The change of R occurs in accordance with the equation

$$\ddot{R} = -\frac{\partial U_{\mathrm{eff}}}{\partial R}, \tag{10}$$

where the effective "potential energy" is

$$U_{\mathrm{eff}} = \frac{1}{2R^2} + \ln R. \tag{11}$$

The plot of U_{eff} presented in Fig. 15, shows that R is a periodical function of time at any value of "full energy"

$$E \equiv \frac{\dot{R}^2}{2} + \frac{1}{2R^2} + \ln R. \tag{12}$$

This means the stability of the mode in question since it is entirely characterized by a single parameter (which may be represented by the radius R). The radius fluctuates between the values $R_{\min}$ and $R_{\max}$ (in Fig. 15, the pairs $R_{\min}^{(1)}$, $R_{\max}^{(1)}$ and $R_{\min}^{(2)}$, $R_{\max}^{(2)}$ correspond to the effective energies equal to E_1 and E_2, respectively).

In conclusion, note that Eq. (10) is a special case (corresponding to $\gamma = 2$) of the equation

$$\ddot{R} = -\frac{1}{R} + R^{1-2\gamma}, \tag{13}$$

which describes expansions–compressions of a uniform gaseous cylinder with the adiabatic index γ.

6. Consider the nonlinear evolution of a perturbation of a circular cylinder to an elliptical cylinder. The system is assumed to be cold in the (x, y) plane of rotation (Antonov [9]).

Solution. The sought "equations of motion" will be derived by the following method. Assume that the Lagrange coordinates of the particles (x, y) are linearly dependent on the initial coordinates x_0, y_0 (i.e., let us restrict ourselves only to the perturbation of such a type):

$$x = \alpha(t)x_0 + \beta(t)y_0, \qquad y = \gamma(t)x_0 + \delta(t)y_0. \tag{1}$$

Notice that the linear coupling in (1) is constant in time (if the initial velocities are subordinated to it), for transformation (1) converts the homogeneous circular cylinder to a homogeneous elliptical cylinder, inside which the field strength depends on the coordinates linearly. Find the semiaxes $a(t)$ and $b(t)$ of the ellipse at the time t. Using $x_0^2 + y_0^2 = r_0^2$, substitute in (1) x_0 and y_0 by x and y:

$$(\gamma^2 + \delta^2)x^2 + (\alpha^2 + \beta^2)y^2 - 2(\alpha\gamma + \beta\delta)xy = r_0^2(\alpha\delta - \beta\gamma)^2. \tag{2}$$

For $a(t)$ and $b(t)$ it yields

$$a^2 + b^2 = (\alpha^2 + \beta^2 + \gamma^2 + \delta^2)r_0^2, \qquad ab = (\alpha\delta - \beta\gamma)r_0^2. \tag{3}$$

The calculation of the gravitational energy per unit length of the elliptical cylinder is nonstandard. Start with the determination of the potential. The potential of the cylinder can be most easily obtained by performing the limit transition $c \to \infty$ in the familiar expression for the potential of the three-axial ellipsoid [148]:

$$\Phi_< = \pi G \rho_0 \left(\sum_{i=1}^{3} A_i x_i^2 - I \right), \tag{4}$$

where $x_1 = x$, $x_2 = y$, $x_3 = z$,

$$A_i = abc \int_0^\infty \frac{du}{\Delta(a_i^2 + u)}, \tag{5}$$

$$I = abc \int_0^\infty \frac{du}{\Delta}, \tag{6}$$

$$\Delta \equiv \sqrt{(a^2 + u)(b^2 + u)(c^2 + u)}.$$

The interval of integration in formulae (5) and (6) should be divided into two regions: (1) $0 < u < c$ and (2) $c < u < \infty$. In the first one, one can expand the terms of the type $(c^2 + u)$, $(c^2 + u)^{1/2}$ with respect to the small ratio u/c^2, while in the second, expressions of the form $(a^2 + u)^{1/2}$, $(b^2 + u)^{1/2}$ etc. are expanded with respect to the small ratios a^2/u, b^2/u. Then, the integrals are reduced to tabulated ones, and we get

$$\Phi_<^{(c)} \simeq \frac{2GM}{a + b}\left(\frac{x^2}{a} + \frac{y^2}{b} \right) + 2GM \ln(a + b) - 2GM \ln 4c \tag{7}$$

(M is the mass per unit length).

As $c \to \infty$, the potential is divergent, and therefore it must be "renormalized" in a reasonable way. This is attained by change of the origin. For instance, the potential of the three-axial ellipsoid can be counted from the potential of the rotation ellipsoid with the same large semiaxis $c_0 = c$, $a_0 = b_0$, $a_0 b_0 = ab$ (in the limit $c \to \infty$ from the circular cylinder):

$$\Phi_< = \frac{2GM}{a + b}\left(\frac{x^2}{a} + \frac{y^2}{b} \right) + 2GM \ln \frac{a + b}{2a_0} - \frac{GM}{a_0^2}r^2. \tag{8}$$

Integrating (8) over the elliptical cross section of the cylinder, let us calculate the potential energy

$$U = GM^2[\ln(a + b) + \text{const}] = \frac{\Omega^2 r_0^2 M}{4}[\ln(\alpha + \delta)^2 + (\beta - \gamma)^2] + \text{const}. \tag{9}$$

The constants independent of a, b are further inessential. Therefore, the action integral is expressed in the following way:

$$S = \frac{Mr_0^2}{4} \int \left\{ \frac{\dot{\alpha}^2 + \dot{\beta}^2 + \dot{\gamma}^2 + \dot{\delta}^2}{2} - \Omega^2 \ln[(\alpha + \delta)^2 + (\beta - \gamma)^2] \right\} dt.$$

Accordingly, the equations of motion are

$$\ddot{\alpha} = \ddot{\delta} = -2\Omega^2(\alpha + \delta)q^{-1}, \qquad \ddot{\beta} = -\ddot{\gamma} = -2\Omega^2(\beta - \gamma)q^{-1}, \tag{10}$$

where

$$q \equiv (\alpha + \delta)^2 + (\beta - \gamma)^2 = \left(\frac{a+b}{r_0}\right)^2. \tag{11}$$

From Eq. (10), first of all, it follows that the combinations $(\alpha - \delta)$ and $(\beta + \gamma)$ are arbitrary linear functions of time:

$$\alpha - \delta = C_1 t + C_2, \qquad \beta + \gamma = C_3 t + C_4 \tag{12}$$

$(C_1, C_2, C_3, C_4$ are constants). However, on the other hand, according to (3), the difference of the semiaxes $(a - b)$ is expressed through these combinations:

$$\frac{(a-b)^2}{r_0^2} = (\alpha - \delta)^2 + (\beta + \gamma)^2; \tag{13}$$

therefore

$$\left(\frac{a-b}{r_0}\right)^2 = (C_1 t + C_2)^2 + (C_3 t + C_4)^2. \tag{14}$$

Thus, any initial impact that does not satisfy the special condition $C_1 = C_3 = 0$ leads ultimately to the singular state with $b = 0$. This is the main result of [9]. It shows that the initial state is unstable, and the instability turns out to be of power type. In linear theory, this conforms, as noted in §4, to the presence of a multiple root $\omega = -1$ for perturbations of the type under consideration.

Using Eqs. (10), it is possible to analyze also the evolution of $q = (a + b)^2/r_0^2$. If we introduce the notations $(\alpha + \delta) \equiv A$, $(\beta - \gamma) \equiv B$, so that $q = A^2 + B^2$, then from (10) it is easy to infer

$$t = \frac{1}{2} \int^q \frac{dq}{\sqrt{C_5 q - C_6 - 4\Omega^2 q \ln q}}, \tag{15}$$

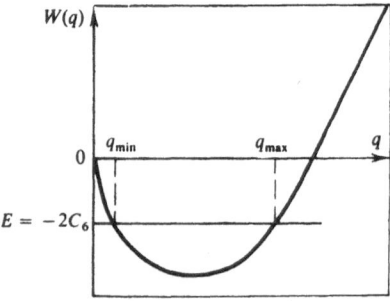

Figure 16. Graph of effective potential energy $W(q)$ for circular cylinder-elliptical cylinder-type perturbations.

where the constants C_5 and C_6 are

$$C_5 = \dot{A}^2 + \dot{B}^2 + 4\Omega^2 \ln q, \qquad (16)$$

$$C_6 = (B\dot{A} - A\dot{B})^2. \qquad (17)$$

Since the width Δq of the interval, where the subroot expression in (15) is positive, tends to zero along with the parameters of the initial impact, at not very large perturbations, the value q [and consequently, also $(a + b)$] changes periodically. This is also illustrated by the plot of the effective potential energy

$$W(q) = 8\Omega^2 q \ln q - 2C_5 q \qquad (18)$$

in Fig. 16. At some fixed value of the "energy" $E \equiv -2C_6$, the value q fluctuates between q_{min} and q_{max}.

CHAPTER III

Equilibrium and Stability of Collisionless Spherically Symmetrical Systems

In dealing with *spherical collisionless* systems we first of all bear in mind the spherical galaxies and globular star clusters. Approximately spherical shape is inherent in elliptical galaxies of small oblateness. The stars are the "particles" in such systems. Compact [146] galactic clusters with a rather large number of members are also referred by us to such systems: In order to study them, one may employ the same mathematical apparatus, only here individual galaxies should be considered as particles. Another example is the system of globular clusters of stars, e.g., in our Galaxy. As is well known, it also has a density distribution close to the spherically symmetrical one, and the symmetry center of this distribution defines the galactic center. The "particles" here are evidently individual stellar clusters. The estimate of the average time of pair collisions made in §1, Chapter IX, shows that the collisionless approximation is rather good for the study of oscillations and stability of all of these systems.

The investigation into spherically symmetrical systems can be divided into the following parts:

1. *Spherically symmetrical systems with isotropic distribution functions of particles in velocities.* At the beginning of §1, devoted to the equilibrium states of spherically symmetrical systems, there are found distribution functions of the simplest form dependent only on the energy of a particle $E = v^2/2 + \Phi_0$ and consequently being isotropic with respect to velocity directions **v**. Their perturbations are investigated in §2, where some general theorems establishing the stability of a broad class of such systems are proved.

2. *Systems with circular orbits* (Einstein model,[1] Fig. 2) are dealt with in §3. The density $\rho_0(r)$ in these models may be arbitrary; in particular, it may be chosen as in real systems rapidly falling toward the edge. However, in reality, the stars in spherical galaxies and globular clusters move mainly along very elongated trajectories. Therefore, the Einstein model, seemingly, is mainly of academic interest. It is one of the simplest models of spherically symmetrical collisionless systems, about which it is convenient to make clear some general properties inherent to a given geometry.

In Section 3.1, the oscillation spectrum of a *homogeneous*-in-density spherically symmetrical system of masses rotating along circular trajectories is found. The oscillations turn out to be neutral [87].

In the same model with a uniform density, the question of the possibility of surface wave excitation due to the change in the spherical shape of the surface without any change in local density of the system (subsection 3.1.3) is also considered. The spectrum of eigenvalues is calculated, among which no complex values occurred [153, 154].

Section 3.4 deals with the stability of the nonuniform density system. Mainly local perturbations are investigated, for which a simple criterion of stability is derived. For stable systems, the results are in many respects similar to those obtained in §7 of the previous chapter for oscillations of the inhomogeneous cylinder.[2]

Section 1 gives an example of the distribution function that describes a system of masses rotating along circular trajectories for the case when at each point of the sphere the number of particles rotating in one direction does not coincide with the number of particles rotating in the opposite direction; therefore, as a whole, the model possesses the zero-different rotation moment. Section 3.3 analyzes the behavior of small perturbations of such a system [125]. The frequencies of all possible oscillations are calculated, and the system stability is proved.

3. *Systems with orbits close to circular.* In §1, the class of stationary functions of distribution for systems with trajectories of particles close to circular (the dispersion of radial velocities is finite but small) [143] is obtained. Construction of such stationary solutions is performed on the basis of the method of expansion in a series of the δ functions and their derivatives (the method of moments). Section 3.2 proves the stability of the functions of distribution obtained with respect to arbitrary small perturbations (for the case of a homogeneous system) [143]. Unlike the model with merely circular orbits of particles, the oscillations turn out to be collective: Due to the presence of the radial dispersion of velocities of the particles, perturbation of

---

[1] In 1939, A. Einstein [155] considered the stability of relativistic clusters of stars moving along circular trajectories.

[2] It is intriguing to note that, in the *relativistic uniform-density* Einstein model, the eigenfrequencies of radial oscillations form a continuous spectrum similar to *nonuniform Newtonian* spheres with circular orbits [153].

one layer is transmitted onto the other layers, thus resulting in a wavelike variation of density inside the sphere. The local disturbances of nonhomogeneous systems with nearly circular orbits of particles are considered in Section 3.4.

4. *Systems with elliptical orbits.* In the spherically symmetrical homogeneous rotating system suggested by Freeman [203], there are only elliptical orbits, whose projections onto the rotational plane are circumferences (along the rotation axis, the particles perform harmonical oscillations) (see §1). Then the movement of matter in the plane parallel to the rotation plane is hydrodynamic in that each element of the body has a definite speed. In §4, the growth rates of surface wave excitation are calculated [91, 96]. It is also shown that, due to the presence of dispersion along the rotation axis, the system turns out to be stable with respect to the flattening into a two-axis ellipsoid. Also, the stability of the distribution function of a homogeneous sphere at rest with all possible elliptical orbits is investigated. The spectrum of oscillations of the model is calculated and its stability shown [112, 115].

5. *Systems with radial trajectories.* In §1, some distribution functions of a nonuniform-density system with merely radial motions of the particles are found. It is easy to see that such systems must be unstable according to Jeans, due to the absence of the stabilizing influence of the velocity dispersion in the transversal directions (see §5). Since the systems with the isotropic velocity distributed are stable, the boundary σ_{cr} that separates the stable and unstable models in the anisotropy degree $(2\overline{v_r^2}/\overline{v_\perp^2}) \equiv \sigma$ lies in the region $\sigma_{cr} > 1$.

6. *Spherical systems of a general type.* In §6 we consider the stability of systems with equilibrium states of a sufficiently general type. Corresponding distribution functions are determined in §1 as series depending on one or a few parameters. Under the variation of these parameters the characteristics of equilibrium states change in wide limits. For instance, Sections 6.2 and 6.3 are devoted $[41^{ad}]$ to stability investigation of two-parameters series of Camm's distribution functions, which cover all the region from the systems with circular orbits to the systems with radial trajectories. The stability criterion with respect to most large-scale perturbations are found. At last, the hypothesis of universality of the obtained stability criterion is discussed.

In the last section of this chapter (§7), a brief summary of the results of investigation of the stability of spherically symmetrical collisionless systems is given.

§ 1 Equilibrium Distribution Functions

The theoretical investigations into equilibrium states of the collisionless spherically symmetrical gravitating systems have been under way for a long time, and the most important results are reflected in Veltmann's review [32].

It seems advisable to consider in this section mainly those distribution functions whose stability will be of further interest to us.

1. The equilibrium distribution functions f_0 in collisionless spherically symmetrical systems must satisfy the equations

$$\mathbf{v} \frac{\partial f_0}{\partial \mathbf{r}} - \frac{\partial \Phi_0}{\partial \mathbf{r}} \frac{\partial f_0}{\partial \mathbf{v}} = 0, \qquad (1)$$

$$\Delta \Phi_0 = 4\pi G \rho_0, \qquad (2)$$

in which Φ_0 (self-consistent potential) and ρ_0 (density) depend only on radius r. The integrals of motion are (the masses of all particles are assumed to be equal and to be unity: $M = 1$) energy $E = v^2/2 + \Phi_0$ and the vector of the angular momentum of particle $\mathbf{L}$. Therefore, the equilibrium distribution function in the general case is $f_0(E, \mathbf{L})$. If the total momentum of rotation of the system is zero, it is usually assumed that $f_0 = f_0(E, L^2)$, where $L^2 = r^2(v_\theta^2 + v_\varphi^2)$ is the square of the angular momentum of an individual particle.

Such systems are generally anisotropic: the components of the "stress" tensor $T_{\alpha\beta} = \int f_0 v_\alpha v_\beta \, d\mathbf{v}$ in different directions will differ. For the distribution function $f_0 = f_0(E, L^2)$ the "pressure" along r, θ, and φ is

$$P_r = T_{rr} = \int f_0 v_r^2 \, d\mathbf{v}, \qquad P_t = T_{\theta\theta} = T_{\varphi\varphi} = \frac{1}{2} \int f_0 v_\perp^2 \, d\mathbf{v}.$$

Strictly speaking, one should, of course, deal here not with "pressure" but with the velocity dispersions in the radial $c_r^2 \sim P_r/\rho_0$ and transversal $c_t^2 \sim P_t/\rho_0$ directions. There are two evident degenerated cases. The first case corresponds to systems with circular orbits of all particles ($v_r = 0$, $P_r = 0$, $P_t \neq 0$), while the second, to systems with purely radial movements $v_\theta = v_\varphi = v_\perp = 0$, $P_t = 0$, $P_r \neq 0$). The stability of these systems is considered in §§3 and 5.

2. "The hydrodynamical pressure" turned out so far to be useful only for the systems with the *isotropic* particle-velocity distribution function, when $f_0 = f_0(E)$, $P_r = P_t \equiv P_0$ (for more detail, see §2). Therefore, the corresponding terminology is used below only for the description of such systems. Using the specified anisotropic distribution function $f_0(E)$, it is easy to determine the density $\rho_0(r)$ and pressure $P_0(r)$:

$$\rho_0 = 4\pi \sqrt{2} \int_{E=\Phi_0}^{\infty} f_0(E)(E - \Phi_0)^{1/2} \, dE,$$

$$P_0 = \frac{8\pi \sqrt{2}}{3} \int_{E=\Phi_0}^{\infty} f_0(E)(E - \Phi_0)^{3/2} \, dE. \qquad (3)$$

By differentiating the latter equality and comparing the result with the former, we will have[3]

$$-\frac{dP_0}{dr} = \rho_0 \frac{d\Phi_0}{dr}. \tag{4}$$

This equation expresses the well-known condition of the hydrodynamical equilibrium as studied, e.g., in the stellar equilibrium theory. To obtain equilibrium configurations, it must be supplemented with the state equation $P_0 = P_0(\rho_0)$. In this case, however, the state equation is represented by relations (3) which parametrically present the dependence $P_0 = P_0(\rho_0)$; in explicit form, the state equation can be derived by eliminating the parameter Φ_0.

The above statement allows a conclusion to be reached about the possibility of using a fairly well-developed theory of equilibrium of individual stars in order to investigate the relevant stationary states of stellar systems. The hydrodynamical analogy thus considered is also used in the study of the stability of spherically symmetrical systems with isotropic distribution functions (see §2). It possesses an obvious advantage of illustrativeness, although its use has seemingly not resulted as yet in any essentially new results.

3. *Examples of isotropic distributions.* Consider some most typical partial examples of isotropic distributions of particles in velocities and their hydrodynamical analogies.

3a. The solutions by Emden (see, e.g., [101]) are well known, for a polytropic gaseous sphere with the pressure $P_0 = k\rho^{1+1/n} \equiv k\rho^\gamma$ (γ is called the adiabatic index, and n the polytropic index). The relevant distribution functions of collisionless systems have, as it appears, the form of

$$f_0 = c[\Phi_0(a) - E]^\alpha \qquad (c = \text{const}). \tag{5}$$

Indeed, substituting (5) into (3), we get

$$\rho_0(r) = c_1[\Phi_0(a) - \Phi_0(r)]^{\alpha + 3/2}, \tag{6}$$

where the new constant c_1 is defined by the expression

$$c_1 = 8\sqrt{2}\pi c \int_0^1 x\,dx \int_0^{\sqrt{1-x^2}} (1 - x^2 - y^2)^\alpha\,dy \qquad (\alpha > -1). \tag{7}$$

Similarly, calculation of the "pressure" $P_0(r)$ yields in this case $P_0(r) \sim [\Phi_0(a) - \Phi_0(r)]^{\alpha + 5/2}$, therefore $P_0 \sim \rho^{(2\alpha + 5)/(2\alpha + 3)}$. Therefore, each α is associated with a definite adiabatic index $\gamma = (2\alpha + 5)/(2\alpha + 3)$, and the polytropic index $n = \alpha + \frac{3}{2}$. Substituting (6) into the Poisson equation, we get

---

[3] A similar equation for any $f_0 = f_0(E, L^2)$ is [101]

$$-\frac{1}{r^2}\frac{d}{dr}(r^2 P_r) + \frac{2P_t}{r} = \rho_0 \frac{d\Phi_0}{dr},$$

or

$$\frac{d(\rho_0 c_r^2)}{dr} + \frac{2\rho_0}{r}(c_r^2 - c_t^2) = -\rho_0 \frac{d\Phi_0}{dr}.$$

the nonlinear equation for the potential Φ_0, which can be reduced to the Emden equation

$$xy'' + 2y' + xy'' = 0, \tag{8}$$

if it is assumed that

$$y \equiv \frac{[\Phi_0(a) - \Phi_0(r)]}{\Phi_*}, \qquad x = r/r_0, \tag{9}$$

$$r_0^{-2} = 4\pi Gc_1\Phi_*^{n+1/2}.$$

The distribution functions of (5) were independently derived by Jeans [241] and Eddington [196, 197]. The properties of the solutions of the Emden equation are thoroughly studied in the theory of internal stellar structure. The value $\alpha = \frac{7}{2}$ is a boundary value; it separates the solutions with finite and infinite radius and mass of the system. The density $\rho_0(r)$ that corresponds to it is the familiar Schuster law

$$\rho_0(r) = \rho_0[1 + r^2/r_0^2]^{-5/2} \qquad (\rho_0, r_0 \text{ constants}). \tag{10}$$

Other solutions in elementary functions are obtained at $\alpha = -\frac{3}{2}, -\frac{1}{2}$. However, the first of these solutions does not exist in the collisionless case, since at $\alpha \le -1$, the distribution function (5) yields the divergent integral in (7), which defines the density ρ_0. Such a situation has, seemingly, a general character: not every motion and not every equilibrium in hydrodynamical systems may coexist in collisionless systems (see also below, item 4 and §2). The limiting case $\alpha = -1$ corresponds to the following [214]:

$$f_0 = \text{const} \cdot \delta(\Phi_0(a) - E). \tag{11}$$

The equation of state of ideal gas $P \sim \rho^{5/3}$ is $f_0(E)$ corresponded to by, as is easy to see, the distribution function of the Fermi-Dirac type (with zero temperature)

$$f_0(E) - \text{const} \cdot \theta(E_0 - E).$$

3b. The Maxwellian distribution function deserves, of course, special attention. In the collisionless case, it has the form

$$f_0(E) = c \cdot e^{-E/\theta} \qquad (\Phi_0 \le E < \infty), \tag{12}$$

where c and θ are constants and θ has the meaning of effective "temperature" of the stellar "gas." The difficulty is in the fact that (12) leads to systems with infinite mass M (and radius R). Such a result is natural. The particles with a fairly high speed, possessing, therefore, a positive full energy, must have escaped from the system of limited size. The density $\rho_0(r)$, corresponding to (12), is $\rho_0(r) = \text{const} \cdot e^{-\Phi_0(r)/\theta}$. Therefore, if it is assumed that the system mass is finite, $M < \infty$, then the potential $\Phi_0(r) \to 0$ at $r \to \infty$ (as $-GM/r$). However, as $\Phi_0 \to 0$, we in this case have $\rho_0(r) \to \text{const}$ and, consequently, $M = \infty$, which contradicts our assumption.

The Poisson equation for the case of interest has the form

$$\frac{d^2\Phi_0}{dr^2} + \frac{2}{r}\frac{d\Phi_0}{dr} = \text{const} \cdot e^{-\Phi_0/\theta}.$$

It has one simple solution in elementary functions,

$$\Phi_0 = 4\pi G\beta \ln r + \text{const}, \tag{13}$$

and $\rho_0(r) = \beta/r^2$. This solution ensues, provided that $4\pi G\beta = 2\theta$.

Note that distribution function (12) can be derived also by the limiting transition from the Emden solution, where it corresponds to the isothermic adiabatic index $\gamma = 1$ ($n \to \infty$). The hydrodynamical equation of state ensuing from (3) yields here $P_0 = \theta\rho_0$, i.e., really $\gamma = 1$. Owing to the difficulty noted above, the "pure" Maxwellian distribution in (12) must be somewhat modified in order to obtain the finite mass and size of the system. For instance, instead of (12), one can consider the "cutoff" of the Maxwellian distribution, assuming $f_0 = 0$ at $E < E_0$. In this case, however, the differential equation for the potential Φ_0 proves to be

$$\Phi_0'' + \frac{2}{r}\Phi_0' + a_1 e^{a_2\Phi_0} \, \text{erf}_2\sqrt{a_3 + a_4\Phi_0} = 0 \tag{14}$$

(a_1, a_2, a_3, a_4 are constants), which may be solved only numerically [47, 352].

As was noted already by Eddington [197], relation (3) for the given density distribution $\rho_0(r)$ [and, consequently, the potential $\Phi_0(r)$] can be considered the integral equation for the distribution function $f_0(E)$. Moreover, it is an integral equation of the familiar type, a generalized Abel equation, which is solved analytically in the general form. In fact, by differentiating Eq. (3) over Φ_0 we arrive at the ordinary Abel equation [132].[4]

4. *Anisotropic distribution functions.* Turn now to the consideration of anisotropic distribution functions. There are two possible approaches to the determination of the equilibrium states.

First. Using the given distribution function of the form $f_0 = f_0(E, L^2)$ the density $\rho_0(r)$ is calculated by double integration:

$$\rho_0(r) = \frac{2\pi}{r^2} \iint f_0(E, L^2)\left[-2\Phi_0(r) + 2E - \frac{L^2}{r^2}\right]^{-1/2} dE \, dL^2. \tag{15}$$

The integration region in the plane E, L^2 is limited [32] by the axis E ($L^2 > 0$), by the line of circular velocities, defined by obvious parametric equations

$$2E = r\Phi_0' + 2\Phi_0, \tag{16}$$

$$L^2 = r^3\Phi_0',$$

as well as by the condition of gravitational coupling $E \le 0$.

---

[4] It should, however, be noted that physically acceptable solutions do not result for all densities $\rho_0(r)$. For instance, at $\rho_0(r) = \text{const} \cdot \theta(R - r)$ (uniform sphere) $f_0(E)$ formally determined from (3) is negative everywhere, apart from the phase region boundary (see Problem 4 of Chapter IV, which deals with a similar point).

Second. The relation of (15) can be considered also as the integration equation for f_0 (E, L^2), if it is assumed that $\rho_0(r)$ and $\Phi_0(r)$ are given. Then, using the function of one variable $\rho_0(r)$ it is necessary, according to (15), to define the function of two variables $f_0 = f_0(E, L^2)$. In this case it is to be expected that the solution of the written integral equation will be ambiguous [32]. Below we give an example of a two-valued solution. The question concerning the finding of all solutions for (15) remains unsolved so far.[5]

Circular orbits. The simplest anisotropic distribution function is the distribution function of particles rotating along circular trajectories (see Fig. 2):

$$f_0 = \frac{\rho_0(r)}{2\pi v_0(r)}\, \delta(v_r)\delta(v_\perp - v_0(r)), \tag{17}$$

where $v_\perp^2 = v_\theta^2 + v_\varphi^2$, $v_0^2 = r\Phi_0'$. The particles are maintained in circular orbits due to the exact equality of the gravitational attraction and centrifugal force. The density $\rho_0(r)$ may apparently be set in an arbitrary way.

Systems with nearly circular orbits. One may suggest [143] a general method of calculation of stationary functions of distribution for models close to the Einstein model (the radial velocity dispersion is finite but small). The essence of the method is as follows. The initial set of Vlasov's equations can be presented in the form

$$\hat{L}f_0 + v_r\left(\frac{\partial f}{\partial r} - \frac{v_\perp}{r}\frac{\partial f}{\partial v_\perp}\right) + \left(\frac{v_\perp^2}{r} - \frac{\partial\Phi}{\partial r}\right)\frac{\partial f}{\partial v_r} - \frac{1}{r}\left(\cos\alpha\,\frac{\partial\Phi}{\partial\theta} + \frac{\sin\alpha}{\sin\theta}\frac{\partial\Phi}{\partial\varphi}\right)\frac{\partial f}{\partial v_\perp}$$

$$+ \frac{1}{rv_\perp}\left(\sin\alpha\,\frac{\partial\Phi}{\partial\theta} - \frac{\cos\alpha}{\sin\theta}\frac{\partial\Phi}{\partial\varphi}\right)\frac{\partial f}{\partial\alpha} = 0, \tag{18}$$

$$\Delta\Phi(\mathbf{r}) = 4\pi G \int f(r, v_r, v_\perp, \alpha)\, dv_r v_\perp\, dv_\perp\, d\alpha. \tag{19}$$

Here we selected the spherical coordinate system, and instead of v_θ and v_φ, the variables $v_\perp$ and α: $v_\perp^2 = v_\theta^2 + v_\varphi^2$; $\alpha = \arctan(v_\varphi/v_\theta)$ are introduced; the operator L denotes

$$\hat{L} = \frac{\partial}{\partial t} + \frac{v_\perp}{r}\left(\cos\alpha\,\frac{\partial}{\partial\theta} + \frac{\sin\alpha}{\sin\theta}\frac{\partial}{\partial\varphi} - \sin\alpha\cot\theta\,\frac{\partial}{\partial\alpha}\right). \tag{20}$$

For the derivation of the kinetic equation in the spherical coordinate system (as in any other orthogonal system) we refer to §1 of the Appendix.

In considering the spherically symmetrical ($\partial/\partial\theta = \partial/\partial\varphi = 0$), stationary ($\partial/\partial t = 0$) solutions of the form $f_0 = f_0(r, v_r, v_\perp)$ (so that also $\partial/\partial\alpha = 0$),

---

[5] The two alternative approaches described in stellar dynamics are called [101] the Jeans problem and the inverse Jeans problem or, respectively, the "inductive" or "deductive" methods [32].

from (18) and (19) we get

$$v_r\left(\frac{\partial f_0}{\partial r} - \frac{v_\perp}{r}\frac{\partial f_0}{\partial v_\perp}\right) + \left(\frac{v_\perp^2}{r} - \frac{\partial \Phi_0}{\partial r}\right)\frac{\partial f_0}{\partial v_r} = 0,$$

$$\frac{\partial}{\partial r}(r^2\Phi_0) = 4\pi G\rho_0 r^2 = 8\pi^2 Gr^2 \int f_0(r, v_r, v_\perp)v_\perp \, dv_\perp \, dv_r.$$

(21)

The solution of (17) trivially satisfies (21). Represent now f_0 in the form of expansion in δ functions and their derivatives:

$$f_0 = [a\delta(v_\perp - v_0) + a_1\delta'(v_\perp - v_0)$$
$$+ a_2\delta''(v_\perp - v_0)]\delta(v_r) + b\delta(v_\perp - v_0)\delta''(v_r),$$

(22)

where a, a_1, a_2, b, v_0 are the functions of the coordinate r; and δ', δ'' are the derivatives of the δ function. We shall call such expansions "δ expansions." Such δ expansions prove to be convenient above all in the investigation of the stability of the relevant equilibrium systems, where they were first used [87] (see §3). After substitution of (22) into (21), with the help of the Poisson equation, we obtain a set of three equations for four unknown functions a, a_1, a_2, b. This permits representation of the general distribution function in (22) in the form dependent, apart from $\rho_0(r)$, also on another arbitrary function $\theta(r)$ having the meaning of a small "temperature":

$$b = \frac{\theta(r)\rho_0}{4\pi v_0}, \qquad a = \frac{\rho_0}{2\pi v_0} + \frac{1}{4\pi v_0^2}\left[\frac{\theta\rho_0(rv_0)'}{2v_0^2} - \left(\frac{r\rho_0\theta}{v_0}\right)'\right],$$

$$a_1 = \frac{1}{4\pi v_0}\left[\frac{\rho_0\theta(rv_0)'}{2v_0^2} - \left(\frac{r\rho_0\theta}{v_0}\right)'\right], \qquad a_2 = \frac{\theta\rho_0(rv_0)'}{8\pi v_0^2}.$$

(23)

In [143], the stability of a homogeneous system described by distribution function (22) at $n_0(r) = $ const and $\theta(r) = $ const is shown, and the frequencies of collective oscillations are calculated (see §3, this chapter).

The "anisotropic generalization" of distribution functions (5) are the following distribution functions derived by Camm [180]:

$$f_0 \sim L^\beta[\Phi_0(a) - E]^\alpha, \qquad \beta > -2, \alpha > -1.$$

(24)

Density ρ_0, corresponding to (24), is

$$\rho_0 = \text{const} \cdot [\Phi_0(a) - \Phi_0(r)]^{\alpha + (\beta + 3)/2}, \qquad a = \text{radius}.$$

Using the Poisson equation, it is further possible to derive the Emden–Fowler equation

$$xy'' + 2y' + x^{\beta+1}y^{\alpha+(\beta+3)/2} = 0, \qquad y \equiv \frac{\Phi_0(a) - \Phi_0(r)}{\Phi_0(a) - \Phi_0(0)}, \qquad x \equiv \frac{r}{a}$$

At $\alpha + (\beta + 3)/2 = 1$, the solutions for this equation can be expressed through the Bessel function [22].

In the limiting case $\alpha = -1$ we have [214]

$$f_0 = \text{const} \cdot \delta[\Phi_0(a) - E]L^\beta.$$

In the other limiting case $\beta = -2$, we get

$$f_0 = \text{const} \cdot [\Phi_0(a) - E]^\alpha \delta(L^2).$$

Calculation of the average squares of the radial and transversal velocities for the distribution function in (24) leads to the relation

$$\frac{2\overline{v_r^2}}{\overline{v_\perp^2}} = \frac{2}{\beta + 2}.$$

This yields that, as $\beta \to -2$, the function in (24) describes configurations with purely radial movement of stars; for $\beta = 0$ the system becomes isotropic, and $\beta \to \infty$ corresponds to the systems with circular orbits.

For all distribution functions in (24) it is typical that on the system boundary the total particle velocity vanishes. Such a requirement is not necessary; it is sufficient to require zero identity at $r = a$ of the radial velocity v_r. For that purpose, the particle energy on the boundary must be equal to the "effective" potential energy at $r = a$ ($U_{\text{eff}} = L^2/2r^2 + \Phi_0$). This requirement is satisfied by the second series of the anisotropic distribution functions [180]

$$f_0 \sim L^\beta \left[\frac{L^2}{2a^2} + \Phi_0(a) - E \right]^\alpha. \tag{25}$$

For the density ρ_0 in this case we have from (25)

$$\rho_0(r) \sim \int L^{\beta+1} \left[\frac{L^2}{2a^2} + \Phi_0(a) - \frac{v_r^2}{2} - \frac{L^2}{2r^2} - \Phi_0(r) \right]^\alpha \frac{dL}{r^2} \, dv_r, \tag{26}$$

where integration is performed over the region

$$0 \le v_r^2 \le \frac{L^2}{a^2} - \frac{L^2}{r^2} + 2[\Phi_0(a) - \Phi_0(r)],$$

$$0 \le L^2 < \frac{2a^2 r^2 [\Phi_0(a) - \Phi_0(r)]}{a^2 - r^2}, \tag{27}$$

where $L_{\max} = a^3 \Phi_0'(a)$. As a result of the integration, we shall have

$$\rho_0(r) = \text{const } r^\beta \frac{[\Phi_0(a) - \Phi_0(r)]^{\alpha + (\beta + 3)/2}}{(a^2 - r^2)^{\beta/2 + 1}}.$$

Substitution into the Poisson equation leads to the following nonlinear differential equation for the potential:

$$\Phi_0'' + \frac{2}{r} \Phi_0' = 4\pi G \text{ const } r^\beta \frac{[\Phi_0(a) - \Phi_0(r)]^{\alpha + (\beta + 3)/2}}{(a^2 - r^2)^{\beta/2 + 1}}. \tag{28}$$

At $\alpha = -\frac{1}{2}, \beta = 0$ we obtain the distribution function of the system with the uniform density ρ_0 [112, 115]

$$f_0 = \frac{\rho_0}{\pi^2 a^2 \Omega_0^2} \theta \left[\frac{L^2}{2a^2} + \frac{\Omega_0^2 a^2}{2} - E \right] \left(\Omega_0^2 (a^2 - r^2) - v_r^2 - \frac{L^2}{r^2} + \frac{L^2}{a^2} \right)^{-1/2}, \tag{29}$$

where $\Omega_0^2 = 4\pi G\rho_0/3$, $\theta(x)$ is the Heaviside function. Another solution for Eq. (28) for $\alpha = \frac{3}{2}$, $\beta = 0$ has been found by Kuz'min and Veltmann [62]. It yields for the density $\rho_0(r)$ the Schuster law (10). This example illustrates the above-mentioned ambiguity of the solution of Eq. (15) for $f_0(E, L^2)$. Still another distribution function yielding the same density as in (10) is the isotropic model (5) with $\alpha = \frac{7}{2}$.

The calculation of the ratio of radial and transversal velocity dispersions leads for (25) to the result

$$\frac{2\overline{v_r^2}}{\overline{v_\perp^2}} = \frac{2(1 - r^2/a^2)}{\beta + 2}. \tag{30}$$

This value, of course, vanishes at $r \to a$; at $\beta \to -2$ the particles move along radial trajectories, and at $\beta \to \infty$, the orbits of stars become circular. At $\beta = 0$, from (30) we get $2\overline{v_r^2}/\overline{v_\perp^2} = 1 - r^2/a^2$, so that near the center $r \to 0$ the system is, in essence, isotropic, while in the vicinity of the boundary $r \to a$ it is similar to the system with circular orbits.

The distribution functions of a homogeneous sphere in (29) contain *all possible elliptical orbits* of the stars admissible in the potential $\Phi_0 \sim r^2/2$. Freeman in [203] suggested, in particular, the sphere model, also with uniform density, the equilibrium in which is provided by the rotation of particles in the plane parallel to (xy) and by the local velocity dispersion along the z-axis. This distribution function has the form

$$f_0 = \frac{\rho_0}{\pi} \frac{\delta(v_x)\delta(v_y)}{\sqrt{1 - r^2 - v_z^2}}, \tag{31}$$

and (31) is written in the frame of reference rotating at an angular velocity $\Omega = (4\pi G\rho_0/3)^{1/2}$; the sphere radius and the value Ω are assumed to be equal to unity. From (31) it is evident that, in this frame of reference, the particles do not migrate in the (x, y) plane but perform harmonical oscillations about the z-axis at a frequency Ω. In the inertial system, the trajectories of particles are ellipses, but only with circular projections onto the (x, y) plane. Rotation of all particles in (31) occurs in one and the same direction so that the moment of rotation of the entire system is different from zero. However, by using superposition of such systems rotating in opposite directions, one seemingly may obtain systems with any moment from zero to some maximum.

We mention here two more interesting series of the distribution functions. One of them was apparently first suggested by Osipkov [27^{ad}]:

$$f_0 = f_0(X_\lambda) = \frac{A}{\pi^2} [(1 + 3\lambda)U^2 X_\lambda^{-1/2}/4 + 4(1 + \lambda)X_\lambda^{1/2}U + 8(1 - \lambda)X_\lambda^{3/2}], \tag{32}$$

where $X_\lambda \equiv -2E - U/2 - \lambda L^2/r_0^2$. A, U, λ, r_0 are parameters; moreover, A, U, r_0 are connected with the relation $4\pi G r_0^2 A U^2 = 1$, and the dimensionless parameter λ is restricted to the interval $(-1/3, 97/45)$. Details concerning properties of these spherical star models can be found in Section 6.1, where we shall give also the results of stability investigation of models (32). Here we write only the density $\rho_0(r)$ and potential $\Phi_0(r)$ corresponding to (32):[5']

$$\rho_0(r) = 2AU^3 \frac{(3 - r^2/r_0^2)}{(1 + r^2/r_0^2)^3}, \qquad (33)$$

$$\Phi_0(r) = -\frac{U}{1 + r^2/r_0^2}. \qquad (34)$$

The second series of distribution functions was given by Kuz'min and Velltmann [18[ad], 19[ad]]:

$$f_0 = f_0(x, \xi) = (1 + p\xi)^{-5/2}\Pi(u),$$

$$\Pi(u) = 2\sqrt{2} \frac{\rho_0}{\pi^2}\left[\lambda + \frac{8}{5}pu + \frac{16}{7}(1 - \lambda)u^2 - \frac{64}{21}pu^3\right]u^{3/2}, \qquad (35)$$

$$u = \frac{x - \lambda\xi}{1 + p\xi}, \qquad x = -E, \qquad \xi = \frac{L^2}{2}.$$

Series (35) depends on two dimensionless parameters: λ and p. As one can see, the distribution (35) gives the phase description of Shuster model (10). In Section 6.3 the distribution functions (35) will be investigated for stability; there one may also obtain detailed representation of the properties of equilibrium states characterizing these distribution functions.

Of special interest are spherical systems with a *purely radial movement* (see Fig. 21). As indicated by Agekyan [2], such systems must be formed at the time of stellar formation in a diffuse cloud or as a result of ejections out of a superdense nucleus. Agekyan studied the partial model of such a system with identical apocentric distances R, i.e.,

$$f_0 \sim \delta(E - E_0)\delta(L^2).$$

The differential equation for the potential has here such a form:

$$x^2y'' + 2xy' + \frac{\mu}{\sqrt{y}} = 0, \qquad (36)$$

where

$$y = \frac{R}{GM}[\Phi_0(R) - \Phi_0(r)], \qquad x = \frac{r}{R},$$

$$\mu = \left[\int_0^1 \frac{dx}{\sqrt{y}}\right]^{-1}.$$

[5'] Relations (33) and (34) were found by Idlis [14[ad]]; and for brevity, we call models described by formulas (32)–(34), Idlis models.

It is possible to find the asymptotic behavior of the solutions of Eq. (36) near the system center. At $x \to 0$, the equation is reduced to the following:

$$x^2 y' \simeq -\frac{\mu}{\sqrt{y}} x,$$

whose solution is

$$y = (-\tfrac{3}{2}\mu \ln cx)^{2/3},$$

where the integration constant c should be found from the exact solution. In the same work by Agekyan, Eq. (36) was integrated numerically. In particular, it is shown that c is close to unity (or is exactly equal to 1). Assuming that $c = 1$, we shall arrive at the following behavior of the density and velocity of particles $v_0 = 2\sqrt{\Phi_0(R) - \Phi_0(r)}$ near the system center:

$$v_0 \simeq \sqrt{\frac{2GM}{R}} (-\tfrac{3}{2}\mu \ln x)^{1/3},$$

$$\rho_0 \simeq \frac{\mu M}{2\pi R^3 x^2 (-\tfrac{3}{2}\mu \ln x)^{1/3}}.$$

The general form of distribution functions of systems with radial trajectories of stars is as follows:

$$f_0 = r^{-2}\delta(v_\theta)\delta(v_\varphi)F(E) \sim \delta(L^2)F(E). \tag{37}$$

For the arbitrary density $\rho(r)$ the problem of definition of $F(E)$ is reduced to the solution of the Abel integral equation [132]. If it is again assumed, similarly to (5) and (24), that

$$F(E) = \begin{cases} \text{const}[\Phi_0(a) - E]^\alpha, & \Phi_0(a) > E, \\ 0, & \Phi_0(a) < E, \end{cases} \tag{38}$$

then for the potential $\Phi_0(r)$ we obtain the differential equation

$$\Phi_0'' + \frac{2}{r}\Phi_0' - \frac{c}{r^2}[\Phi_0(a) - \Phi_0(r)]^{\alpha + 1/2} = 0 \qquad (c > 0). \tag{39}$$

In particular, at $\alpha = -\tfrac{1}{2}$ we get $\Phi_0 = 4\pi G\beta(\ln r/a - 1)$, $\rho_0 = \beta r^{-2}$. We write in the explicit form the relevant distribution function [112]

$$f_0 = \frac{\beta}{\pi r^2} \delta(v_\theta)\delta(v_\varphi)\left(8\pi G\beta \ln \frac{a}{r} - v_r^2\right)^{-1/2}. \tag{40}$$

We make here another remark: In the case of systems with purely radial stellar movement, solutions are lacking which correspond to systems with uniform density. It may readily be shown in solving the Abel equation for the sought-for phase density $F(E)$:

$$\rho_0 = \text{const} = \frac{1}{2r^2} \int F(E) \frac{dE}{\sqrt{E - \Phi_0}}. \tag{41}$$

The solution of this equation shows that there is some finite range of values E in which $F(E) < 0$.

From qualitative considerations it is easy to see that the clusters with radial stellar motion [corresponding to $\beta \to -2$, see (24)] are unstable owing to the absence of the stabilizing influence of the stellar speed dispersion in the tangential direction.

Eddington [196, 197] derived the general form of distribution functions of "Maxwellian" type:

$$f_0 = \frac{\rho_0}{(1 + \alpha r^2)^2} q^{\mu} \exp[-\tfrac{1}{2}(p^2 + q)], \tag{42}$$

where

$$p = av_r^2, \qquad q = bv_{\perp}^2, \qquad a = \text{const}, \qquad b = a/(1 + \alpha r^2).$$

This distribution function may be derived from Eq. (21), if the latter is solved by the method of separation of variables [143].

Assuming that

$$f_0(v_r, v_{\perp}, r) = g(v_r)h(v_{\perp}, r), \tag{43}$$

we obtain

$$g(r) = A \exp(-v_r^2/2\theta), \tag{44}$$

where $1/\theta$ is the separation parameter.

For $h(v_{\perp}, r)$, we have the equation

$$\left(\frac{\partial h}{\partial r} - \frac{v_{\perp}}{r} \frac{\partial h}{\partial v_{\perp}}\right) - \frac{1}{\theta}\left(\frac{v_{\perp}^2}{r} - \frac{\partial \Phi_0}{\partial r}\right)h = 0, \tag{45}$$

which, after separation of $h(v_{\perp}, r) = a'(v_{\perp})b'(r)$, yields

$$b' = Br^c \exp\left(\frac{-\Phi_0}{\theta}\right), \qquad a' = Dv_{\perp}^c \exp\left(\frac{-v_{\perp}^2}{2\theta}\right), \tag{46}$$

where c is the new separation parameter.

Now, substituting f_0 into the Poisson equation, we get

$$\frac{d^2\Phi_0}{dr^2} + \frac{2}{r}\frac{d\Phi_0}{dr} - 4\pi GFr^c \exp\left(\frac{-\Phi_0}{\theta}\right) = 0. \tag{47}$$

The single solution of this equation in elementary functions is $\Phi_0 = \ln r$, and $\rho(r) \sim \beta/r^2$. In this case, the partial solution is written in the form [196, 197]

$$f_0(v_r, v_{\perp}, r) = \frac{\beta}{r} \frac{v_{\perp}^{2(4\pi G\beta/2\theta - 1)}}{2\pi\sqrt{2\pi\theta}\,(2\theta)^{4\pi G\beta/2\theta}\Gamma(4\pi G\beta/2\theta)} \exp\left(-\frac{v_r^2 + v_{\perp}^2}{2\theta}\right). \tag{48}$$

Formula (48) coincides with (42). For the density $\rho_0(r)$, the distribution function in (42) leads to the coupling

$$\rho_0 = \frac{b}{1 + \alpha r^2} \exp\left(\frac{\Phi_0 - \Phi}{h^2}\right), \qquad h = \text{const}. \tag{49}$$

It turns out that such models always possess infinite mass and radius and cannot therefore correspond to a real physical system. At $\alpha = 0$ and $\mu = 0$, (42) goes over to the Maxwellian function of (12). The case $\alpha = 0$, $\mu \neq 0$ was considered in detail in [22]. It is interesting in that it admits the relativistic generalization. Then

$$\rho_0 \sim \beta r^{-2}, \qquad f_0 \sim r^{-2} v_\perp^{2(4\pi G/2\theta - 1)} \exp\left(\frac{-v^2}{2\theta}\right). \tag{50}$$

The δ-like character of function (48) at the point $v_\perp = (4\pi G\beta)^{1/2} \equiv v_0$ shows that the limit of f_0 at small θ will be a certain series of δ function and its derivatives.

The expansion is performed in a standard way:

$$g(x) = \sum_n \gamma_n \delta^{(n)}(x - x_0),$$

where

$$\gamma_n = \frac{(-1)^n}{n!} \int g(x)(x - x_0)^n \, dx.$$

The degree of deviation from circular orbits is convenient to characterize by the small parameter, the dispersion of kinetic energy of radial movement of volume unit:

$$\int v_r^2 f_0(v_r, v_\perp, r) \, dv_r v_\perp \, dv_\perp = \rho_0(r) \frac{\theta}{2}.$$

Here θ is the small parameter having dimension v_r^2.

Expansion of $g(v_r)$ gives

$$(2\pi\theta)^{-1/2} \exp(-v_r^2/2\theta) = \delta(v_r) + \tfrac{1}{2}\theta \delta''(v_r) + o(\theta).$$

A more cumbersome calculation of the "transversal" function leads to the expansion

$$\frac{v_\perp^{2(v_0^2/2\theta - 1)}}{(2\theta)^{v_0^2/2\theta}\Gamma(v_0^2/2\theta)} \exp\left(\frac{-v_\perp^2}{2\theta}\right) = \frac{1}{v_0}\left(1 + \frac{3}{8x}\right)\delta(v_\perp - v_0) + \frac{3}{8x}\delta'(v_\perp - v_0)$$

$$+ \frac{v_0}{8x}\delta''(v_\perp - v_0) + o\left(\frac{1}{x}\right), \tag{51}$$

where $x = v_0^2/2\theta$. For f_0, within an accuracy of up to θ, we have

$$f_0 = \left[\frac{1}{v_0}\delta_r\delta_\perp\left(1 + \frac{3}{8x}\right) + \frac{\theta}{2v_0}\delta_r''\delta_\perp + \frac{3}{8x}\delta_r\delta_\perp' + \frac{v_0}{8x}\delta_r\delta_\perp''\right]\frac{\beta}{2\pi r^2},$$

$$\delta_r \equiv \delta(v_r), \qquad \delta_\perp \equiv \delta(v_\perp - v_0), \tag{52}$$

Above we have already seen how the expansion over low "temperature" is performed in the case of arbitrary density.

Let us in conclusion consider the problem of *the rotation* of collisionless spherically symmetrical systems. This problem has apparently first been raised by Lynden-Bell [279], who showed a possibility of rotation in principle. In fact, it is clear that, starting from a nonrotating sphere with $f_0 = f_0(E, L)$, one can always obtain a multitude of distribution functions corresponding to rotating systems (parameter $0 \leq \alpha \leq 1$):

$$f_0(E, L) \rightarrow f_0(E, L, L_z) = \alpha f_0(E, L)\theta(L_z) + (1 - \alpha) f_0(E, L)\theta(-L_z) \quad (53)$$

[here $\theta(L_z)$ is the step function]. This transformation means that we have divided the particles into groups with $L_z > 0$ and $L_z < 0$ and reversed the movement in one of the groups. Then it is easy to see that neither the density $\rho_0(r)$ nor the potential $\Phi_0(r)$ changes [49]. As an example, we shall consider in more detail the system of particles rotating in circular orbits. Refer to Fig. 2. The distribution function in (17), as considered earlier, corresponds to the case when, in the plane tangent to an arbitrary sphere at any point of it, all the vectors of particle speeds totaled a zero vector, i.e., there was no rotation of the sphere as a whole. We single out conventionally the z-axis, and in the equatorial plane let the number of particles rotating in one direction exceed the number of particles rotating in the opposite direction. Let, besides, such a difference be lacking on the pole (it diminishes on moving away from the equator). The system thus constructed possesses a nonzero rotational moment.

In the equilibrium case $\partial/\partial t = \partial/\partial \varphi = 0$ and $\Phi_0 = \Phi_0(r)$, from (18) we have

$$\frac{v_\perp}{r}\left(\cos\alpha\,\frac{\partial}{\partial\theta} - \sin\alpha\cot\theta\,\frac{\partial}{\partial\alpha}\right)f_0 + v_r\left(\frac{\partial f_0}{\partial r} - \frac{v_\perp}{r}\frac{\partial f_0}{\partial v_\perp}\right) + \left(\frac{v_\perp^2}{r} - \frac{\partial\Phi_0}{\partial r}\right)\frac{\partial f_0}{\partial v_r} = 0.$$

$$(54)$$

It is easy to see that this equation is satisfied, for example, by the function[6] [124]

$$f_{0\mu} = \frac{\rho_0}{2\pi v_0}\,\delta(v_r)\delta(v_\perp - v_0)(1 + \mu\sin\theta\sin\alpha) \qquad (|\mu| \leq 1). \quad (55)$$

The distribution function of (55) describes the system with the finite total moment

$$L_z = \int f_0 r\sin\theta v_\perp \sin\alpha v_\perp\, dv_\perp\, dv_r\, d\alpha\, r^2\, dr\, d\Omega \neq 0$$

and with the spherically symmetrical distribution of density (and potential).

Consider for the sake of definiteness a sphere with $\rho_0 = \text{const}$. Calculate the angular speed of rotation of such a sphere. We have [124]

$$\langle v_\varphi \rangle = \frac{1}{\rho_0}\int f_0 v_\varphi\, d^3v = \tfrac{1}{2}\mu\Omega r\sin\theta, \qquad \Omega_{\text{rot}} = \frac{\mu\Omega}{2}. \quad (56)$$

---

[6] The general solution of (54) corresponding to circular orbits is the function

$$f_0 = \frac{\rho_0}{2\pi v_0}\,\delta(v_r)\delta(v_\perp - v_0)[1 + F(\sin\theta\sin\alpha)]$$

with arbitrary F that satisfies $\int F\, d\alpha = 0$. However, it may be shown that (55) is the unique distribution function describing a uniformly rotating (at the angular speed $\mu\Omega/2$) sphere.

Here Ω is the frequency of a star's revolution in circular orbit about the gravity center, Ω_{rot} is the rotational frequency of the sphere about the z-axis. It is evident that the uniform sphere (55) rotates as a solid body.

§ 2 Stability of Systems with an Isotropic Particle Velocity Distribution

The *stability* of spherically symmetrical stellar clusters with the isotropic (in the inertial frame of reference) velocity distribution is most thoroughly investigated; in this case some general theorems have also been proved. The main results were first attained by Antonov [4, 6], and later "rediscovered" [7] in works of many authors (at least, in the most essential features).

2.1 The General Variational Principle for Gravitating Systems with the Isotropic Distribution of Particles in Velocities $(f_0 = f_0(E), f'_0 \equiv df_0/dE \leq 0)$

Below, we derive the stability criterion valid for systems of *any geometry* with the distribution function $f_0 = f_0(E)$, decreasing monotonically with the energy: $f'_0 \leq 0$.

We start with the derivation of "equations of motion" describing the evolution of perturbations in systems with the isotropic velocity distribution of particles; first of all, we shall not assume that $f'_0 \leq 0$.

By denoting the linear operator $\hat{D}$

$$\hat{D} = \mathbf{v}\frac{\partial}{\partial \mathbf{r}} - \frac{\partial \Phi_0}{\partial \mathbf{r}}\frac{\partial}{\partial \mathbf{v}}, \tag{1}$$

we write the perturbed kinetic equation in the form

$$\frac{\partial f_1}{\partial t} + \hat{D}f_1 = \mathbf{v}\frac{df_0}{dE}\frac{\partial \Phi_1}{\partial \mathbf{r}}. \tag{2}$$

Now we divide (2) by df_0/dE and take into account that $\hat{D}$ acting on any function of integrals of motion yields zero. In particular, $\hat{D}(1/f'_0) = 0$, and therefore

$$\frac{\hat{D}f_1}{f'_0} = \hat{D}\frac{f_1}{f'_0}. \tag{3}$$

The Φ_1 function is dependent only on $\mathbf{r}$, so that one can write

$$\mathbf{v}\frac{\partial \Phi_1}{\partial \mathbf{r}} = \hat{D}\Phi_1. \tag{4}$$

---

[7] One of the recent investigations of such a kind is that of Katz [75^{ad}].

Thus, reduce Eq. (2) to the form

$$\frac{\partial f_1}{f_0' \, \partial t} + \hat{D}\left(\frac{f_1}{f_0'} - \Phi_1\right) = 0. \tag{5}$$

On the other hand, the gravity law gives the following coupling between the distribution function f_1 and the potential Φ_1:

$$\Phi_1(\mathbf{r}, t) = -G \iint \frac{f_1(\mathbf{r}', t, \mathbf{v}') \, d\mathbf{v}' \, d\mathbf{r}'}{|\mathbf{r} - \mathbf{r}'|}, \tag{6}$$

where the integral refers to the entire phase volume occupied by the system. From (5) and (6) it follows that the initial system of equations is identical with the following equation:

$$\frac{\partial f_1/\partial t}{\partial f_0/\partial E} + \hat{D}\left[\frac{f_1}{\partial f_0/\partial E} + G \iint \frac{f_1(\mathbf{r}', \mathbf{v}', t) \, d\mathbf{r}' \, d\mathbf{v}'}{|\mathbf{r} - \mathbf{r}'|}\right] = 0, \tag{7}$$

or, for the eigenmode $f_1 = \varphi e^{-i\omega t}$,

$$-\frac{i\omega\varphi}{\partial f_0/\partial E} + \hat{D}\left[\frac{\varphi}{\partial f_0/\partial E} + G \iint \frac{\varphi(\mathbf{r}', \mathbf{v}') \, d\mathbf{r}' \, d\mathbf{v}'}{|\mathbf{r} - \mathbf{r}'|}\right] = 0. \tag{7'}$$

It is somewhat surprising that the equation of motion (7) thus derived is the first-order equation. In fact, from physical considerations it is clear that the perturbed cluster can either oscillate or contract, while all motions of this kind are described by second-order differential equations (of hyperbolic type). In reality, equations of just this type are implicitly contained in (7); it can be inferred in the following way [4]. We split the function $f(t, \mathbf{r}, \mathbf{v})$ into the even and odd parts: $f = f_+ + f_-$, where

$$f_+ = \tfrac{1}{2}[f(+\mathbf{v}) + f(-\mathbf{v})], \qquad f_- = \tfrac{1}{2}[f(+\mathbf{v}) - f(-\mathbf{v})].$$

The even part f_+ does not change under reflections in the velocity space, while the odd part in such a reflection changes its sign. The even part f_+ defines the perturbed density of particles, "tension" in clusters of particles, and the rest of the even momenta in phase space. The odd part f_- defines the flux of particles, the energy flux, and like odd momenta of f. The operator $\hat{D}$ is odd; it changes the parity of the function, on which it acts. Bearing in mind the above, we split Eq. (7') into the even and odd parts in the following way:

$$-i\omega\varphi_+ + \hat{D}\varphi_- = 0, \tag{8}$$

$$-\frac{i\omega\varphi_-}{\partial f_0/\partial E} + \hat{D}\left[\frac{\varphi_+}{\partial f_0/\partial E} + G \iint \frac{\varphi_+(\mathbf{r}', \mathbf{v}')}{|\mathbf{r} - \mathbf{r}'|} \, d\mathbf{r}' \, d\mathbf{v}'\right] = 0. \tag{9}$$

Equation (8) expresses φ_+ through φ_-. By eliminating φ_+ from the system of (8), (9), we shall get the second-order equation for one odd part of the perturbed distribution function φ_-:

$$\frac{\omega^2\varphi_-}{(-f_0')} = \hat{K}\varphi_-, \qquad \frac{1}{(-f_0')}\frac{\partial^2 f_-}{\partial t^2} = -\hat{K}f_-, \tag{10}$$

where the operator $\hat{K}$ acts as

$$\hat{K}\varphi_- = \frac{\hat{D}^2\varphi_-}{f_0'} + G\hat{D}\iint\frac{\hat{D}\varphi_-}{|\mathbf{r} - \mathbf{r}'|}\,d\mathbf{r}'\,d\mathbf{v}'. \tag{11}$$

One can easily make sure that this operator is a self-adjoint one on a class of continuous functions, which vanish outside some restricted phase volume:

$$\iint g\hat{K}h\,d\mathbf{r}\,d\mathbf{v} = \iint h\hat{K}g\,d\mathbf{r}\,d\mathbf{v}. \tag{12}$$

Equation (11) is similar to the ordinary equation of oscillation, where $\hat{K}$ plays the role of the "elasticity coefficient" and $1/(-f_0')$ is the "mass."

The results obtained so far are valid for any isotropic distribution function. Let us now make an assumption important for further calculations that the $f_0(E)$ function decreases monotonically with energy: $f_0' \leq 0$ everywhere. With such a restriction it is possible just to apply the property of self-adjointness of (12) and prove that the squares of all eigenfrequencies ω^2 are real. It is performed in a usual manner. We turn, in (10), to complex-conjugate quantities:

$$\frac{-\overline{\omega^2}\,\overline{\varphi}_-}{\partial f_0/\partial E} = \hat{K}\overline{\varphi}_-, \tag{13}$$

Multiplying (10) by $\overline{\varphi}_-$, and (13) by φ_+, subtracting and integrating over the total phase space, we get

$$(\omega^2 - \overline{\omega^2})\iint\frac{|\varphi_-|^2}{\partial f_0/\partial E}\,d\mathbf{r}\,d\mathbf{v} = 0.$$

Since, according to the assumption, f_0' is negative everywhere, so $\omega^2 = \overline{\omega^2}$. Therefore, ω can be real or purely imaginary but cannot be complex. By multiplying (10) by $\overline{\varphi}_-$ and integrating over phase space of the system, we find the following expression for ω^2:

$$\omega^2 = \frac{\iint \overline{\varphi}_- \hat{K}\varphi_-\,d\mathbf{r}\,d\mathbf{v}}{\iint |\varphi_-|^2/(-f_0')\,d\mathbf{r}\,d\mathbf{v}}. \tag{14}$$

From (14) follows the general criterion of stability, which can be formulated in the following way.

The necessary and sufficient condition for the stability of any collisionless configuration characterized by the isotropic distribution function $f_0 = f_0(E)$, $f_0'(E) \leq 0$ is the positive definiteness of the operator $\hat{K}$:

$$K(\psi, \psi) \equiv \iint \psi\hat{K}\psi\,d\mathbf{r}\,d\mathbf{v} > 0 \tag{15}$$

for all ψ. More precisely, it should be noted that the condition in (15) guarantees the lack of solutions growing faster than t. For example, there is an obvious growing solution corresponding to the uniform motion of the cluster as a whole. Actually, let us turn from a fixed system of reference to one moving along the x-axis at a speed ε with respect to the former. Then the new distribution function is

$$f(x, y, z, v_x, v_y, v_z, t) = f_0(x + \varepsilon t, y, z, v_x + \varepsilon, v_y, v_z)$$

$$= f_0 + \varepsilon\left(t\frac{\partial f_0}{\partial x} + \frac{\partial f_0}{\partial v_x}\right) + \cdots, \qquad (16)$$

i.e., the perturbation is proportional to

$$t\frac{\partial f_0}{\partial x} + \frac{\partial f_0}{\partial v_x}, \qquad (17)$$

and grows in a powerlike manner ($\sim t$). Therefore, using the stability criterion formulated, one should not include in the number of trial functions $\partial f_0/\partial x$, $\partial f_0/\partial y$, $\partial f_0/\partial z$ and their linear combinations. The condition $K(\psi, \psi) \geq 0$ also does not ensure the lack of other solutions growing in a similar manner. We therefore shall not consider the question of the linearly growing solutions for (10). In talking about the stability below, we shall, as a rule, bear in mind the nonnegativity of $K(\psi, \psi)$. The question concerning the solutions corresponding to $\omega = 0$ (but only radial), which may increase in a powerlike manner, are considered in [239]. Formula (15) may be reduced to the form

$$K(\psi, \psi) \equiv \Gamma(\beta, \beta) = -\iint \frac{\beta^2}{f_0'} d\mathbf{r}\, d\mathbf{v}$$

$$-G \iiiint \frac{\beta(\mathbf{r}, \mathbf{v})\beta(\mathbf{r}', \mathbf{v}')}{|\mathbf{r} - \mathbf{r}'|} d\mathbf{r}\, d\mathbf{r}'\, d\mathbf{v}\, d\mathbf{v}', \qquad (18)$$

where the notation $\beta \equiv \hat{D}\psi$ is introduced and ψ is assumed to be a continuous function equal to zero on the phase volume boundary and outside it, but not equal to zero identically.

The stability criterion of (15) refers to arbitrary perturbations and is valid for stellar systems with arbitrary geometry, not necessarily spherically symmetrical, it is only required that the equilibrium distribution function be isotropic, $f_0 = f_0(E)$, and decreasing, $df_0/dE < 0$.

2.2 Sufficient Condition of Stability

Example: stability of polytropic models with $f_0'(E) \leq 0$. The above-formulated general criterion of stability was applied by Antonov [6] for the investigation of the stability of spherically symmetrical systems. Instead of the general criterion in (18) which is the necessary and sufficient condition for stability, it proved to be convenient to use a simpler sufficient criterion of stability. By utilizing the Cauchi–Bunyakovsky inequality

$$\left(\int f_1 f_2\, d\mathbf{v}\right)^2 \leq \left(\int f_1^2\, d\mathbf{v}\right)\left(\int f_2^2\, d\mathbf{v}\right), \qquad (19)$$

with

$$f_1 = \sqrt{|f_0'|}, \qquad f_2 = \frac{\beta}{\sqrt{|f_0'|}},$$

we shall have

$$\int \frac{\beta^2}{|f_0'|} \, d\mathbf{v} \geq \frac{(\int \beta \, d\mathbf{v})^2}{\int |f_0'| \, d\mathbf{v}}. \tag{20}$$

Further, from the equation

$$\rho_0 = \int f_0(\tfrac{1}{2}v^2 + \Phi_0) \, d\mathbf{v}, \tag{21}$$

it follows that [at $f_0(E_{\max}) \neq \infty$]

$$\frac{d\rho_0}{d\Phi_0} = \int f_0'\left(\frac{v^2}{2} + \Phi_0\right) d\mathbf{v} = -\int \left|\frac{df_0}{dE}\right| d\mathbf{v}. \tag{22}$$

Thus, taking into account (22) and (20), we can state that stability will be proved if we prove the positive definiteness of the expressions

$$Q(\alpha, \alpha) \equiv -\int \frac{\alpha^2}{(d\rho_0/d\Phi_0)} \, d\mathbf{r} - G\iint \frac{\alpha(\mathbf{r})\alpha(\mathbf{r}')}{|\mathbf{r} - \mathbf{r}'|} \, d\mathbf{r} \, d\mathbf{r}' \geq 0, \tag{23}$$

where $\alpha(\mathbf{r}) = \int \beta \, d\mathbf{v}$. Antonov [6] investigated the function $\alpha(\mathbf{r})$, meeting the requirement

$$\int \alpha(\mathbf{r}) \, d\mathbf{r} = 0 \qquad \text{or} \qquad \iint \beta \, d\mathbf{r} \, d\mathbf{v} = 0, \tag{24}$$

which is satisfied by a broader class of functions than that for the previous condition:

$$\beta = \hat{D}\psi, \qquad \psi(\mathbf{r}, -\mathbf{v}) = -\psi(\mathbf{r}, \mathbf{v}). \tag{25}$$

For functions of the type (25), equalities (24) are, of course, also fulfilled.

Prove, by using (23), the stability of polytropic models in (5) with $\alpha < \tfrac{3}{2}$ ($n < 3$). Let us consider the radial perturbations. Note that the double integral in (23) can be interpreted as the doubled energy of the "charge" field distributed with density $\alpha(r)$. It may be represented in another form more convenient to us [6]:

$$2 \cdot \frac{1}{8\pi} \int |\nabla\psi_\alpha|^2 \, d\mathbf{r} = \int_0^\infty \left|\frac{d\psi_\alpha}{dr}\right|^2 r^2 \, dr, \tag{26}$$

where ψ_α is the potential (respectively $-\nabla\psi_\alpha \equiv \mathbf{E}_\alpha$ is the strength) of the field corresponding to the charge $\alpha(r)$:

$$\frac{d\psi_\alpha}{dr} = \frac{4\pi}{r^2} \tau, \qquad \tau \equiv \int_0^r r^2\alpha \, dr. \tag{27}$$

The first integral in (23) may also be rewritten in terms of τ for, according to (27),

$$\alpha = \frac{1}{r^2} \frac{d\tau}{dr}.$$

(28)

Therefore, we finally obtain instead of (23) the following expression:

$$4\pi \int \frac{(d\tau/dr)^2 \, dr}{r^2(-d\rho_0/d\Phi_0)} - 16\pi^2 G \int \frac{\tau^2 \, dr}{r^2}.$$

(29)

The boundary conditions follow from (24) and (27): τ must vanish at the boundary (and outside it) and at the center must have the third-order zero.

As a result, we have the variational problem involving the proof for the fact that the minimum of the ratio of the first and second terms of (29) is more than unity:

$$\min \frac{\displaystyle\int \frac{(d\tau/dr)^2 \, dr}{r^2(-d\rho_0/d\Phi_0)}}{4\pi G \int \tau^2 \, dr/r^2} \equiv \lambda > 1.$$

(30)

The solution is yielded by the differential equation

$$\frac{d}{dr}\left(\frac{d\tau/dr}{r^2(-d\rho_0/d\Phi_0)}\right) + 4\pi G\lambda \frac{\tau_0}{r^2} = 0,$$

(31)

where λ is the first eigenvalue of (31) at $\tau(0) = \tau(a) = 0$ (a is the radius).

We take the comparison function $\tau_0(r)$ satisfying the boundary condition $\tau_0(0) = 0$ and the differential equation

$$\frac{d}{dr}\left(\frac{d\tau_0/dr}{r^2(-d\rho_0/d\Phi_0)}\right) + 4\pi G \frac{\tau_0}{r^2} = 0.$$

(32)

It turns out that for ρ_0 and Φ_0, corresponding to the Emden equation, the $\tau_0(r)$ function may be found in the explicit form

$$\tau_0 = -(n - 3)r^2 \frac{d\Phi_0}{dr} + 4\pi G(n - 1)r^3 \rho_0.$$

(33)

Now, multiplying (32) by τ and (31) by τ_0, let us subtract and integrate from 0 to a:

$$\int_0^a \left\{\tau_0 \frac{d}{dr}\left[\frac{d\tau}{dr}\bigg/ r^2\left(-\frac{d\rho_0}{d\Phi_0}\right)\right] - \tau \frac{d}{dr}\left(\frac{d\tau_0/dr}{r^2(-d\rho_0/d\Phi_0)}\right)\right\} dr$$
$$+ 4\pi G(\lambda - 1) \int_0^a \frac{\tau\tau_0}{r^2} \, dr = 0.$$

(34)

We integrate (34) by parts, taking into account that τ_0 at $r = 0$ has the third-order zero:

$$\left[\frac{\tau_0 \, d\tau/dr}{r^2(-d\rho_0/d\Phi_0)}\right]_{r=a} + 4\pi G(\lambda - 1) \int_0^a \frac{\tau\tau_0}{r^2} \, dr = 0.$$

(35)

The $\tau(r)$ function within the interval $(0, a)$ retains the one and same sign as the first (nodeless) eigenfunction of equation (31). For $\tau > 0$, there must then be $d\tau/dr\,|_{r=a} < 0$. The function $\tau_0 > 0$ at $n < 3$, as follows from (33). Finally, the derivative $(-d\rho_0/d\Phi_0)$ is positive. Taking into account all the above, from (35) we get that at $n < 3$ ($\alpha < \frac{3}{2}$) the value $\lambda > 1$, i.e., the system is stable.

2.3 Other Theorems about Stability. Stability with Respect to Nonradial Perturbations

Using the sufficient criterion of stability in (23)–(24)[8] Antonov, by means of a set of artificial mathematical techniques, was able to prove some theorems concerning the stability of clusters of particles with $f'_0(E) \leq 0$. First, he obtained the following sufficient condition of stability (at $f'_0 < 0$):

$$\frac{d^3\rho_0}{d\Phi_0^3} < 0 \tag{36}$$

at all r. This condition, in particular, is satisfied, according to (6), by the bounded ($\alpha < \frac{7}{2}$) polytropic models of (5) with $\alpha > \frac{1}{2}$ as well as the isochronic Henon model [214], for which (in corresponding units)

$$\rho_0 = \frac{(1 - \Phi_0)^4(3 + \Phi_0)}{(1 + \Phi_0)^3} \qquad (0 < |\Phi_0| < 1).$$

Secondly, as we have seen, particularly for polytropic models of (5), §1, their stability at $\alpha < \frac{3}{2}$ is proved. Thus, all polytropic bounded models with the decreasing distribution function prove to be stable, i.e., at $0 < \alpha < \frac{7}{2}$. We shall not go into details of these proofs here.

Consider only another important theorem proved by Antonov [6]. It turns out that for stability of isotropic spherical systems with $f'_0 < 0$, generally only radial perturbations may be dangerous, i.e., nonradial oscillations are always stable. We shall prove this theorem. Expression (23) can be rewritten in a somewhat different manner [6, 145].

$$(\alpha, \hat{Q}\alpha) \equiv Q(\alpha, \alpha) = \int \left[\frac{\alpha(\mathbf{r})}{d\rho_0/d\Phi_0} + 4\pi G\Delta^{-1}\alpha(\mathbf{r}) \right] \alpha(\mathbf{r})\, d\mathbf{r}, \tag{37}$$

where Δ^{-1} is the operator inverse to the Laplace operator. Thus, for stability it is sufficient that $Q(\alpha, \alpha)$ be nonnegative, i.e., that the integral operator

$$\hat{Q} \equiv \frac{1}{d\rho_0/d\Phi_0} + 4\pi G\Delta^{-1} \tag{38}$$

be nonnegative: $\hat{Q} \geq 0$.

---

[8] In [145], a similar criterion is obtained as a condition of positiveness of the form $\Gamma(\beta, \beta)$ for arbitrary β. Here $Q(\alpha, \alpha)$ appears to be the solution of the variational problem for the minimum $\Gamma(\beta, \beta)$ at a fixed $\alpha(\mathbf{r}) = \int \beta(\mathbf{r}, \mathbf{v})\, d\mathbf{v}$. However, restrictions in (24) and (25) are essential.

Further, we shall follow the work of Khazin and Shnol' [145], which gives a somewhat simpler proof in comparison to the original work of Antonov [6]. Thus, assume that in the expansion $\alpha(\mathbf{r})$ over spherical harmonics there is no zero term:

$$\alpha(\mathbf{r}) = \sum_{\substack{l=1 \\ |m| < l}}^{\infty} \alpha_{lm}(r) Y_{lm}(\theta, \varphi). \tag{39}$$

It is to be proved that then $\Gamma(\beta, \beta) > 0$. For proof, we substitute (39) into expression (37) for $Q(\alpha, \alpha)$:

$$Q(\alpha, \alpha) = \sum_{\substack{l=1 \\ |m| < l}}^{\infty} Q_l(\alpha_{lm}, \alpha_{lm}), \tag{40}$$

where

$$Q_l(\alpha, \alpha) = \int \left(\frac{\alpha(r)}{a_0(r)} + 4\pi G \Delta_l^{-1} \alpha \right) \alpha(r) \, d\mathbf{r}, \qquad a_0 = \frac{d\rho_0}{d\Phi_0}, \tag{41}$$

$$\Delta_l \equiv \frac{1}{r^2} \frac{d}{dr} \left(r^2 \frac{d}{dr} \right) - \frac{l(l+1)}{r^2}. \tag{42}$$

The expression for $Q(\alpha, \alpha)$ is split, as was to be expected, into a sum over separate spherical harmonics (see §2, Appendix). At $l > 1$, the obvious inequality $\Delta_l \leq \Delta_1$ takes place. But for any pair[9] of the self-adjoint operators $\hat{P}$ and $\hat{Q}$ it follows, from $\hat{P} > \hat{Q} > 0$, that $\hat{P}^{-1} < \hat{Q}^{-1}$. Therefore, $\Delta_l^{-1} > \Delta_1^{-1}$, so that $Q_l(\alpha, \alpha) > Q_1(\alpha, \alpha)$. Thus, one has to prove the nonnegativity of the operator

$$\frac{1}{a_0(r)} + 4\pi G \Delta_1^{-1} \geq 0. \tag{43}$$

Inequality (43) is equivalent to the following:

$$\hat{P}_1 \equiv -\Delta_1 - 4\pi G a_0(r) \geq 0. \tag{44}$$

Actually, we rewrite (43) as

$$\hat{P} \equiv \frac{1}{a_0} \geq -4\pi G \Delta_1^{-1} \equiv \hat{Q},$$

where $\hat{Q} > 0$ since $(\alpha, \hat{Q}\alpha)$ is proportional to the minus gravitational energy for a mass distributed with a density $\alpha(\mathbf{r})$. Therefore, $\hat{P}^{-1} < \hat{Q}^{-1}$, which is identical to (44). It can easily be verified that the function

$$\alpha_0(r) = \frac{1}{r^2} \int_0^r \rho_0(r) r^2 \, dr \tag{45}$$

---

[9] This is seen, for instance, from the following equalities [145]:

$$\hat{P} - \hat{Q} = \hat{P}^{1/2}(1 - \hat{C})\hat{P}^{1/2}, \qquad \hat{P}^{-1} - \hat{Q}^{-1} = \hat{P}^{-1/2}(1 - \hat{C}^{-1})\hat{P}^{-1/2},$$
$$\hat{C} = \hat{P}^{-1/2}\hat{Q}\hat{P}^{-1/2}.$$

satisfies the equation $\hat{P}_1 \alpha = 0$. Since $\alpha_0(r)$ has no nodes at $r > 0$, then due to familiar oscillative theorems [41, 55] the spectrum λ of the operator $\hat{P}_1$ is nonnegative. Then $\lambda = 0$ is its eigenvalue [accordingly $\alpha_0(r)$ is the eigenfunction], corresponding to the "lowest level", $\hat{P}_1 \geq 0$. The theorem is proved.

2.4 Variational Principle for Radial Perturbations

In works following [4, 6], i.e., [239, 288], a great complexity of the variational principle of (15) is emphasized, and attempts are made to somewhat simplify the stability analysis. First of all, quite natural simplification is achieved, if one restricts oneself to the consideration of only radial oscillations. With due regard for the theorem just proved, such a restriction must not affect the ultimate inferrences of the stability. In this case one can obtain equations of motion, and the variational principle to be simpler as compared to (7), (15) [239].

The linearized kinetic equation will be written in the form

$$\frac{\partial f_1}{\partial t} + \hat{D}f_1 - \frac{\partial f_0}{\partial E} v_r \frac{\partial \Phi_1}{\partial r} = 0. \tag{46}$$

Disturbances are assumed to be radial; however, on the other hand, the distribution function is not necessarily to be considered as isotropic, $f_0 = f_0(E, L^2)$. Equation (46) must be, as usual, supplemented by the equation for Φ_1, expressed through f_1. Let us write it in the following form:

$$\frac{\partial^2 \Phi_1}{\partial t \, \partial r} = \frac{G}{r^2} \frac{\partial M(r)}{\partial t} = \frac{G}{r^2}(-4\pi r^2)j(r) = -4\pi G \int v_r f_1 \, d\mathbf{v}, \tag{47}$$

where $M(r)$ is the mass inside the radius r and $j(r)$ is the mass flow in the radial direction; (46) and (47) represent the initial equations of motion.

We divide f_1 again into even and odd parts, $f_1 = f_+ + f_-$, and obtain, instead of (46) and (47), the following system of equations:

$$\frac{\partial f_+}{\partial t} + \hat{D}f_- = 0, \tag{48}$$

$$\frac{\partial f_-}{\partial t} + \hat{D}f_+ - \frac{\partial f_0}{\partial E} v_r \frac{\partial \Phi_1}{\partial r} = 0, \tag{49}$$

$$\frac{\partial^2 \Phi_1}{\partial t \, \partial r} = -4\pi G \int v_r f_- \, d\mathbf{v}. \tag{50}$$

Eliminating f_+, we have the equation for f_-

$$\frac{1}{f_0'} \frac{\partial^2 f_-}{\partial t^2} = \hat{T}f_-, \tag{51}$$

where the operator $\hat{T}$ is

$$\hat{T}f_- = \frac{\hat{D}^2 f_-}{f_0'} - 4\pi G v_r \int v_r f_- \, dv. \tag{52}$$

Self-adjointness of this operator is proved elementarily. Equation (51) can be inferred also from the variational principle

$$\delta \int \left[\frac{(\partial f_-/\partial t)^2}{-df_0/dE} - f_- \hat{T}f_- \right] dv \, d\mathbf{r} \, dt. \tag{53}$$

The expression for the operator $\hat{T}$ really has in this case a simpler structure than the operator on the right-hand side of (18). Accordingly, the form of the stability criterion is simpler: for $df_0/dE < 0$ the system is stable with respect to radial perturbations then and only then, where the operator $\hat{T}$ is positively defined for all functions $f(\mathbf{r}, \mathbf{v}) = \varphi(\mathbf{r}, \mathbf{v})$ dependent on $|\mathbf{r}|$, $\mathbf{v}$, and restricted in phase space. From (51), it is possible to prove conservation of the value similar to the pulsation energy [288, 239]:

$$H = \int \left[\left(\frac{\partial f_-}{\partial t} \right)^2 \Big/ |f_0'| + f_- \hat{T}f_- \right] dv \, d\mathbf{r} = \text{const}, \tag{54}$$

or in terms of the complete f,

$$H = \int \left[\left(\frac{\partial f}{\partial t} \right)^2 \Big/ |f_0'| \right] d\mathbf{r} \, dv - 4\pi G \int \left(\int v_r f \, dv \right)^2 d\mathbf{r}. \tag{54'}$$

In conclusion, note that the results, similar to those obtained here and in the previous section, are valid not only for $f_0 = f_0(E)$ (as in §2) and not only for radial perturbations of spherical systems (as here) but also for perturbations of any cluster which do not violate the spatial symmetry of the equilibrium configuration.[10]

2.5 Hydrodynamical Analogy

Even such simplified variational principles as (53) are nonetheless more difficult to apply than, for instance, the relevant results in the theory of gaseous spheres, where these variational principles involve functions of only one coordinate r. This defines the importance of the so-called hydrodynamical analogy actually established by Antonov [6], but explicitly formulated by Lynden-Bell [284, 288].

In the previous section it was already noticed that the search for stationary states of stellar systems in some cases is equivalent to solving the corresponding problem in the equilibrium theory of gaseous spheres. It turns

---

[10] For instance, for all axial-symmetrical perturbations of rotating axial-symmetrical systems.

out that the problem of the stability of collisionless system models with isotropic distribution functions can also be reduced to the problem of the stability of a certain hydrodynamical system. The possibility of such a transition from the six-dimensional phase space to the ordinary three-dimensional one, i.e., to the solution of a mathematically far more straight-forward problem just constitutes the composition of the hydrodynamical analogy.[11] The exact formulation of the theorem [288, 239] for spherically symmetrical[12] clusters is the following. Consider the bounded spherically symmetrical stellar system with the isotropic distribution function $f_0(E)$ and $df_0/dE < 0$. It is stable with respect to radial perturbations if the gaseous sphere with the same density distribution $\rho_0(r)$ in the radius,

$$\rho_0(r) = 4\pi\sqrt{2} \int_{E=\Phi_0}^{\infty} f_0(E)(E - \Phi_0)^{1/2} \, dE \qquad (55)$$

and with the pressure

$$P_0(r) = \frac{8\pi\sqrt{2}}{3} \int_{E=\Phi_0}^{\infty} f_0(E)(E - \Phi_0)^{3/2} \, dE, \qquad (56)$$

is stable with respect to radial perturbations, for which the "adiabatic index" is

$$\gamma = \frac{\rho_0}{P_0} \frac{dP_0/dr}{d\rho_0/dr}. \qquad (57)$$

To prove this theorem, first of all note that according to Lynden-Bell [284] the necessary and sufficient condition for stability of a gaseous sphere (and, consequently, the sufficient condition for stability of a stellar cluster) is the positive definiteness of the operator

$$\hat{P}_1 = -\Delta - 4\pi G\rho \frac{d\rho}{dP} \qquad (58)$$

in space of all functions of r, vanishing on infinity $r = \infty$:

$$\int \psi \hat{P}_1 \psi \, dr > 0 \qquad (59)$$

for all $\psi(r)$.

But this condition is equivalent to (38), which is proved similarly to the proof of the inequality $\Delta_l < \Delta_1$ in Section 2.3.

---

[11] The possibility of such a change was also suggested earlier [101]. However, then it was merely a hypothesis which was supported by some other statistical considerations. It proves, however, that it can be strictly substantiated.

[12] In reality, this theorem is given by Lynden-Bell [284] in a somewhat more general context. There are also some equivalent formulations [145, 239]; we give the formulation from [239].

The authors in [239, 284, 288] somewhat generalize the results inferred earlier in [6] in that part which refers to the stability of polytropic models: they admit arbitrary barotropic equations of state $P = P(\rho)$.

Such theorems all establish only sufficient conditions of stellar cluster stability. Inverse statements are generally not true, so that stellar clusters are "more stable" in comparison with gaseous ones. Thus, in [288], there are several examples of stable stellar clusters, to which correspond unstable gaseous spheres. Such a situation is, in particular, due to the fact that perturbation created in the collisionless system does not have to maintain itself (e.g., travelling like a wave): it may vanish if the particles "have run away" from the region of perturbation.

2.6 On the Stability of Systems with Distribution Functions That Do Not Satisfy the Condition $f'_0(E) \leq 0$

Until now we have dealt with only distribution functions satisfying the condition $f'_0 \leq 0$. Seemingly, they all are stable. Investigation into system stability, in which $f'_0 > 0$ in some region of phase space,[13] is more complex. The condition $f'_0 < 0$ does not define the boundary between the stable and unstable solutions. For example, stable one-dimensional systems are known, in which this condition is violated. Perpendicular oscillations of the homogeneous flat layer (see Chapter I, §5) may serve as an example.

Henon [217] investigated the stability of polytropic models in (5), §1, with $\alpha < 0$, for which $f'_0 > 0$, however, only with respect to radial perturbations. Unlike previous papers, the study in [217] is performed not by the analytical, but by numerical methods. The system was represented by 1000 spherical layers each being thought as a *set* of stars having at a given time the same distances r from the center, the radial v_r, and transversal $v_\perp$ velocities. The distribution of the number of layers with such values $(r, v_r, v_\perp)$ was specified in accordance with $f_0(E)$. Due to discreteness and random selection of the location and velocities, the "numerical" $f_0(E)$ fluctuated near the average theoretical value, and these fluctuations were to "trigger" any instability, if any.

In stable cases (for instance $n > 1$), fluctuations are purely statistical. In the case $n = \frac{1}{2}(\alpha = -1)$, collective oscillations of the spatial structure, which were lacking at $n > 1$, as well as "bulging" out of the boundary of the phase (r, v_r) region, which has a trend toward slow growth with time, are noticeable. Therefore, the case $n = \frac{1}{2}$ appears to be slightly unstable.

Henon estimates the boundary between the stable and unstable configurations to be $n = 0.5$–0.6. However, it is possible that the critical value is exactly $n = \frac{1}{2}$, so that all polytropic models with $f_0 = f_0(E)$ are stable.

---

[13] For real systems with confined phase volume, f'_0 evidently cannot be positive everywhere.

§ 3 Stability of Systems of Gravitating Particles Moving on Circular Trajectories

3.1 Stability of a Uniform Sphere [87]

Consider the simplest example of a system of particles with homogeneous density and show its stability with respect to arbitrary small perturbations.

Subsection 3.1.1 deals with the method of the solution for the kinetic equation similar to the familiar method of integration over trajectories in plasma physics [138]. Subsection 3.1.2 gives the spectrum of eigenfrequencies of the system that is a discrete set of real numbers. Subsection 3.1.3 presents the finding of the spectrum of surface oscillations which also turns out to be stable.

3.1.1. Derivation of the Equation of Eigenoscillations. Let us employ the spherical coordinates r, θ, φ and characterize the velocity through the values v_r, $v_\perp = (v_\theta^2 + v_\varphi^2)^{1/2}$, $\alpha = \arctan(v_\varphi/v_\theta)$. In these variables, the kinetic equation is written in the form

$$\hat{L}f + v_r\left(\frac{\partial f}{\partial r} - \frac{v_\perp}{r}\frac{\partial f}{\partial v_\perp}\right) + \left(\frac{v_\perp^2}{r} - \frac{\partial \Phi}{\partial r}\right)\frac{\partial f}{\partial v_r} - \nabla_\perp\Phi\frac{\partial f}{\partial v_\perp} = 0, \qquad (1)$$

where

$$L = \frac{\partial}{\partial t} + \frac{v_\perp}{r}\left[\cos\alpha\frac{\partial}{\partial\theta} + \frac{\sin\alpha}{\sin\theta}\frac{\partial}{\partial\varphi} - \sin\alpha\cot\theta\frac{\partial}{\partial\alpha}\right], \qquad (2)$$

$$\nabla_\perp\Phi\frac{\partial}{\partial\mathbf{v}_\perp} = \frac{1}{r}\left(\cos\alpha\frac{\partial\Phi}{\partial\theta} + \frac{\sin\alpha}{\sin\theta}\frac{\partial\Phi}{\partial\varphi}\right)\frac{\partial}{\partial v_\perp} - \frac{1}{rv_\perp}\left(\sin\alpha\frac{\partial\Phi}{\partial\theta} - \frac{\cos\alpha}{\sin\theta}\frac{\partial\Phi}{\partial\varphi}\right)\frac{\partial}{\partial\alpha}. \qquad (3)$$

The equilibrium distribution function is defined by the expression

$$f_0 = \frac{\rho_0}{2\pi v_0}\delta(v_r)\delta(v_\perp - v_0). \qquad (4)$$

The unperturbed potential is

$$\Phi_0 = \frac{\Omega^2 r^2}{2} + \text{const}, \qquad (5)$$

where

$$\Omega^2 = \frac{4\pi G\rho_0}{3}. \qquad (6)$$

ρ_0 is the constant density, and $v_0 = \Omega r$.

Linearize Eqs. (1)–(3) by denoting the equilibrium values by the index 0 and the perturbed values by the index 1.

In the linear approximation it follows from (1) that

$$\hat{L}f_1 + v_r\left(\frac{\partial f_1}{\partial r} - \frac{v_\perp}{r}\frac{\partial f_1}{\partial v_\perp}\right) + \left(\frac{v_\perp^2}{r} - \frac{\partial \Phi_0}{\partial r}\right)\frac{\partial f_1}{\partial r}$$

$$= \frac{\partial \Phi_1}{\partial r}\frac{\partial f_0}{\partial v_r} + \frac{1}{r}\left(\cos\alpha\,\frac{\partial \Phi_1}{\partial \theta} + \frac{\sin\alpha}{\sin\theta}\frac{\partial \Phi_1}{\partial \varphi}\right)\frac{\partial f_0}{\partial v_\perp}. \quad (7)$$

Taking into account that $f_0 \sim \delta(v_r)\cdot\delta(v_\perp - v_0)$, we find that (7) is satisfied at f_1 of the form[14]

$$f_1 = \delta(v_r)[A\delta(v_\perp - v_0) + B\delta'(v_\perp - v_0)] - C\delta'(v_r)\delta(v_\perp - v_0), \quad (8)$$

where the prime corresponds to the derivative over the argument. From (7) we find that the A, B, and C functions satisfy the equations

$$\hat{L}^0 A - \frac{1}{r\Omega}\left(\hat{L}^0 - \frac{\partial}{\partial t}\right)B + \frac{1}{r}\frac{\partial}{\partial r}(rC) = 0, \quad (9)$$

$$\hat{L}^0 B - 2\Omega C = \frac{\rho_0}{2\pi\Omega^2 r^2}\left(\hat{L}^0 - \frac{\partial}{\partial t}\right)\Phi_1, \quad (10)$$

$$\hat{L}^0 C + 2\Omega B = -\frac{\rho_0}{2\pi\Omega r}\frac{\partial\Phi_1}{\partial r}. \quad (11)$$

The operator $\hat{L}^0$ differs from the operator $\hat{L}$ in the change of $v_\perp/r$ for Ω. At f_1 of the form of (8) the density perturbation is

$$\rho_1 = \int_0^{2\pi}(r\Omega A - B)\,d\alpha. \quad (12)$$

The way to solve Eqs. (9)–(11) and seek ρ_1 is the following. Multiplying both sides of Eq. (10) by the operator $\hat{L}^0$ and expressing, with the help of (11), the value $\hat{L}^0 C$ by B and Φ_1, we obtain

$$(\hat{L}^0 + 2i\Omega)(\hat{L}^0 - 2i\Omega)B = \frac{\rho_0}{2\pi\Omega^2 r^2}\left[\left(\hat{L}^0 - \frac{\partial}{\partial t}\right)\hat{L}^0\Phi_1 + 2\Omega^2 r\frac{\partial\Phi_1}{\partial r}\right]. \quad (13)$$

Hence

$$B = \frac{\rho_0}{2\pi\Omega^2 r^2}(\hat{L}^0 - 2i\Omega)^{-1}(\hat{L}^0 + 2i\Omega)^{-1}\left[\left(\hat{L}^0 - \frac{\partial}{\partial t}\right)\hat{L}^0\Phi_1 + 2\Omega^2 r\frac{\partial\Phi_1}{\partial r}\right],$$

$$(14)$$

where the power (-1) denotes the inverse operator, whose action will be explained further.

We express the A function using Eqs. (9) and (11) through B and Φ_1:

$$A = \frac{(\hat{L}^0)^{-1}}{r\Omega}\left[\left(\hat{L}^0 - \frac{\partial}{\partial t}\right)B - \frac{1}{2}\frac{\partial}{\partial r}(r\hat{L}^0 B) + \frac{\rho_0}{4\pi\Omega^2}\left(\hat{L}^0 - \frac{\partial}{\partial t}\right)\frac{\partial}{\partial r}\left(\frac{\Phi_1}{r}\right)\right].$$

$$(15)$$

---

[14] The method of expansion over δ functions and their derivatives used below is a version of the familiar method of momenta convenient for calculations.

Substitute this result into (12):

$$\rho_1 = - \int_0^{2\pi} \left\{ \frac{\partial}{\partial t} (\hat{L}^0)^{-1} \left[B + \frac{\rho_0}{2\pi\Omega^2} \frac{\partial}{\partial r} \left(\frac{\Phi_1}{r} \right) \right] + \frac{1}{r} \frac{\partial}{\partial r} \left(rB - \frac{\rho_0 \Phi_1}{2\pi r \Omega^2} \right) \right\} d\alpha.$$

(16)

Equation (16) along with (14) yields the required coupling of ρ_1 with Φ_1 necessary for the self-consistency of perturbations by the equation

$$\Delta\Phi_1 = 4\pi G \rho_1(\Phi_1).$$

(17)

Let us now give the form of the operator $(\hat{L}^0)^{-1}$. Let the function $X = X(t, \theta, \varphi, \alpha)$ satisfy the equation

$$\hat{L}^0 X = a(t, \theta, \varphi, \alpha),$$

(18)

where a is some known function. Since $\hat{L}^0$ is the operator of the differentiation of d/dt along the unperturbed (in this case, circular) path of a particle, we may write

$$X = (\hat{L}^0)^{-1} a = \int_{-\infty}^{t} a\{t', \theta_0(t', \theta, \varphi, \alpha), \dots, \dots\} \, dt',$$

(19)

where the dots denote the initial φ_0, α_0 expressed through $t, \theta, \varphi, \alpha$. In a similar way, we find

$$(\hat{L}^0 \pm 2i\Omega)^{-1} a = \int_{-\infty}^{t} e^{\mp 2i\Omega(t - t')} \cdot a(t') \, dt'.$$

(20)

Let us now describe the specific vein of the calculation of integrals of the type (19) and (20). Let the potential perturbation have the form

$$\Phi_1 = \chi_l(r) \Phi_1^l(t, \theta, \varphi),$$

(21)

where

$$\Phi_1^l = e^{-i\omega t} Y_m^l(\theta, \varphi),$$

$$Y_m^l(\theta, \varphi) \equiv e^{-im(\pi/2 - \varphi)} \cdot P_{m0}^l(\cos \theta).$$

(22)

$P_{m0}^l(\cos \theta)$ are the functions accurate to coefficients coinciding with Legendre polynomials (see the book of Vilenkin [33]). All normalizing coefficients are included in $\chi_l(r)$.

When a is equal to the right-hand side of (22), Eqs. (19) and (20) are written as

$$(\hat{L}^0 + iq\Omega)^{-1} a = e^{-i\omega t} \int_0^{\infty} e^{-i(\omega - q\Omega)\tau} Y_m^l[\theta(t - \tau), \varphi(t - \tau)] \, d\tau$$

$$(q = 0, \pm 2); \quad (23)$$

the variable of integration t' is substituted for $\tau = t - t'$. From the equation of motion we find

$$\cos \theta(t) = \cos \gamma \sin \psi(t),$$

$$\tan [\tilde{\varphi}_0 - \varphi(t)] = \cot \psi(t)/\sin \gamma, \tag{24}$$

$$\cot \alpha(t) = -\cot \gamma \cos \psi(t),$$

where

$$\psi(t) = \psi_0 + \Omega(t - t_0). \tag{25}$$

Here the constants $\tilde{\varphi}_0$, ψ_0, γ are introduced, which can be expressed in terms of θ_0, φ_0, α_0 by considering (24) at $t = t_0$.

Taking into account (24) and utilizing the addition theorem, let us represent Y_m^l in the form of a sum of three-index functions [33].

$$Y_m^l[\theta(t - \tau, \tilde{\varphi}_0, \gamma, \psi_0); \varphi(t - \tau, \tilde{\varphi}_0, \gamma, \psi_0)]$$

$$= \sum_{s=-l}^{l} T_{ms}^l\left[\frac{\pi}{2} - \tilde{\varphi}_0, \gamma - \frac{\pi}{2}, \frac{\pi}{2} - \psi(t)\right] P_{s0}^l(0) \int_0^\infty e^{i[\omega - (q+s)\Omega]\tau} \, d\tau. \tag{26}$$

Here the function

$$T_{ms}^l\left[\frac{\pi}{2} - \tilde{\varphi}, \theta, \frac{\pi}{2} - \varphi_2\right] = e^{im\varphi_1 + is\varphi_2 - i\pi(m-s)/2} P_{ms}^l(\cos \theta). \tag{27}$$

Using (27), we transform the right-hand side of (23):

$$(\hat{L}^0 + iq\Omega)^{-1}a = e^{-i\omega t}\sum_{s=-l}^{l} T_{ms}^l\left[\frac{\pi}{2} - \tilde{\varphi}_0, \gamma - \frac{\pi}{2}, \frac{\pi}{2} - \psi(t)\right] P_{s0}^l(0)$$

$$\times \int_0^\infty e^{i[\omega - (q+s)\Omega]\tau} \, d\tau. \tag{28}$$

The integral over t is calculated with due regard for the Landau bypass rule $(\omega \to \omega + i\Delta, \Delta \to +0)$:

$$\int_0^\infty \exp\{i[\omega - (q + s)\Omega]\tau\} \, d\tau = \frac{i}{\omega - (q + s)\Omega}. \tag{29}$$

Thereafter, using the transformation (that is, as in (26), a consequence of the addition theorem)

$$T_{ms}^l\left[\frac{\pi}{2} - \tilde{\varphi}_0, \gamma - \frac{\pi}{2}, \frac{\pi}{2} - \psi(t)\right]$$

$$= \sum_{s'=-l}^{l} T_{ms'}^l\left[\frac{\pi}{2} - \varphi(t), \theta(t), \frac{\pi}{2} - \alpha(t)\right] \cdot e^{-is\pi} P_{s's}^l(0) \tag{30}$$

and relations between θ_0, φ_0, α_0 and θ, φ, α, we switch from the variables $\tilde{\varphi}_0$, γ, ψ_0 to the variables θ, φ, α. This completes the calculation of the B function.

We substitute the result into (16) and, integrating over the angle α, obtain the expression for the perturbed density:

$$\rho_1 = e^{-i\omega t} Y_m^l(\theta, \varphi) \frac{\rho_0}{\Omega^2} \left[\left(\frac{d^2 \chi_l}{dr^2} + 2 \frac{d}{dr} \frac{\chi_l}{r} \right) \sum_{s=-l}^{l} |P_{s0}^l(0)|^2 \right.$$

$$\times \frac{-\Omega^2}{(\omega - s\Omega)^2 - 4\Omega^2} + \left. \frac{\chi_l}{r^2} \sum_{s=-l}^{l} |P_{s0}^l(0)|^2 \frac{2\omega\Omega^2 + s\Omega\omega(s\Omega - \omega)}{(s\Omega - \omega)[(\omega - s\Omega)^2 - 4\Omega^2]} \right],$$

$$(31)$$

where

$$|P_{s0}^l(0)|^2 = \begin{cases} \dfrac{(l+s)!(l-s)!}{\left[\left(\dfrac{l+s}{2} \right)! \left(\dfrac{l-s}{2} \right)! 2^l \right]^2}, & \text{if } (l+s) \text{ is even,} \quad (32) \\[4ex] 0, & \text{if } (l+s) \text{ is odd.} \quad (33) \end{cases}$$

We transform the right-hand side of (31) by expanding the frequency functions ω in simple fractions. In this way, Poisson equation (17) is reduced to the form

$$(1 + a_l)\Delta\chi_l(r) = 0. \tag{34}$$

Here

$$\Delta = \frac{\partial^2}{\partial r^2} + \frac{2}{r} \frac{\partial}{\partial r} - \frac{l(l+1)}{r^2}, \tag{35}$$

$$a_0 = \frac{3\Omega^2}{\omega^2 - 4\Omega^2}, \tag{36}$$

$$a_1 = \frac{(\omega^2 - 3\Omega^2)3\Omega^2}{(\omega^2 - \Omega^2)(\omega^2 - 9\Omega^2)}, \tag{37}$$

$$a_2 = \frac{3\Omega^2(\omega^2 - 7\Omega^2)}{(\omega^2 - 16\Omega^2)(\omega^2 - 4\Omega^2)}, \tag{38}$$

$$a_l = \sum_{s=-(l+2)}^{l+2} \frac{\alpha_s^l}{(\omega/\Omega) - s} \qquad (l = 3, 4, 5, \ldots). \tag{39}$$

Summation in (39) is performed over s provided the number $(l+s)$ is even:

$$\alpha_s^l = \tfrac{3}{4}[|P_{s-2}^l(0)|^2 - |P_{s+2}^l(0)|^2] \qquad (|s| \le l - 2), \tag{40}$$

$$\alpha_{\pm l}^l = \mp\tfrac{3}{4}|P_{l-2}^l(0)|^2, \tag{41}$$

$$\alpha_{\pm (l+2)}^l = \mp\tfrac{3}{4}|P_l^l(0)|^2. \tag{42}$$

Equation (34) is satisfied at an arbitrary radial dependence Φ_1, if

$$1 + a_l = 0. \tag{43}$$

This is the required dispersion equation of eigenoscillations of the homogeneous sphere. The case of surface oscillations $\Delta\chi_l = 0$ is treated below (see subsection 3.1.3).

Equation (43) can be transformed to the form

$$a_l = \sum \alpha_{ls} \frac{3}{(\omega - s)^2 - 4} = -1, \tag{44}$$

where

$$\alpha_{ls} \equiv |P_s^l(0)|^2 \quad \text{and we put } \Omega = 1. \tag{45}$$

Consider some particular cases. Perturbations with $l = 0$ corresponding to radial oscillations of the sphere have a frequency $\omega^2 = 1$. Perturbations with $l = 1$ have frequencies $\omega_1^2 = 0$, $\omega_2^2 = 7$. The case of $l = 1$ corresponds to dipole perturbation. For perturbations of the quadrupole type $(l = 2)$, the squares of eigenfrequencies are $\omega_0^2 = 0$, $\omega_{1,2}^2 = (17 \pm \sqrt{117})/2$, all these frequencies are not negative. It turns out (see next subsection) that this is true also in the general case, at any l.

3.1.2. Proof for the Realness of Eigenfrequencies. It is convenient to represent characteristic equation (44) in the following "multiplicative" form:

$$\frac{3\omega^2(\omega^2 - 4)\cdots[\omega^2 - (l - 3)^2][\omega^2 - (l^2 + l + 1)]}{(\omega^2 - 1)(\omega^2 - 9)\cdots[\omega^2 - (l + 2)^2]} = -1 \quad (l \text{ odd}), \tag{46}$$

$$\frac{3(\omega^2 - 1)(\omega^2 - 9)\cdots[\omega^2 - (l - 3)^2][\omega^2 - (l^2 + l + 1)]}{(\omega^2 - 4)(\omega^2 - 16)\cdots[\omega^2 - (l + 2)^2]} = -1 \quad (l \text{ even}), \tag{47}$$

The equivalence of (46) and (47) and the initial form of equation (44) can be proved by expanding the proper fraction in the left-hand side of (46) or (47) in simple summands and making sure that this expansion coincides with the left-hand side of (44). This is performed elementarily, and we do not consider it.

Now, using the multiplicative form of characteristic equations (46) and (47), it is easy to prove the realness of all eigenfrequencies ω, i.e., the lack of exponential instabilities.

First let l be odd. We construct, assuming ω to be real, a tentative plot of the $a_l(\omega)$ function (Fig. 17). Figure 17 shows a part of this plot corresponding to $\omega > 0$ [the $a_l(\omega)$ function is even]. The eigenfrequencies lie on the intersections of the curves $a_l = a_l(\omega)$ and the straight line $a_l = -1$. Find the number of intersections. If it is coincident with the power of characteristic equation (46) [equal to $(l + 3)$], then it will be this that will signify the realness of all eigenfrequencies. We make sure that this is the case.

First of all from Fig. 17 it is easy to see that the intervals

$$(1 - 3), (3 - 5), \ldots, (l - 4, l - 2), (l, l + 2) \tag{48}$$

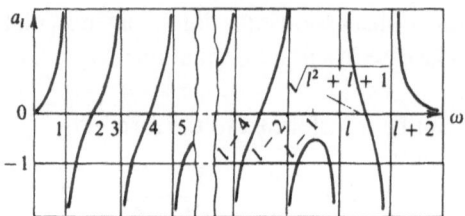

Figure 17. Graphic definition of eigenfrequencies of "internal" oscillations of a homogeneous sphere with circular particle orbits [Eq. (46)] (l is odd).

each contain one frequency. This yields $l - 1$ frequencies (with due regard for negative). It remains to show that within the $(l - 2, l)$ interval, there are two frequencies. But this follows just from the estimate of a_l in the middle of this interval

$$-a_l(l - 1) = \frac{3l(l - 1)(2l - 4)!!}{(4l^2 - 1)(2l - 3)!!} < \frac{3l(l - 1)}{(4l^2 - 1)} < 1. \tag{49}$$

The proof for realness of all eigenfrequencies at any odd l is thus completed. The case of the even l is considered in a quite similar way and leads to the same result.[15]

3.1.3. Stability with Respect to Surface Perturbations [154]. Equation (34) has a structure identical to the equation for potential perturbations of the electric potential of homogeneous plasma, which gives the dispersion equation of Langmuir oscillations [86]. However, unlike the homogeneous infinite plasma, allowance for the confined nature of the gravitating system under consideration leads to the possibility of the existence of oscillations of the surface with $\Delta\chi_l = 0$.

Naturally, the question arises regarding the stability of spherically symmetrical systems with respect to surface perturbations. The specific character of surface oscillations with $\Delta\chi_l = 0$ lies in the fact that, according to the Poisson equation, the perturbed density of the system, performing such oscillations, is zero. Thus, one may indicate an analogy with oscillation of a sphere of incompressible liquid. Along with "local" perturbations, studied above, these perturbations constitute the complete spectrum of possible oscillations of the model in question.

Below we give the derivation of the corresponding characteristic equation somewhat different from the one used in the original work [154]. Using the method that coincides completely with that described in subsection 3.1.1, one can infer the differential equation for the potential $\Phi_1^{(l)} = \chi_l(r)Y_{lm}(\theta, \varphi)$ in the

---

[15] Note that we have neglected above the possibility of integer frequencies (in Ω units), including the zero one; see, e.g., formula (34), subsection 3.1.1. The last frequency ($\omega^2 = 0$) is really present; this is revealed in analysis of system oscillations with weak inhomogeneity (see Problem 1).

case of the arbitrary (not necessarily uniform) density $\rho_0(r)$. It has the form [105, 125].

$$\frac{d}{dr}\left[r^2 A_l(r)\frac{d\chi_l}{dr}\right] - B_l(r)\chi_l = 0, \tag{50}$$

where the following notations are introduced:

$$A_l(r) = 1 + \omega_0^2 \sum_{s=-l}^{l} \frac{\alpha_{ls}}{(\omega - s\Omega_0)^2 - \Omega_0^2 - \omega_0^2}, \tag{51}$$

$$B_l(r) = l(l+1) + \sum_{s=-l}^{l} \alpha_{ls}$$

$$\times \left(r^2 \frac{d}{dr}\left\{\frac{\omega_0^2}{r} \cdot \frac{2s\Omega_0}{(\omega - s\Omega_0)[(\omega - s\Omega_0)^2 - \Omega_0^2 - \omega_0^2]}\right\}\right.$$

$$+ \omega_0^2 \left\{\frac{4s\Omega_0(\omega - s\Omega_0) + s^2[(\omega - s\Omega_0)^2 - \omega_0^2 + 3\Omega_0^2]}{(\omega - s\Omega_0)^2[(\omega - s\Omega_0)^2 - \Omega_0^2 - \omega_0^2]}\right.$$

$$\left.\left.+ \frac{(l+s+1)(l-s)}{(\omega - s\Omega_0)(\omega - s\Omega_0 - 2\Omega_0)}\right\}\right), \tag{52}$$

$$\omega_0^2 = 4\pi G\rho_0(r) = 3\Omega_0^2(r) + r\frac{d\Omega_0^2}{dr}, \tag{53}$$

$$\Omega_0^2 = \frac{4\pi G}{r^3}\int_0^r \rho_0(r)r^2\,dr. \tag{54}$$

From Eq. (50), by integrating over the "transition layer" at the boundary $r = R$, we get the following boundary condition for the oscillations of the homogeneous sphere with a radius $R = 1$:

$$\left[1 + \omega_0^2 \sum_{s=-l}^{l} \frac{\alpha_{ls}}{(\omega - s\Omega_0)^2 - \Omega_0^2 - \omega_0^2}\right]\frac{d\chi_l}{dr}\Big|_{1-0}^{1+0}$$

$$= \omega_0^2 \sum_{s=-l}^{l} \frac{2s\Omega_0\alpha_{ls}}{(\omega - s\Omega_0)[(\omega - s\Omega_0)^2 - \Omega_0^2 - \omega_0^2]}\chi_l\Big|_{1-0}^{1+0}. \tag{55}$$

The boundary condition in (55) denoting the discontinuity of the first derivative of the potential, evidently corresponds to the appearance, at the perturbed boundary, of an effective simple layer with the surface density $\sigma = \rho_0\Delta R$. The second boundary condition is the requirement of the continuity of the potential itself:

$$\chi_l\big|_{1-0}^{1+0} = 0. \tag{56}$$

The boundary conditions of (55) and (56) must, of course, also be satisfied for the above perturbations with $\Delta\chi_l \neq 0$, so that the χ_l function is not quite arbitrary. It is necessary to require that the sphere boundary not be displaced, while the redistribution of the density ρ_1 corresponding to the potential χ_l

yields the zero 2^l-pole moment[16]

$$\int_0^R \rho_1 r^{l+2} \, dr = \int_0^R \Delta\chi_l r^{l+2} \, dr = 0. \tag{57}$$

Such a perturbation does not produce any potential outside the sphere:

$$\chi_{l>} \equiv 0, \qquad \frac{d\chi_{l>}}{dr} \equiv 0. \tag{58}$$

If one also imposes the requirement [not contradicting (57)]

$$\chi_{l<}(R) = 0, \tag{59}$$

then with due regard for dispersion equation (43), the boundary conditions in (55) and (56) will be satisfied.

Since for the surface oscillations of interest

$$\chi_l(r > 1) = r^{-(l+1)}, \qquad \chi_l(r < 1) = r^l, \tag{60}$$

then using (55) we get the required characteristic equation

$$a_l \equiv \frac{3}{2l+1} \sum_{s=-l}^{l} \frac{\alpha_{ls}}{(\omega - s)^2 - 4}\left(l - \frac{2s}{\omega - s}\right) = -1. \tag{61}$$

The dispersion relation in (61) substitutes for (43) for the case $\Delta\chi_l = 0$.

At $l = 1$, we have $\omega^2 = 0$. This is an obvious result, implying displacement of the sphere as a whole. At $l = 2$, $\omega^2 = 14/5$, subsequent l are satisfied by a larger number of eigenfrequencies; however, they are all real.

To prove this, we represent again the characteristic equation in the multiplicative form:

$$a_l = \frac{3l}{2l+1} \cdot \frac{(\omega^2 - 1)(\omega^2 - 9)\cdots[\omega^2 - (l-3)^2]}{(\omega^2 - 4)(\omega^2 - 16)\cdots(\omega^2 - l^2)} = -1 \qquad (l \text{ even}), \tag{62}$$

$$a_l = \frac{3l}{2l+1} \cdot \frac{\omega^2(\omega^2 - 4)\cdots[\omega^2 - (l-3)^2]}{(\omega^2 - 1)(\omega^2 - 9)\cdots(\omega^2 - l^2)} \qquad (l \text{ odd}). \tag{63}$$

A tentative plot of the $a_l(\omega)$ function for the case of even l (for odd l, the consideration is similar) is given in Fig. 18. The proof is performed in the same way as in the previous subsection. The intervals

$$(2, 4), (4, 6), \ldots, (l - 4, l - 2) \tag{64}$$

each contain one root, as well as symmetrical intervals at negative ω; the total number of the roots is $(l - 4)$ here. To them we must add each two roots within the intervals $(l - 2, l)$ and $(-l + 2, -l)$, since for $a_l(l - 1)$ in this case we have

---

[16] We have already encountered a similar condition in the cylinder case [see (11) in §7, previous chapter]. Note that condition (57) is a natural generalization for the case of perturbations with $l > 1$ of the condition $\int_0^R \rho_1(r)r^2 \, dr = 0$, following from the law of conservation of momentum.

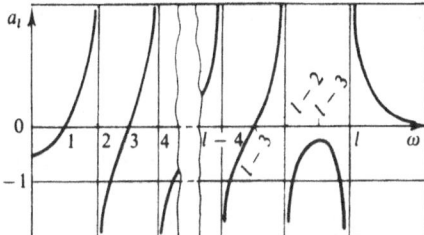

Figure 18. Diagram of function $a_l(\omega)$ for "surface" oscillations of a homogeneous Einstein model [see Eq. (62); l is even].

the same estimate as earlier:

$$-a_l(l-1) = \frac{3l(l-1)\,(2l-4)!!}{(4l^2-1)\,(2l-3)!!} < \frac{3l(l-1)}{4l^2-1} < 1. \tag{65}$$

The complete number of roots l is coincident with the degree of the characteristic equation, which does mean stability.

The surface oscillations of the collisionless sphere considered in this item are similar to sphere oscillations of incompressible liquid in the sense that here we do not also have perturbation of the local density: $\rho_1 = 0$. We recall the expression for oscillation frequencies of the incompressible sphere (that of Kelvin)

$$\omega^2 = \frac{2l(l-1)}{2l+1}. \tag{66}$$

It is easy to see that (66) and (61) have little in common. For example, for $l = 2$ from (66) we obtain $\omega^2 = \frac{4}{5}$ while from (61) $\omega^2 = \frac{14}{5}$; for $l = 0$ from (66) $\omega^2 = \frac{12}{7}$ and from (61) we get two different frequencies ω_1^2, ω_2^2 (which are defined by the equation $7\omega^4 - 61\omega^2 + 63 = 0$), etc.

3.2 Stability of a Homogeneous System of Particles with Nearly Circular Orbits [143]

In the previous section, the paths of particles were assumed circular. The case of circular orbits is degenerate in that the frequency of collective oscillations is independent of the radial wave number. Proceeding from the similarity with plasma media, it may be suggested that degeneration is canceled if the particles have a radial velocity (the case of noncircular trajectories). The problem of the equilibrium distribution function of such a system is considered by us in §1 (item 4). Below it is shown that degeneration really is canceled in the case when the paths of the particles deviate from circular ones. For the sake of simplicity, we shall consider oscillations of the homogeneous sphere. The unperturbed distribution function, according to §1, is of the form

$$f_0 = \frac{\rho_0}{2\pi v_0}\left[\delta_r\delta_\perp\left(1 + \frac{\theta}{2v_0^2}\right) + \delta_r\delta'_\perp\frac{\theta}{2v_0} + \frac{\theta}{2}\delta_r\delta''_\perp + \frac{\theta}{2}\delta''_r\delta_\perp\right], \qquad (1)$$

where $\delta_\perp = \delta(v_\perp - v_0)$, $\delta_r = \delta(v_r)$, and θ is the mean square of the radial velocity.

We seek the perturbed f_1 function in the form

$$f_1 = A\delta_r\delta_\perp + B\delta_r\delta'_\perp - C\delta'_r\delta_\perp + \theta[C_1\delta'_r\delta'_\perp + C_2\delta_r\delta''_\perp$$
$$+ C_3\delta'_r\delta''_\perp + C_4\delta_r\delta'''_\perp + C_5\delta'''_r\delta_\perp + C_6\delta''_r\delta_\perp + C_7\delta''_r\delta_\perp]. \qquad (2)$$

For coefficients $A, B, C, C_1, \ldots, C_7$ we obtain the system of ten equations, and by solving these one can, after rather cumbersome calculations, arrive at the expression for ρ_1 of the form

$$\rho_1 = \rho_0\left[\sum_{s=-l}^{l}|P_{s0}^l(0)|^2\frac{\Delta\Phi_1}{4\Omega^2 - \omega_s^2} + \sum_{s=-l}^{l}\frac{3\theta|P_{s0}^l(0)|^2\Delta(\Delta\Phi_1)}{(4\Omega^2 - \omega_s^2)(16\Omega^2 - \omega_s^2)}\right]. \qquad (3)$$

This provides a possibility of obtaining the equation for the perturbed density

$$\frac{\theta}{\alpha}\Delta\rho_1 + \rho_1 = 0, \qquad (4)$$

where

$$\alpha = \frac{1}{3}\frac{\sum_{s=-l}^{l}|P_{s0}^l(0)|^2(\omega_s^2 - \Omega^2)/(\omega_s^2 - 4\Omega^2)}{\sum_{s=-l}^{l}|P_{s0}^l(0)|^2(3\Omega^2)/(\omega_s^2 - 4\Omega^2)(16\Omega^2 - \omega_s^2)}. \qquad (5)$$

For the case $l = 0$ we get $(k^2 = \alpha/\theta)$

$$\omega^2 = \Omega^2 + \tfrac{3}{5}k^2v_T^2, \qquad (6)$$

where $v_T^2 = \theta$, $k^2 = k_\perp^2 + k_r^2$, $k_\perp^2 = l(l + 1)/r^2$.

For $l = 2$

$$\omega_1^2 = 3.1\Omega^2 + 7.4k^2v_T^2, \qquad \omega_2^2 = 13.9\Omega^2 + 0.5k^2v_T^2, \qquad \omega_3^2 = 0, \qquad (7)$$

etc.

Thus, we see that allowance for the radial velocity dispersion does not lead to violation of the homogeneous sphere stability.

3.3 Stability of a Homogeneous Sphere with Finite Angular Momentum [124]

In previous subsections of this section, we have dealt with the stability of the model of the cluster in the form of a spherically symmetrical system of masses rotating in circular (and close to circular) orbits (Fig. 2). The total moment of rotation of such a system was zero. However, in §1 (see item 4), we have seen that this is generally not necessary: One can construct models of rotating spherical clusters. Therefore, the question naturally arises as to whether the spherically symmetrical rotating configurations are stable. Turning to the plasma stability theory [86], we shall note that practically all plasma instabilities are due to the presence of the stationary magnetic field. As was already repeatedly stressed, the kinetic equation for the distribution function of plasma particles in the magnetic field is coincident with the kinetic equation for "cold" rotating gravitating configurations in the proper frame of reference. There is a familiar set of instabilities in plasma [86] consuming for its development the energy of the stationary magnetic field. For this reason, the aim of this section is to clear up the possibility of the gravitational instability, which would result, e.g., in slowing down the sphere rotation as a whole.

Consider oscillations of a homogeneous model, described by the distribution function in (55) §1:

$$f_0 = \frac{\rho_0}{2\pi v_0} \delta(v_r)\delta(v_\perp - v_0)(1 + \mu \sin \theta \sin \alpha) \qquad (|\mu| \leq 1). \qquad (1)$$

We investigate the stability of the above system as in Section 3.1. The solution of the linearized kinetic equation, according to Section 3.1, is required in the form

$$f_1 = A\delta(v_r)\delta(v_\perp - v_0) + B\delta(v_r)\delta'(v_\perp - v_0) - C\delta'(v_r)\delta(v_\perp - v_0). \qquad (2)$$

To determine A, B, C, we get the equations

$$\hat{L}^0 A - \frac{1}{r\Omega}\left(\hat{L}^0 - \frac{\partial}{\partial t}\right)B + \frac{1}{r}\frac{\partial}{\partial r}(rC) = -\mu \sin \theta \cos \alpha \frac{\rho_0}{2\pi v_0} \hat{K}^0\Phi_1, \qquad (3)$$

$$\hat{L}^0 B - 2\Omega C = \frac{\rho_0}{2\pi\Omega^2 r^2}\left(\hat{L}^0 - \frac{\partial}{\partial t}\right)\Phi_1(1 - \mu \sin \theta \sin \alpha), \qquad (4)$$

$$\hat{L}^0 C + 2\Omega B = -\frac{\rho_0}{2\pi\Omega r}\Phi_1'(1 + \mu \sin \theta \sin \alpha). \qquad (5)$$

Here

$$\Phi_1' = \frac{\partial \Phi_1}{\partial r}, \qquad \hat{K}^0 = \frac{v_0}{r}\left(\sin \alpha \frac{\partial}{\partial \theta} - \frac{\cos \alpha}{\sin \theta}\frac{\partial}{\partial \varphi}\right),$$

$$\hat{L}^0 = \frac{\partial}{\partial t} + \Omega \cos \alpha \frac{\partial}{\partial \theta} + \frac{\sin \alpha}{\sin \theta}\frac{\partial}{\partial \varphi} - \sin \alpha \cot \theta \frac{\partial}{\partial \alpha}.$$

Transforming the expression $\cos \alpha \sin \theta \, \hat{K}^0\Phi_1$, we have

$$\cos \alpha \sin \theta \, \hat{K}^0\Phi_1 = \sin \alpha \sin \theta\left(\hat{L}^0 - \frac{\partial}{\partial t}\right)\Phi_1 - \Omega \frac{\partial \Phi_1}{\partial \varphi}. \qquad (6)$$

We present A, B, C in the form

$$A = A_0 + A_1\mu \sin \alpha \sin \theta, \qquad B = B_0 + B_1\mu \sin \alpha \sin \theta,$$

$$C = C_0 + C_1\mu \sin \alpha \sin \theta.$$

For A_0, B_0, C_0, we shall infer the system of equations similar to the system for A, B, C in Section 3.1.

Noting that

$$\hat{L}^0 \begin{Bmatrix} B_1 \\ C_1 \end{Bmatrix} \sin \alpha \sin \theta = \sin \alpha \sin \theta \hat{L}^0 \begin{Bmatrix} B_1 \\ C_1 \end{Bmatrix},$$

we have $B_1 = B_0$, $C_1 = C_0$. According to (6), we transform Eq. (3):

$$\hat{L}^0 A_1 - \frac{1}{r\Omega} \left(\hat{L}^0 - \frac{\partial}{\partial t} \right) B_1 + \frac{1}{r} \frac{\partial}{\partial r} r C_1$$

$$= \frac{1}{\Omega r^2} \left[- \left(\hat{L}^0 - \frac{\partial}{\partial t} \right) \Phi_1 \frac{\rho_0}{2\pi v_0} + \frac{\rho_0/2\pi v_0}{\sin \alpha \sin \theta} \Omega \frac{\partial \Phi_1}{\partial \varphi} \right]. \qquad (7)$$

Calculating from (7) A_1, for the perturbed density ρ_1, we shall obtain

$$\rho_1 = \int_0^{2\pi} (A_0 v_0 - B_0) \, d\alpha + \mu \int_0^{2\pi} (A_1 v_0 - B_1) \sin \alpha \sin \theta \, d\alpha. \qquad (8)$$

The calculation of the second integral in (8) is reduced to the calculation of the integral (see Section 3.1)

$$\int_0^{2\pi} \sum_{s=-l}^{l} T_{ms'}^l \left(\frac{\pi}{2} - \varphi, \theta, \frac{\pi}{2} + \alpha \right) e^{-is\pi} P_{s's}^l(0) \sin \alpha \sin \theta \, d\alpha = I. \qquad (9)$$

Using the recurrent formulae for $P_{mn}^l(\cos \theta)$ [33],

$$\sin \theta \frac{dP_{mn}^l(\cos \theta)}{d \cos \theta} + \frac{m - n \cos \theta}{\sin \theta} P_{mn}^l(\cos \theta) = i\alpha_n P_{m, n-1}^l(\cos \theta),$$

$$\sin \theta \frac{dP_{mn}^l(\cos \theta)}{d \cos \theta} - \frac{m - n \cos \theta}{\sin \theta} P_{mn}^l(\cos \theta) = i\alpha_{n+1} P_{m, n+1}^l(\cos \theta),$$

$$\alpha_n = \sqrt{(l + n)(l - n + 1)},$$

we have that the integral in (9) is

$$I = Y_m^l(\theta, \varphi) P_{s0}^l e^{is\pi} \frac{2\pi m s}{l(l + 1)}. \qquad (10)$$

For ρ_1 we ultimately get

$$\rho_1 = -\rho_0 \sum_{s=-l}^{l} |P_{s0}^l(0)|^2 \frac{\Delta \Phi_1}{(\omega - s\Omega)^2 - 4\Omega^2}$$

$$- \rho_0 \sum_{s=-l}^{l} |P_{s0}^l(0)|^2 \frac{s}{(\omega - s\Omega)^2 - 4\Omega^2} \frac{m\mu}{l(l + 1)} \Delta \Phi_1. \qquad (11)$$

Substituting (11) into the linearized Poisson equation, we obtain the following dispersion equation:

$$\psi(x) \equiv 1 + \sum_{s=-l}^{l} \frac{3}{(x - s)^2 - 4} \left(1 + \frac{\mu m s}{l(l + 1)} \right) |P_{s0}^l(0)|^2 = 0 \qquad \left(x = \frac{\omega}{\Omega} \right). \qquad (12)$$

This equation also is reduced to the multiplicative form

$$a_{lm} = \frac{3(\omega^2 - 1)(\omega^2 - 9)\cdots[\omega^2 - (l - 3)^2]}{\omega(\omega^2 - 4)(\omega^2 - 16)\cdots[\omega^2 - (l + 2)^2]} \{\omega[\omega^2 - (l^2 + l + 1)]$$

$$+ \mu m[\omega^2 - (l^2 + l - 2)]\} = -1 \qquad (l \text{ even}) \qquad (13)$$

$$a_{lm} = \frac{3\omega^2(\omega^2 - 4)\cdots[\omega^2 - (l - 3)^2]}{\omega(\omega^2 - 1)(\omega^2 - 9)\cdots[\omega^2 - (l + 2)^2]} \{\omega[\omega^2 - (l^2 + l + 1)]$$

$$+ \mu m[\omega^2 - (l^2 + l - 2)]\} = -1 \qquad (l \text{ odd}). \qquad (14)$$

The proof of stability also in this case is most readily made by constructing plots of the a_{lm} functions (Fig. 19). A small complication in comparison with the equations considered earlier is due to the presence of the "nonstandard" cubic polynomial in the expression for α_{lm}:

$$F_\alpha(\omega) = \omega^3 - \alpha\omega^2 - (l^2 + l + 1)\omega - \alpha(l^2 + l - 2), \qquad (15)$$

where α denotes the parameter $\alpha = \mu m$; it is easy to see that $-l < \alpha < l$, but it will be enough to consider the case for positive α: $0 < \alpha < l$. It is easy to verify that $F_\alpha(\omega)$ always contains one root each within the intervals $(l, l + 2)$ and $(-l - 2, -l)$ (respectively, ω_1 and ω_2). These are the fixed roots in Fig. 19. The third root ω_3 is more movable. With an increase in α from 0 to l, it changes from small negative values $\omega_3 \simeq -\alpha(l^2 + l - 2)/(l^2 + l + 1)$ at $\alpha \to 0$ to the value $\omega_3 = -(l - 1)$ at $\alpha = l$. It turns out, however, that the qualitative picture $\alpha_{lm}(\omega)$ is independent of the position of ω_3 in the region $(-l + 1, 0)$. The plot of $\alpha_{lm}(\omega)$ in Fig. 19 corresponds to the value ω_3 within the interval $(-l + k, -l + k + 1)$.

The realness of all the roots is readily proved from the consideration of this plot and the following estimates:

$$-a_{lm}(l - 1) = \frac{(2l - 4)!!}{(2l - 3)!!} \frac{3(l - 1)(l + \alpha)}{(4l^2 - 1)} < \frac{(2l - 4)!!}{(2l - 3)!!} \cdot \frac{3}{2} < 1, \qquad (16)$$

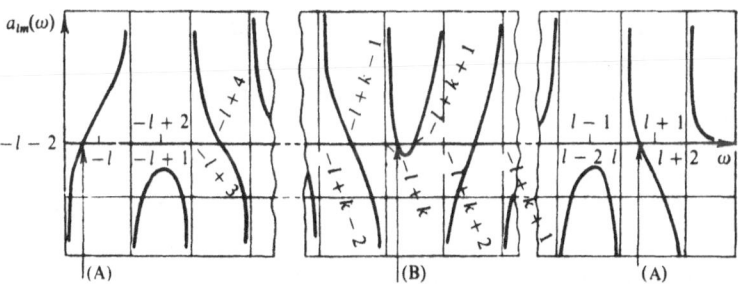

Figure 19. Diagram of the function $a_{lm}(\omega)$ for "internal" oscillations of homogeneous, solid-rotating spheres with circular orbits: (A) "fixed" roots; (B) "mobile" root.

if $l \geq 4$;

$$-a_{lm}(l-1)|_{l=2} = \frac{2+\alpha}{5} < 1 \qquad (\alpha < l < 3), \tag{17}$$

$$-a_{lm}(-l+1) = \frac{(2l-4)!!}{(2l-3)!!} \frac{3(l-1)(l-\alpha)}{(4l^2-1)} < \frac{3(l-1)(l-\alpha)}{(4l^2-1)} < 1. \tag{18}$$

From the plot in Fig. 19 it is seen that one root is contained in each of the intervals

$$(-l+2, -l+4), \ldots, (-l+k-2, -l+k), \tag{19}$$

$$(-l+k+2, -l+k+4), \ldots, (-2, 0), \tag{20}$$

$$(0, 2), \ldots, (l-4, l-2), \tag{21}$$

$$(-l-2, -l), (l, l+2), \tag{22}$$

and from the estimates of (16)–(18) it follows that the other two roots each lie within the intervals $(-l, -l+2)$ and $(l-2, l)$. In the series (19) there are $(k/2 - 1)$ roots, in (20), $(l - k - 2)/2$ roots, and in (21), $-(l/2 - 1)$ roots. It is necessary to add to them two roots (22) and four roots in the intervals $(-l, -l+2), (l-2, l)$. The total number of roots is coincident with the power of the characteristic equation, being $(l + 3)$. This is what proves the realness of all frequencies of oscillations of the system in question.

It is easy to obtain [154] the characteristic equation also for oscillations of the "surface" type $(\Delta \chi_l = 0)$

$$\frac{3}{2l+1} \sum_{s=-l}^{l} \frac{\alpha_{ls}}{(\omega-s)^2 - 4} \left(-l + \frac{2s}{\omega-s}\right) \left[1 + \frac{\mu m s}{l(l+1)}\right] = 1. \tag{23}$$

The multiplicative form in this case is as follows: $a_{lm} = -1$,

$$a_{lm} = \frac{3l}{2l+1} \frac{(\omega^2-1)\cdots[\omega^2-(l-3)^2]}{\omega(\omega^2-4)\cdots(\omega^2-l^2)}\left(\omega + \mu m \frac{l-1}{l}\right) \qquad (l \text{ even}) \tag{24}$$

$$a_{lm} = \frac{3l}{2l+1} \frac{\omega^2 \cdots [\omega^2-(l-3)^2]}{\omega(\omega^2-1)\cdots(\omega^2-l^2)}\left(\omega + \mu m \frac{l-1}{l}\right) \qquad (l \text{ odd}). \tag{25}$$

The realness of all the roots is proved in the same way, as in the previous case [even somewhat easier; the "movable" root here is $\omega_0 = -\mu m(l-1)/l$].

Consider in more detail the case of weak rotation (small μ). Rewriting Eq. (12) in the form

$$1 + \sum_{s=-l}^{l} \frac{3\alpha_{ls}}{(\omega-s)^2-4} = -\frac{3\mu m}{l(l+1)} \sum_{s=-l}^{l} \frac{s\alpha_{ls}}{(\omega-s)^2-4}; \tag{26}$$

let us solve it by subsequent approximations:

$$\omega = \omega_0 + \mu\omega_1. \tag{27}$$

Here ω_0 is some frequency of the sphere at rest, defined only by the main index l of the spherical harmonic $Y_l^m(\theta, \varphi)$; $\mu\omega_1$ is the additive due to rotation, Substituting (27) into (26) we obtain

$$\mu\omega_1 = \frac{\mu m a(\omega_0)}{2}, \tag{28}$$

where

$$a(\omega_0) \equiv \frac{\sum_{s=-l}^{l} s\alpha_{ls}/[(\omega_0 - s)^2 - 4]}{l(l+1) \sum_{s=-l}^{l} (\omega_0 - s)\alpha_{ls}/[(\omega_0 - s)^2 - 4]^2}.$$

For perturbations of the surface type there ensues the expression similar to (28).

We have arrived at the natural result. A slow rotation cancels the above-noted $(2l + 1)$-fold degeneration of frequencies, which are split according to (28) similarly to the splitting of the levels of the quantum-mechanical system in the magnetic field (the Zeeman effect). The role of the Lorentz force in this case is played by the Coriolis force. Similar splitting of frequencies may be traced also in other examples (for oscillations of nearly spherical ellipsoids, gaseous spheres, etc.). In reality, it represents the general law that is the consequence of the group theory.

In conclusion it should be emphasized that we have proved above the (exponential) stability of merely *uniformly* rotating spheres. Possibly, among more general systems with circular orbits there are unstable ones (see Section 3.2, Chapter IV, where this point is discussed in more detail).

3.4 Stability of Inhomogeneous Systems

3.4.1. The Case of Circular Orbits [105, 106]. The equation describing small oscillations of the spherically symmetrical system of masses rotating in circular trajectories with arbitrary dependence of the density on the radius has already been given in Section 3.1 [see (50)]. It is possible to investigate this equation in the same way as was done in §7, Chapter II, in studying the flute-like oscillations of a related system—a cylinder with circular orbits of particles.

The ensuing results are in many respects identical.

For radial perturbations we have a continuous spectrum of the frequencies $\omega^2 = \Omega^2(r_0)$, $0 < r_0 < R$, to which correspond oscillations of individual infinitely thin layers, $\Phi_1' \sim \delta(r - r_0)$.

The local dispersion relations in this case can be derived in the same way, as in §7, Chapter II. They define the set of continuous spectra of the form $[\omega_s \equiv \omega - s\Omega(r)]$:

$$\omega_s(r) = 0, \tag{1}$$

$$\omega_s^2(r) - \varkappa^2(r) = 0, \tag{2}$$

originating from the poles of the $B_l(r, \omega)$ function and of the continuous spectrum defined by the equation

$$A_l(r, \omega) = 0, \qquad (3)$$

where the spectra in (1) and (2) correspond to stable oscillations [and (2) is similar to (24) in §7, Chapter II]. Of more interest in Eq. (3), which can describe both stable and unstable localized perturbations (see Problems below).

One can make sure of the existence of continuous spectra of frequencies of eigen (singular) oscillations by solving (by the method similar to that used in §7, Chapter II) the problem of the evolution of initial perturbations localized in some narrow region. For locally stable systems [the roots of equation (3) are real], one defines at the same time also the law of *damping* in time of such perturbations ($\sim t^{-2}$ or t^{-3}). There is nothing new to reveal about this point in comparison with the case of the nonhomogeneous cylinder. For locally unstable systems [when there are imaginary roots of Eq. (3)], perturbations of this type *increase* according to the (mainly) exponential law.

For each fixed r (and a given l) Eq. (3) has the finite number of roots ω. Remember that in the cylinder case we had one root ($\omega^2 = 2\Omega^2$). This difference seems to be associated with the presence in the spherically symmetrical systems with circular orbits of the velocity dispersions v_θ^2 and v_φ^2. Additional degrees of freedom (as compared with the flute-like oscillations of the cold cylinder, which bear a purely hydrodynamic character) correspond to them.

Analysis of the location of roots in Eq. (3) [106] (see Problem 1) leads to the conclusion about the (local) instability of systems with $d\Omega^2/dr > 0$. However, if the contrary condition ($d\Omega^2/dr < 0$) is fulfilled, so the system turns out to be stable with respect to perturbations of this kind.

The last condition, however, is far from signifying the complete stability, for, in principle, even provided this condition is met, nonlocalized perturbations may be unstable. The presence of such perturbations is easily established, for example, from analysis of small oscillations of a nearly homogeneous sphere, which can be performed in the same way as in the case of a nearly uniform cylinder (§7, Chapter II). In the limit of the homogeneous sphere, these oscillations correspond to "surface" modes as considered in subsection 3.1.3.[17] For the sphere with the arbitrary degree of inhomogeneity one can in a comparatively simple way obtain the dispersion equation that describes similar "global" perturbations (according to classification in §7, Chapter II) only at $l \gg 1$, where these perturbations also become localized (near the sphere surface $r = R$). In this case perturbations, rapidly decreasing as they move off from the boundary $r = R$, may be required in the form

$$\chi_l \sim e^{-k(R-r)} \qquad (kR \gg 1). \qquad (4)$$

---

[7] Also not excepted is the existence of a discrete spectrum of "internal" oscillations. This question, however, remains uninvestigated as yet (as in the case of cylinder oscillations).

The dispersion equation is as follows:

$$l[1 + \sqrt{|A_l|}\ \text{sgn}\ A_l]_R = 2\Omega\omega_0^2 \sum_{s=-l}^{l} \frac{s\alpha_{ls}}{\omega_s(\omega_s^2 - \varkappa^2)}. \tag{5}$$

Nontrivial solutions of (5) at $l \gg 1$ correspond to $A_l \approx -1$, or

$$2 + \omega_0^2 \sum_{s=-l}^{l} \frac{\alpha_{ls}}{\omega_s^2 - \varkappa^2} = 0. \tag{6}$$

For the homogenous sphere this equation, as it should, transforms into Eq. (61), Section 3.1 (at $l \gg 1$).

The manner of arrangement of the roots of Eq. (6) can be ascertained in the same way as for Eq. (3). Analysis (see Problem 1) shows that perturbations of the form (4) do not lead to any new instabilities [in comparison with local oscillations of the "internal" type, which are described by Eq. (3)].

3.4.2. Localized Perturbations in Spherical Systems with Nearly Circular Orbits may also be considered in a general form. Since in the case of purely circular orbits the mode $l = 2$ is most unstable (it begins to grow at the most "soft" condition: $d\Omega^2/dr > 0$—see Problem 1), it is interesting to investigate the influence of nonzero radial velocity dispersion just on this mode [42^{ad}].

It is convenient to use the procedure of separation of angles described in detail in §2 of the Appendix. For given l, the potential perturbation is proportional to certain spherical harmonic $Y_l^m(\theta, \varphi)$. Let us present the perturbation of the distribution function as

$$f_1 = \sum_s \lambda_s(v_\perp, v_r, r, t)T_{ms}^l(\varphi, \theta, \alpha),$$
$$T_{ms}^l(\varphi_1, \theta, \varphi_2) = e^{-im\varphi_1 - is\varphi_2}P_{ms}^l(\cos \theta), \tag{7}$$

where the P_{ms}^l are three-index functions [33]. Functions $P_{m0}^l(\cos \theta)$ are proportional to associated Legendre functions; therefore, we assume

$$\Phi_1 = \chi(r, t)T_{m0}^l. \tag{8}$$

Then for unknown functions λ_s ($s = 0, 1, 2, \dots, l; \lambda_{-s} = \lambda_s$) one can obtain (see §2 of the Appendix) the set of equations which can be reduced to the following, in the case of mode $l = 2$ *interesting* for us:

$$\hat{D}\lambda_0 - i\Omega\alpha_0 \lambda_1 = \frac{\partial \chi}{\partial r}\frac{\partial f_0}{\partial v_r}, \tag{9}$$

$$\hat{D}\lambda_1 - i\frac{\Omega}{2}(\alpha_0 \lambda_0 + \alpha_2 \lambda_2) = \frac{\chi\alpha_1}{2ir}\frac{\partial f_0}{\partial v_\perp}, \tag{10}$$

$$\hat{D}\lambda_2 - i\frac{\Omega}{2}\alpha_2 \lambda_1 = 0, \tag{11}$$

where

$$\hat{D} = \frac{\partial}{\partial t} + v_r \frac{\partial}{\partial r} - \frac{v_r v_\perp}{r} \frac{\partial}{\partial v_\perp} + \left(\frac{v_\perp^2}{r} - \frac{\partial \Phi_0}{\partial r}\right) \frac{\partial}{\partial v_r}, \tag{12}$$

$$\alpha_0 = \alpha_1 = \sqrt{6}, \qquad \alpha_2 = 2, \qquad r\Omega^2 = \frac{\partial \Phi_0}{\partial r};$$

f_0 is the equilibrium distribution function. For the system with nearly circular orbits

$$\hat{D} = \frac{\partial}{\partial t} + \varkappa\left(i\alpha v_x + v_y \frac{\partial}{\partial v_x} - v_x \frac{\partial}{\partial v_y}\right), \tag{13}$$

where we introduce the notations:

$$v_x \equiv v_r \cdot \frac{\varkappa}{2\Omega} \cdot \frac{1}{r\Omega}, \qquad v_y = \frac{v_\perp'}{r\Omega}, \qquad v_\perp = v_\perp' + r\Omega,$$

$$\frac{\varkappa^2}{2\Omega} = \frac{\partial(r\Omega)}{\partial r} + \Omega, \qquad \alpha = \frac{2\Omega^2}{\varkappa^2} kr$$

and also used WKB-substitution $\partial/\partial r \to ik$ (k is the wave number of the perturbation). For the distribution function $f_0 = f_0(v_x^2 + v_y^2)$ we have finally the set of equations:

$$\hat{D}\lambda_0 - i\Omega\alpha_0 \lambda_1 = i f_0' \frac{\varkappa}{\Omega^2 r^2} kr\chi v_x, \tag{9'}$$

$$\hat{D}\lambda_1 + i\frac{\Omega}{2}(\alpha_0 \lambda_0 + \alpha_2 \lambda_2) = -i\frac{\chi\alpha_1}{r^2\Omega} f_0' v_y \tag{10'}$$

$$\hat{D}\lambda_2 - i\frac{\Omega}{2}\alpha_2 \lambda_1 = 0, \tag{11'}$$

where the operator $\hat{D}$ acts according to (13); it may be written as

$$\hat{D} = -i(\omega - \varkappa\alpha v \cos \varphi) - \varkappa\frac{\partial}{\partial \varphi}, \tag{13'}$$

ω = frequency, $v_x = v \cos \varphi$, $v_y = v \sin \varphi$.

We solve the set of equations (9')–(11') as follows. First of all from (9') and (11') we obtain the equation for the quantity

$$\Lambda \equiv \alpha_2 \lambda_0 - 2\alpha_0 \lambda_2:$$

$$\hat{D}\Lambda = \frac{\varkappa\alpha_2}{\Omega^2 r^2} f_0' ikr\chi v \cos \varphi. \tag{14}$$

Hence we find Λ (see for comparison §5, Chapter II):

$$\Lambda = \frac{iA}{\alpha}\left[1 - Q \frac{1}{2\pi} \int_{-\pi}^{\pi} dx \, e^{i[vx + \alpha v_x \sin x + \alpha v_y(1 + \cos x)]}\right], \tag{15}$$

$$Q \equiv \frac{v\pi}{\sin v\pi}, \qquad A = \alpha_2 C, \qquad C = -\frac{1}{\Omega^2 r^2} f_0' ikr\chi, \qquad v \equiv \frac{\omega}{\varkappa}.$$

Summing Eqs. (9')–(11') with weights $b_1 = \alpha_0$, $b_2 = 2\alpha_2$, $b_3 = \alpha_2$, we obtain for the function $\Lambda_1 \equiv b_1\lambda_0 + b_2\lambda_1 + b_3\lambda_2$ the following equation:

$$i(v_1 - \alpha v \cos \varphi)\Lambda_1 + \frac{\partial \Lambda_1}{\partial \varphi} = B \cos \varphi \left(\text{where } v_1 = v + \mu,\ \mu = \frac{2\Omega}{\varkappa},\ B = b_1 C \right),$$

(16)

which has the solution

$$\Lambda_1 = \frac{iB}{\alpha}\left[1 - Q_1 \frac{1}{2\pi} \int_{-\pi}^{\pi} e^{i[v_1 x + \alpha v_x \sin x + \alpha v_y(1 + \cos x)]}\, dx \right],$$

$$Q_1 \equiv \frac{v_1 \pi}{\sin v_1 \pi}.$$

(17)

After this we obtain for λ_0 the equation

$$i(v_0 - \alpha v \cos \varphi)\lambda_0 + \frac{\partial \lambda_0}{\partial \varphi} = C \cos \varphi - E_1\Lambda_1 - E\Lambda;$$

$$v_0 \equiv v - \mu, \qquad E_1 = \frac{i\alpha_0 \mu}{8}, \qquad E = \frac{i\alpha_2 \mu}{16}.$$

(18)

Let us solve this equation:

$$\lambda_0 = \lambda_0^{(1)} + \lambda_0^{(2)} + \lambda_0^{(3)},$$

(19)

$$\lambda_0^{(1)} = \frac{iC}{\alpha}\left[1 - Q_0 \frac{1}{2\pi} \int_{-\pi}^{\pi} e^{i[v_0 x + \alpha v_x \sin x + \alpha v_y(1 + \cos x)]}\, dx \right],$$

(20)

$$\lambda_0^{(2)} = C_0 \frac{e^{i\alpha v \sin \varphi}}{e^{2iv_0\pi} - 1} \int_0^{2\pi} e^{i[v_0 x - \alpha v \sin(x + \varphi)]}\, dx,$$

(21)

$$\lambda_0^{(3)} = \frac{e^{i\alpha v \sin \varphi}}{e^{2iv_0\pi} - 1} \frac{1}{2\pi\alpha} \int_{-\pi}^{\pi} dx \int_0^{2\pi} dy\, e^{i[v_0 y - \alpha v \sin(y + \varphi)]} \cdot (iBE_1 Q_1 e^{iv_1 x}$$

$$+ iAEQ e^{iv x})e^{i\alpha \sin x\, v \cos(y + \varphi)} + i\alpha(1 + \cos x)v \sin(y + \varphi),$$

(22)

where $\lambda_0^{(1)}$ corresponds to the first term on the right-hand side of Eq. (18), $\lambda_0^{(2)}$ corresponds to that part of a sum $-(E_1\Lambda_1 + E\Lambda) \equiv C_0 = 8C\Omega/\alpha\varkappa b_2$ which has no dependence on φ, and $\lambda_0^{(3)}$ corresponds to retaining part of this sum.

Disturbed density may be calculated by the formula (see §2, Appendix)

$$\rho_1 = 2\pi \int \lambda_0 v_\perp\, dv_\perp\, dv_r.$$

(23)

For the distribution function f_0 of Schwarzschild type,

$$f_0 = \frac{\rho_0}{2\pi^2 \varkappa r c^2} \exp\left[-\frac{V_1^2}{2c^2}(v_x^2 + v_y^2) \right] \qquad \left(V_1 \equiv \frac{2\Omega^2 r}{\varkappa} \right),$$

(23')

we obtain after integration (23) the required dispersion relation in the form

$$x \frac{\varkappa^2}{4\pi G \rho_0} = 1 - \frac{\nu}{\nu_0} Q_0 \frac{1}{2\pi} \int_{-\pi}^{\pi} ds \cos \nu_0 s\, e^{-x(1 + \cos s)}$$

$$+ \frac{\mu}{\nu_0} Q_0 \frac{1}{16\pi^2} \int_{-\pi}^{\pi} ds\, (3Q_1 e^{i\nu_1 s} + Q e^{i\nu s}) \int_{-\pi}^{\pi} dy\, e^{i\nu_0 y} e^{-x[1 - \cos(s + y)]}, \quad (24)$$

where we used the WKB relation between χ and ρ_1

$$-k^2 \chi = 4\pi G \rho_1, \quad (25)$$

and denoted

$$x = \frac{k^2 c^2}{\varkappa^2}, \qquad Q_0 = \frac{\nu_0 \pi}{\sin \nu_0 \pi}, \qquad Q = \frac{\nu \pi}{\sin \nu \pi}, \qquad Q_1 = \frac{\nu_1 \pi}{\sin \nu_1 \pi}.$$

Double integrals in Eq. (24) may be easily reduced to single ones.

Let us consider, for example, one of two double integrals which involves Eq. (24):

$$I = \frac{\mu}{16} \frac{1}{\sin \nu_0 \pi \sin \nu_1 \pi} \int_{-\pi}^{\pi} ds\, e^{i\nu_1 s} \int_{-\pi}^{\pi} dy\, e^{i\nu_0 y + x \cos(s + y)} \quad (26)$$

Expanding $e^{x \cos(s + y)}$ into Fourier series and calculating simple integrals, we find

$$I = \frac{\mu}{4} \sum_n \frac{\alpha_n(x)}{(\nu_1 + n)(\nu_0 + n)} = \frac{1}{8} \left(\sum_n \frac{\alpha_n(x)}{\nu_0 + n} - \sum_n \frac{\alpha_n(x)}{\nu_1 + n} \right).$$

On the other hand, it is easily seen that the sums in the brackets can express through the single integrals:

$$\frac{\sum_n \alpha_n(x)}{(\nu + n)} = \frac{1}{2 \sin \nu \pi} \int_{-\pi}^{\pi} ds\, e^{i\nu s - x \cos s}.$$

As a result of such calculations, the dispersion relation (24) reduces to the following form which is convenient for analyses [42^{ad}]:

$$xu = 1 - \frac{1}{16\pi} [3Q_0 F_x(\nu_0) + 3Q_1 F_x(\nu_1) + 2QF_x(\nu)], \quad (27)$$

where

$$u = \frac{\varkappa^2}{4\pi G \rho_0} = \frac{4}{(4 - \mu^2)}, \qquad F_x(\nu) = \int_{-\pi}^{\pi} ds \cos \nu s\, e^{-x(1 + \cos s)}. \quad (28)$$

In the limit of the cold sphere ($x \to 0$) we have from Eq. (27) the equation

$$4u = 4 + \frac{\frac{5}{2} - 4\nu_0^2}{\nu_0^2 - 1} - \frac{\frac{3}{2}}{\nu_1^2 - 1} - \frac{1}{\nu^2 - 1}, \quad (29)$$

which may obtain also directly from Eq. (3) for $l = 2$.

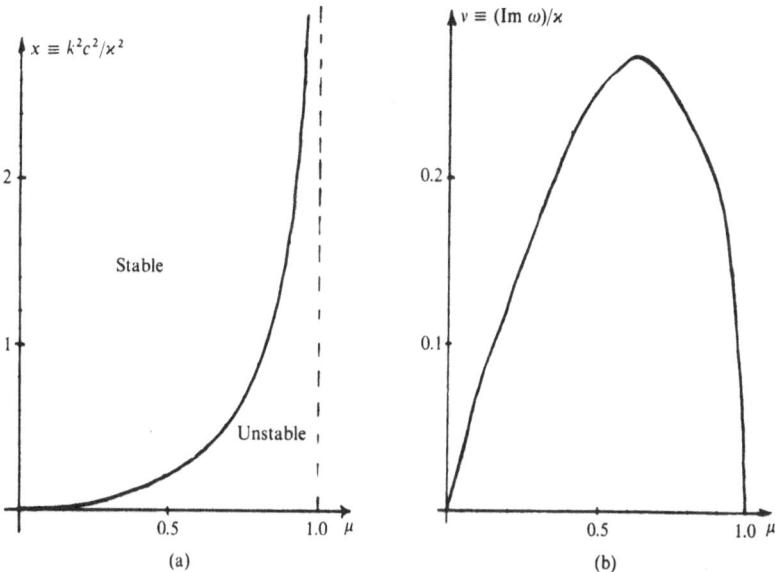

Figure 20. Stability of spherical systems with circular and nearly-circular orbits: (a) Curve of the marginal stability for the spheres with nearly-circular orbits. (b) Dependence of the instability growth rate on the parameter $\mu \equiv 2\Omega/\varkappa$ for the spheres with circular orbits ($l = 2$).

Equation (27) may be explicitly solved with respect to the square of dimensionless frequency[18] v^2 in two limiting cases: (1) $|\mu - 1| \ll 1$ and (2) $\mu \ll 1$. In both cases $v^2 \ll 1$; in the more interesting first limit we obtain ($x \gg 1$)

$$v^2 \approx \delta\left(\delta - \frac{a}{x^{3/2}}\right), \tag{30}$$

where $a = \text{const} = 9\sqrt{\pi/2}/16\pi$, $\delta = 1 - \mu$. The instability corresponds to $v^2 < 0$, e.g., in this limit it occurs if $a/x^{3/2} < \delta < 0$. The condition $\delta < 0$ means $\mu < 1$, which coincides with the instability condition of a cold sphere. The stability boundary ($v^2 = 0$) corresponds to $\delta = 0$ or $\delta = a/x^{3/2}$. The last formula yields the asymptotic behavior for $\mu \to 1$ of the marginal stability curve (of the stability boundary) for the systems under consideration. The instability region is infinitely extending if $\mu \to 1$, but the growth rates of perturbations are simultaneously going to zero.

For $\mu \ll 1$ ($x \ll 1$, $v \ll 1$) we have from Eq. (27)

$$v^2 \approx \tfrac{1}{2}(\tfrac{3}{2}x - \mu^2), \tag{31}$$

so that at the stability boundary $\mu^2 = 3x/2$.

For an arbitrary value of μ the marginal curve $x = x(\mu)$ is determined from Eq. (27) (with $v = 0$) by numerical methods. We represent this curve in Fig. 20(a). In the next, Fig. 20(b), we give the approximate curve of dependence of the instability increment of a cold sphere on the value of parameter μ.

---

[18] It is easily seen that Eq. (27) determines just the square of the frequency, v^2.

§4 Stability of Systems of Gravitating Particles Moving in Elliptical Orbits [112, 114]

4.1 Stability of a Sphere with Arbitrary Elliptical Particle Orbits

Assuming in formula (29), §1, $\Omega = 1$, $a = 1$, we rewrite the distribution function under investigation in the form

$$f_0 = \frac{\rho_0}{\pi^2} \frac{\theta[(1 - r^2)(1 - v_\perp^2) - v_r^2]}{\sqrt{(1 - r^2)(1 - v_\perp^2) - v_r^2}}. \tag{1}$$

The unperturbed potential of the homogeneous sphere is

$$\Phi_0 = \frac{r^2}{2} + \text{const.} \tag{2}$$

To investigate small perturbations, let us utilize the method of variation of the phase volume of the system (Chapter I, §6). Perturbation theory is constructed in a standard way. We write [111]

$$f = \frac{\rho_0}{\pi^2} \frac{\theta[(1 - r^2)(1 - v_\perp^2) - v_r^2 - \varepsilon\chi]}{\sqrt{(1 - r^2)(1 - v_\perp^2) - v_r^2 - \varepsilon\chi}}, \tag{3}$$

where $\varepsilon\chi(t, r, v)$ is the required perturbation which is the deviation of the argument of the distribution function from the stationary value. The linearized kinetic equation then has the form

$$\frac{d\chi}{dt} = 2v_r \frac{\partial \Phi_1}{\partial r} + \frac{2}{r}\left(\cos \alpha \frac{\partial \Phi_1}{\partial \theta} + \frac{\sin \alpha}{\sin \theta} \frac{\partial \Phi_1}{\partial \varphi}\right)(1 - r^2)v_\perp, \tag{4}$$

where $\tan \alpha = v_\varphi/v_\theta$. By performing, under the integral in the expression for complete density,

$$\rho = \frac{\rho_0}{2\pi^2} \int \frac{\theta[\varkappa^2(1 - v_\perp^2) - v_r^2 - \varepsilon\chi]}{\sqrt{\varkappa^2(1 - v_\perp^2) - v_r^2 - \varepsilon\chi}} \, dv_r \, dv_\perp^2 \, d\alpha \qquad (\varkappa^2 \equiv 1 - r^2), \tag{5}$$

the substitution of variables

$$-\varkappa_1^2 v_{\perp 1}^2 = -\varkappa^2 v_\perp^2 - \varepsilon\chi(0, v_\perp), \tag{6}$$

$$-v_{r1}^2 = -v_r^2 - \varepsilon\chi_1, \tag{7}$$

where

$$\varkappa_1^2 = 1 - r^2 - \varepsilon\chi_0,$$

$$\chi_0 \equiv \chi(v_r = 0, v_\perp = 0), \qquad \chi(0, v_\perp) \equiv \chi_2 = \chi(v_r = 0, v_\perp) - \chi_0,$$

$$\chi_1 = \chi - \chi_0 - \chi_2, \tag{8}$$

we shall obtain the formula for boundary displacement

$$\Delta R = -\tfrac{1}{2}\varepsilon\chi_0|_{r=1},\tag{9}$$

and the recipe for calculation of the perturbed density

$$\rho_1 = \frac{\rho_0}{2\pi^2}\int\frac{dv_r\,dv_\perp^2\,d\alpha}{\sqrt{(1-r^2)(1-v_\perp^2)-v_r^2}}$$

$$\times\left[\frac{1}{2}\frac{\partial}{\partial v_r}\left(\frac{\chi_1}{v_r}\right)+\frac{(\partial/\partial v_\perp^2)(\chi_0 v_\perp^2+\chi_2)}{1-r^2}\right].\tag{10}$$

The angular dependence of perturbation of the potential Φ_1 in the case of interest, as for any spherically symmetrical systems with equilibrium distribution functions, depending only on E and L^2, $f_0 = f_0(E, L^2)$, can be selected to be proportional with an individual spherical harmonic (see §2, Appendix)

$$\Phi_1 = \chi_l(r)Y_l^m(\theta, \varphi).\tag{11}$$

For homogeneous spheres, the radial part of the potential χ_l is represented in the form of a polynomial of the power n $(n \geq l)$:

$$\chi_l^{(n)}(r) = (r^n + C_2 r^{n-2} + \cdots + C_n)r^l \qquad (n \text{ even}),$$

which is the partial manifestation of the general result, according to which the potential perturbation of the systems with the quadratic potential similar to (1) can be sought for in the form of polynomials in degrees of Cartesian coordinates x, y, z.

Consider first of all the radial perturbations; then

$$\Phi_1^{(n)} = r^n + C_2 r^{n-2} + \cdots + C_n \qquad (n \text{ even}).\tag{12}$$

The kinetic equation

$$\frac{d\chi}{dt} = -2v_r\frac{\partial\Phi_1}{\partial r} = 2\frac{\partial\Phi_1}{\partial t} - 2\frac{d\Phi_1}{dt}\tag{13}$$

is solved by the method of integration over trajectories:

$$\chi(r, v_r, v_\perp) = -2i\omega\int_{-\infty}^{0}\Phi_1(r')e^{-i\omega t'}\,dt' - 2\Phi_1(r),\tag{14}$$

where

$$r'^2 = r^2\cos^2 t + v^2\sin^2 t + rv_r\sin 2t.\tag{15}$$

Substituting (12) and (15) into (14), one can obtain the expression for the perturbation $\chi(r, v_r, v_\perp)$ in the case of the arbitrary radial mode n. The function $\chi(r, v_r, v_\perp)$ thus obtained should be presented in the form (8), and then, using formula (10), we calculate the perturbed density ρ_1, as well as the potential $\Phi_1(r)$ corresponding to this density ρ_1 [it is evident that for radial perturbations the boundary displacement in (9) is not essential]. Comparison of the calculated potential with the initial expression in (12) yields the spectrum of system oscillations (and as always, to obtain the spectrum it is enough to retain in calculations only the term with the highest degree r).

The calculations described in the general case lead to a cumbersome expression for the oscillation spectrum. Therefore, below we shall restrict ourselves only to the calculation of frequencies for some "lower" (the most large-scale) modes ($n = 2, 4$).

For $n = 2$, we have

$$\Phi_1 \sim r^2; \tag{16}$$

therefore from (12)–(15) we get

$$\chi \sim -\frac{4r^2}{\omega^2 - 4} + \frac{4v^2}{\omega^2 - 4}. \tag{17}$$

We decompose χ to formulae (8):

$$\chi_0 = -\frac{4r^2}{\omega^2 - 4}, \qquad \chi_2 = \frac{2v_1^2}{\omega^2 - 4}, \qquad \chi_1 = \frac{4v_r^2}{\omega^2 - 4}. \tag{18}$$

Let us now find the perturbed density [according to formula (10)]

$$\rho_1 \sim \rho_0 \frac{6}{\omega^2 - 4}, \tag{19}$$

so that the perturbed potential must be

$$\Phi_1 \sim \frac{3}{\omega^2 - 4} r^2. \tag{20}$$

Hence, comparing with (16), we get $\omega^2 = 1$.

For $n = 4$, the calculations are similar. We assume that $\Phi_1 \sim r^4$; then

$$\chi \sim 4r^4 \left(\frac{1}{\omega^2 - 16} + \frac{1}{\omega^2 - 4} \right) + 4v^4 \left(\frac{1}{\omega^2 - 16} - \frac{1}{\omega^2 - 4} \right) - 16$$

$$\times \frac{r^2 v_r^2 - r^2 v^2 / 2}{\omega^2 - 16},$$

$$\rho_1 \sim \rho_0 \left(\frac{\frac{33}{2}}{\omega^2 - 16} + \frac{\frac{5}{2}}{\omega^2 - 4} \right) r^2, \qquad \Phi_1 \sim \frac{3r^4}{20} \left(\frac{\frac{35}{2}}{\omega^2 - 16} + \frac{\frac{5}{2}}{\omega^2 - 4} \right). \tag{21}$$

The oscillation spectrum for the $n = 4$ mode is obtained in the form

$$\frac{7}{\omega^2 - 16} + \frac{1}{\omega^2 - 4} = \frac{8}{3}; \tag{22}$$

hence we determine the squares of frequencies

$$\omega^2 = \tfrac{1}{2}(23 \pm \sqrt{207}). \tag{23}$$

All frequencies in (23) are real, which corresponds to oscillation stability.

Consider now perturbations that bend the system boundary. Among them, seemingly, the most "dangerous" for stability are perturbations that imitate oscillations of incompressible liquid, i.e., those corresponding to the case where the perturbed density $\rho_1 = 0$. We derive the general formula, describing the spectrum of these perturbations (as we shall see, it has a rather simple form). The perturbed potential in this case is written as:

$$\Phi_1^{(l,\,m)} = r^l P_l^m(\cos\theta)e^{im\varphi}. \tag{24}$$

It is evident, however, that the oscillation frequencies of the system in question are independent of the azimuthal number m; therefore, we deal with perturbations as

$$\Phi_1^{(l,\,0)} = r^l P_l(\cos\theta) \equiv F_l(x, y, z) = z^l + \cdots, \tag{25}$$

where $F_l(x, y, z)$ is the harmonic polynomial of the power l. It may be shown that in order to determine the spectrum it is sufficient to restrict oneself to the higher degree z in the expression for Φ_1 in the calculations. The χ function is determined from the kinetic equation by integrating over trajectories (it is more convenient in Cartesian coordinates). For the boundary displacement, we obtain

$$\Delta R = \frac{1}{2^l} \sum_{k=0}^{l-1} C_{l-1}^k \left(\frac{1}{\omega - 2k + l - 2} - \frac{1}{\omega - 2k + l} \right) P_l(\cos\theta). \tag{26}$$

The spectrum is determined from the condition

$$\frac{\partial\Phi_1}{\partial r}\bigg|_{r=1+0} - \frac{\partial\Phi_1}{\partial r}\bigg|_{r=1-0} = 4\pi G\sigma = 4\pi G\rho_0 \Delta R, \tag{27}$$

and can be presented as

$$1 = -\frac{3l}{2^{l-1}(2l+1)} \sum_{k=0}^{l-1} C_{l-1}^k \frac{l-2k}{\omega^2 - (l-2k)^2}. \tag{28}$$

At $l = 1$, we infer a natural result: $\omega^2 = 0$. The quadrupole perturbation $l = 2$ has $\omega^2 = \frac{14}{5}$ (a more detailed calculation of this mode is given in [112]).

For subsequent l, we shall have a larger number of eigenmodes; they all are stable. Indeed, let, e.g., l be odd: $l = 2N + 1$ (the case for even l is treated in a similar way). Then Eq. (28) can be reduced to the form

$$-1 = \sum_{p=-N}^{N} \frac{\gamma_p}{(\omega - 2p)^2 - 1}, \qquad \gamma_p \equiv \frac{3(2N+1)}{2^{2N+2}(4N+3)} \frac{(2N)!}{(N+p)!(N-p)!}, \tag{29}$$

which is equivalent to the standard form in (68), §6, Chapter I. The stability follows from the easily provable inequality

$$\gamma_p = \frac{3}{4} \frac{2N+1}{4N+3} \cdot \frac{(2N)!}{2^{2N}(N+p)!(N-p)!} < 1. \tag{30}$$

(in the form of (30) each of three co-factors is less than 1).

4.2 Instability of a Rotating Freeman Sphere

Freeman [203] suggested collisionless models of rotating homogeneous ellipsoids (with arbitrary semiaxes a, b, c). They are dealt with in the next chapter. In the limit $a = b = c \equiv R$ we get a sphere, the equilibrium in which is ensured by rotation of particles in the (x, y) plane and by the local velocity dispersion on the z-axis, so that at each point of the sphere we have only one-dimensional dispersion. The model in question rotates about the z-axis with the maximally possible angular velocity $\Omega = (4\pi G\rho_0/3^{1/2})$.

The distribution function (in a rotating frame of reference) has the form

$$f_0 = \frac{\rho_0}{\pi} \frac{\delta(v_x)\delta(v_y)}{\sqrt{1 - r^2 - v_z^2}}, \tag{1}$$

where the radius R of the sphere and the value Ω are assumed to be 1.

We apply the perturbation theory [91, 96] combined from the method of δ expansion and from the method of variation of phase volume. We write the perturbed distribution function as

$$f = \frac{\rho_0}{\pi} \frac{\delta(v_x)\delta(v_y)}{\sqrt{1 - r^2 - v_z^2 - \varepsilon\chi}} + \frac{\varepsilon\rho_0}{\pi} B \frac{\delta'(v_x)\delta(v_y)}{\sqrt{1 - r^2 - v_z^2}} + \frac{\varepsilon\rho_0}{\pi} C \frac{\delta(v_x)\delta'(v_y)}{\sqrt{1 - r^2 - v_z^2}}, \tag{2}$$

where new unknown functions

$$\chi = \chi(t, r, v_z), \qquad B = B(t, r, v_z), \qquad C = C(t, r, v_z)$$

are introduced.

Substituting f into the kinetic equation (it is convenient to write it here in Cartesian coordinates)

$$\frac{\partial f}{\partial t} + v_x \frac{\partial f}{\partial x} + v_y \frac{\partial f}{\partial y} + v_z \frac{\partial f}{\partial z} + 2v_x \frac{\partial f}{\partial v_y} - 2v_y \frac{\partial f}{\partial v_x} + \left(x - \frac{\partial\Phi}{\partial x}\right) \frac{\partial f}{\partial v_x}$$

$$+ \left(y - \frac{\partial\Phi}{\partial y}\right) \frac{\partial f}{\partial v_y} - \frac{\partial\Phi}{\partial z} \frac{\partial f}{\partial v_z} = 0, \tag{3}$$

and linearizing it, we arrive at the following system of equations for the χ, B, C functions:

$$\frac{\partial\chi}{\partial t} + v_z \frac{\partial\chi}{\partial z} - z \frac{\partial\chi}{\partial v_z} - 2(1 - r^2 - v_z^2)\left(\frac{\partial B}{\partial x} + \frac{\partial C}{\partial y}\right) - 2(Bx + Cy) = v_z \frac{\partial\Phi_1}{\partial z}, \tag{4}$$

$$\frac{\partial B}{\partial t} + v_z \frac{\partial B}{\partial z} - z \frac{\partial B}{\partial v_z} - 2C = \frac{\partial\Phi_1}{\partial x}, \tag{5}$$

$$\frac{\partial C}{\partial t} + v_z \frac{\partial C}{\partial z} - z \frac{\partial C}{\partial v_z} + 2B = \frac{\partial\Phi_1}{\partial y}. \tag{6}$$

We show that the system under consideration is unstable with respect to surface perturbations. We select the simplest type of surface bendings, at which the potential perturbation has the form

$$\Phi_1 = \varepsilon(x + iy)^l \equiv \varepsilon A r^l Y_l^l(\theta, \varphi) \tag{7}$$

(ε is the parameter of smallness); then from (5) and (6) we get

$$B = \frac{l(x + iy)^{l-1}}{\omega - 2}, \qquad C = -\frac{l(x + iy)^{l-1}}{\omega - 2}. \tag{8}$$

Substitution of (8) into (4) yields

$$\chi = -2l \frac{(x + iy)^l}{\omega(\omega - 2)}. \tag{9}$$

From (9) it follows that the density perturbation is really equal to zero. The boundary displacement ΔR can be determined from the equation of the perturbed sphere surface

$$1 - r^2 + 2l\varepsilon \frac{(x + iy)^l}{\omega(\omega - 2)} = 0,$$

so that

$$\Delta R = \varepsilon \frac{l}{\omega(\omega - 2)} (x + iy)^l|_{r=1} = \varepsilon \frac{l}{\omega(\omega - 2)} A Y_l^l(\theta, \varphi). \tag{10}$$

The displacement ΔR refers to the surface density $\sigma = \rho_0 \Delta R$. The potential produced by the simple layer with the surface density σ, inside and outside the sphere, respectively:

$$\Phi_< = D_1 r^l Y_l^l(\theta, \varphi), \tag{11}$$

$$\Phi_> = D_2 r^{-l-1} Y_l^l(\theta, \varphi), \tag{12}$$

D_1 and D_2 are determined from the equations:

$$\Phi_< = \Phi_>|_{r=1}, \qquad \frac{\partial \Phi_>}{\partial r} - \frac{\partial \Phi_<}{\partial r}\bigg|_{r=1} = 4\pi G\sigma.$$

Hence we determine the coefficient

$$D_1 = -\varepsilon \frac{3l}{2l + 1} \frac{A}{\omega(\omega - 2)}. \tag{13}$$

Comparing (13) with (7), we arrive at the following equation for the determination of the eigenfrequencies ω:

$$\omega^2 - 2\omega + \frac{3l}{2l + 1} = 0; \tag{14}$$

hence

$$\omega = 1 \pm \sqrt{1 - \frac{3l}{2l + 1}}. \tag{15}$$

This equation shows that at all l, apart from $l = 1$ (the case corresponding to sphere displacement as a whole) instability occurs. Its growth rate with increasing l increases and asymptotically approaches the value $\gamma_\infty = 1/\sqrt{2}$.

Note that with respect to simple axial-symmetrical perturbations the system proves to be stable. Thus, for the mode that corresponds to isotropic extension or contraction of the sphere, we get $\omega^2 = 1$, while for the case of perturbation $\Phi_1 \sim r^2 P_2(\cos \theta)$ converting the sphere to the ellipsoid of rotation, $\omega^2 = 14/5$.

Freeman's sphere in (1) is the simplest model for the system with elliptical trajectories of particles. Not all possible elliptical trajectories are represented here but only those whose projections onto the (x, y) plane yield circumferences. The movement of all particles along these circumferences occurs in one direction, so that the sphere, as a whole, happens to rotate with the maximum speed. Instability is explained by just this fast rotation of the sphere (as a whole), due to which the effective gravity force on its surface vanishes. The system that is the *superposition* of two identical Freeman's spheres rotating in opposite directions (and consequently, as a whole, being at rest) is stable with respect to perturbations under consideration. Actually, quite similarly to the previous case, we can derive the following equation[19] for frequencies of eigenoscillations (7):

$$\frac{3l}{2l + 1} \left[\frac{1}{(\omega + l)(\omega + l - 2)} + \frac{1}{(\omega - l)(\omega - l + 2)} \right] + 2 = 0. \quad (16)$$

This is the biquadratic equation for ω, all solutions of which, as is easy to prove, are real.

As far as flattening along the z-axis into a ellipsoid of revolution is concerned, all such systems are stable: the velocity z-component dispersion proves to be sufficient for their stabilization.

The Camm model considered in Section 4.1, which presents all possible elliptical orbits, proved, to be stable as we have seen.

It is interesting to compare the form of eigenfunctions in the two cases considered. Let us give for Freeman's sphere (1) an example of the mode that changes not only the location of the boundary but also the local density:

$$\Phi_1 = [br^4 P_4^2(\cos \theta) + r^4 P_2^2(\cos \theta) + ar^2 P_2^2(\cos \theta)]e^{2i\varphi}. \quad (18)$$

It is easy to see that it is a definite combination of spherical harmonics; individual harmonics will not in this case be eigenfunctions. The reason evidently is the lack of spherical symmetry of the initial state of the system *movement*. For the (single) stable root $\omega_7 = 0.64 + 0.22i$; the coefficients b and a are $b = 0.0089 - 0.00511i$ and $a = -1.1052 + 0.02543i$. The characteristic

---

[19] If the relative weights of oppositely rotating components are α and β ($\alpha + \beta = 1$), then Eq. (16) is replaced by

$$\frac{3l}{2l + 1} \left[\frac{\alpha}{(\omega + l)(\omega + l - 2)} + \frac{\beta}{(\omega - l)(\omega - l + 2)} \right] + 1 = 0. \quad (17)$$

equation for oscillation frequencies has the form

$$\omega^8 + 8\omega^7 + \frac{37}{3}\omega^6 - \frac{608}{21}\omega^5 - \frac{986}{21}\omega^4 + \frac{1012}{21}\omega^3$$

$$+ \frac{88}{21}\omega^2 - \frac{1599}{147}\omega - \frac{1312}{49} = 0.$$

The remaining roots of this equation are: $\omega_1 = 1.74$, $\omega_2 = -0.25$, $\omega_3 = -1.13$, $\omega_4 = -1.70$, $\omega_5 = -3.93$, $\omega_6 = 0.64 - 0.22i$, $\omega_8 = -4.0$.

The "next" mode also has a similar appearance here:

$$\Phi_1 = [cr^3 P_3^1(\cos\theta) + br^3 P_1^1(\cos\theta) + arP_1^1(\cos\theta)]e^{i\varphi},$$

where the constants c, b, a as well as eigenfrequencies of oscillations are calculated in [91].

Unlike the Freeman sphere, the eigenfunctions of the Camm aniso-tropic sphere are proportional to individual spherical harmonics:

$$\Phi_1 \sim (r^l + ar^{l-2} + \cdots)P_l^m(\cos\theta)e^{im\varphi},$$

where the oscillation frequencies are independent of m. This is a natural consequence of a *complete* spherical symmetry of the equilibrium state.

§ 5 Stability of Systems with Radial Trajectories of Particles [50, 11, 43[ad]]

5.1 Linear Stability Theory

As is known, the stellar orbits in spherical galaxies and spherical clusters are very much elongated along their radius. Due to this, the question arose of the stability of systems with motion of particles close to radial.

For purely radial movements, the answer to this question appears to be rather obvious [50, 111]: such systems must be unstable. Actually, if, e.g., in the movement in circular orbits there is some analog of elastic forces, then at purely radial movements nothing prevents "sticking" of neighboring orbits: The velocity dispersion of the particles in the tangential direction, which would have stabilized the system, is lacking in this case.

Consider now a narrow cone in such a sphere (AOB in Fig. 21) and imagine that at the initial time we have compressed it to produce the cone $A'OB'$ by approaching the generatrices $AO \rightarrow A'O$, $BO \rightarrow B'O$. The inevit-ability of the further Jeans "flattening" is then evident, if one further takes into account that the particles cannot escape from the perturbation region[20] (in the linear regime).

---

[20] This is not the case where the initial perturbation is located in the region restricted in radius. Then the particles "run away" from the perturbed region, and the perturbation dimin-ishes in part. at least at the beginning. Instability here may have an oscillative character.

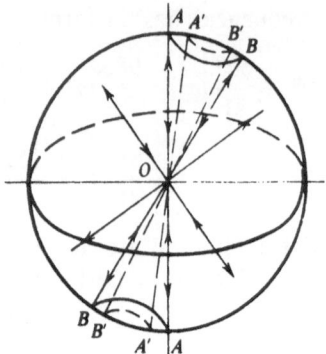

Figure 21. Spherically symmetrical systems with purely radial particle motions.

Due to the importance of principle of this question, let us nonetheless give a strict mathematical proof for the instability [11].

Since $f_0 \sim \delta(v_\theta)\delta(v_\varphi)$, we expand the perturbed distribution function f_1 into the series in δ functions of v_θ and v_φ and their derivatives:

$$f_1 = A\delta(v_\theta)\delta(v_\varphi) + B\delta'(v_\theta)\delta(v_\varphi) + C\delta(v_\theta)\delta'(v_\varphi), \tag{1}$$

where A, B, C are the $v_r, r, \theta, \varphi, t$ functions. The perturbed density ρ_1, θ and φ components of the matter flow, j_θ and j_φ, are represented, respectively, by the integrals over v_r from A, B, and C:

$$\rho_1 = \int A \, dv_r, \tag{2}$$

$$j_\theta = - \int B \, dv_r, \tag{3}$$

$$j_\varphi = - \int C \, dv_r. \tag{4}$$

Substituting (1) into kinetic equation (1), Section 3.1, we obtain the system of equations [$f_0 = \delta(L^2)\varphi_0(E)$]:

$$\frac{\partial A}{\partial t} + \hat{D}A + \frac{2v_r}{r} A - \frac{1}{r}\left(\frac{\partial B}{\partial \theta} + B \cot \theta\right) - \frac{1}{r \sin \theta}\frac{\partial C}{\partial \varphi} = \frac{\partial \Phi_1}{\partial r}\frac{v_2}{r^2}\frac{\partial \varphi_0}{\partial E}, \tag{5}$$

$$\frac{\partial B}{\partial t} + \hat{D}B + \frac{3v_r}{r} B = \frac{1}{r}\frac{\partial \Phi_1}{\partial \theta}\frac{1}{r^2}\varphi_0, \tag{6}$$

$$\frac{\partial C}{\partial t} + \hat{D}C + \frac{3v_r}{r} C = \frac{1}{r \sin \theta}\frac{\partial \Phi_1}{\partial \varphi}\frac{1}{r^2}\varphi_0, \tag{7}$$

where $\hat{D}$ is the following operator:

$$\hat{D} = v_r \frac{\partial}{\partial r} - \frac{\partial \Phi_0}{\partial r}\frac{\partial}{\partial v_r}. \tag{8}$$

The Poisson equation has the form

$$\frac{1}{r^2}\frac{\partial}{\partial r}\left(r^2\frac{\partial \Phi_1}{\partial r}\right) + \frac{1}{r^2 \sin \theta}\frac{\partial}{\partial \theta}\left(\sin \theta \frac{\partial \Phi_1}{\partial \theta}\right) + \frac{1}{r^2 \sin^2 \theta}\frac{\partial^2 \Phi_1}{\partial \varphi^2} = 4\pi G \int A \, dv_r.$$

(9)

For $\Phi_1 = \chi(r)Y_l^m(\theta, \varphi)$, from Eqs. (5)–(9), it is easy to see that there should be

$$B = F_1(r, v_r, t)\frac{\partial}{\partial \theta}Y_l^m, \qquad \sin \theta C = F_1(r, v_r, t)\frac{\partial}{\partial \varphi}Y_l^m, \qquad (10)$$

$$A = a_1(r, v_r, t)Y_l^m,$$

so that instead of (5)–(9) we shall have

$$\frac{1}{r^2}\frac{d}{dr}\left(r^2\frac{d\chi}{dr}\right) - \frac{l(l+1)}{r^2}\chi = 4\pi G \int a_1 \, dv_r, \qquad (11)$$

$$\frac{\partial a_1}{\partial t} + \hat{D}a_1 + \frac{2v_r}{r}a_1 + \frac{l(l+1)}{r}F_1 = \frac{\partial \chi}{\partial r}\frac{v_r}{r^2}\frac{\partial \varphi_0}{\partial E}, \qquad (12)$$

$$\frac{\partial F_1}{\partial t} + \hat{D}F_1 + \frac{3v_r}{r}F_1 = \frac{\chi}{r^2}\varphi_0. \qquad (13)$$

The natural substitution of $a_1 r^2 = a$ and $F_1 r^3 = F$ somewhat simplifies the system (11)–(13), by reducing it to the following:

$$\frac{d}{dr}\left(r^2\frac{d\chi}{dr}\right) - l(l+1)\chi = 4\pi G \int a \, dv_r, \qquad (14)$$

$$\frac{\partial a}{\partial t} + \hat{D}a + \frac{l(l+1)}{r^2}F = v_r \frac{d\chi}{dr}\frac{d\varphi_0}{dE} = \frac{d\chi}{dr}\frac{\partial \varphi_0}{\partial v_r}, \qquad (15)$$

$$\frac{\partial F}{\partial t} + \hat{D}F = \chi \varphi_0. \qquad (16)$$

We represent the distribution function $\varphi_0(E)$ in the form

$$\varphi_0(E) = \int \varphi_0(E_0)\delta(E - E_0) \, dE_0, \qquad (17)$$

and solve first of all Eqs. (15) and (16) for one "monochromatic" component of (17):

$$\delta(E - E_0) = \frac{1}{2|v_{E_0}|}[\delta(v_r - v_{E_0}) + \delta(v_r + v_{E_0})], \qquad v_{E_0} \equiv \sqrt{2[E_0 - \Phi_0(r)]},$$

(18)

In this case, one can further perform the δ expansion over $(v_r \pm v_{E_0})$:

$$a = a_+ \delta(v_r - v_0) + a_- \delta(v_r + v_0) + b_+ \delta'(v_r - v_0) + b_- \delta'(v_r + v_0), \qquad (19)$$

$$F = F_+ \delta(v_r - v_0) + F_- \delta(v_r + v_0), \qquad (20)$$

where the coefficients $a_\pm, b_\pm, F_\pm$ are the r, t (and E_0) functions; $v_0 \equiv v_{E_0}$. The plus and minus signs distinguish the two halves of the flow with opposite velocities. After substitution of (19) and (20) into (15) and (16) we have

$$\frac{\partial a_\pm}{\partial t} \pm v_0 \frac{\partial a_\pm}{\partial r} \pm v_0' a_\pm - \frac{\partial b_\pm}{\partial r} + \frac{l(l+1)}{r^2} F_\pm = 0, \tag{21}$$

$$\frac{\partial b_\pm}{\partial t} \pm v_0 \frac{\partial b_\pm}{\partial r} \pm 2v_0' b_\pm = \frac{1}{2v_0} \frac{\partial \chi}{\partial r}, \tag{22}$$

$$\frac{\partial F_\pm}{\partial t} \pm v_0 \frac{\partial F_\pm}{\partial r} \pm v_0' F_\pm = \frac{1}{2v_0} \chi. \tag{23}$$

This system will be simplified if the substitution

$$X_- = v_0 a_\pm, \qquad Y_\pm = v_0^2 b_\pm, \qquad Z_\pm = v_0 F_\pm \tag{24}$$

is made. Then we shall have instead of (21)–(23)

$$\hat{D}_\pm X_\pm - \frac{1}{v_0} \frac{\partial Y_\pm}{\partial r} + \frac{2v_0'}{v_0^2} Y_\pm + \frac{l(l+1)}{r^2} Z_\pm = 0, \tag{25}$$

$$\hat{D}_\pm Y_\pm = \frac{v_0}{2} \frac{\partial \chi}{\partial r}, \tag{26}$$

$$\hat{D}_\pm Z_\pm = \tfrac{1}{2}\chi, \tag{27}$$

where

$$\hat{D}_\pm = \frac{\partial}{\partial t} \pm v_0 \frac{\partial}{\partial r} = \left(\frac{d}{dt}\right)_\pm \tag{28}$$

are the differentiation operators in time along unperturbed radial trajectories of the flow particles. The closed system of equations will ensue if the Poisson equation is added to (25)–(27). Since the contribution of an individual component with the energy E_0 to the perturbed density is

$$\rho_{1E_0} = (a_+ + a_-)\frac{\varphi_0(E_0)}{r^2} = \frac{\varphi_0(E_0)(X_+ + X_-)}{v_0 r^2} = \rho_{E_0}(X_+ + X_-), \tag{29}$$

the total perturbed density

$$\rho_1 = \int dE_0 \, \rho_{1E_0}, \tag{30}$$

and the Poisson equation will be of the form

$$\frac{1}{r^2}\frac{d}{dr}\left(r^2 \frac{d\chi}{dr}\right) - \frac{l(l+1)}{r^2}\chi = 4\pi G \int dE_0 (X_+ + X_-)\rho_{E_0} \tag{31}$$

[for the one-component system, on the right-hand side of (31), simply for $4\pi G(X_+ + X_-)/r^2 v_0 = \omega_0^2(X_+ + X_-)$, where $\omega_0^2 = 4\pi G\rho_0(r)$ is the local Jeans frequency].

The problem of searching for the eigenmodes $(\sim e^{-i\omega t})$ of the system of equations (25)–(27), (31) is very complicated, in any event for the analytical solution. But to prove instability, it is sufficient to restrict oneself to perturbations of any partial form. Since the system, in the transversal direction, is cold, then it is clear beforehand that perturbations with whatever small "transversal" (with respect to radius) wavelength (among others) should be unstable. In the approximate equations describing the evolution of perturbations, one then has to take into account only transversal forces, which contribute to the joining of orbits. This circumstance strongly simplifies the formal analysis.

Thus, let us make use of the approximation of the quasiclassical type, but select the direction of fast change of "humps" and "hollows" of perturbed values, this time, to be perpendicular with the radii (unlike, e.g., the works devoted to short wavelength stability of cold disk systems; see Section 2.2, Chapter V).

The Poisson equation in this approximation will be obtained from (31) by neglecting the terms containing derivatives over r:

$$-\frac{l^2 \chi}{r^2} = 4\pi G \int dE_0 \, (X_+ + X_-)\rho_{E_0}. \tag{32}$$

From (26) and (27) it follows that $Y_\pm$ and $Z_\pm$ are the quantities of one order of magnitude; therefore, the equations for $X_\pm$ are written thus:

$$\hat{D}_\pm X_\pm \approx -\frac{l^2}{r^2} Z_\pm, \tag{33}$$

From (27) we express χ in terms of $Z_\pm$, $\chi = 2\hat{D}_\pm Z_\pm$ and substitute into (32):

$$-\frac{l^2}{r^2} \hat{D}_\pm Z_\pm = 2\pi G \int dE_0 \, (X_+ + X_-)\rho_{E_0}.$$

Introducing the new variables

$$\mathscr{I}_\pm = -v_0 F_\pm = -Z_\pm, \tag{34}$$

$$\xi_\pm = \frac{1}{l^2} r v_0 a_\pm = \frac{1}{l^2} r X_\pm, \tag{35}$$

we finally obtain the following symmetrical system of equations [11]:

$$\hat{D}_\pm \xi_\pm = \frac{\mathscr{I}_\pm \pm v_0 \xi_\pm}{r}, \tag{36}$$

$$\hat{D}_+ \mathscr{I}_+ = \hat{D}_- \mathscr{I}_- = 2\pi Gr \int dE_0 \, \rho_{E_0}(\xi_+ + \xi_-). \tag{37}$$

Instability is proved [11] by constructing the Lyapunov functional

$$V(t) = \int dE_0 \int \frac{\xi_+ \mathscr{I}_+ + \xi_- \mathscr{I}_-}{r} \, dm_{E_0}, \qquad dm_{E_0} = r^2 \rho_{E_0} \, dr, \tag{38}$$

where the integrals over m_{E_0} are taken between the turning points of the corresponding flows.

Differentiating (38) and taking into account the invariance of dm_{E_0}, we shall obtain, after some calculations, an explicitly positive-definite form:

$$\frac{dV}{dt} = \int dE_0 \int \frac{(\mathscr{I}_+)^2 + (\mathscr{I}_-)^2}{r^2} dm_{E_0}$$

$$+ 2\pi Gr^2 \int dr \left[\int dE_0 \, \rho_{E_0}(\xi_+ + \xi_-) \right]^2 \geq 0. \tag{39}$$

This means instability (see, e.g., [156]). The meaning of the values $\xi^\pm$ and $\mathscr{I}^\pm$ is readily cleared up; it is the following:

$$\xi^\pm = \frac{r}{l^2} \frac{\rho_1^\pm}{\rho_0}, \tag{40}$$

$$\mathscr{I}^\pm = r\tilde{v}_\theta = r\tilde{v}_\varphi, \tag{41}$$

where $\tilde{v}_\theta = \tilde{v}_\varphi$ are the amplitudes of the average θ and φ components of velocity of the flow in question. The quantities $\xi_\pm$ may also be given another interpretation. For this purpose, let us write the continuity equation for an individual flow with allowance for the smallness of the terms containing the derivatives over r, in comparison with the derivatives over θ and φ:

$$\frac{\partial \rho_1^\pm}{\partial t} + \rho_0^\pm \left(\frac{1}{r} \frac{\partial v_\theta^\pm}{\partial \theta} + \frac{1}{r \sin \theta} \frac{\partial v_\varphi^\pm}{\partial \varphi} \right) = 0. \tag{42}$$

In switching over to linear displacements ζ_θ, ζ_φ ($v_\theta = \partial \zeta_\theta / \partial t$, $v_\varphi = \partial \zeta_\varphi / \partial t$) here, from (42) we shall obtain

$$\rho_1^\pm = -\rho_0^\pm \left(\frac{1}{r} \frac{\partial \zeta_\theta^\pm}{\partial \theta} + \frac{1}{r \sin \theta} \frac{\partial \zeta_\varphi^\pm}{\partial \varphi} \right). \tag{43}$$

Since the transversal displacements $(\zeta_\theta, \zeta_\varphi) \sim \xi \cdot r \cdot \text{grad } Y_l^m(\theta, \varphi)$, then from (43) it follows that

$$\rho_1^\pm = \rho_0^\pm \zeta l(l + 1) \simeq \rho_0^\pm \zeta^\pm l^2. \tag{44}$$

From comparison of (44) and (40) it is easy to see that $\xi^\pm = \zeta^\pm$. $\xi^\pm$ in this approximation coincides with the amplitude of the linear displacement in the transversal direction (without taking into account the angular dependence).

Just this interpretation was followed by Antonov, who considered perturbations in the form proportional with the sectorial harmonics, $\Phi_1 = Y_l^l(\theta, \varphi)\chi(r) \sim \sin^l \theta \cos l\varphi \cdot \chi(r)$, that is, $l = m$, $l \gg 1$. Then Φ_1 is essentially different from zero in the narrow zone near the equatorial plane

$$|\theta - \tfrac{1}{2}\pi| \sim l^{-1/2}, \tag{45}$$

in which the displacement in φ is much greater than the displacement in θ due to the angular dependence of Φ_1. The quantities $\xi^\pm$ and $\mathscr{I}^\pm$ in this case should be thought of correspondingly as the linear displacement in the equatorial plane in the φ azimuth and the angular momentum (rv_φ) of the particle.

5.2 Simulation of a Nonlinear Stage of Evolution

In the previous subsection we stated that spherical systems with purely radial motions of stars are unstable in a linear approximation. It means that initially small disturbances will then increase. It is interesting to follow an evolution of such systems in a nonlinear stage. This problem was solved in $[43^{ad}]$ by a "method of particles" for the simplest model of a stellar sphere with the equilibrium distribution function (40), §1:

$$f_0 = \frac{\beta}{\pi r^2} \delta(v_\theta)\delta(v_\varphi)\left(8\pi G\beta \ln \frac{a}{r} - v_r^2\right)^{-1/2}. \tag{1}$$

5.2.1. On the Simulation of a Collisionless System by a "Method of Particles."
Since here we encounter this method for the first time, let us give a brief description (for technical details, cf. §3, Appendix).

A real system [for example, a stellar sphere, described by the distribution function $f_0(E, L)$] is represented in the form of the system N of points interacting with each other. Since actually N cannot be taken large, $N_{max} \simeq 300$–400,[21] and, in such a model system with a purely Coulomb law of interaction, the role of "collisions" for close passages of particles would be highly overestimated, then for the force of interaction between two particles with the masses m_1 and m_2 one takes the Newton law "cutoff" at small distances in the simplest way[22]

$$F_i = -G \frac{m_1 m_2 (x_{1i} - x_{2i})}{(r_{12}^2 + c^2)^{3/2}}, \tag{2}$$

where F_i is the ith component of force ($i = x, y, z$), r_{12} is the separation between the particles, c is the cutoff radius. Normally, c is chosen of the order of several percent from the radius of the system R. It is clear that such a cutoff should not essentially affect the correctness of the description of the large-scale processes of interest (in particular, of the long-wave instability), with characteristic dimensions of the order of R.

Each of the particles of the model system moves according to the second Newtonian law under the action of the forces of the form (2) produced by all the other particles,

$$\frac{d^2 x_i^{(n)}}{dt^2} = -G\tilde{m}^2 \sum_{m \neq n} \frac{(x_i^{(n)} - x_i^{(m)})}{(r_{mn}^2 + c^2)^{3/2}} \qquad (n = 1, 2, \ldots, N), \tag{3}$$

where it is assumed that individual masses of particles are identical and are equal to $\tilde{m}$. If one assumes a certain initial state, i.e., initial coordinates and

---

[21] For a larger number of particles N, other methods of simulation $[294, 67^{ad}]$ are applied. The configuration space of the system is divided into rather small regions; then the number of particles in each of these regions is counted, the average density is found, and the Poisson equation is solved by yielding the self-consistent potential, etc.

[22] If one assumes in (2) that there is, for example, exponential cutoff, then for the calculation of the exponent one would spend all the counting time.

velocities N of particles, then, by solving by computer the system of equations of second order (3), we may, in principle, determine all the further evolution of the model system. Here, of course, one has to bear in mind that any computation of this kind will not possess any property of reproductivity, implying that by assuming another initial state with locations of particles in the phase space of the system close to the first, we shall finally arrive at quite different co-ordinates and velocities of individual particles (deviations small at the initial time will become large in the course of time). Therefore, on the one hand, it is really possible to speak only about some average statistical characteristics of the system, and, on the other hand, it is necessary to duplicate the count by different methods, by attaining independence of these average characteristics of a concrete method of calculations.

This is the general ideology, which was, for example, followed by Ostriker and Peebles [301] in the problem of the instability of the form of galactic disk models (cf. Section 3.2, Chapter IV). It also remains in force without changes in the approach discussed below.

The differences are associated with the method of assuming the initial state. In [301, 294] dealing with spiral galactic simulation, the construction of the initial state of the system consisted of successive stages: (1) distribution of particles in the galactic plane approximately corresponding to the previously assumed law of variation of the surface density with the distance r from the center [for example, $\sigma_0(r) \sim 1/r$], (2) calculation of the velocities of the circular, regular rotation, and (3) the addition to the regular rotation of small peculiar motions (normally, with a Maxwellian velocity distribution of particles).

It is evident that, for our purposes, such an approach is inapplicable. For example, in the case of interest here, one needs to construct the initial state corresponding to the distribution function (7). We, in fact, shall "play" the positions and velocities of particles in such a way that in the limit as $N \to \infty$, the model system obtained will exactly correspond to $f_0(E, L)$. However, even for the real $N \sim 100$–300, the system "played" is close to the equilibrium one (1) (of course, the similarity grows with increasing N). The model system obtained may thus be considered as the one representing an equilibrium state with superposed small initial perturbations. In the system in itself, these perturbations will increase (in the case of instability). Owing to the statistical character of the "playing" of the distribution function, the typical scale of the initial perturbations $l \sim R/\sqrt[3]{N}$ (R is the size of the system); however, there are also larger scales, including $\sim R$. (The latter may be enhanced by imposing, in addition perturbations ("artificial") of a corresponding scale). In spherical systems, perturbations with a wavelength $\lambda \sim l$ are normally stable so that by solving the equations of motion in (3) we shall be able to follow the evolution of perturbations with $\lambda \sim R$, i.e., in essence, the variation of the form. It seems on the face of it that the systems with trajectories close to radial ones are the exception. However, as we shall see below, in this case an evolution involves also not only "transversal heating" of the system but an evident change of its form as well (see Fig. 23). The most interesting, large-scale disturbances may also be selected in an artificial way, if we suppress a growth of short waves (for example, by a choice of a sufficiently large cutoff radius c).

The distribution function of the system is usually normalized for the mass of the system. For statistical simulation, one must switch over to the distribution function $F(\varepsilon, \mu)$, normalized for 1, such that the differential

$$dw = F(\varepsilon, \mu) \, d\varepsilon \, d\mu \qquad \left(\int dw = 1 \right) \qquad (4)$$

will be equal to the probability of the fact that the particle has a dimensionless energy and angular momentum within the interval, $(\varepsilon, \varepsilon + d\varepsilon)$ and $(\mu, \mu + d\mu)$. It is easy to see that the required function

$$F(\varepsilon, \mu) = \frac{16\pi^2}{M} \frac{\mu f_0(\varepsilon, \mu)}{\Omega_1(\varepsilon, \mu)}, \qquad (5)$$

where $\Omega(\varepsilon, \mu)$ is the frequency of radial oscillations of the particle.

In accordance with the distribution function in (5), we "play" the energies ε_i and the angular momenta μ_i of the N particles ($i = 1, 2, \ldots, N$). For that purpose, one may make use of any of the standard methods for statistical simulation [$3^{ad}, 9^{ad}$]; some details regarding specific distributions are given in §3, Appendix.

Our next aim is to play the coordinates $(x_i, y_i, z_i)\, N$ of particles corresponding to the energies ε_i and momenta μ_i obtained earlier. The particle with given $(\varepsilon, \mu) = (\varepsilon_i, \mu_i)$ moves in a certain orbit (of the roselike type) in some certain plane. The law of motion along the radius r is described by the formula (25), §4, Appendix, which, for the sake of convenience, we rewrite here as

$$w_1(x, \varepsilon, \mu) = \begin{cases} v_1 \dfrac{\partial \tilde{I}}{\partial \varepsilon} \equiv w_1^+ & (0 \le w_1 \le \pi), \\[3mm] 2\pi - v_1 \dfrac{\partial \tilde{I}}{\partial \varepsilon} = 2\pi - w_1^+ \equiv w_1^- & (\pi \le w_1 \le 2\pi), \end{cases} \qquad (6)$$

where x is the dimensionless radius, w_1 is the angular variable (conjungate to the variable of radial action I_1), the signs $(+)$ and $(-)$ correspond to the "direct" (from x_{min} to x_{max}) and "inverse" (from x_{max} to x_{min}) movements of the particle, and the notation for the integral

$$\tilde{I}(x, \varepsilon, \mu) = \int_{x_{min}}^{x} dx' \left(2\varepsilon - 2\varphi(x') - \frac{\mu^2}{x'^2} \right)^{1/2} \qquad (7)$$

is introduced. The particles in an equilibrium system are uniformly distributed in angular variables w_i ($i = 1, 2, 3$), including w_1 [within $(0, 2\pi)$]. Accordingly, playing first w_1, then the radius x will be found by solving Eq. (6) with respect to x. This problem is very simply solved by the usual numerical methods. Then we turn to the determination of the angular spatial coordinates

θ, φ of the particle. With respect to the angle θ, it moves in accordance with the formula

$$\cos\theta = \cos\theta_0 \cos\left(w_2 - \frac{\partial S_1}{\partial I_2}\right), \tag{8}$$

derived in §4, Appendix. Here $\sin\theta_0 = \mu_z/\mu$ (μ_z is the z-component of the angular momentum),

$$\frac{\partial S_1}{\partial I_2} = \begin{cases} v_2 \dfrac{\partial \check{I}}{\partial\varepsilon} + \dfrac{\partial \check{I}}{\partial\mu}, & 0 < w_1 < \pi, \\[3mm] -\left(v_2 \dfrac{\partial \check{I}}{\partial\varepsilon} + \dfrac{\partial \check{I}}{\partial\mu}\right), & \pi < w_1 < 2\pi. \end{cases} \tag{9}$$

Playing uniformly in $(-1, 1)$ $\cos\theta_z$, where θ_z is the angle between the vector of angular momentum μ and the axis z, we find $\mu_z = \mu\cos\theta_z$; simultaneously, also $\cos\theta_0$ in (8) is determined. Then we play [uniformly in $(0, 2\pi)$] the angular variable w_2; then by formulae (8) and (9) we shall find $\cos\theta$. The angle φ is played in a similar way. According to formula (6), §4, Appendix:

$$\varphi = w_3 - \frac{\partial S_1}{\partial I_3} - \frac{\partial S_2}{\partial I_3}, \tag{10}$$

$$\frac{\partial S_1}{\partial I_3} = \frac{\partial S_1}{\partial I_2} \cdot \operatorname{sgn}\mu_z \tag{11}$$

$$\frac{\partial S_2}{\partial I_3} = \operatorname{sgn}\mu_z \cdot \arccos\frac{\cos\theta}{\cos\theta_0} - \operatorname{arccot}\left[\frac{\sin\theta_0 \cos\theta}{\sqrt{\cos^2\theta_0 - \cos^2\theta}}\right]. \tag{12}$$

Again we play, uniformly in $(0, 2\pi)$, the last angular variable w_3; thereafter from (10)–(12) we find φ. Thereby, the coordinates of the particles are determined.

The next stage is to compute the velocities of the particle. But they are defined unambiguously from the already known ε, μ, μ_z, $r = x$, θ, φ:

$$v_\perp = \frac{\mu}{x}, \qquad |v_r| = \sqrt{2\varphi(x) + 2\varepsilon - v_\perp^2},$$

$$v_\varphi = \frac{\mu_z}{x\sin\theta}, \qquad |v_\theta| = \sqrt{v_\perp^2 - v_m^2}. \tag{13}$$

The sign of v_r is positive if $0 < w_1 < \pi$ (the movement from $x_{\min}$ to $x_{\max}$), and negative for $(\pi < w_1 < 2\pi)$ (the inverse movement from $x_{\max}$ to $x_{\min}$). In the same way, we find also the signs of v_θ: $v_\theta > 0$ for $\sin\alpha < 0$ ($\alpha \equiv w_2 - \partial S_1/\partial I_2$); $v_\theta < 0$ for $\sin\alpha > 0$.

The procedure thus described is performed for each of N particles; thereafter the initial state of the model system will be determined completely. One has only to switch over to the Cartesian x, y, z, v_x, v_y, v_z, and then one may start to solve numerically the equations of motion (3).

Using the above method, initial states of any integrable [69] systems (or a system with such equilibrium potentials, which admit a full separation of the variables in the usual sense [69]) are constructed. For spherically symmetrical systems, one may suggest also another somewhat less formal manner of playing the equilibrium state.

5.2.2. Simulation of a Spherical System with Radial Trajectories of Stars.
The distribution function (1) corresponds to the density

$$\rho_0(r) = \frac{\beta}{r^2} \tag{14}$$

and the potential

$$\Phi_0(r) = 4\pi G\beta \left(\ln \frac{r}{a} - 1 \right). \tag{15}$$

The energy of the particle is

$$E = \frac{v_r^2}{2} + 4\pi G\beta \left(\ln \frac{r}{a} - 1 \right). \tag{16}$$

The period of oscillations of a particle with an energy E in the potential (15) is

$$T = 2 \int_0^{r_{max}} \frac{dr}{v^r} = \sqrt{\frac{2\pi}{E_0'}} \, a e^{1 + E/E_0'}, \tag{17}$$

$$E_0' \equiv 4\pi G\beta.$$

Therefore, the probability density, for the particle, to have an energy E is

$$F(E) = \frac{1}{\sqrt{\pi}} \frac{e^{-(E_0 - E)/E_0}}{\sqrt{(E_0 - E)/E_0}}, \tag{18}$$

$$\int_{-\infty}^{\infty} F(E) \, dE = 1,$$

where $E_0 = \Phi_0(a) = -4\pi G\beta \, (E_0 < 0)$ is the maximum energy, $E \in (-\infty, E_0)$. Thus, we see that in this case the distribution of stars in the variable

$$s = \sqrt{(E_0 - E)/E_0}$$

is normal:

$$F(s) = \frac{2}{\sqrt{\pi}} e^{-s^2}, \qquad s \in (0, \infty). \tag{19}$$

The dimensionless values are easily introduced in the following way. Let the number of stars in the model system be N. If the mass of an individual particle is assumed to be unity, then the total mass $M = N = 4\pi\beta R$. Assume further $E_0' \equiv 4\pi G\beta = 1$, $R = 1$; then the gravitational constant $G = 1/N$.

Of course, in this case the general scheme of the construction of the initial state of the model system, described in subsection 5.2.1, simplifies strongly due to the one-dimensional motion of particles. To begin with, one may play the direction of motion of a star (i.e., for example, the angular location of the point on a sphere of a unit radius, which with identical probability may be in any place). Then the value s is played according to the normal law of (19) and, consequently, the energy $E = -1 - s^2$. Further, we find the initial radius of the particle with the energy E by resolving, with respect to r, the equation

$$w_1 = \begin{cases} w_1^+ \equiv 2\sqrt{\pi} \displaystyle\int_{v(r)}^{\infty} e^{-u^2}\, du, & w_1 < \pi, \\[2mm] w_1^- = 2\pi - w_1^+, & w_1 > \pi, \end{cases} \tag{20}$$

where $v(r) = \sqrt{1 + E - \ln r}$ and w_1 is a radial angular variable which is uniformly distributed within the interval $(0, 2\pi)$.

Let us direct the velocity of the particle along the ray determined at the first stage of playing; a value of the velocity for given E and r is $|v_r| = \sqrt{2E + 2 - 2\ln r}$, while its direction is outward from the center for $w_1 < \pi$ and toward the center for $w_1 > \pi$.

In one typical model realization we assumed $N = 200$, $c = 0.1$, and then obtained the system with a kinetic energy $T = T_r = 94.41$, a potential energy $W = -187.9$. Distributions of particle numbers into spherical layers for the given realization (n) and for the exact equilibrium state (n_0) are compared in Fig. 22 (in the case $n_0 = N/K = $ const, where K is the number of layers). It is seen that, indeed, there occurs a fairly close resemblance. The distribution of particles over "the energy parameter" s turns out to be close to normal.

In Fig. 23(a, b, c, d) we represent projections (in a dimetry projection) of point stars of the system at different moments of evolution of an initial state described. One step in this example corresponds to a time interval $\Delta t = T/50$. In Fig. 23(a) a time elapsed equal to only $T/5$, so the system at this moment

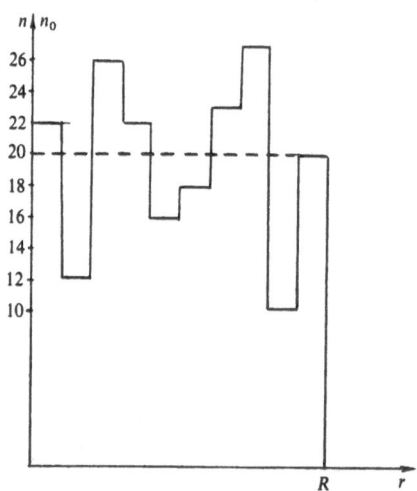

Figure 22. Numbers of particles n (solid line) in the spherical layers for one of the model (40), §1 (trajectories are purely radial). Numbers of particles n_0 for the exact equilibrium are shown by dotted line.

(a) Step 10

(b) Step 30

(c) Step 50

(d) Step 60

Figure 23. Nonlinear evolution of the model (40), §1 of the spherical system with radial trajectories.

is in the state which differs only slightly from the initial one. Deformation of the system's form begins to be considerable with the 40th step; it is already very clear in Fig. 23(c) and 23(d) (50th and 60th steps).

Since a deformation is obviously elliptical, it is natural to follow an evolution of components of the quadrupole momenta tensor of the system

$$d_{ik} = \frac{1}{N} \sum_{1}^{N} (3x_i x_k - r^2 \delta_{ik}). \tag{21}$$

It is also useful to calculate semiaxes a, b, c of the "equivalent" homogeneous ellipsoid for which (by definition)

$$d_1 = \tfrac{1}{5}(2a^2 - b^2 - c^2), \qquad d_2 = \tfrac{1}{5}(2b^2 - a^2 - c^2),$$
$$d_3 = \tfrac{1}{5}(2c^2 - a^2 - b^2), \tag{22}$$

where d_1, d_2, d_3 are principal values of the quadrupole momenta tensor of the system under consideration. From Eqs. (22) we may obtain

$$a^2 = \tfrac{5}{3}(\overline{r^2} + d_1), \qquad b^2 = \tfrac{5}{3}(\overline{r^2} + d_2), \qquad c^2 = \tfrac{5}{3}(\overline{r^2} + d_3), \tag{23}$$

where $\overline{r^2}$ is a mean square of radii of particles [for the homogeneous ellipsoid $\overline{r^2} = \tfrac{1}{5}(a^2 + b^2 + c^2)$]. In Table II we represent the values d_1, d_2, d_3, a, b, c, calculated from the written formulae. In the last column we give also the values of the rms radius of the system's particles

$$(\overline{r^2})^{1/2} = \left[\frac{1}{N} \sum_{1}^{N} (x^2 + y^2 + z^2) \right]^{1/2}.$$

It is seen that, on the background of a slow expansion of the system as a whole (due to the approximateness of the realization of the equilibrium state), there is a very fast amplification of the rather small ellipticity occurring at the initial moment.

Computations similar to ones described above were performed also for the other realizations of the model (1). Namely, we varied the particles' number N, the cutoff radius c, the time step of integration Δt and, finally, the sequences of random numbers under statistical playing of the equilibrium state. Apart from this, the solution of the set of equations (3) is performed by two different methods (one of which is the well-known Runge–Kutta method [7^{ad}]). These computations confirm that the picture of an evolution presented

Table II

Step	d_1	d_2	d_3	a	b	c	$(\overline{r^2})^{1/2}$
0	−0.055	−0.042	0.097	0.66	0.68	0.83	0.56
10	−0.067	−0.043	0.110	0.67	0.70	0.87	0.58
20	−0.085	−0.036	0.121	0.69	0.75	0.91	0.61
30	−0.138	−0.033	0.171	0.67	0.79	0.98	0.64
40	−0.197	−0.031	0.228	0.64	0.83	1.06	0.67

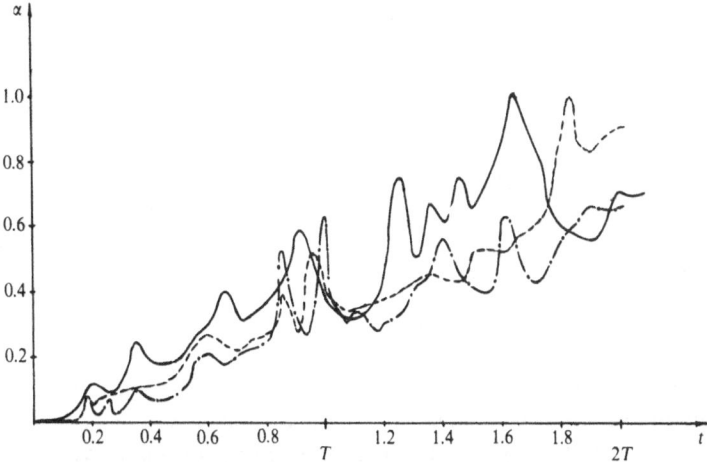

Figure 24. Time-dependence of anisotropy $\alpha = T_\perp / 2T_r$ for a number of realizations of the spherical system with radial trajectories.

above on one example was really typical. In particular, the main feature of this evolution was always obviously manifested: namely, the elliptical deformation of the system. Detailed dynamics of the nonlinear stage of development of initial disturbances may be rather complicated, even nonregular especially for smaller N when the statistical fluctuations are large. For the sake of illustration we give in Fig. 24 graphs of dependence on the time t of global anisotropy[23] $\alpha = T_\perp / 2T_r$ for three values of cutoff radius: $c = 0.01, 0.02, 0.05 (N = 100)$. One may see that on the average the pictures of evolution are approximately the same for all three cases, though there is considerable nonregularity (the latter is greater for smaller values of c).

§ 6 Stability of Spherically Symmetrical Systems of General Form [44[ad]]

Above, in the investigation of stability of stellar spheres we have imposed essential restrictions on the form of the distribution function (as in §2, particularly devoted to isotropic distributions), or on the admissible character of the orbits of particles; moreover, orbits of the simplest form, circular (in §3), purely radial (in §5), and, finally, elliptical orbits (in §4, where we have dealt with the systems homogeneous in density), have been considered. This allowed us in each case considered either to give a full analytical solution of the stability problem or in any event (as, e.g., for the systems with circular orbits) to advance considerably the process of analytical solution.

Although the restraints mentioned are highly artificial, nevertheless, the results of all the investigations already described, taken altogether, allow us

---

[23] It is necessary to bear in mind that here parameter α doesn't determine relative degree of "transversal heating" of the system since in the case T_r and $T_\perp$ involve also the energy of macroscopic contraction (apart from thermal energy).

to arrive at some general conclusions. The main one of these is the following: among the models of spherically symmetrical stellar systems there are both stable and unstable ones, and the boundary separating these and those models must fall somewhere between the models with isotropic distributions and the models with radial trajectories of all the particles. Unfortunately, the approach adopted in the previous treatment (the study of the systems of definite classes) does not allow us to go significantly farther than such qualitative inferences; in practice, we have already completely exhausted its possibilities.

To obtain quantitative criteria of stability, it is necessary to study more general models. Below, we investigate the stability of several series of distribution functions described briefly in §1. All these distribution functions are dependent on some parameters [α, β in Camm series in (24), (25), §1; λ in the Idlis series (32), §1, etc.], which may vary continually within a certain range, so that the models describe, for the corresponding values of the parameters, both isotropic velocity distributions of particles and the anisotropic distributions, including the fairly extended ones in the radial direction.

Analytical investigation of stability of such systems seems to be impossible. We use two numerical methods: the method of matrix equation [34[ad], 35[ad]] and, the most suitable for our aims, modification of the method of simulation of the stellar system by the totality of N point masses interacting with each other according to the Newtonian law [43[ad]]; above (in §5) we already applied "the method of particles" for the investigation of nonlinear stage of an evolution of spherical systems with the radial orbits of stars.

Both methods have a rather common character; they are applicable not only for the investigation of the spherical systems but also for systems with other geometry.[24] Further (for example, in the next chapter) we shall consider the corresponding applications.

The arrangement of the material in this section is as follows. In Section 6.1, the Idlis model (32), §1, is considered; Section 6.2 is concerned with the first Camm's distribution functions (24), §1; Section 6.3 considers the stability of a spherical system with the phase density distribution of Kuzmin and Veltmann (35), §1. The stability investigation of the Idlis model and the Kuzmin–Veltmann model is carried out only by the matrix method; of the Camm model, by the two methods, the matrix method and the N-body method.

At the beginning of each subsection, a more detailed, as compared to §1, representation about the corresponding equilibrium model is given. In particular, the boundaries of the phase region of the system are outlined, and data are given on the radii of circular orbits, on the minimum and maximum distance from the center of the particles with different values of energy and angular momentum, frequency of oscillation of the particles, the radius, the kinetic (separately, the radial and "transversal") and potential energies of the

---

[24] It should be noted, in particular, that the matrix method similar to that used here for stellar spheres, was applied by Kalnajs [74[ad]] for the formulation of the general statement of the problem of the disk system stability (cf. §5, Appendix).

system, etc. Further, also the peculiarities of the computational algorithm adopted is emphasized for a given system. Finally, in conclusion, we represent the results of the investigation of the stability of the model (by one of the two methods mentioned above). The general conclusions following from the results attained in this and the preceding sections are discussed in §7.

6.1 Series of the Idlis Distribution Functions [44[ad]]

6.1.1. Equilibrium State. Consider a stellar sphere described by the distribution function (32), §1:

$$f_0 = f_0(x_\lambda) = \frac{A}{\pi^2} [\tfrac{1}{4}(1 + 3\lambda)U^2 x_\lambda^{-1/2} + 4(1 + \lambda)Ux_\lambda^{1/2} + 8(1 - \lambda)x_\lambda^{3/2}],$$

(1)

where A and U are the dimensional and λ is the dimensionless parameters of the model,

$$x_\lambda = -2E - \frac{U}{2} - \frac{\lambda L^2}{r_0^2},$$

(2)

E and L are the energy and the angular momentum of the particle (per the unit mass):

$$E = \frac{v_r^2}{2} + \frac{v_\perp^2}{2} + \Phi_0(r), \qquad L = rv_\perp;$$

(3)

r_0 is the parameter with dimension of length.

It can easily be tested that (1) indeed yields a self-consistent model with the density

$$\rho_0(r) = \int f_0 \, dv = AU^3 \frac{2(3 - r^2/r_0^2)}{(1 + r^2/r_0^2)^3}$$

(4)

and the potential

$$\Phi_0(r) = -\frac{U}{1 + r^2/r_0^2}.$$

(5)

Thus, we have a system of finite radius $R = \sqrt{3}r_0$, with a reasonable (everywhere decreasing) behavior of the density $\rho_0(r)$. From (5), the meaning of the parameter U introduced in (1),

$$U = -\Phi_0(0),$$

is clear. The "full" gravitational potential $\Phi(r)$ satisfying the matching condition with the outer (Coulomb) potential $\Phi_> = -GM/r$ (M is the mass of the system) differs from (5) by a constant value:

$$\Phi = \Phi_0(r) + C, \qquad C = \frac{U}{4} - \frac{GM}{r_0\sqrt{3}}.$$

(6)

It is convenient to write the energy of particles in the form of (3) by changing the origin accordingly.

Substituting (5) and (4) into the Poisson equation

$$\frac{1}{r^2}\frac{d}{dr}r^2\frac{d\Phi_0}{dr} = 4\pi G\rho, \tag{7}$$

we find the connection between the parameters A, U, and r_0:

$$4\pi G r_0^2 A U^2 = 1. \tag{8}$$

The mass of the system is

$$M = 4\pi \int_0^R \rho(r)r^2\,dr = \frac{\pi A U^3 r_0^3 3\sqrt{3}}{2}. \tag{9}$$

The "de-dimensioning" necessary below is readily performed here by the natural method, by assuming that

$$4\pi G = 1, \qquad U = 1, \qquad A = 1. \tag{10}$$

Then the length unit will be, according to (8), r_0: $r_0 = 1$. In these units, the distribution function

$$f_0(x_\lambda) = \frac{1}{\pi^2}\left[\tfrac{1}{4}(1 + 3\lambda)x_\lambda^{-1/2} + 4(1 + \lambda)x_\lambda^{1/2} + 8(1 - \lambda)x_\lambda^{3/2}\right], \tag{11}$$

$$x_\lambda = -2\varepsilon - \tfrac{1}{2} - \lambda\mu^2, \tag{12}$$

where ε and μ are the dimensionless energy and the angular momentum of the particle; the density

$$\rho_0(x) = \frac{2(3 - x^2)}{(1 + x^2)^3}, \tag{13}$$

where $x = r/r_0$ is the dimensionless radius; the potential

$$\Phi_0(x) = -\frac{1}{1 + x^2}. \tag{14}$$

It is easy to see that the series of models under consideration is indeed one-parametric: it is completely determined by the value λ. The mass of the system M in our units is a universal constant, independent of the parameters:

$$M = \frac{3\pi\sqrt{3}}{2} = 8.1621. \tag{15}$$

The gravitational energy of the system

$$W = \tfrac{1}{2}\int_0^{\sqrt{3}} \rho(x)\Phi(x)4\pi x^2\,dx \tag{16}$$

[where $\Phi(x) = \Phi_0(x) + C$, $C = \tfrac{1}{4} - M/4\pi\sqrt{3}$] is also a universal constant:

$$W = -3.17533. \tag{17}$$

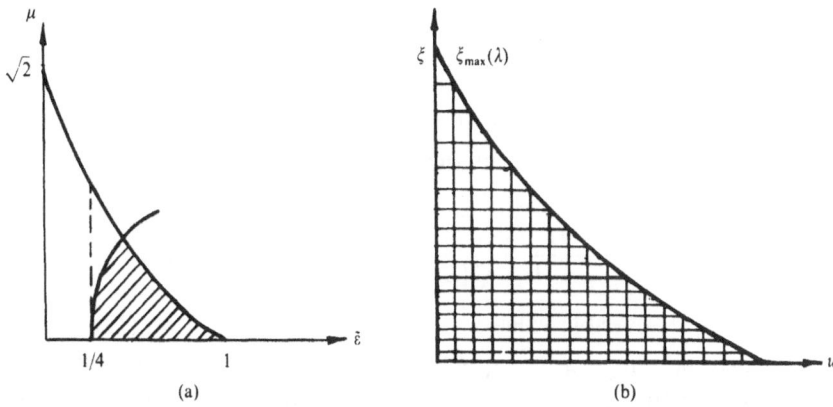

Figure 25. Phase region for the models of the Idlis series: (a) in plane $(\tilde{\varepsilon}, \mu)$; (b) in plane (u, ξ).

The same may be said also about the total kinetic energy of the system, which according to the virial theorem, is

$$T = -\tfrac{1}{2}W = 1.58766. \tag{18}$$

The phase region of the system on the plane $(\tilde{\varepsilon} = -\varepsilon, \mu)$ is limited [cf. Fig. 25(a)]: from below by the abscissa axis $\mu = 0 \,(\mu > 0)$, from the left, by the parabola

$$\tilde{\varepsilon} = \frac{1}{4} + \frac{\lambda\mu^2}{2}, \tag{19}$$

expressing the condition $x_\lambda > 0$ necessary in connection with the presence of radicals in expression (12) for $f_0(x_\lambda),$[25] and finally, from above, by the line of circular orbits. The equation of the latter (cf. §1):

$$\mu^2 = x^3 \frac{\partial\Phi_0}{\partial x}, \qquad \varepsilon = -\tilde{\varepsilon} = \Phi_0 + \frac{x}{2}\frac{\partial\Phi_0}{\partial x}. \tag{20}$$

In the given case, for $\Phi_0(x)$ from (5)

$$\tilde{\varepsilon} = \frac{1}{(1+x^2)^2}, \qquad \frac{\mu}{\sqrt{2}} = \frac{x^2}{1+x^2}. \tag{21}$$

The parameter x (the radius of the circular orbit of the particle with given ε and μ) is readily eliminated from (21), and we obtain the equation of the line of circular orbits in the form of a parabola

$$\tilde{\varepsilon} = \left(1 - \frac{\mu}{\sqrt{2}}\right)^2. \tag{22}$$

---

[25] The condition $x_\lambda > 0$ replaces in this case a more usual condition of the "gravitational bound" of particles in the system. The latter requires that the energy ε of the particle with a given angular momentum μ be no more than the value of the effective potential energy $U_{\text{eff}} = \Phi_0(x) + \mu^2/2x^2$ on the boundary $x = R$. For $\Phi_0(x) = -1/(1+x^2)$ and $R = \sqrt{3}$ this yields: $\tilde{\varepsilon} > \tfrac{1}{4} - \mu^2/6$. This boundary parabola coincides with (19) only for $\lambda = -\tfrac{1}{3}$. For all $\lambda > -\tfrac{1}{3}$ the condition $x_\lambda > 0$ automatically produces a "gravitational bound." This, in particular, means that the boundary $x = R$ are reached only by the particles with the zero angular momentum $\mu = 0$.

Further, we shall use instead of $\tilde{\varepsilon}$ and μ the practically more convenient variables $u = \sqrt{x_\lambda}$ and $\xi = \mu/\sqrt{2}$. In these variables, the phase region appears to be significantly simpler [cf. Fig. 25(b)]. Both from below and now on the left, it is bounded by the straight lines ($\xi = 0$ and $u = 0$) and from above, again by the line of circular orbits, the equation of which in the new variables[26] is

$$\xi = -\frac{1}{\lambda - 1} + \sqrt{\frac{1}{(\lambda - 1)^2} + \frac{1}{2(\lambda - 1)}} \left(\tfrac{3}{2} - u^2\right). \tag{23}$$

The maximally possible value $x_\lambda = \tfrac{3}{2}$, which is evident from (12) and Fig. 25(a), correspondingly $u_{max} = \sqrt{\tfrac{3}{2}}$. The same value ensues also from Eq. (23): $u_{max} = u (\xi = 0)$; u_{max} is independent of λ. The maximum value of the variable ξ is also found from (23): $\xi_{max}(\lambda) = \xi(0) = \tfrac{3}{2}/(2 + \sqrt{3\lambda + 1})$. The condition of positiveness of the distribution function f_0 yields (for $\lambda > 1$)

$$x_\lambda < (x_\lambda)_c = \frac{2(\lambda + 1) + \sqrt{4(\lambda + 1)^2 + 2(\lambda - 1)(3\lambda + 1)}}{8(\lambda - 1)}. \tag{24}$$

Hence one may infer that $(x_\lambda)_c = (x_\lambda)_{max} = \tfrac{3}{2}$ for $\lambda = \lambda_{max} = 97/45$; for $\lambda > \lambda_{max}$ the phase density of the system $f_0(x_\lambda)$ would be negative near $x_\lambda = \tfrac{3}{2}$. Therefore, we shall restrict ourselves to the consideration of the models with $\lambda < 97/45$. On the other hand, as is evident from (11), for $\lambda < -\tfrac{1}{3}$, the distribution function $f_0 < 0$ for sufficiently small x_λ; in order to avoid this, one should assume that $\lambda > -\tfrac{1}{3}$.

Our main purpose is the investigation of the models for stability described herein, which in turn is defined first of all by the ratio of the velocity dispersions of the particles in the radial and tangential (transversal to the radius) directions. The squares of the velocity dispersions are, according to the definition,

$$\overline{c_r^2} = \frac{\int v_r^2 f \, dv}{\rho(x)}, \qquad \overline{c_\perp^2} = \frac{\tfrac{1}{2} \int v_\perp^2 f \, dv}{\rho(x)}. \tag{25}$$

The integrals in (25) with the distribution function (11) are easily calculated. For c_r^2, we get

$$c_r^2 = \frac{\varkappa^4}{p^2 \rho(x)} \left(\frac{1 + 3\lambda}{16} + \frac{1 + \lambda}{6} \varkappa^2 + \frac{1 - \lambda}{8} \varkappa^4\right), \tag{26}$$

[where $\varkappa^2 = (3 - x^2)/(2 + 2x^2)$, $p^2 = (1 + \lambda x^2)$], and

$$c_\perp^2 = \frac{c_r^2}{p^2} = \frac{c_r^2}{1 + \lambda x^2}. \tag{27}$$

Figure 26 shows the graphs $c_r(x)$ and $c_\perp(x)$ for $\lambda = 0$ and $\lambda = \lambda_c = 1.737$. The ratio of the squares of the velocity dispersions is, according to (27),

$$\sigma = \frac{c_r^2}{c_\perp^2} = 1 + \lambda x^2, \tag{28}$$

---

[26] Equation (23) is valid for $\lambda > 1$; just such λ are mainly of interest to us.

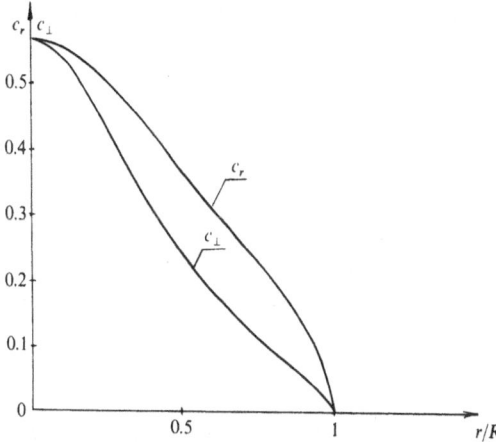

Figure 26. Dependence of velocity dispersions C_r, $C_\perp$ on the radius for one of the Idlis models.

so that $\lambda = 0$ corresponds to the isotropic case; for $\lambda < 0$ the transversal dispersion is more than the radial one (this case, being obviously stable, is of relatively little interest), while for $\lambda > 0$ the anisotropy $\sigma > 1$, i.e., the radial dispersion exceeds the transversal one. In the last case, one may expect, for rather great values of the parameter λ, that instability appears. Indeed, as we shall see later (in 6.1.3), for λ, close to $\lambda_{max} = 97/45 = 2.1556$, the system becomes unstable; the critical value of the parameter where instability first appears, $\lambda = \lambda_c = 1.737$. The instability growth rate increases monotonically with varying λ from λ_c to λ_{max}, remaining small, however, so that one may say that the system remains (even at the maximum λ) near the stability boundary.

It is clear (and is confirmed by calculations (cf. 6.1.3) that at the stability boundary, there is excited the most large-scale nonradial mode having no nodes in the radius r. Therefore, it is logical to suggest that the stability criterion must be defined by some characteristics averaged over the system — of the type of ratio of the kinetic radial and transversal energies. Table III lists the results of the calculation of some macrocharacteristics of the Idlis models of (11) for different values of the parameter λ.

Table III

λ		T_r	$T_\perp$	$u_r \equiv -T_r/W$	$u_\perp \equiv -T_\perp/2W$	$v \equiv 2T_r/T_\perp$
$\lambda_{max} =$	2.155	0.729	0.429	0.229	0.135	1.699
	1.906	0.714	0.436	0.224	0.137	1.635
$\lambda_c =$	1.737	0.703	0.442	0.221	0.139	1.590
	1.408	0.680	0.473	0.214	0.142	1.499
	1.010	0.648	0.469	0.204	0.147	1.379
	0.761	0.625	0.481	0.196	0.151	1.298
	−0.034	0.523	0.532	0.164	0.167	0.983
$\lambda_{min} = -\frac{1}{3}$		0.460	0.563	0.145	0.177	0.817

It is, furthermore, interesting to follow the change of the size of the region, in which circular orbits are possible, with the varying parameter λ. Since the radius of the circular orbit $x_k = \sqrt{\xi/(1-\xi)}$ increases monotonically together with ξ, the maximally possible for a given λ, x_k is reached at $\xi = \xi_{max}$; hence

$$\frac{(x_k^2)_{max}}{R^2} = [2\sqrt{3\lambda+1}+1]^{-1/2}. \tag{29}$$

This formula shows that the circuit orbits may indeed not exist in the whole sphere, but only in its part: $(x_k)_{max}/R < 1$ (except for the marginal case $\lambda = -\frac{1}{3}$, when $(x_k)_{max} = R$). For $\lambda = 0$, i.e., for the isotropic model, $(x_k)_{max}/R \simeq 0.58$. The region, where circular orbits are possible, with increasing λ, is "hidden" still deeper in the system; the size of this region also is a characteristic of the system stability degree. The minimal $(x_k)_{max} \simeq 0.4R$ ensues for $\lambda = \lambda_{max} = 97/45$.

6.1.2. Some Peculiarities of the Computing Algorithm. The instability of the Idlis models was investigated [44^{ad}] with the help of the matrix equation derived in §4, Appendix. For the sake of convenience we write this equation here:

$$\sum_{m=1}^{\infty} (M_{mn}^{(l)} + \delta_{mn}) a_m = 0, \qquad n = 1, 2, 3, \ldots . \tag{30}$$

Here the a_m are the coefficients of expansion of the perturbed potential Φ_1 and the density σ_1 (dimensionless) over the functions that are terms of whichever biorthonormal system $\{\chi_k^{(l)}(x)\}, \{\sigma_k^{(l)}(x)\}$:

$$\Phi_1 = \Phi_1(x, \theta) = P_l(\cos \theta) \sum_k a_k \chi_k^{(l)}(x), \tag{31}$$

$$\sigma_1 = \sigma_1(x, \theta) = P_l(\cos \theta) \sum_k a_k \sigma_k^{(l)}(x). \tag{32}$$

Biorthonormality means that

$$\int_0^R \sigma_k^{(l)}(x)\chi_n^{(l)}(x)x^2 \, dx = -\delta_{nk}. \tag{33}$$

In the Appendix, we give two examples of biorthonormal systems of functions, from which in this case the simplest one is employed:

$$\chi_n^{(l)} = \frac{\sqrt{2}}{\alpha_n} \frac{1}{\sqrt{x}|J_{l+1/2}(\alpha_n)|} J_{l+1/2}\left(\alpha_n \frac{x}{R}\right), \tag{34}$$

$$\sigma_n^{(l)} = -\frac{\sqrt{2}}{R^2} \frac{\alpha_n}{\sqrt{x}|J_{l+1/2}(\alpha_n)|} J_{l+1/2}\left(\alpha_n \frac{x}{R}\right), \tag{35}$$

where $J_\nu(\alpha)$ is the Bessel function and α_n must be defined from the equation

$$J_{l-1/2}(\alpha_n) = 0. \tag{36}$$

The matrix elements $M_{mn}^{(l)}$ in (30) are defined by formulae (20) §4, Appendix. The time for computer computations may be essentially reduced if $M_{m,n}^{l}$ is reduced to the form depending on ω^2:

$$M_{mn}^{(l)} = 2 \sum_{l_1 = 0}^{\infty} \sum_{l_2 = -l}^{l} D_{l_1 l_2} \int \frac{d\varepsilon\, \mu\, d\mu}{v_1(\mu, \varepsilon)} S(l_1)(\chi_n)_{l_1 l_2} (\chi_m)_{l_1 l_2}$$

$$\times \frac{\Omega_{l_1 l_2}}{\omega^2 - \Omega_{l_1 l_2}^2} \left[(l_1 v_1 + l_2 v_2) \frac{\partial f_0}{\partial \varepsilon} + l_2 \frac{\partial f_0}{\partial \mu} \right], \tag{37}$$

$$S(l_1) = 1 \quad \text{at } l_1 \neq 0, \qquad S_1(0) = \tfrac{1}{2}, \qquad \Omega_{l_1 l_2} = l_1 v_1 + l_2 v_2,$$

$D_{l_1 l_2}$ are the numbers defined by expression (20a), §4, Appendix; $v_1(\varepsilon, \mu)$, $v_2(\varepsilon, \mu)$ are the oscillation frequencies of the particles in the radius r and in the angle θ, equal to [69]

$$v_1 = \frac{\partial \varepsilon(I_i)}{\partial I_1}, \qquad v_2 = \frac{\partial \varepsilon(I_i)}{\partial I_2}. \tag{38}$$

$\varepsilon(I_i)$ is the energy of the particle considered as the function of the action variables I_j ($j = 1, 2, 3$); I_1 and I_2 are the actions corresponding to the motion in r and θ. Finally, the functions $(\chi_m^{(l)})_{l_1 l_2}$ may be calculated by the formulae (also derived in the Appendix):

$$(\chi_m^{(l)})_{l_1 l_2}(\varepsilon, \mu) = 2v_1 \int_{x_{\min}}^{x_{\max}} \frac{\chi_m^{(l)}(x)\, dx}{\sqrt{2\varepsilon - \mu^2/x^2 - 2\varphi(x)}} \cdot \cos \psi_{l_1 l_2}(x, \varepsilon, \mu), \tag{39}$$

$$\psi_{l_1 l_2}(\varepsilon, \mu) = \left[(l_1 v_1 + l_2 v_2) \frac{\partial}{\partial \varepsilon} + l_2 \frac{\partial}{\partial \mu} \right] \tilde{I}(x, \varepsilon, \mu), \tag{40}$$

$$\tilde{I}(x, \varepsilon, \mu) = \int_{x_{\min}}^{x} dx' \left[2\varepsilon - 2\varphi(x') - \frac{\mu^2}{x'^2} \right]^{1/2}, \tag{41}$$

where $x_{\min}(\varepsilon, \mu)$ and $x_{\max}(\varepsilon, \mu)$ are the minimal and maximal radii of the particle with given ε, μ. Integration in (37) is carried out over the phase region of the system [Fig. 25(a)].

The potential perturbation Φ_1 can be taken in the form (31) without loss of generality, due to a total[27] spherical symmetry of the system in question. Both radial ($l = 0$) and nonradial (with $l = 1$ and $l = 2$) perturbations have been considered [44^{ad}].

To practically solve the problem on computer, it is necessary to switch over from the infinite matrix equation in (30) to some finite system of equations, i.e., to perform the "truncation" procedure. We truncate the initial system in the simplest way, by separating from the full (infinite) determinant of the system the left-hand upper angle sized $N \times N$. With such a way of

---

[27] Corresponding symmetry is possessed not only by the density distributions ρ_0 and potential distributions Φ_0, but also the distribution function f_0, which does not single out any direction (unlike, e.g., the Freeman model considered in Section 4.2). This, in particular, means independence of eigenfrequencies on the azimuthal number m in the angular dependence of the perturbed potential of the general form $P_l^m(\cos \theta)e^{im\varphi}$.

"truncation," those terms are eliminated which will not be essential for the large scale modes of interest (cf. expression (34) for $\chi_n^{(l)}$). It turns out also that the main results (as for instance the value of the growth rate of instability, the critical value of the parameter $\lambda = \lambda_c$ in the Idlis series, etc.) are weakly dependent on N for $N > 2$; in the computations described below [44^{ad}], the maximum dimension of the determinant was 5×5.

Turn now to the description of the computing algorithm adopted in [44^{ad}]. The phase region of the system on the plane (u, ξ) was replaced by a discrete set of points (u_i, ξ_i) in such a way, as shown in Fig. 25(d). The maximum size of the network (supplemented to give a square) was 51×51. Then, in this set, all the values incorporated in the matrix equation of (30) were calculated.

Since in the case under consideration we have an explicit (and very simple) expression of (14) for the equilibrium potential, a significant part of the preliminary work necessary to calculate the eigenfrequencies is carried out analytically. For example, limiting (maximal and minimal) radii of the orbit with given ε, μ are calculated. From the condition of vanishing of the radial velocity of the particle, we have

$$\tilde{\varepsilon} + \frac{\xi^2}{x^2} - \frac{1}{1 + x^2} = 0;$$

therefore

$$x^2_{\substack{\max \\ \min}} = \frac{1}{2\tilde{\varepsilon}} \{1 - \xi^2 - \tilde{\varepsilon} \pm \sqrt{[1 - (\xi + \sqrt{\tilde{\varepsilon}})^2][1 - (\xi - \sqrt{\tilde{\varepsilon}})^2]}\}. \quad (42)$$

Let us have some further discussion about the calculation [by formulae (38)] of the frequencies v_1 and v_2. Since (cf. §4, Appendix)

$$I_1 = \frac{I}{\pi}, \qquad I = I(\varepsilon, \mu) = \int_{x_{\min}}^{x_{\max}} dx \left(2\varepsilon - 2\varphi(x) - \frac{\mu^2}{x^2}\right)^{1/2}, \quad (43)$$

$$\mu = I_2 + |I_3|, \quad (44)$$

we shall find, by differentiating (43) over I_1 and I_2 and considering I as a complex function $I(\varepsilon(I_1, I_2), \mu(I_2, I_3))$:

$$v_1 = \pi \left/ \left(\frac{\partial I}{\partial \varepsilon}\right)\right., \qquad v_2 = -\left(\frac{\partial I}{\partial \mu}\right) \left/ \left(\frac{\partial I}{\partial \varepsilon}\right)\right.. \quad (45)$$

Instead of v_2, one can easily calculate the relation

$$\frac{v_2}{v_1} = -\frac{1}{\pi}\frac{\partial I}{\partial \mu}. \quad (46)$$

For arbitrary u, ξ, the determination of the frequencies is reduced to the calculation of simple integrals

$$\frac{\partial I}{\partial \varepsilon} = \frac{1}{\sqrt{2\tilde{\varepsilon}}} \int_0^{\pi/2} d\varphi \sqrt{1 + x^2_{\min} + (x^2_{\max} - x^2_{\min}) \sin^2 \varphi}, \quad (47)$$

$$\frac{\partial I}{\partial \mu} = -\frac{\mu}{\sqrt{2\bar{\varepsilon}}} \int_0^{\pi/2} d\varphi \frac{\sqrt{1 + x_{\min}^2 + (x_{\max}^2 - x_{\min}^2)\sin^2 \varphi}}{x_{\min}^2 + (x_{\max}^2 - x_{\min}^2)\sin^2 \varphi}. \tag{48}$$

In particular, in circular orbits

$$v_1 = (2 - 2\xi)^{3/2}, \qquad \frac{v_2}{v_1} = \frac{1}{2\sqrt{1 - \xi}}, \tag{49}$$

and for radial trajectories ($\mu \to 0$) we have the universal ratio $v_2/v_1 = \frac{1}{2}$.

First, on the set (u_i, ξ_i), we calculate [44^{ad}] the massives of the frequencies $v_1(u_i, \xi_i)$ and $v_2(u_i, \xi_i)$; then data sets for functions $\partial\tilde{I}(x_\alpha, u_i, \xi_i)/\partial\varepsilon$ and $\partial\tilde{I}(x_\alpha, u_i, \xi_i)/\partial\mu$ incorporating expression (40) for $\psi_{l_1 l_2}$. The main computer time was spent on the calculation and accumulation on magnetic tape of data sets $\chi_{l_1 l_2}^{(n)}(u_i, \xi_i)$ for $n = 1, 2, \ldots, 5, l_1 = 0, 1, \ldots, 4; l_2 = -2, 0, 2$ (for $l = 2$) or $l_2 = -1, 1$ (for $l = 1$). The characteristic frequencies ω were determined then from the zero-equality condition of the determinant of the "truncated" system (30).

6.1.3. Main Results. Figure 27 represents the instability growth rates of the Idlis models obtained by the above method for several values of the parameter $\lambda = \lambda_{\max}(=2.1556)$, 1.9, 1.8, 1.5, 1.01 for the mode $l = 2$. The three points corresponding to $\lambda = \lambda_{\max}$, 1.9, 1.8, for which the instability was revealed with confidence, lie on the straight line; for $\lambda = 1.01$, instability is apparently absent; the small growth rate $\gamma \simeq 0.002$ for $\lambda = 1.5$ seems to be wholly due to the "noise" effect (the accuracy of computation was insufficient for reliable revealing such a weak instability). The maximum value of the instability growth rate $\gamma_{\max}(\lambda_{\max}) = 0.045$ is much less than the frequencies of oscillations of the particles v_1 and v_2 (which in the adopted units are ~ 1). This illustrates that, even at $\lambda = \lambda_{\max}$, the system is in the vicinity of the stability boundary. Extrapolation of the straight line passing

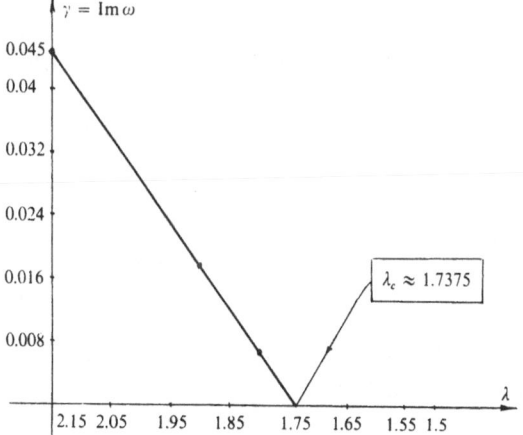

Figure 27. Growth rates of the instability for the Idlis models.

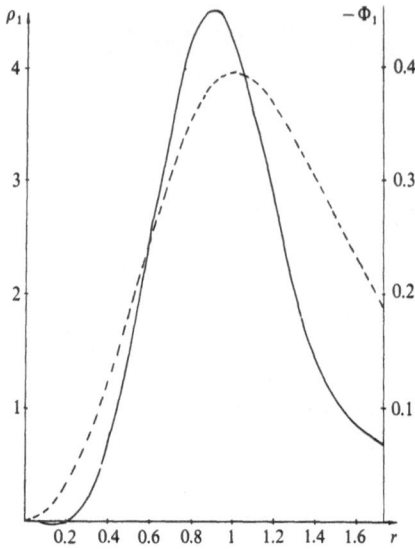

Figure 28. Eigenfunctions ρ_1 and $(-\Phi_1)$ for disturbances of quadrupole type (the Idlis model).

through the three upper points leads to a critical value $\lambda = \lambda_c = 1.7375$. In Table III, a separate line gives the values of the macroscopic parameters characterizing the Idlis model for $\lambda = \lambda_c$.

The graphs of the eigenfunctions $\Phi_1(x)$ (nodeless) and $\rho_1(x)$ for the mode $l = 2$ under consideration at $\lambda = 97/45$ and $\lambda = 1.9$ are given in Figure 28; they confirm that we do indeed deal with the most large-scale mode.

For $l = 1$, there is a root $\omega^2 = 0$ corresponding to the displacement of the system as a whole along the x-axis. The eigenfunctions Φ_1 and ρ_1 computed for this root have within a very good accuracy coincided, respectively, with $\partial\Phi_0/\partial x$ and $\partial\rho_0/\partial x$. None of the other roots of ω^2 for $l = 1$ in [44^{ad}] have been found (neither stable nor unstable).

The radial perturbations ($l = 0$) were investigated with the aim of making sure there is stability of the models in question with respect to such perturbations, in spite of the presence of the term $\sim x_\lambda^{-1/2}$, increasing with energy, in the distribution function f_0. Indeed, no roots of ω^2 which were unstable were found. For $l = 0$ and $l = 2$ we have the real frequencies: $\omega \simeq 0.733$ ($l = 0$) and $\omega \simeq 0.896$ ($l = 2$) (for $\lambda = \lambda_{\max}$). The real frequencies may evidently lie at "nonresonance" points, $\omega \neq \Omega_{l_1 l_2}(\varepsilon, \mu) = l_1 v_1(\varepsilon, \mu) + l_2 v_2(\varepsilon, \mu)$. Therefore, the search for real frequencies was [44^{ad}] carried out in the following ("purposeful") way. First, the maximum and minimum values of the combinations $\Omega_{l_1 l_2}(\varepsilon, \mu)$ were calculated for various pairs of (l_1, l_2). Here, on the real axis ω, "gaps" were sought, in which none of $\Omega_{l_1 l_2}(\varepsilon, \mu)$ was available: for example, the region $0.971 < \omega < 2.828$ for $l = 0$, or the region $0.335 < \omega < 0.917$ for $l = 2$ (at $\lambda = \lambda_{\max}$). Just in these ranges, real frequencies were then sought for. It turned out that the location of the real roots is weakly dependent on λ: for example, for $l = 2$ and $\lambda = 1.9$, $\omega \simeq 0.894$; for $\lambda = 1.8$, $\omega \simeq 0.893$, etc. For $l = 1$, the "gap" lies in $-0.46 < \omega < 0.46$.

6.2 First Series of Camm Distribution Functions (Generalized Polytropes)[27'] [41[ad]]

6.2.1. Equilibrium State. If the Camm distribution function of the first series (24), §1, is written in the form

$$f_0 = f_0(E, L) = CL^\beta[\Phi_0(a) - E]^\alpha, \tag{1}$$

where C is a constant, then the density corresponding to it will be

$$\rho_0 = \rho_0(r) = kr^\beta[\Phi_0(a) - \Phi_0(r)]^{\alpha + (\beta + 3)/2}, \tag{2}$$

and the new constant k is related to C by the relation

$$k = 2\pi C2^{(\beta + 1)/2}B\left(\frac{\beta + 3}{2}, \alpha + 1\right)B\left(\frac{1}{2}, \frac{\beta}{2} + 1\right), \tag{3}$$

$[B(a, b)$ is the Euler integral of the first kind]. Integration of $\rho_0 = \int f_0 \, dv$, yielding (2), has meaning only for $\alpha > -1, \beta > -2$.

Let us turn immediately to the dimensionless values, assuming that

$$4\pi G = 1, \qquad k = 1, \qquad \Phi_0(a) - \Phi_0(0) \equiv U = 1. \tag{4}$$

Thereby we define also the measurement units of all the other values, including the length l, mass m, and time τ. The Poisson equation in these units written for

$$y(x) = \frac{\Phi_0(a) - \Phi_0(r)}{\Phi_0(a) - \Phi_0(0)} \qquad \left(x = \frac{r}{l}\right), \tag{5}$$

takes on the form of the standard Emden–Fowler equation:

$$xy'' + 2y' + x^{\beta + 1}y^{\alpha + (\beta + 3)/2} = 0, \tag{6}$$

where $y(0) = 1$. The distribution function in (1) is rewritten as

$$f_0(\varepsilon, \mu) = c\mu^\beta[\varphi_0(R) - \varepsilon]^\alpha, \tag{7}$$

where ε and μ are the notations normally used by us for dimensionless energy and angular momentum, while the dimensionless coefficient of proportionality c is found by means of (3):

$$c = \frac{\Gamma(\alpha + (\beta + 5)/2)}{(2\pi)^{3/2}2^{\beta/2}\Gamma(\alpha + 1)\Gamma(\beta/2 + 1)} \tag{8}$$

(Γ is the Euler integral of the first kind). The dimensionless density is expressed by a simple formula

$$\sigma(x) = x^\beta[y(x)]^{\alpha + (\beta + 3)/2}. \tag{9}$$

The Emden–Fowler equation (6) is integrated analytically only in few special cases; we are, however, aware [55] of some qualitative properties of

---

[27'] Results of this section (and also of the next section, 6.3) were obtained by one of the authors (V.L.P.)

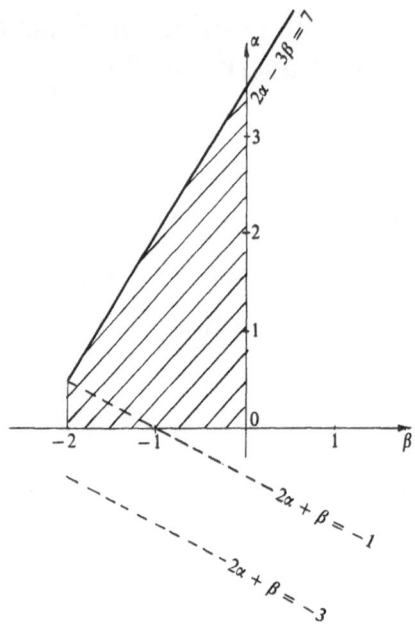

Figure 29. Region of parameters α, β (dashed) corresponding to "reasonable" models of the first Camm series.

the solutions of this equation depending on the values of the parameters α and β. A representation of these properties may be obtained from Figure 29, where the range of parameters of interest for us in the future is dashed. The models corresponding to this range describe finite systems with normal (decreasing) anisotropic distributions $f_0(\varepsilon, \mu)$, extended in the radial direction. The straight line separating the finite systems ($R < \infty$) from infinite ones is

$$2\alpha - 3\beta = 7; \tag{10}$$

below the straight line

$$2\alpha + \beta = -3 \tag{11}$$

the density ρ_0 at the "boundary" $r = R$ does not vanish; finally, on the straight line

$$2\alpha + \beta = -1 \tag{12}$$

the points (α, β) corresponding to the simplest solutions of Eq. (6) in the form of the Bessel functions are located.

The numerical solution for the Emden–Fowler equation does not constitute any problem if one notes that there is a solution analytical with respect to the variable $z = x^{\beta+2}$ ($\beta + 2 > 0$). Switching over in (6) from x to z and substituting into the resulting equation the series

$$y = \sum_{n=0}^{\infty} C_n z^n, \tag{13}$$

one may by turns determine all the coefficients C_n through C_0, which, according to the definition (5) of y is

$$C_0 = y(0) = 1. \tag{14}$$

In particular, we get

$$C_1 = \frac{dy}{dz}\bigg|_{z=0} = -\frac{1}{(\beta + 2)(\beta + 3)}, \tag{15}$$

$$2C_2 = \frac{d^2y}{dz^2}\bigg|_{z=0} = \frac{\alpha + (\beta + 3)/2}{(\beta + 3)(2\beta + 5)(\beta + 2)^2}. \tag{16}$$

The numerical integration of Eq. (6) with initial conditions (14)–(16) in $z = 0$ can be performed simply by any of standard methods (for example, by the Runge–Kutta method [7^{ad}]). As a result, the $y(x) = \varphi(R) - \varphi(x)$ function is calculated. Several typical graphs for these functions and for the dimensionless density $\sigma(x)$ are given in Fig. 30. They, in particular, confirm the increase in the system radius as one approaches the upper boundary of the region, dashed in Fig. 29. Figure 30 shows also how, with decreasing value of β, the degree of concentration of the mass of the system in the central region $(x \to 0)$ grows. The radii $R = R(\alpha, \beta)$ of the models for different values of (α, β) are also given in one of the columns of Tables IV and V where, in addition, there are listed some other macroscopic characteristics: the mass M of the system, the radial T_r and transversal $T_\perp/2$ kinetic energies, the total kinetic energy T_{tot}, the potential energy W, the ratios $u_r = T_r/(-W)$, $u_\perp = T_\perp/(-W)$, the "global" anisotropy $2T_r/T_\perp$, which define the stability of the modesls (cf. 6.1.3). In addition, in the tables, for each pair (α, β), are listed: the maximum radius

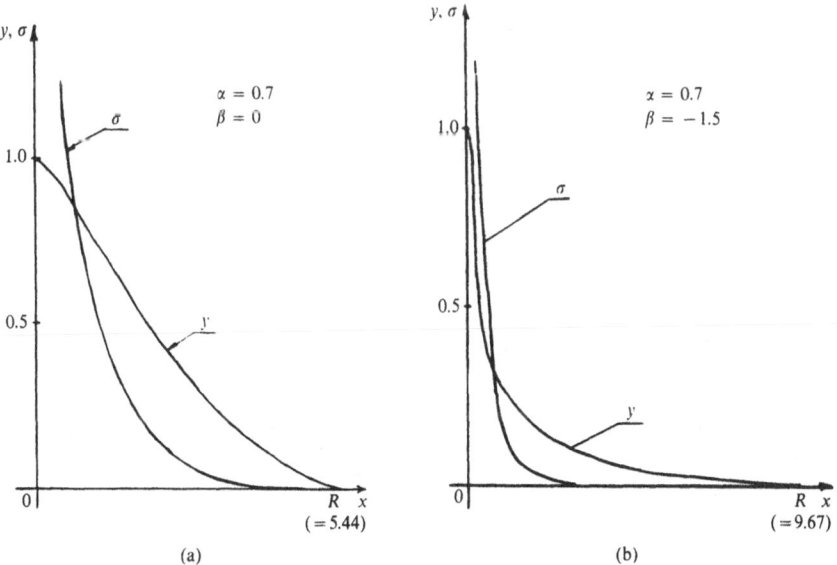

Figure 30. Dimensionless potentials $y(x)$ and densities $\sigma(x)$ for two models of the first Camm series. (a) $\alpha = 0.7$, $\beta = 0$; (b) $\alpha = 0.7$, $\beta = -1.5$.

Table IV. $\alpha = 0.3$.

β	M	R	$(x_k)_{\text{max}}$	$(x_k)_{\text{max}}/R$	T_r	$T_\perp$	T_{tot}	W	$u_r = -T_r/W$	$u_\perp = -T_\perp/2W$	$\xi = 2T_r/T_\perp$	$\bar{E}$	$\bar{\mu}$	μ_{max}	$\bar{\mu}/\mu_{\text{max}}$
0	31.68	4.05	2.41	0.595	3.05	6.10	9.15	−18.41	0.166	0.166	1	0.252	0.99	2.10	0.47
−0.15	30.34	4.16	2.46	0.591	2.93	5.41	8.34	−16.77	0.174	0.161	1.08	0.252	0.95	2.09	0.45
−0.30	28.81	4.28	2.52	0.589	2.78	4.72	7.49	−15.07	0.184	0.157	1.18	0.252	0.90	2.07	0.435
−0.45	27.06	4.40	2.58	0.586	2.59	4.02	6.62	−13.30	0.195	0.151	1.29	0.251	0.84	2.04	0.41
−0.60	25.05	4.52	2.63	0.582	2.40	3.35	5.75	−11.51	0.208	0.146	1.43	0.249	0.76	1.99	0.38
−0.75	22.70	4.62	2.67	0.578	2.14	2.67	4.81	−9.62	0.222	0.139	1.60	0.246	0.69	1.92	0.36
−0.90	19.96	4.67	2.68	0.574	1.83	2.01	3.84	−7.70	0.238	0.131	1.82	0.240	0.60	1.82	0.33
−1.05	16.76	4.65	2.65	0.570	1.48	1.41	2.88	−5.77	0.256	0.122	2.11	0.229	0.49	1.67	0.29
−1.20	13.06	4.47	2.52	0.564	1.08	0.87	1.95	−3.90	0.278	0.111	2.50	0.216	0.38	1.45	0.26
−1.35	8.91	4.00	2.24	0.560	0.67	0.44	1.11	−2.22	0.303	0.098	3.08	0.148	0.25	1.14	0.22
−1.50	4.65	3.04	1.69	0.556	0.30	0.15	0.45	−0.90	0.333	0.083	4.00	0.169	0.13	0.72	0.18
−1.65	1.26	1.45	0.80	0.552	0.06	0.02	0.08	−0.16	0.370	0.065	5.71	0.124	0.034	0.26	0.13

Table V. $\alpha = 2$.

β	M	R	$(x_k)_{\text{max}}$	$(x_k)_{\text{max}}/R$	T_r	$T_\perp$	T_{tot}	W	$u_r = -T_r/W$	$u_\perp = -T_\perp/2W$	$\xi = 2T_r/T_\perp$	$\bar{E}$	$\bar{\mu}$	μ_{max}	$\bar{\mu}/\mu_{\text{max}}$
0	23.75	9.54	4.88	0.511	1.50	3.01	4.51	−9.12	0.165	0.165	1	0.389	0.93	2.99	0.311
−0.15	23.11	11.07	5.63	0.509	1.44	2.66	4.11	−8.29	0.174	0.161	1.08	0.393	0.91	3.18	0.286
−0.30	22.47	13.33	6.76	0.507	1.37	2.33	3.70	−7.47	0.184	0.156	1.18	0.375	0.89	3.44	0.259
−0.45	21.86	17.00	8.58	0.505	1.30	2.01	3.30	−6.66	0.195	0.151	1.29	0.364	0.883	3.84	0.230
−0.60	21.38	23.89	12.02	0.503	1.21	1.70	2.91	−5.85	0.207	0.145	1.43	0.348	0.884	4.50	0.196
−0.75	21.20	40.70	20.41	0.501	1.12	1.41	2.53	−5.08	0.221	0.138	1.60	0.325	0.92	5.86	0.157
−0.90	21.90	121.24	60.68	0.500	1.04	1.14	2.18	−4.35	0.238	0.131	1.82	0.289	1.05	10.28	0.102

of circular orbits $(x_k)_{max}$, the ratio $(x_k)_{max}/R$, the mean energy $\bar{E}$, and the mean angular momentum $\bar{\mu}$ of the star, the maximum angular momentum μ_{max}, and the ratio $\bar{\mu}/\mu_{max}$. The tables were compiled in order to have orientation of how the main characteristics of the models vary with varying parameters α, β. Therefore, the accuracy of computations here was not very high, which is evident, for example, from the degree of satisfaction of the virial theorem $2T_{tot} = -W$; in computations associated with the investigation of the stability (6.2.2, 6.2.3), the accuracy was significantly improved. The tables, in particular, show that the "dimensionlessing" introduced by us is natural (apart from the fact that it leads to the simplest form of the Poisson equation) also in the sense that, in the units obtained, all the values $(M, T_r, W, \text{etc.})$ characterizing the system on the average have an order of ~ 1. But this simultaneously implies that the units introduced by the method described above vary during transition from one model to the other. Therefore, it is evident that by comparing the models with different values of (α, β) one has to compare the ratios of the type of u_r, $u_\perp$, $\bar{E}$, $\bar{\mu}/\mu_{max}$, etc.), which are independent of the measurement units rather than the absolute values of the quantities. Note that, in "absolute" units, the mass of the system M_a, may, of course, be arbitrary even for fixed values of α, β: assignment of a certain M_a fixes (also in absolute units) the scale factors k and U.

The ratio of the radial T_r to the transversal $T_\perp$ kinetic energies increases as the parameter β approaches $\beta_{min} = -2$, in accordance with the fact that the ratio of the squares of velocity dispersions for the series in question

$$\frac{2c_r^2}{c_\perp^2} = \frac{2}{\beta + 2} \to \infty \qquad \text{as } \beta \to -2. \tag{17}$$

The behavior of c_r^2 (and $c_\perp^2$) with the radius x is coincident with the behavior of the function $y(x)$ since, for example,

$$c_r^2(x) = \frac{y(x)}{\alpha + (\beta + 5)/2} \sim y(x). \tag{18}$$

Consequently, the graphs $y(x)$ for different (α, β), given in Fig. 30, may at the same time be considered also as the graphs c_r^2 and $c_\perp^2$ (in corresponding scales).

The phase regions of the models in question, on the plane $(\tilde{\varepsilon} \equiv \varphi(R) - \varepsilon, \mu)$, have, as seen from Fig. 31, a simple form. Above, as always, they are limited by the line of circular orbits. However, here this line is not defined analytically, as was in the case of the Idlis models (cf. 6.1.1); however, the numerical calculation is not complicated. It is easy to see that as $\beta \to \beta_{min}$ the phase region gets concentrated near the abscissa axis $\mu = 0$.

Figure 32 shows [for the two models: (1) $\alpha = 0.7$, $\beta = -0.75$ and (2) $\alpha = 0.7$, $\beta = -1.65$] the radii of circular orbits of particles x_k for different $\tilde{\varepsilon}$. For all models $(x_k)_{max} < R$; the boundaries of the system $x = R$ are reached only by the particles with $\mu = 0$. Very typical is the deformation of the curve $x_k = x_k(\tilde{\varepsilon})$ with decreasing β; the larger part of the stars is here near the

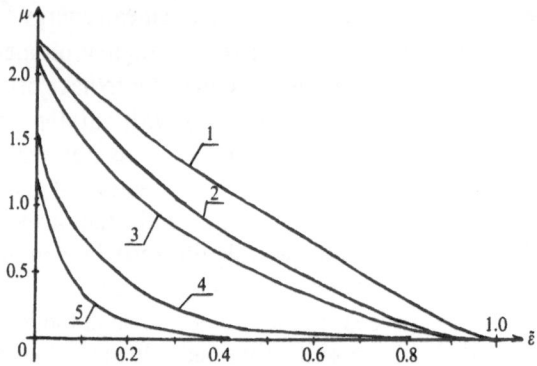

Figure 31. Phase region $(\tilde{\varepsilon}, \mu)$ for a number models of the first Camm series. (1) $\alpha = 0.7$, $\beta = 0$; (2) $\alpha = 0.7$, $\beta = -0.75$; (3) $\alpha = 0.7$, $\beta = -1.05$; (4) $\alpha = 0.7$, $\beta = -1.5$; (5) $\alpha = 0.7$, $\beta = -1.65$.

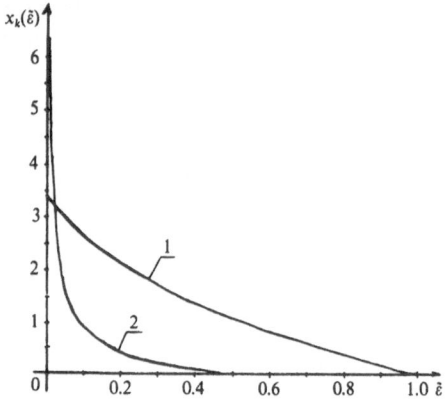

Figure 32. Radii of circular orbits for two models of the first Camm series. (1) $\alpha = 0.7$, $\beta = -0.75$; (2) $\alpha = 0.7$, $\beta = -1.65$.

center. The degree of this deformation [together with the value of the ratio $(x_k)_{max}/R$, which decreases monotonically together with β, cf. tables] also are indices of stability or instability of the system.

The last figure, Fig. 33 gives, for a typical model, $\alpha = 0.7$ and $\beta = -0.75$, the minimum x_{min} and maximum x_{max} radii of the orbits, corresponding to the maximum energy ε for different values of the angular momentum $\mu(\tilde{\varepsilon} = 0)$. The curves $x_{min}(\mu)$ and $x_{max}(\mu)$ merge at the point $\mu = \mu_{max}$ corresponding to the circular orbit $x_k = 3.38$.

6.2.2. Some Peculiarities of the Computing Algorithm. The stability of the models of the Camm first series is investigated both by the "method of particles" and by the matrix method. The scheme of calculations by the matrix method in general features is coincident with the scheme adopted for the Idlis models that was described in the previous section. But there are two complicating points. Firstly, in this case, the calculation of all the values incorporated in the matrix elements M_{mn}, including the line of circular orbits, $x_{min}(\tilde{\varepsilon}_i, \mu_i)$, $x_{max}(\tilde{\varepsilon}_i, \mu_i)$ on the set $(\tilde{\varepsilon}_i, \mu_i)$, etc. was to be per-

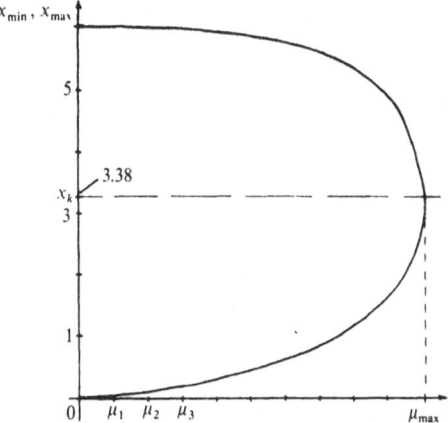

Figure 33. Minimum (x_{min}) and maximum (x_{max}) radii of star orbits for the minimum energy ε with different values of the angular momentum μ (first Camm series).

formed by numerical methods. Secondly, the matrix elements themselves if written in a standard form of (20) or (21), §4, Appendix, for the distribution function of (7) with $\beta < -1$, contain formally divergent integrals. The regularization necessary in this case is made in §4, Appendix. From the two biorthonormalized systems (the potential–density) given therein here the general system of (42) is used, which is more flexibly appropriated for the investigation of oscillations of the models with any α, β (for details cf. Appendix).

The method of particles, more exactly, the variety of this method used in [43^{ad}], was described above in Section 5.2 and also in §3, Appendix, in which item 1 deals with some peculiarities of the simulation of models of the first Camm series.

6.2.3. Main Results. We present in Fig. 34 graphs of dependence of the instability increment on the parameter β, for three sequences of models of the

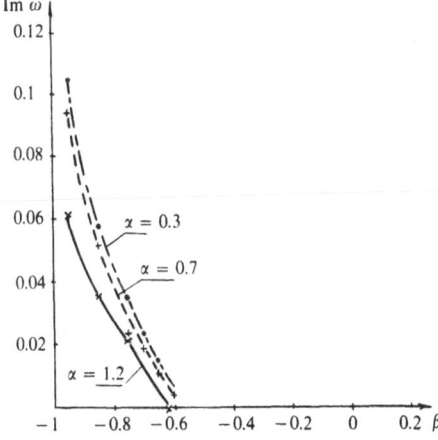

Figure 34. Dependence of the instability growth rate on the parameter β for three sequences of models of the first Camm series ($\alpha = 0.3, 0.7, 1.2$).

first Camm series, each of which corresponds to some fixed value of a second parameter α ($\alpha = 0.3, 0.7, 1.2$). The most essential conclusion, which one may see from this figure, is the fact that the stability boundaries (i.e., the positions of those points $\beta_c \simeq 0.6$ where Im $\omega = 0$) depend only very weakly on a value of the second parameter α.[28] This agrees with the fact that for the models of the series considered the ratio of the local stellar dispersion in the radial direction to one in the transversal direction (and, so, global aniso-tropy of the system ξ, on the value of which depends probably the stability or instability of the system) is determined by only parameter β: $2c_r^2/c_\perp^2 = 2T_r/T_\perp \equiv \xi = 2/(\beta + 2)$. The critical value $\beta = -0.6$ corresponds to the anisotropy $\xi = 1.43$. Recall that for Idlis models considered in the previous subsection the critical anisotropy had a similar value, $\xi_c \simeq 1.59$.

In all the cases the most unstable mode is the nodeless one, $l = 2$ (the elliptical deformation): just this mode excites at the stability boundary. In the "apparently unstable" region (i.e., for sufficiently large $\xi > \xi_c$ or for $\beta < \beta_c$) the smaller-scale modes may also be excited due to the instability, in particular with the same angular dependence $l = 2$. For example, for the model with $\alpha = 0.7$, $\beta = -0.95$ there are the three unstable roots with increments: Im $\omega_1 \simeq 0.094$, Im $\omega_2 \simeq 0.033$, Im $\omega_3 \simeq 0.008$. The eigen-functions, corresponding to these roots, are represented in Figs. 35(a), (b), and (c), respectively. It is seen that the most unstable solution [Fig. 35(a)] hasn't any nodes in a radius; the next mode [Fig. 35(b)] has one node; and, finally, the solution with minimum increment (in the given case), Fig. 34(c), has two radial nodes.

For the sake of a control of computations we also computed, in particular, the eigenfunctions ($\delta\rho$, $\delta\Phi$) for the most large-scale mode with $l = 1$. As was required, these functions were coincident with ρ_0', Φ_0' (within the limits of accuracy of computations).

Note also that for the models considered here (unlike, for example, the models of the Idlis series, see subsection 6.1) there are none of the "gaps" into the distribution of the resonance frequencies $\omega = l_i\Omega_i$; the latter occupy all the real axis ω. Though it still does not prohibit the presence of the real eigenfrequencies in the spectrum of the system's oscillations, however, it makes its search more difficult.

Now we turn to the description of some results obtained by the "method of particles."

Figure 36 represents distributions of the numbers of particles into the spherical layers for one of realizations of the model, with $\alpha = 0.3$, $\beta - 0.95$ (n)

---

[28] Computations were also performed for a few other values of α (in particular, for $\alpha = 2$, 2.4); we didn't present these so that the figure would not be too encumbered. This figure at a glance already represents the stability picture of the models under consideration with a sufficient clarity. Note at the same time that Im ω in Fig. 34 are given in the units (defined above), which vary from one model to the other. Divergence of the curves for $\beta < \beta_c$ will be less if one will calculate Im ω in the units of some characteristic frequency for the system, for example into the units of $\bar{v}_1$. The computation shows that, for example, for $\beta = -0.95$, $\bar{v}_1$ decreases from $\bar{v}_1 \simeq 0.63$ to $\bar{v}_1 \simeq 0.32$ as α increases from $\alpha = 0.3$ to $\alpha = 2.0$.

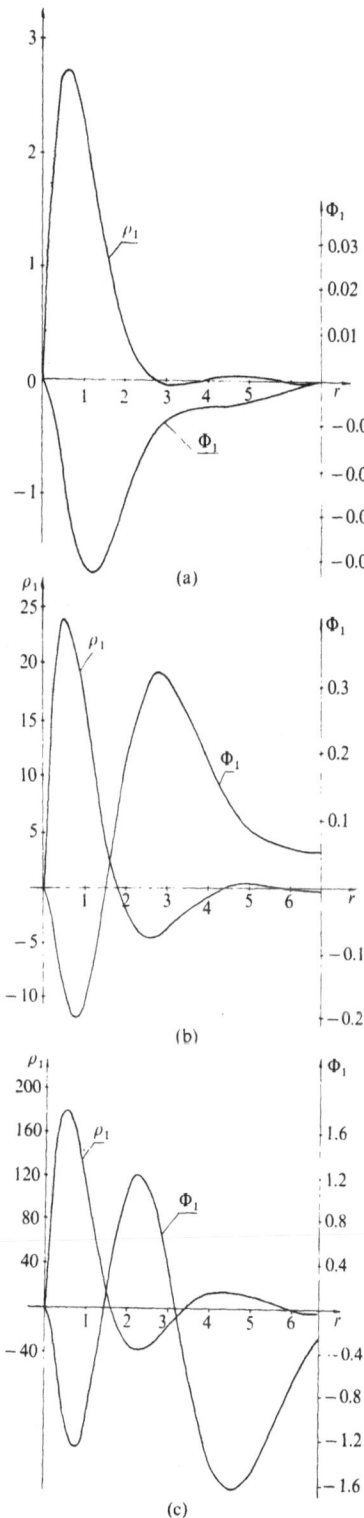

Figure 35. Eigenfunctions for disturbances (mode $l = 2$) in one of unstable models of the first Camm series.

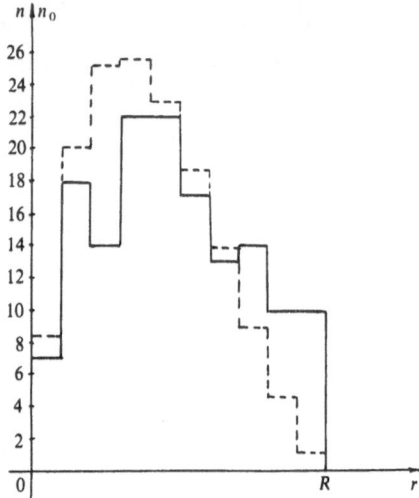

Figure 36. Particle numbers (n) in spherical layers (solid line) for one of realization of the model $\alpha = 0.3$, $\beta = -0.95$ (first Camm series) and for the exact equilibrium (n_0) (dotted line).

Table VI

	T_r	$T_\perp$	T_{tot}	W	E_{tot}	$u_r = -T_r/W$
Realization	1.76	2.14	3.90	-8.16	-4.26	0.22
Exact equilibrium	1.72	1.81	3.53	-7.05	-4.52	0.24

	$u_\perp = -T_\perp/W$	$\zeta = 2T_r/T_\perp$	$\bar{E}$	E_{max}	μ_{max}	L_z
Realization	0.13	1.64	0.28	0.97	1.70	-0.4
Exact equilibrium	0.13	1.905	0.24	1.00	1.78	0

and of the exact equilibrium (n_0) for $N = 150$, $c = 0.1$. The other macroscopic parameters of the model system obtaining for this realization correspond to the first line of Table VI (values of the same parameters for the exact equilibrium are given in the second line). From Fig. 36 and from Table VI one may judge likeness of the model realization to the true equilibrium state. The most essential difference occurs in values of global anisotropies (1.640 and 1.905). The anisotropy $\xi = 1.64$ corresponds to "effective" $\beta = -2 + 2/\xi \simeq -0.78$. As is seen from Fig. 34, for $\beta = -0.78$ (and $\alpha = 0.3$), the system must be unstable but with the very small increment Im $\omega = 0.04$ (or ≈ 0.065 if we express Im ω by $\bar{v}_1 \simeq 0.62$). Such weak instability simply hasn't sufficient time to become considerable (during the times of numerical runnings $t \sim 2T$, $T \sim 1/\bar{v}_1$). Indeed, from the projections (in a dimetry projection) of point-stars of the model system at different moments (Fig. 37) one cannot see the appearance of any ellipticity. This confirms also Table VII, in which we represent the principal values of the quadrupole

Table VII

Step	d_1	d_2	d_3	a	b	c	$(\overline{r^2})^{1/2}$
0	−0.290	0.038	0.252	2.14	2.26	2.34	1.73
10	−0.262	0.051	0.210	2.16	2.28	2.34	1.75
20	−0.237	0.050	0.187	2.20	2.31	2.36	1.78
30	−0.209	0.034	0.175	2.26	2.35	2.40	1.81
40	−0.187	0.014	0.175	2.32	2.39	2.46	1.85
75	−0.163	0.0005	0.163	2.59	2.64	2.69	2.04
100	−0.216	0.027	0.189	2.79	2.87	2.91	2.21

momenta tensor, values a, b, c of semiaxes of the "equivalent" homogeneous ellipsoid, and rms radii $(\overline{r^2})^{1/2}$ (definitions of all the values listed are exactly coincident with ones described in subsection 5.2). It is seen that at the stage of an evolution, which is reflected by this table, the initial small ellipticity becomes even smaller during the time. In order to reveal for certain, for the chosen values of the model parameters ($\alpha = 0.3$, $\beta = -0.95$), the instability, which must be relatively weak (Im $\omega \simeq 0.105$, see Fig. 34), it turns out to be necessary to run the evolutions of many different realizations (i.e., accumulate the sufficient statistics) and then carry out a relevant statistical processing of the results. Such a work was actually made for the values α, β given above. Among different realizations for this α, β, we find those which correspond to the larger "effective" β (the example of just such a realization we had above), as well those corresponding to the smaller "effective" β (i.e., more unstable). In most realizations, actually, the evolution revealed the instability: the principal values of the quadrupole moment tensor and a degree of the system's ellipticity had a trend to some growth. At the same time, plotter pictures, for the most realizations in the case, only slightly differed from Fig. 37. On the other hand, for models with β corresponding to a sufficiently strong instability (according to the linear theory; for example, for $\beta = -1.5$, $\alpha = 0.3$ when Im $\omega \sim \bar{v}_1$) the elliptical deformation is clearly manifested in the plotter pictures; moreover, these pictures are very similar to Fig. 23, reflecting an evolution of spheres with purely radial star motions (which, of course, is quite natural).

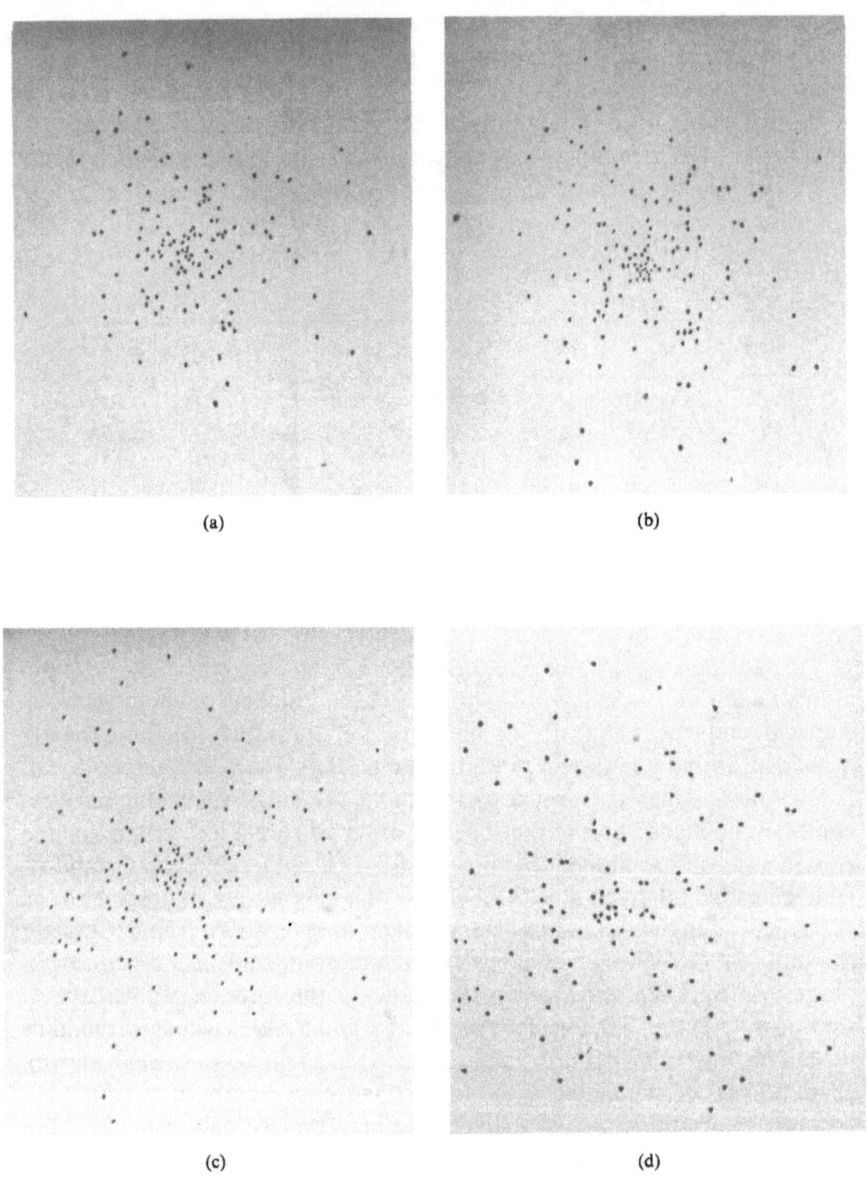

Figure 37. Projections of point-stars locations for the computer model (first Camm series) at different times. (a) step 11, (b) step 31, (c) step 51, (d) step 76.

6.3. Shuster's Model in the Phase Description

6.3.1. Equilibrium State. In this section, we shall investigate the stability of distribution functions (35), §1, which are suggested by Kuzmin and Veltmann [18^{ad}, 19^{ad}]:

$$f_0(\tilde{\varepsilon}, \xi) = (1 + p\xi)^{-5/2} \cdot \Pi(u),$$

$$\Pi(u) = 2\sqrt{2}\,\pi^{-2}\rho_0[\lambda + \tfrac{8}{5}pu + \tfrac{16}{7}(1 - \lambda)u^2 - \tfrac{64}{21}pu^3]u^{3/2}, \qquad (1)$$

$$u = \frac{\tilde{\varepsilon} - \lambda\xi}{1 + p\xi}, \qquad \tilde{\varepsilon} = -E, \qquad \xi = \frac{\mu^2}{2},$$

where E and μ are an energy and an angular momentum of a particle; p and λ are dimensionless parameters of the model.[29]

We recall that the phase density (1) corresponds to the volume star density

$$\rho_0(r) = \frac{\rho_0}{(1 + r^2/r_0^2)^{5/2}} \qquad (2)$$

and the gravitational potential

$$\Phi_0(r) = -\frac{\Phi_0}{(1 + r^2/r_0^2)^{1/2}}, \qquad (3)$$

where the parameters ρ_0, Φ_0, r_0 are connected (as it follows from the Poisson's equation) by the relation

$$\rho_0 = \frac{3\Phi_0}{r_0^2}. \qquad (4)$$

The values of the parameters λ and p are not quite arbitrary; the restrictions arise [5^{ad}] in the connection with the requirements of nonnegativeness of the phase density f_0: $\Pi(1) \geq 0$, of decreasing f_0 with the energy E: $\Pi'(1) \geq 0$ and, finally, of "the radial prolateness" of the velocity distribution $(\partial f_0/\partial \xi < 0)$

$$5p\Pi(1) + 2(\lambda + p)\Pi'(1) \geq 0.$$

There is the last restriction from above, which means

$$0 < 136p + 91\lambda \leq 112. \qquad (5)$$

For stability investigation it is natural to introduce the dimensionless quantities involving the problem so that the expression for the potential gets maximally simplified; in the given case we assume $\Phi_0 = 1$, $r_0 = 1$, then $\rho_0 = 3$. It is also natural to assume $4\pi G = 1$.

---

[29] We note here [18^{ad}, 19^{ad}] that the function $\Pi(u)$ satisfies Abel's equation

$$\rho_0(r) = \frac{4\sqrt{2}\,\pi}{pr^2\Phi + \lambda r^2 + 1}\int_0^\Phi du\sqrt{\Phi - u}\,\Pi(u)\,du,$$

$(\Phi(r) \equiv -\Phi_0(r))$ and, consequently, $\Pi(u)$ may be obtained as the solution of this equation [for $\rho_0(r)$ in the form (2)].

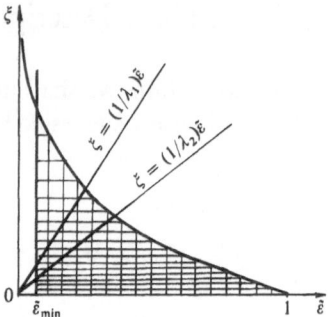

Figure 38. Phase regions $(\tilde{\varepsilon}, \zeta)$ for the models of the Kuzmin–Veltmann series.

As $r \to \infty$, $\Phi_0(r) \simeq -1/r$, but, on the other hand, the potential must evidently approach Newton's one: $\Phi_0(r) \to -GM/r$, where M is the mass of the system. Since in the units assumed $G = 1/4\pi$, we obtain, for the mass, $M = 4\pi$.

The line of circular orbits at the plane $(\tilde{\varepsilon}, \zeta)$, for the Shuster model, is determined parametrically (the parameter is r, the radius of the corresponding orbit) by the following two equations:

$$\zeta = \frac{r^4}{2(1 + r^2)^{3/2}}, \qquad \tilde{\varepsilon} = \frac{2 + r^2}{2(1 + r^2)^{3/2}}. \tag{6}$$

Near the center $(r \to 0)$ (6) gives $\zeta \approx \frac{1}{2}(1 - x)^2$, while for the large distances from the center $(r \to \infty)$ one can obtain $\zeta = 1/4x$. The phase region of the system at the plane $(\tilde{\varepsilon}, \zeta)$ is bounded, apart from the line of circular orbits (6), by two straight lines: $\zeta = 0$ $(\zeta > 0)$ and $\tilde{\varepsilon} = \lambda\zeta$ $(\tilde{\varepsilon} > \lambda\zeta)$. The last condition is connected with the presence of the radical into expression (1) for the phase density; only this condition distinguishes phase regions of the models with different values of the parameter λ (see Fig. 38). The form of the phase region at the plane $(\tilde{\varepsilon}, \zeta)$ is not dependent of p.

As for previous series of distribution functions investigated against the stability in Sections 6.1 and 6.2, here we also give tables (Table VIII) in which some essential macroscopic characteristics of models (1) for different values of parameters λ, p[30] are represented. From the tables one may see that these characteristics for the models under consideration vary within sufficiently wide ranges; moreover, dependence on λ is considerably stronger than on p. Note at the same time that dependence of the global anisotropy parameter of the system $\zeta = 2T_r/T_\perp$ on p (for a fixed value of λ) is almost exactly linear.

---

[30] Potential energy in the units assumed is $W = 3\pi^2/8 \approx -3.701$, a total kinetic energy $T_r + T_\perp \simeq 1.851$ (in accordance with the virial theorem), an average particle energy $\bar{E} \simeq 0.442 = \text{const}$, an average angular momentum $\bar{\mu}$ decreases smoothly (for each fixed λ) with increasing p: for $\lambda = 0.01$, from $\bar{\mu} = 0.540$ for $p = 0.001$ to $\bar{\mu} = 0.435$ for $p = 0.817$, or, for $\lambda = 0.256$, from $\bar{\mu} = 0.448$ for $p = 0.001$, to $\bar{\mu} = 0.401$ for $p = 0.652$, and so on.

Table VIII

$\lambda = 0.010$				$\lambda = 0.256$				$\lambda = 0.492$			
p	T_r	$T_\perp$	$\xi = 2T_r/T_\perp$	p	T_r	$T_\perp$	$\xi = 2T_r/T_\perp$	p	T_r	$T_\perp$	$\xi = 2T_r/T_\perp$
0.001	0.626	1.224	1.026	0.001	0.757	1.095	1.382	0.001	0.827	1.024	1.615
0.137	0.668	1.184	1.128	0.131	0.778	1.073	1.449	0.083	0.837	1.014	1.651
0.273	0.701	1.150	1.219	0.261	0.797	1.054	1.513	0.165	0.847	1.004	1.686
0.409	0.730	1.211	1.302	0.392	0.815	1.036	1.574	0.248	0.856	0.995	1.720
0.545	0.755	1.096	1.380	0.522	0.832	1.019	1.632	0.330	0.865	0.986	1.754
0.681	0.778	1.073	1.452	0.652	0.847	1.004	1.689	0.412	0.873	0.977	1.787
0.817	0.799	1.052	1.520					0.494	0.822	0.969	1.820

$\lambda = 0.738$				$\lambda = 0.984$				$\lambda = 1.800$			
p	T_r	$T_\perp$	$\xi = 2T_r/T_\perp$	p	T_r	$T_\perp$	$\xi = 2T_r/T_\perp$	p	T_r	$T_\perp$	$\xi = 2T_r/T_\perp$
0.001	0.881	0.970	1.83	0.001	0.923	0.923	1.990	0.073	1.026	0.825	2.487
0.067	0.887	0.964	1.84	0.042	0.927	0.924	2.004	0.187	1.032	0.819	2.519
0.132	0.893	0.958	1.86	0.083	0.930	0.921	2.019	0.300	1.036	0.813	2.551
0.198	0.899	0.952	1.89								
0.264	0.905	0.946	1.91								
0.330	0.911	0.940	1.93								

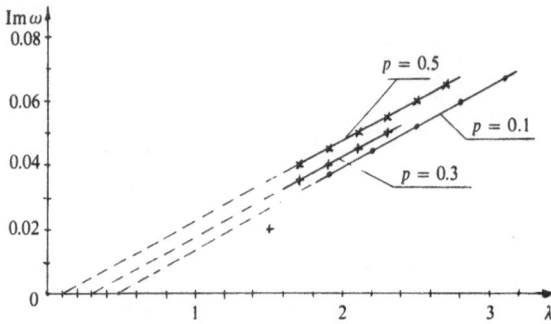

Figure 39. Dependence of the instability growth rate on the parameter λ for three sequences of the models of the Kuzmin–Veltmann series ($p = 0.1, 0.3, 0.5$).

6.3.2. Peculiarities of the Computing Algorithm were connected in particular with the infinity (unlimitedness) of Shuster's models. So we introduced some maximum radius $R \gg r_0$ and, thus, bounded the phase space of the system for small $\tilde{\varepsilon}$ (see Fig. 38). In other respects given series are similar to Idlis' series (Section 6.1) (first of all due to the fact that we also have here explicit and very simple analytical expressions for the equilibrium potential and density). If one used the variables u, ζ, we came to the computation scheme which is practically identical to one used in the case of Idlis' models.

It is more convenient, however, to employ the variables $\tilde{\varepsilon}$, ζ since with these variables [unlike (u, ζ)] the equation for the line of circular orbits doesn't depend on values of parameters (λ, p) of the concrete model. So, at the plane $(\tilde{\varepsilon}, \zeta)$, one may introduce some universal network (see Fig. 38), compute all the necessary quantities on this network, and throw away those points of the network which don't hit the phase region of the system. For such a choice of variables only one computation of data sets[31] $\chi_{l_1 l_2}^{(n)}(\tilde{\varepsilon}_i, \zeta_i)$ (see sub-section 6.1) is necessary.

6.3.3. Main Results. Similar to Fig. 34 in the previous subsection, in this case we present in Fig. 39 graphs of dependence of instability increment on the parameter λ, for three sequences of Kuzmin–Veltmann models for fixed values of the second parameter p ($p = 0.1, 0.3, 0.5$). It is seen that for not very small Im ω the dependence Im $\omega(\lambda)$ is close to the linear one. At the same time, for small ω ($=$ Im ω) computations in a given case are not exact. The error is due to small denominators in the expressions for matrix elements M_{mn}. Stars, which have energies close to the maximum energy, move during an essential part of a total time in a near-Coulomb potential $\sim 1/r$. But all[32]

---

[31] In the variables (u, ζ) one must repeat the computation of these data sets for each new model, i.e., for each pair of parameters (λ, p). Note that for the one-parametric series of Idlis' models the utilization of variables (ε, ζ) permitted computer time to diminish considerably as well, though this was less important there than in the case of (two-parametric!) models investigated in this subsection.

[32] So the described effect, for the nonfinite models under consideration, is far more important than for the finite systems investigated earlier.

these stars are resonant at $\omega = 0$ for the corresponding l_1, l_2, l_3 since in the Coulomb potential $\Omega_1 = \Omega_2$ [see formula (37), Section 6.1]. If we increase the exactness of computations [in particular, by using a more "dense" network on the plane (ε, μ)], curves Im $\omega(\lambda)$ become more straight. In connection with the above, for the estimate of the critical values according to data given, we shall extrapolate the linear parts of curves Im $\omega(\lambda)$ determined with confidence to the left up to its intersections with the axis of abscissae. The values λ_c corresponding to stability boundaries Im $\omega = 0$, determined by this way, are the following: $\lambda \approx 0.47$ for $p = 0.1$; $\lambda \approx 0.30$ for $p = 0.3$ and $\lambda \approx 0.12$ for $p = 0.5$. These critical values of the parameters (λ, p) correspond to anisotropies $\xi_c = 2T_r/T_\perp$, which equal 1.60, 1.57, 1.50, respectively. These values ξ_c are as near to each other and also to the values ξ_c determined earlier (Sections 6.1, 6.2) for other spherically symmetric stellar systems.

The data given above refer to the nodeless mode $l = 2$ which is the most unstable one (as well as for the models investigated in the previous Sections 6.1 and 6.2).

A general discussion of the stability problem for spherically-symmetrical stellar systems is given in the next section, while some applications of stability theory developed above are considered in Chapter VIX.

§ 7 Discussion of the Results

Let us draw some conclusions about the investigation into stability of collisionless spherically symmetrical systems. In §2, the stability of distribution functions of a very wide class (isotropic and decreasing with energy) is shown. It is most probable that all isotropic distribution functions with $f'_0(E) < 0$ are stable. Also (§3) the stability of all models with circular orbits of particles is provided with respect to localized perturbations for $d\Omega^2/dr < 0$. It may be suggested that such systems are stable also for arbitrary perturbations.

The suggestion seems very probable that any distribution functions, intermediate between circular and isotropic, are also stable. An exact investigation in §4 (Section 4.1) of one such "intermediate" system (Camm model) which showed its stability supports this suggestion.

On the other hand, in §5 instability of models with purely radial trajectories of particles is proved. And, finally, the investigation of general models of spherically symmetrical systems, in §6, allows us to formulate the universal stability criterion. It turns out that the stability depends on the value of the global anisotropy of the system

$$\xi = \frac{T_r}{(T_\perp/2)}, \tag{1}$$

i.e., from the ratio of the total kinetic energies of radial and transversal star motions. The critical values of the parameter ζ for the models, considered in §6, are within the interval

$$\zeta_c = 1.70 \pm 0.25. \tag{2}$$

For $\zeta > \zeta_c$ the system is unstable, and for $\zeta < \zeta_c$ it is stable. The interval of values ζ_c is rather narrow, though the models considered are very different from each other. This just implies the universality of the criterion obtained. The degree of the universality may be seen from (2).

Apart from the universality of the obtained criterion it is interesting that, for the instability, the anisotropy must not be too large: for $\zeta > 2$, the system is apparently unstable. By the way, one sometimes considers [61[ad]] the models with mean anisotropy $\zeta \sim 5\text{--}8$ (for more detail see §8, Chapter IX).

Thus, we now have a sufficiently general picture of the stability of spherically symmetrical stellar systems which are schematically represented in Fig. 40. In this figure we take on the axes the radial and transversal kinetic energies of the system. The dotted line means the boundary separating stable systems from unstable ones. As was shown in §6, the instability begins with the elliptical deformation of the sphere if the dependence of the distribution function on the energy E and on the angular momentum L has a "natural" character, i.e., $f'_E < 0, f'_L < 0$.

If the latter condition is not satisfied, the radial oscillations may also be unstable. Henon [217] investigated, by numerical methods, the stability of models (24), §1 (first Camm series), for $\alpha < 0$ ($f'_{0E} < 0$) with respect to the

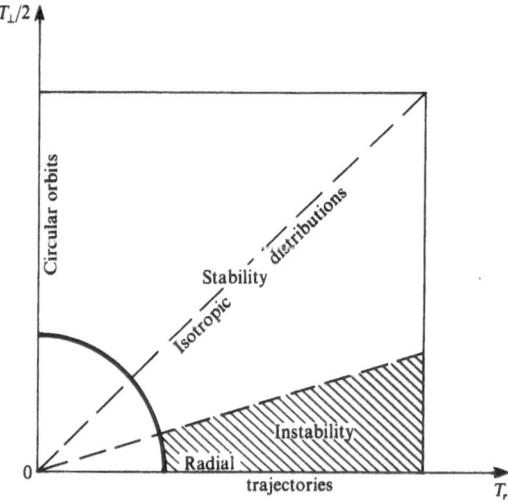

Figure 40. Probable locations in plane $(T_r, T_\perp/2)$ of stable and unstable collisionless spherically symmetrical systems.

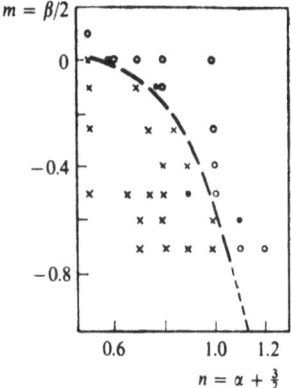

Figure 41. Approximate boundary of stability for models (24), §1, to radial oscillations [217].

radial disturbances. The results are summed in Fig. 41, where in the (α, β) $(m \equiv \beta/2, n \equiv \alpha + \frac{3}{2})$ plane of the parameters, the circles indicate stable models, the crosses, unstable, and the dots correspond to uncertain cases close to the boundary between stable and unstable solutions; the latter is roughly represented by the dashed line. The instability region, as is easy to see, is located below the line $\beta = 0$, corresponding to (isotropic) polytropes. In particular, it may be noticed that the model with $f_0 \sim \delta(E_0 - E)\delta(L^2)$, corresponding, in (24), §1, to $\alpha = -1$, $\beta = -2$, should already be unstable with respect to radial perturbations. If the dashed curve in Fig. 41 is approximated up to $\beta = -2$ (which is possible with greater uncertainty) we shall obtain the boundary value $\alpha \simeq (-0.3)–(-0.4)$. The models for clusters with radial trajectories ($\beta = -2$) with larger α are stable, and unstable with smaller ones. In particular, all "nonpathological" models with $f'_{0E} < 0$ (i.c., $\alpha > 0$) are most probably stable with respect to radial perturbations. Instability is due evidently to a too rapid growth f_0 with energy. With respect to nonradial perturbations, such systems, as was seen in (§5) are unstable, and this, according to continuity, should be true also for sufficiently close distribution functions.

Let us also make some comments concerning the stability of rotating collisionless spheres. We have considered two such examples, and, in one case, the system happened to be stable (models with circular orbits, §3), while in the other, unstable (Freeman's sphere model, §4). Of course, the existence of rotating collisionless spherically symmetrical systems is hardly justified evolutionarily. Nevertheless, these results allow one to question the presence of necessary correlation between the degree of oblateness and the velocity rotation of the system suggested from the classical equilibrium figure theory. The ways of the evolution of collisionless systems (in particular, stellar ones) may be diversiform.

Problems

1. Investigate the stability of arbitrary spherically symmetrical systems with circular orbits with respect to localized perturbations (Pal'chik et al. [106]).

Solution. As indicated in §3, this problem is reduced to the investigation of the position of the roots ω of the equation

$$A_l(r, \omega) = 0. \tag{1}$$

Introduce the notations

$$x = \frac{\omega^2}{\Omega^2(r)}, \qquad \varkappa^2 = \frac{\omega_0^2}{\Omega^2} + 1, \qquad \alpha_s = \varkappa + s, \qquad \beta_s = \varkappa - s. \tag{2}$$

In these notations, the $A_l(r, \omega)$ function will take the form

$$A_l(r, x) = 1 + 2(\varkappa^2 - 1) \sum_{s=1}^{l} \alpha_{ls} \frac{x - \alpha_s \beta_s}{(x - \alpha_s^2)(x - \beta_s^2)} + (\varkappa^2 - 1) \frac{\alpha_{l0}}{x - \varkappa^2}. \tag{3}$$

Let us restrict ourselves to consideration of only unstable (see below) solutions. Namely, for each given l we shall investigate such distributions of the initial density $\rho_0(r)$, for which the condition

$$\varkappa^2 > l^2 \qquad (l \geq 2) \tag{4}$$

is satisfied. First of all let l be odd. Then

$$A_l(r, x) = 1 + 2(\varkappa^2 - 1) \sum_{s=1}^{l}{}' \alpha_{ls} \frac{x - \alpha_s \beta_s}{(x - \alpha_s^2)(x - \beta_s^2)}, \tag{5}$$

where $\sum'$ means that summing is performed for odd s. This sum contains $(l + 1)/2$ summands; consequently, Eq. (1) has $(l + 1)$ roots. All these roots x are real. Qualitatively, the form of the $A_l(x)$ function is given in Fig. 42.[33] Between each pair of neighboring poles of $A_l(x)$ there is one root. The sum total of such roots is l, they are real and

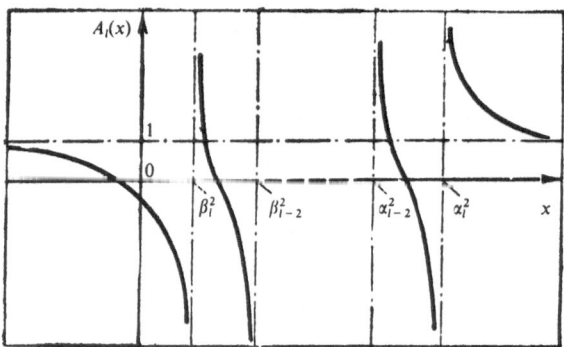

Figure 42. Graphic representation of localized dispersion equation $A_l(x) = 0$ with $\varkappa^2/\Omega^2 > l^2$ (unstable mass distributions) [106].

---

[33] To construct the plot in Fig. 42, it is sufficient to analyze the behavior of $A_l(x)$ near the poles $x = \mu_s$ ($\mu_s = \alpha_s^2$ or β_s^2); under the condition in (4) it happens that $A_l(x \approx \mu_s) \sim a/(x - \mu_s)$ ($a > 0$). It is readily proved also that the $A_l(x)$ function is monotonic in the case under consideration [106].

positive. The last $(l + 1)$th root lies evidently somewhere within the interval $-\infty < x < \beta_l^2 = (\varkappa - l)^2$, where the $A_l(x)$ function varies from 1 to $(-\infty)$. Clarify the sign of this root. For this purpose, calculate the value of $A_l(r, x)$ at $x = 0$

$$A_l(r, 0) = 1 - 2(\varkappa^2 - 1) \sum_{s=1}^{l} \frac{\alpha_{ls}}{\alpha_s \beta_s} = 1 - 2 \sum_{s=1}^{l}{}' |P_{s0}^l(0)|^2 \frac{\varkappa^2 - s^2 + s^2 - 1}{\varkappa^2 - s^2}$$

$$= - \sum_{s=-l}^{l}{}' \frac{s^2 - 1}{\varkappa^2 - s^2} |P_{s0}^l(0)|^2. \tag{6}$$

In the derivation of (6), the formula $\sum_{s=-l}^{l} |P_{s0}^l(0)|^2 = 1$ has been used. Since $s^2 \geq 1$ and $\varkappa^2 > l^2$, we have $A_l(r, 0) < 0$. Consequently, the root in question is negative (see Fig. 42).

In a similar way, the case of even l is investigated. The $A_l(r, x)$ function then has the form

$$A_l(r, x) = 1 + \frac{\alpha_{l0}}{x - \varkappa^2}(\varkappa^2 - 1) + 2(\varkappa^2 - 1) \sum_{s=2}^{l}{}'' \alpha_{ls} \frac{x - \alpha_s \beta_s}{(x - \alpha_s^2)(x - \beta_s^2)}, \tag{7}$$

where $\sum''$ means that summing is performed for even s. In the same way as in the case of odd l, the $A_l(r, x)$ function has $(l + 1)$ roots, from which l roots are positive [provided the condition (4) is satisfied]. To determine the sign of the $(l + 1)$th root, make an estimate of $A_l(r, 0)$:

$$A_l(r, 0) = 1 - \alpha_{ls} \frac{\varkappa^2 - 1}{\varkappa^2} 2(\varkappa^2 - 1) \sum_{s=2}^{l}{}'' \frac{\alpha_{ls}}{\varkappa^2 - s^2}$$

$$= - \sum_{s=-l}^{l}{}' \frac{s^2 - 1}{\varkappa^2 - s^2} \alpha_{ls} < - \sum_{s=2}^{l}{}'' \frac{s^2 - 1}{\varkappa^2} \alpha_{ls} = \frac{1}{\varkappa^2}\left(1 - \sum_{s=-l}^{l}{}'' \alpha_{ls} s^2\right) < 0,$$

so that this root is negative.

Thus it is shown that, under the condition in (4), the (local) instability takes place.

It is easy to calculate [106], e.g., the growth rate of instability for the systems with density, slowly increasing toward an edge so that $\varkappa^2/\Omega^2 = 4 + \varepsilon(r)$, $0 < \varepsilon(r) \ll 1$. For (unstable) even harmonics we have

$$-\omega^2 = \varepsilon(r)\Omega^2(r) \frac{3\alpha_{12}}{13\alpha_{12} + 2\alpha_{10} + 16\sum_{s=4}^{l}[(s^2 - 1)/(s^2 - 4)]\alpha_{ls}},$$

This formula, in particular, shows the presence of solutions with the zero frequency (at any even l) within the limit of a homogeneous sphere (see remark in the footnote at the end of subsection 3.1.2).

The instability increment for the mode $l = 2$ obtained numerically is represented in Fig. 20(a). A somewhat longer consideration [106] shows that when the condition opposite to that in (4) is satisfied, the system is stable. An approximate form of the $A_l(x)$ function for this case is given in Fig. 43.

Note that with respect to perturbations with $l = 0, 1$, the system is stable at any initial distribution. Actually, in the case of radial perturbations, $l = 0$, Eq. (1) has a single positive root $x_0 = 1$, and two roots: $x_1 = 0$, $x_2 = 3 + \varkappa^2$ ($x_2 > 0$) for dipole perturbations ($l = 1$).

The following (quadrupole) harmonic ($l = 2$) leads to excitation of the system provided that $\varkappa^2 > 4$, or which is equivalent,

$$\frac{d\Omega^2}{dr} > 0. \tag{8}$$

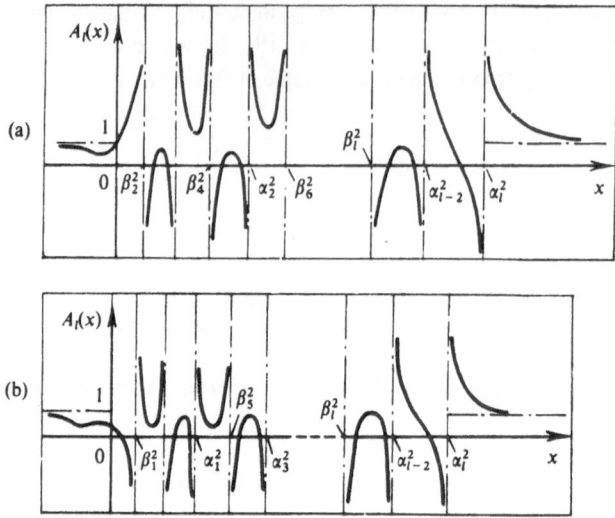

Figure 43. Function $A_l(x)$ for stable density distribution $\rho_0(r)$ $(x^2/\Omega^2 < l^2)$ [106]: (a) even l; (b) odd l.

The manner of arrangement of the roots in Eq. (6), §3, 4, describing "global" perturbations for $l \gg 1$ (when they are localized near the sphere boundary $r = R$) can be determined, as above. The distributions of density $\rho_0(r)$ satisfying the inequalities

$$l^2 < x^2 < x_c^2 \qquad (9)$$

happen to be unstable (at given l). Here x_c^2 is found from the equation

$$\sum_{s=-l}^{l} \frac{s^2 - 1}{x_c^2 - s^2} \alpha_{ls} = 1 \qquad (10)$$

in the case of odd l and a similar equation for even l.

2. Investigate the nonlinear stability of the collisionless Camm sphere [(29), §1] with respect to radial perturbations retaining spatial homogeneity of density (Antonov and Nuritdinov [12]).

Solution. Write the equilibrium distribution function in the form

$$f_0 = \text{const}\{1 + [\mathbf{r}_0, \mathbf{v}_0]^2 - r_0^2 - v_0^2\}^{-1/2} \qquad \left(\text{const} - \frac{\rho_0}{\pi^2}\right). \qquad (1)$$

Consider the linear transformation linking the current $\mathbf{r}$, $\mathbf{v}$ and the initial $\mathbf{r}_0$, $\mathbf{v}_0$ Lagrange coordinates and speeds of particles:

$$\mathbf{r} = \alpha(t)\mathbf{r}_0 + \beta(t)\mathbf{v}_0, \qquad \mathbf{v} = \dot{\alpha}(t)\mathbf{r}_0 + \dot{\beta}(t)\mathbf{v}_0. \qquad (2)$$

Invert this transform:

$$\mathbf{r}_0 = \dot{\beta}(t)\mathbf{r} - \beta(t)\mathbf{v}, \qquad \mathbf{v}_0 = -\dot{\alpha}(t)\mathbf{r} + \alpha(t)\mathbf{v}, \qquad (3)$$

and substitute (3) into (1); as a result we obtain the distribution function f of the perturbed state

$$f = \text{const}\{1 + [\mathbf{r}\mathbf{v}]^2 - (\dot{\alpha}^2 + \dot{\beta}^2)r^2 + 2(\alpha\dot{\alpha} + \beta\dot{\beta})\mathbf{r}\mathbf{v} - (\alpha^2 + \beta^2)v^2\}^{-1/2}. \qquad (4)$$

By integrating (4) over velocities **v** we again arrive at the density ρ, independent of spatial coordinates. Consequently, formulae (2) describe the required perturbation.

Let us consider the changes in radius of the sphere $R(t)$. To determine the radius, write the equation of boundary in phase space as

$$(\dot{\alpha}^2 + \dot{\beta}^2)r^2 - 2(\alpha\dot{\alpha} + \beta\dot{\beta})\mathbf{rv} + (\alpha^2 + \beta^2)v^2 - [\mathbf{rv}]^2 = 1. \tag{5}$$

The radius $R(t)$ is evidently the maximum of $|\mathbf{r}|$, at which **v** may still be the real vector. Transform (5):

$$(\alpha^2 + \beta^2)\left(v_2 - \frac{\alpha\dot{\alpha} + \beta\dot{\beta}}{\alpha^2 + \beta^2}r\right)^2 + (\alpha^2 + \beta^2 - r^2)v_\perp^2 = 1 - \frac{r^2}{\alpha^2 + \beta^2}, \tag{6}$$

where the Jacobian of transformation (2) is

$$\alpha\dot{\beta} - \beta\dot{\alpha} = 1. \tag{7}$$

Hence we find that the maximum r is attained at

$$v_r - \frac{\alpha\dot{\alpha} + \beta\dot{\beta}}{\alpha^2 + \beta^2}r = v_\perp = 0; \tag{8}$$

it is equal to

$$R(t) = \sqrt{\alpha^2 + \beta^2}. \tag{9}$$

It is easy to see that the density ρ is proportional to R^{-3},

$$\rho(t) = \rho_0(\alpha^2 + \beta^2)^{-3/2}. \tag{10}$$

Therefore, the equation of motion of an individual star is

$$\ddot{r} = -\frac{r}{(\alpha^2 + \beta^2)^{3/2}}. \tag{11}$$

In comparing with the first equation (2), we obtain

$$\ddot{\alpha} = -\frac{\alpha}{(\alpha^2 + \beta^2)^{3/2}}, \tag{12}$$

$$\ddot{\beta} = -\frac{\beta}{(\alpha^2 + \beta^2)^{3/2}}. \tag{13}$$

Equations (12) and (13) have the form of equations of the problem of two bodies, which are attracted according to the Coulomb law. Respectively, the equation of motion for $R = (\alpha^2 + \beta^2)^{1/2}$ is

$$\ddot{R} = -\frac{\partial U_{\text{eff}}}{\partial R}, \qquad U_{\text{eff}} = \frac{1}{2R^2} - \frac{1}{R}. \tag{14}$$

In the equilibrium state $R = 1$. Small oscillations with respect to this state occur at the frequency $\omega = 1$. The change of the radius $R(t)$ has a character of oscillations, if the "energy"

$$E \equiv \frac{\dot{R}^2}{2} + \frac{1}{2R^2} - \frac{1}{R} \tag{15}$$

is negative: $E < 0$. If, however, the initial perturbation is so strong that $E > 0$, the system ultimately must dissipate (see Fig. 44, where $E_1 < 0, E_2 > 0$).

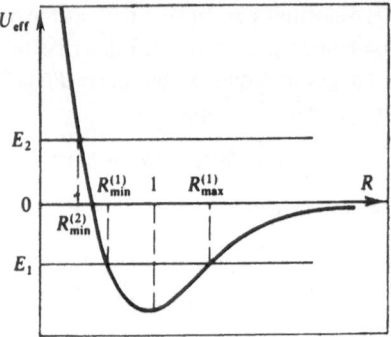

Figure 44. Effective potential energy, U_{eff}, for radial oscillations of a homogeneous Camm sphere.

3. Solve the problem of the development of perturbations in the homogeneous isotropic collisionless universe (Bisnovatyi-Kogan and Zeldovich [24]).

Solution. The problem must be solved at the background of an accurate nonstationary solution, corresponding to expansion or contraction of the homogeneous and isotropic universe. The latter will be considered in the framework of the Newtonian cosmology, The correctness of such consideration is proved (see, e.g., [48]). The problem may be of interest for the description of agglomeration of spherical stellar clusters (in the scheme by Peebles and Dicke [302]), of galaxies or galactic clusters into larger formations.

Find first of all the self-consistent solution of the unperturbed kinetic equation

$$\frac{\partial f_0}{\partial t} + v \frac{\partial f_0}{\partial r} - \frac{\partial \Phi_0}{\partial r} \frac{\partial f_0}{\partial v} = 0 \tag{1}$$

and the Poisson equation

$$\Delta \Phi_0 = 4\pi G \rho_0. \tag{2}$$

We shall seek the solution corresponding to the Newtonian universe with the critical density $\rho_0(t)$. In this case, as is known [48],

$$\rho_0(t) = \frac{1}{6\pi G t^2}, \tag{3}$$

$$\Phi_0(t, r) = \tfrac{2}{3}\pi G \rho(t) r^2, \qquad \frac{\partial \Phi_0}{\partial r} = \frac{2}{9} \frac{\mathbf{r}}{t^2}. \tag{4}$$

The characteristics of kinetic equation (1) are the following (in the Cartesian coordinates $x_i = x, y, z$ at $i = 1, 2, 3$):

$$dt = \frac{dx_i}{v_i} = -\frac{dv_i}{\tfrac{2}{9} x_i / t^2}. \tag{5}$$

The integrals of the system of equations for characteristics in (5) are thus:

$$c_{1i} = 3 v_i t^{1/3} - x_i t^{-2/3}, \tag{6}$$

$$u_i \equiv c_{2i} = v_i t^{2/3} - \tfrac{2}{3} x_i t^{-1/3}. \tag{7}$$

Consequently, the kinetic equation in (1) is satisfied by any function of c_{1i}, c_{2i}, but it is necessary further to ensure the condition of self-consistency:

$$\int f_0 \, dv = \rho_0(t) = \frac{1}{6\pi G t^2}. \tag{8}$$

It is directly seen that this condition is satisfied by the arbitrary function of $c_{2i} \equiv u_i$:

$$f_0 = f_0(u_i). \tag{9}$$

This solution may be given a more illustrative interpretation, if u_i is represented in the form

$$u_i = a(t)(v_i - Hr_i), \tag{10}$$

where the $a(t)$ function is proportional to the radius of the expanding universe of A. A. Fridman, while $H(t)$ is the Hubble "constant." For the case of the critical density $H(t) \sim 1/t, a(t) \sim t^{2/3}$ and (10) coincides with (7). In reality, however, the solution of the form $f_0(u_i)$, with the arbitrary function f_0, remains valid for any density, not only critical density. In the uniform isotropically expanding universe the solution of the kinetic equation therefore will be the arbitrary function[34] of velocities, decreasing according to the adiabatic law (i.e., according to the general expansion of the universe, following this expansion). The term Hr in (10) takes into account the Hubble expansion of the universe with the speed $v_0 = Hr$. Therefore, it is quite natural that the argument of the distribution function involves the difference $v - v_0 = v - Hr$, which corresponds just to chaotic velocities of particles.

Assume for further calculations the Maxwellian distribution function to be

$$f_0 = \frac{1}{6\pi G m} \frac{1}{(\pi \theta)^{3/2}} \exp\left(-\sum_i \frac{u_i^2}{\theta}\right). \tag{11}$$

The effective temperature $T(t) \sim \theta/a^2$ corresponding to (11) decreases with time, i.e., the matter is cooled down at expansion of the world. Consider now small perturbations of such a homogeneous nonstationary background. The linearized kinetic equation will be solved by the trajectory-integration method.

The equations of trajectories are coincident with the equations of characteristics in (4). Both coordinates and velocities of the particle at a given time moment $t'(x', v'_x, \ldots)$ is easily expressed through coordinates and velocities at the time t $(x, v_x, \ldots)$ using the integrals from (6) and (7):

$$x' = \left(3v_x t^{1/3} - \frac{x}{t^{2/3}}\right)t'^{2/3} + \left(2\frac{x}{t^{1/3}} - 3v_x t^{2/3}\right)t'^{1/3},$$

$$v'_x = \frac{2}{3}\left(3v_x t^{1/3} - \frac{x}{t^{2/3}}\right)(t')^{-1/3} + \frac{1}{3}\left(2\frac{x}{t^{1/3}} - 3v_x t^{2/3}\right)(t')^{-2/3} \tag{12}$$

In the unperturbed solution of (12) all spatial scales grow proportionally with $t^{2/3}$; therefore, also the wavelength of perturbation grows $\sim t^{2/3}$ as well. Accordingly, the eigenfunctions are of the form

$$\Phi = e^{ik\xi}\varphi(t), \qquad f = e^{ik\xi}f(t), \tag{13}$$

---

[34] Strictly speaking, this is true only in Newtonian cosmology; in relativistic theory, this arbitrariness somewhat diminishes. The kinetic relativistic theory of the expanding universe is considered in [163] by Bel.

where $\xi = x/t^{2/3}$ (so that the dimension of k is here not coincident with the dimension of the "ordinary" wave number!). Substituting (13) into the linearized kinetic equation and allowing for (2) and (12), we obtain, by integrating in parts,

$$f = ik \frac{\partial f_0}{\partial u_1} \int_0^t \varphi(t')e^{3iku_1[t^{-1/3}-(t')^{-1/3}]}dt'. \tag{14}$$

The Poisson equation is written as

$$\Delta\Phi = -\frac{k^2\varphi(t)}{t^{4/3}} = e^{ik\xi}\int f\, dv. \tag{15}$$

Finally, integrating (14) over v and substituting the result into (15), we arrive at the following integral equation for φ:

$$\varphi(t) + \frac{2}{t^{2/3}}\int_0^t \varphi(t')\tau e^{-(3k^2\theta/4)\tau^2}\, dt' = 0, \tag{16}$$

where $\tau \equiv t^{-1/3} - (t')^{-1/3}$. For long-wave perturbations or for long times when $9k^2\theta\tau^2/4 \ll 1$ and the exponent in (16) can be substituted for unity, the initial integral equation becomes simpler:

$$\varphi(t) + \frac{2}{t^{2/3}}\int_0^t \varphi(t')\tau\, dt' = 0. \tag{17}$$

By further integrating, (17) can be reduced to a simple differential equation

$$t^2\varphi'' + \tfrac{8}{3}t\varphi' = 0. \tag{18}$$

This equation describes the familiar hydrodynamical solution for dust [48]: $\varphi_1 = c_1$, $\varphi_2 = c_2 t^{-5/3}$, which corresponds to $\delta\rho/\rho \sim t^{2/3}$ for φ_1 and $\delta\rho/\rho \sim t^{-1}$ for φ_2. The first solution is unstable: $\delta\rho/\rho \to \infty$ at $t \to \infty$.

Quite a different picture, which does not resemble the hydrodynamical one at all, results at short wavelengths. In [24], the solution is reached under condition

$$\frac{9k^2\theta}{4}t^{-2/3} \gg 1$$

when the integral in (16) can be calculated by the steepest descent method. Let $\lambda \equiv 9k^2\theta/4$; then, assuming $\varphi(t')$ to be a slow function, and $\tau e^{-9k^2\theta\tau^2/4}$ a function rapidly vanishing at large λ, we arrive at the following first-order differential equation:

$$\varphi + 9\sqrt{\frac{\pi}{2e}}\frac{t^2}{\lambda^{3/2}}\varphi' = 0. \tag{19}$$

The solution for (19) is thus

$$\varphi = \text{const}\cdot\exp\left(\frac{1}{9}\sqrt{\frac{2e\lambda^3}{\pi}}\frac{1}{t}\right). \tag{20}$$

Consequently, the solution decreases with the characteristic time T of damping, from the time t_0, equal to

$$T = 9\sqrt{\frac{\pi}{2e\lambda^3}}t_0^2. \tag{21}$$

It may be expressed in the following manner:

$$T \sim t_\lambda \left(\frac{t_\lambda}{t_0}\right)^2,$$ (22)

where t_λ is the time of passage of one wavelength λ, by particle, at the time t_0.

For ideal gas, we seemingly would have had in this case nondamping acoustic oscillations. Such a striking (at first sight) contradiction to hydrodynamical theory is easily understood already in hydrodynamical terms. The hydrodynamical approximation corresponds to the vanishingly small free path length, $l \rightarrow 0$; accordingly, to zero viscosity (coefficient of viscosity $v = lv_T$ where v_T is the thermal speed of particles). At the same time, the "free path length" of collisionless particles, and consequently, the "viscosity" and damping should be definitely large.

We have considered two limiting cases and showed that at $k^2\theta\tau^2 \ll 1$, perturbations are unstable, while at $k^2\theta\tau^2 \gg 1$ they are damping, which means stability. Hence it is easy to see that the boundary separating the stable and unstable solutions is somewhere in the middle, i.e., at $k^2\theta\tau^2 \sim 1$. One can readily make sure that this condition can be otherwise written in the form

$$\left(\frac{2\pi}{\lambda_{cr}(t)}\right)^2 = k_{cr}^2(t) \approx \frac{G\rho_0(t)}{v_T^2(t)},$$

which coincides with the condition obtained by Jeans from consideration of the stability of the homogeneous and *stationary* background (within the framework of hydrodynamical approximation[35]).

---

[35] For the collisionless case, the same was performed by Lynden-Bell [281] and Maksumov [80, 81].

CHAPTER IV

Equilibrium and Stability of Collisionless Ellipsoidal Systems

It is well known that many galaxies as well as individual elements of galaxies, e.g., the bars of SB-galaxies, have a shape close to elliptical. Therefore, it is interesting to construct exact self-consistent models for collisionless ellipsoids.

In classical investigations concerning equilibrium figures and stability of uniform incompressible liquid, which was mentioned by us in the Introduction, most attention was paid just to uniform *ellipsoidal* systems. Apparently, this in many respects is explained by the presence of accurate models convenient for theoretical analysis: ellipsoids of rotation (spheroids) by Maclaurin, three-axis ellipsoids by Jacobi, and so on (see [148]).

In the collisionless case, we now also know some exact stationary models of ellipsoids; here they may play a role, in some respect similar to Maclaurin and Jacobi ellipsoids. We bear in mind first of all the familiar models of Freeman [203], considered in detail below (§1). Physically, they have, as we shall see, little in common with incompressible ellipsoids. This is, of course, not occasional but implies a deep difference in matter states in the form of incompressible liquid and of collisionless gas. This is obviously responsible for the absence of collisionless analogs of Maclaurin and Jacobi ellipsoids. An attempt to detect some relevant distribution functions leads to solutions that have no physical meaning[1] (see Problem 5, this chapter).

Thus, there is no possibility of automatically transferring to the collisionless case the results attained in classical works.

---

[1] True, only among distribution functions of the simplest type; this statement has no general proof so far.

However, on the other hand, there frequently remains one important mathematical property here (which is important for the stability theory): the eigenfunctions describing the potential *perturbation* inside a uniform collisionless ellipsoid may be presented in the form of finite polynomials in degrees of Cartesian coordinates (see §2). This fact is not quite trivial, in particular, because, in the collisionless case, generally not only is the boundary surface of the ellipsoid (as in the incompressible liquid model) perturbed but also the local density of matter.

The fact mentioned means the possibility of analytically solving the problem of stability and even, moreover, of explicitly calculating all eigenfrequencies of oscillations and relevant eigenfunctions of such systems.

Thus, we have here a rather intriguing situation, where a whole large class of systems possessing a quite new physics and having obvious prototypes in nature (galaxies, etc.) can be studied in a unified way, and it is possible to derive accurate spectra of small (and in some cases also nonlinear) oscillations. Idealizations adopted in such an approach are, of course, very essential: the isolated position of the system, its homogeneity, a sharp boundary, etc. Some of them are explicitly nonrealistic, and in a future, more complete theory they will have to be removed.

However, in the region of interest, the accurate mathematical investigations of real systems are difficult or even impossible (we do not speak here about the numerical calculations performed with computer). There are therefore two veins. One of them is to restrict oneself only to consideration (qualitative) of phenomena and to estimates of the characterizing parameters in order of magnitude.

The other method consists of investigation of such simple (e.g., uniform in density) models, which admit an accurate mathematical solution. They are the probe stone for the test of future *a priori* approximate theories belonging to real objects.

This field of research is only in its first stages so far [96], and some partial results have been achieved. Here, however, it is quite possible to realize the general program of a full classification of all possible equilibrium states of uniform ellipsoidal systems and of investigation of their oscillations (at least, in the spirit and volume of recent investigations carried out by Chandrasekhar with co-workers, who are systematizing the older investigations concerning incompressible ellipsoids).

At the beginning of this chapter (§1) we consider the equilibrium distribution functions of ellipsoidal systems. Accurate models of three-axis uniform ellipsoids were found by Freeman [203], who utilized them to study the evolution of the bars of SB-galaxies. With the same purpose, Freeman [202, 204] investigated other nonaxial-symmetrical models: elliptical cylinders and elliptical disks. They have very much in common with ellipsoids; therefore, we consider them also in §1 of this chapter.

There also is given the distribution function of "hot" in the rotational plane, collisionless spheroids [117].

At the end of the section, we discuss the general method of Lynden-Bell [280] of seeking distribution functions of the form $f_0(E, L_z)$ using the known density distribution $\rho_0(r, z)$ (as well as Hunter's method [235] that is a certain modification of Lynden-Bell's method). These methods are illustrated by the examples of accurate self-consistent models of elliptic galaxies suggested by Lynden-Bell and Hunter.

In §2, the stability of Freeman's three-axis ellipsoids is studied with respect to perturbations of the ellipsoid–ellipsoid type. The boundary that separates the regions of stable and unstable solutions is found. The growth rates of the increase of unstable solutions are calculated. We shall consider in more detail the question of the stability of strongly prolated ellipsoids.

In this section, also, the reasons for the polynomial form of eigenfunctions of ellipsoidal systems (incompressible and collisionless) are discussed. At the end of the section, we give some results of the study of stability of elliptic disks [335].

Section 3 deals with the study of stability of rotational ellipsoids. At the beginning (Section 3.1), the theory of perturbations is constructed, which allows one in principle to find all eigenfunctions and eigenfrequencies of Freeman's two-axis ellipsoid, in which the particles, in the rotational plane, move along the circumference. Explicitly, the frequencies of oscillations of the ellipsoid boundary are found in which the local density is not perturbed. Also, the example of the simplest non-axial-symmetrical mode which, apart from the boundary, also changes the local density of the ellipsoid. For this mode, the eigenfunctions and characteristic frequencies are obtained. The lines of identical perturbed potential and of surface density provide a picture of slightly leading spirals.

At the end of Section 3.1, the dispersion equation for arbitrary Jeans oscillations of Freeman's rotational ellipsoid of small thickness is given.

In Section 3.2 we derive the general restrictions due to the condition of the stability of collisionless circular ellipsoids with respect to the large-scale instabilities of three main types (Jeans, fire-hose, and barlike). Then we discuss the question of the universality of corresponding stability criteria. The observational data concerning elliptical galaxies and the other ellipsoidal steller systems are given in §2, Chapter X, where we discuss also some applications of the theory developed in this chapter.

§ 1 Equilibrium Distribution Functions

1.1 Freeman's Ellipsoidal Models

Freeman [203] constructed the models of uniform-density collisionless three-axis ellipsoids called "balanced" by him: for them the gravitational and centrifugal forces along the larger axis compensate exactly for each other. The task was to design a uniform ellipsoid with a boundary, which is un-altered in the frame of reference x, y, z, rotating about the z-axis with an angular speed $-\Omega$:

$$\frac{x^2}{a^2} + \frac{y^2}{b^2} + \frac{z^2}{c^2} = 1. \tag{1}$$

Assume for concreteness that $a > b > c$. To solve the problem, it is necessary to find a stationary solution for the set of Vlasov's equations which here comprises the kinetic equation *in a rotating system*

$$\mathbf{v}\frac{\partial f_0}{\partial \mathbf{r}} + \left(\Omega^2 \mathbf{R} + 2[\Omega \mathbf{v}] - \frac{\partial \Phi_0}{\partial \mathbf{r}}\right)\frac{\partial f_0}{\partial \mathbf{v}} = 0, \tag{2}$$

and the Poisson equation

$$\Delta \Phi_0 = 4\pi G \int f_0 \, d\mathbf{v} = \rho_0 \theta \left[1 - \left(\frac{x^2}{a^2} + \frac{y^2}{b^2} + \frac{z^2}{c^2}\right)\right] \qquad [\mathbf{R} = (x, y, 0)] \tag{3}$$

The potential of the uniform ellipsoid is well known [64, 148]:

$$\Phi_0 = \text{const} + \tfrac{1}{2}(A^2 x^2 + B^2 y^2 + C^2 z^2), \tag{4}$$

where the coefficients A^2, B^2, C^2 are equal to

$$A^2 = \frac{3MG}{k^2(a^2 - c^2)^{3/2}}[F(\omega_0, k) - E(\omega_0, k)], \tag{5}$$

$$B^2 = \frac{3MG}{k^2(a^2 - c^2)^{3/2}}\left[-F(\omega_0, k) + \frac{E(\omega_0, k)}{1 - k^2}\right] - \frac{3MGc}{ab(b^2 - c^2)}, \tag{6}$$

$$C^2 = -\frac{3MGE(\omega_0, k)}{(a^2 - c^2)^{3/2}(1 - k^2)} + \frac{3MGb}{ac(b^2 - c^2)}, \tag{7}$$

$$\text{const} = -\frac{3MGF(\omega_0, k)}{2(a^2 - c^2)^{1/2}}, \tag{8}$$

where M is the mass of the ellipsoid: $M = \tfrac{4}{3}\pi abc\rho_0$,

$$\sin^2 \omega_0 = \frac{a^2 - c^2}{a^2}, \qquad k^2 \equiv \frac{a^2 - b^2}{a^2 - c^2}, \tag{9}$$

and F, E are incomplete elliptic integrals.

The arbitrary rotating ellipsoid is characterized by three independent dimensionless parameters:

$$\frac{b}{c}, \quad \frac{c}{a}, \quad \frac{G\rho_0}{\Omega^2}. \tag{10}$$

Freeman introduces one essential restriction in considering the system balanced along the x-axis:

$$A_1^2 \equiv A^2 - \Omega^2 = 0, \tag{11}$$

so that the ensuing family of ellipsoids is biparametric. We introduce the quantity $\Phi_1 = \Phi_0 + \tfrac{1}{2}\Omega^2 R^2$; then

$$\nabla \Phi_1 = (0, -B_1^2 y, -C^2 z) \tag{12}$$

$(B_1^2 \equiv B^2 - \Omega^2)$, and the equations of individual particles in the system of coordinates rotating along with the ellipsoid will be

$$\ddot{x} + 2\Omega\dot{y} = 0, \qquad \ddot{y} - 2\Omega\dot{x} = -B_1^2 y, \qquad \ddot{z} = -C^2 z. \qquad (13)$$

The general solution of the system in (13) is thus

$$x - x_0 = vt + \frac{2\Omega}{\beta} A_\beta \cos(\beta t + \varepsilon_\beta),$$

$$y = \frac{2\Omega}{B_1^2} v + A_\beta \sin(\beta t + \varepsilon_\beta), \qquad (14)$$

$$z = A_\gamma \sin(\gamma t + \varepsilon_\gamma),$$

where

$$\beta^2 = B_1^2 + 4\Omega^2; \qquad \gamma^2 = C^2. \qquad (15)$$

In (14), $x_0, v, A_\beta, \varepsilon_\beta, \varepsilon_\gamma$ are the arbitrary constants of integration. The total motion described by these formulae can be represented as a superposition of the subsequent simple constituents (Fig. 45):

(1) of drift of the guiding center with a constant speed v along $y = 2\Omega v/B_1^2$,
(2) of rotation, with angular frequency β, round the guiding center, and in the direction opposite to the rotation of the ellipsoid boundary;
(3) of harmonic oscillations, at a frequency γ, near the equatorial plane $z = 0$.

The particles having a nonzero drift velocity v will necessarily be lost by the ellipsoid; therefore for the construction of self-consistent models we use only particles with $v = 0$.

Write down the integrals of motion of an individual star. In the general case, where the ratio of oscillation frequencies in the (x, y) plane and along the z-axis (β/γ) is irrational, there are four such integrals:

$$x_0 = x - \frac{2\Omega}{\beta^2}\dot{y}, \qquad (16)$$

$$I_2 = \dot{x} + 2\Omega y \equiv 0, \qquad (17)$$

$$E_\beta = \tfrac{1}{2}\dot{y}^2 + \tfrac{1}{2}\beta^2 y^2 = \tfrac{1}{2}\beta^2 A_\beta^2, \qquad (18)$$

$$E_\gamma = \tfrac{1}{2}\dot{z}^2 + \tfrac{1}{2}\gamma^2 z^2 = \tfrac{1}{2}\gamma^2 A_\gamma^2. \qquad (19)$$

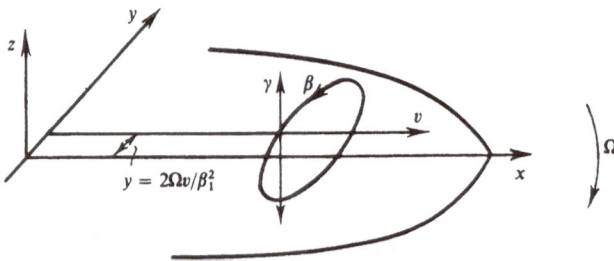

Figure 45. A particle orbit in the Freeman ellipsoid represented as superposition of three elementary motions [203].

Owing to (17), the distribution function of the system must have the form

$$f = f(I_i) = F(x_0, E_\beta, E_\gamma)\delta(I_2). \tag{20}$$

Let us seek the F function in the form $F = F(\mathcal{J})$, where $\mathcal{J} = \mathcal{J}(X_0, E_\beta, E_\gamma) = 1$ is the condition that the orbit touch the ellipsoid surface.

From (16)–(19) it follows that in the general case (when β/γ is irrational) the star orbit, at $\beta t, \gamma t \to \infty$, covers the surface of the elliptical cylinder:

$$\frac{\beta^2}{4\Omega^2}(x - x_0)^2 + y^2 = A_\beta^2, \qquad -A_\gamma \le z \le A_\gamma. \tag{21}$$

The orbit characterized by the integrals (x_0, E_β, E_γ) touches the ellipsoid in (1), if

$$\mathcal{J} = \frac{x_0^2}{a^2 - 4\Omega^2 b^2/\beta^2} + \frac{2E_\beta}{\beta^2 b^2} + \frac{2E_\gamma}{\gamma^2 c^2} = 1. \tag{22}$$

In such a form $\mathcal{J}$ is represented explicitly as a function of integrals of motion. The expression for $\mathcal{J}$ can be rewritten in the following form:

$$\mathcal{J}(\mathbf{r}, \mathbf{v}) = \frac{c_y^2}{\beta^2 b^2 \mu^2} + \frac{c_z^2}{\gamma^2 c^2} + \frac{x^2}{a^2} + \frac{y^2}{b^2} + \frac{z^2}{c^2}, \tag{23}$$

where the notations

$$c_y = \dot{y} - \frac{2\Omega b^2 x}{a^2}, \qquad \mu^2 = 1 - \frac{4\Omega^2 b^2}{\beta^2 a^2} \tag{24}$$

are introduced. Hence it is easy to see that possible values of $\mathcal{J}$ for orbits traversing the point $\mathbf{r}$ inside the ellipsoid satisfy the inequality

$$\frac{x^2}{a^2} + \frac{y^2}{b^2} + \frac{z^2}{c^2} \le \mathcal{J}(\mathbf{r}) \le 1. \tag{25}$$

Find now the $F(\mathcal{J})$ function which must satisfy the equation

$$\rho_0\theta\left[1 - \left(\frac{x^2}{a^2} + \frac{y^2}{b^2} + \frac{z^2}{c^2}\right)\right] = \int F(\mathcal{J})\delta(I_2)\, d\mathbf{v}. \tag{26}$$

Using (25) and transforming in (26) to cylindrical coordinates in the velocity space (q, θ)

$$c_y = \beta b \mu q \sin\theta, \qquad c_z = \gamma c q \cos\theta,$$

it is easy to obtain

$$F(\mathcal{J}) = \frac{\rho_0}{\pi\beta\gamma b c \mu}\delta(\mathcal{J} - 1), \tag{27}$$

so that the ultimate distribution function has the form

$$f_0 = \frac{\rho_0}{\pi\beta\gamma\mu b c}\delta(v_x + 2\Omega y)\delta\left(\frac{c_y^2}{b^2\beta^2\mu^2} + \frac{v_z^2}{\gamma^2 c^2} + \frac{x^2}{a^2} + \frac{y^2}{b^2} + \frac{z^2}{c^2} - 1\right). \tag{28}$$

Consider the motion of matter in such an ellipsoid. Along the z-axis, the particles perform harmonical oscillations at a frequency γ. Of more interest is the motion in the rotational plane (x, y).

Calculate the average macroscopic velocity of matter

$$\langle \mathbf{v} \rangle = \frac{1}{\rho_0} \int \mathbf{v} f_0 \, d\mathbf{v}. \tag{29}$$

It happens to be

$$\langle \mathbf{v} \rangle = \left(-2\Omega y, \frac{2\Omega b^2 x}{a^2}, 0 \right), \tag{30}$$

i.e., the lines of flux in the rotating system are ellipses similar to and concentric with the cross section $z = \text{const}$ of the ellipsoid. The circulation of the matter occurs against the direction of boundary rotation. In particular, the surface of the ellipsoid also is covered by lines of flux of the average circulation.

Lines of flux for the mean velocity $\langle \mathbf{v} \rangle$ evidently do not coincide with trajectories of the particles themselves, despite the fact that the system is stationary in the rotating system.

Calculate now the tensor of velocity dispersion

$$\sigma_{ij} = \frac{1}{\rho_0} \int (v_i - \langle v_i \rangle)(v_j - \langle v_j \rangle) f_0 \, d\mathbf{v} \qquad (i, j = 1, 2, 3). \tag{31}$$

The solely diagonal components σ_{yy} and σ_{zz} happen to be nonzero:

$$\sigma_{xx} = 0,$$

$$\sigma_{yy} = \tfrac{1}{2} b^2 \beta^2 \left(1 - \frac{4\Omega^2 b^2}{\beta^2 a^2} \right) \left(1 - \frac{x^2}{a^2} - \frac{y^2}{b^2} - \frac{z^2}{c^2} \right),$$

$$\sigma_{zz} = \tfrac{1}{2} \gamma^2 c^2 \left(1 - \frac{x^2}{a^2} - \frac{y^2}{b^2} - \frac{z^2}{c^2} \right). \tag{32}$$

Using (30)–(32), it is possible to verify that, as was to be expected, the hydrodynamical Jeans equation ($\mathbf{c} \equiv \langle \mathbf{v} \rangle$)

$$\text{div } \sigma + \mathbf{c}\nabla\mathbf{c} + 2[\Omega\mathbf{c}] - \frac{\partial \Phi_1}{\partial \mathbf{r}} = 0 \tag{33}$$

is satisfied.

The average velocity in the inertial frame of reference is defined by the equation

$$R\langle v_\varphi \rangle = \Omega^2 (x^2 + y^2) - \langle v_y \rangle x + \langle v_x \rangle y = \Omega \left[x^2 \left(1 - \frac{2b^2}{a^2} \right) - y^2 \right]. \tag{34}$$

Hence it follows that in the region $y^2 > (1 - 2b^2/a^2) x^2$ the mean rotational velocity is opposite to the direction of rotation of the frame of reference (i.e., to the direction of rotation of the main axes of the ellipsoid), while in the

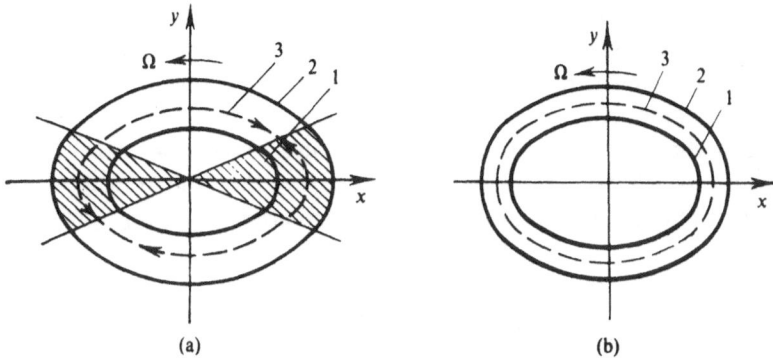

Figure 46. The macroscopic motion of the matter in the plane of rotation of the Freeman triaxial ellipsoid: (a) the case of counterflows: the ellipsoid is sufficiently highly flattened in the plane of rotation, $b/a < 1/\sqrt{2}$ (rotation referred to an inertial frame in the shaded area is in the opposite sense to that in the unshaded area); (b) the case with no counterflows $(b/a > 1/\sqrt{2})$; (1) lines of flux of the motion referred to the rotating frame; (2) the ellipsoid boundary; (3) the sense of the mean rotational motion referred to an inertial frame.

region $y^2 < (1 - 2b^2/a^2) \cdot x^2$ it coincides with it. It is seen that the "counterflows" exist only for ellipsoids sufficiently flattened in the plane of rotation: $2b^2 < a^2$ (Fig. 46). It is interesting to note that such macroscopic movements were described by Freeman and Vaucouleurs [206] in the galaxy NGC 4027 (for more detail see §3, Chapter X). The angular momentum of the Freeman ellipsoid is

$$L_{\text{tot}} = \int \rho[\Omega R^2 - \langle v_y \rangle x + \langle v_x \rangle y] \, dv = \tfrac{1}{5} M\Omega(a^2 - 3b^2). \qquad (35)$$

It is seen that, for $a^2 < 3b^2$, the directions of the angular velocity of the boundary and the complete momentum are opposite, while, for $a^2 > 3b^2$, these directions coincide.

The full kinetic energy equal to the sum of the internal kinetic energy (chaotic movement of stars) and of the energy of "organized" averaged rotation is

$$E_T = \tfrac{1}{2} \iint [(v_x + \Omega y)^2 + (v_y - \Omega x)^2 + v_z^2] \; f_0(\mathbf{r}, \mathbf{v}) \, d\mathbf{r} \, d\mathbf{v}$$

$$= \tfrac{3}{10} \frac{M^2 GF(\omega_0, k)}{(a^2 - c^2)^{1/2}}. \qquad (36)$$

The potential energy is calculated as

$$V = \tfrac{1}{2} \int \rho_0 \Phi_0 \, d\mathbf{r} = -\tfrac{3}{5} \frac{M^2 GF(\omega_0, k)}{(a^2 - c^2)^{1/2}}. \qquad (37)$$

From (36) and (37) it follows, in particular, that

$$2E_T + V = 0, \qquad (38)$$

i.e., validity of the virial theorem.

The stability of three-axis Freeman's ellipsoids with respect to the largest-scale perturbations is investigated in §2 [91, 96]. It turns out that, in the region of oblateness observed in real galactic bars, stability takes place (see also §3, Chapter X).

The partial case of the three-axis ellipsoid in (28) is the two-axis Freeman's ellipsoid, which can be obtained from (28) by the limit transition $b \to a$. The distribution function in the frame of reference rotating together with the ellipsoid is

$$f_0 = \frac{\rho_0}{\pi \gamma c} \delta(v_x) \delta(v_y) \left(1 - \frac{x^2 + y^2}{a^2} - \frac{z^2}{c^2} - \frac{v_z^2}{\gamma^2 c^2} \right)^{-1/2}. \tag{39}$$

The paths of the particles in the stationary system have the following form: their projections onto the (x, y) plane are circumferences, while along the z-axis the particles perform harmonic oscillations with the frequency γ.[2]

A detailed investigation into stability of these ellipsoids is contained in §3, this chapter [91, 93]. They are unstable at any value of oblateness.

Along with the distribution function of (39) where all the particles migrate in the same direction, one may [21] consider the superposition of two such models with opposite rotational direction (and any relative weight).

The models of rotating three-axis stellar ellipsoids in (28) have been applied by Freeman for the description of some properties of bridges ("bars") of spiral galaxies of the type SB (see §3, Chapter X). Using the method described above, Freeman obtained also exact distribution functions for other "barlike" models: of the elliptical cylinder [202] and elliptical disk [204].

Assume that the system (disk or cylinder) rotates with an angular speed $(-\Omega)$ about the z-axis. The gravitational potential in the internal region is expressed by the formula

$$\Phi_0 = \text{const} + \tfrac{1}{2} A^2 x^2 + \tfrac{1}{2} B^2 y^2, \tag{40}$$

where in the case of cylinder

$$A^2 = \frac{4\pi G \rho_0 b}{a + b}, \qquad B^2 = \frac{4\pi G \rho_0 a}{a + b}, \tag{41}$$

and in the case of disk

$$A^2 = \frac{3GM[K(m) - E(m)]}{m^2 a^3},$$

$$B^2 = \frac{3GM[E(m) - b^2 K(m)/a^2]}{m^2 a b^2}, \tag{42}$$

$m^2 \equiv 1 - b^2/a^2$; E and K are the complete elliptical integrals. The major semiaxis is directed along the x-axis, with the small one along the y-axis:

---

[2] There are no non-axial-symmetrical ellipsoids ($a \neq b$) with a similar picture of movement of particles in the plane of rotation (elliptical orbits identical and concentric with the boundary ellipse) (see Problem 6).

$a > b$. Then, as follows from (41) and (42) (and apparently from physical considerations), $B^2 > A^2$. The equations of motion of stars in the potential of (40) in the frame of reference rotating with an angular speed $-\Omega$ are thus:

$$\ddot{x} + 2\Omega\dot{y} = (\Omega^2 - A^2)x,$$
$$\ddot{y} - 2\Omega\dot{x} = (\Omega^2 - B^2)y. \tag{43}$$

The general solution for this system of equations is

$$x = A_\alpha \sin(\alpha t + \varepsilon_\alpha) + A_\beta \sin(\beta t + \varepsilon_\beta),$$
$$y = k_\alpha A_\alpha \cos(\alpha t + \varepsilon_\alpha) - k_\beta A_\beta \cos(\beta t + \varepsilon_\beta), \tag{44}$$

where $(A_\alpha, A_\beta, (\varepsilon_\alpha, \varepsilon_\beta)$ and $(k_\alpha A_\alpha, k_\beta A_\beta)$ are the constant real amplitudes and phases, while the dimensionless values k_α and k_β are

$$k_\alpha = \frac{A^2 - \Omega^2 - \alpha^2}{2\alpha\Omega}, \qquad k_\beta = \frac{\Omega^2 + \beta^2 - A^2}{2\beta\Omega}. \tag{45}$$

Assume A_α and A_β to be positive. The frequencies of oscillation α and β are determined as roots of the fourth-order equation

$$F(\omega) = \omega^4 - (A^2 + B^2 + 2\Omega^2)\omega^2 + (A^2 - \Omega^2)(B^2 - \Omega^2) = 0. \tag{46}$$

Let α and β be the positive roots of (46) and $\beta > \alpha$ for definiteness.

The distribution functions are represented by the following formulae (see Problems 1 and 2 of this chapter):

$$f_0 = \frac{\rho_0 ab\Lambda^2\delta(1 - I)}{\pi k_\alpha k_\beta}, \qquad \text{for cylinder,} \tag{47}$$

$$f_0 = \frac{3M\Lambda^2(1 - I)^{-1/2}}{4\pi^2 k_\alpha k_\beta}, \qquad \text{for disk,} \tag{48}$$

where the notations

$$\Lambda^2 = \frac{4\Omega^2(k_\beta^2 - k_\alpha^2)^2}{(a^2 k_\beta^2 - b^2)(b^2 - a^2 k_\alpha^2)(\beta^2 - \alpha^2)^2}, \tag{49}$$

$$I = (k_\beta^2 - k_\alpha^2)\left[\frac{A_\alpha^2}{\alpha^2 k_\beta^2 - b^2} + \frac{A_\beta^2}{b^2 - a^2 k_\alpha^2}\right]$$

are used.

The structure of "barlike" models, described first by Freeman, was later analyzed in great detail by Hunter, who had made an attempt to depict the complete system of such models [234]. By the way, it should be mentioned at once that he did not reveal any new solutions in comparison with those found by Freeman. Of most interest in the mentioned work by Hunter is likely to be the study of the question concerning the uniqueness of the solutions inferred. Strictly, however, the uniqueness is proved only in some particular cases; namely, for the distribution functions of the elliptic disk or cylinder at rest (see Problem 3, this chapter). Physically acceptable solutions for rotating ellipsoids are achieved only in two particular cases, considered by Freeman: for the disk limit $c \to 0$ and for an ellipsoid rotating with a maximally possible angular velocity $\Omega = A$ (see Problem 5).

1.2 "Hot" Models of Collisionless Ellipsoids of Revolution

The above models of Freeman's spheroids in (28) have none of the velocity dispersion of particles in the (x, y) rotational plane and in this respect are "cold."

If, however, one bears in mind applications to elliptical galaxies, then more satisfactory are "hot" models possessing a finite velocity dispersion in the rotational plane. The attempts to find physically admissible "hot" models of ellipsoidal systems have been fruitless for a long time. We have already noted Hunter's study [234], in which, apart from old Freeman's models [203], only completely "pathological" solutions were found, which had a negative distribution function in some finite region of the phase space. The search for some of the simplest collisionless analogs of liquid Maclaurin and Jacobi ellipsoids (see Problem 5) also lead to similar difficulties. Besides, there is the theorem [179] which states that there is no finite self-gravitating stationary and axial-symmetrical stellar system with an ellipsoidal[3] velocity distribution function and nonzero velocity dispersion and mean movement in the rotational plane.

Below we describe [117] the models of "hot" uniform collisionless spheroids with distribution functions in velocities of a more complex form (not ellipsoidal).

The equation of the boundary of the ellipsoid of revolution in question will be

$$\frac{x^2 + y^2}{a^2} + \frac{z^2}{c^2} = \frac{r^2}{a^2} + \frac{z^2}{c^2} = 1, \tag{50}$$

the density is ρ_0, while the gravitational potential at the inner point (r, z) will be written in the form

$$\Phi_0(r, z) = \frac{\Omega^2 r^2}{2} + \frac{\omega_0^2 z^2}{2}. \tag{51}$$

Let us first of all seek the distribution functions of spheroids at rest, depending on the following three integrals of motion: energy of a particle in the rotational plane (x, y)

$$E_\perp = \tfrac{1}{2} v_r^2 + \tfrac{1}{2} v_\varphi^2 + \tfrac{1}{2} \Omega^2 r^2$$

(v_r and v_φ are, respectively, the radial and azimuthal velocities), the energy of oscillations in the perpendicular direction (z) $E_z = \tfrac{1}{2} v_z^2 + \tfrac{1}{2} \omega_0^2 z^2$ and the

---

[3] I.e., depending on the quadratic form of components of velocity of the particle

$$\sum \alpha_{ik} v_i v_k (+ \sum \beta_i v_i).$$

angular momentum $L \equiv L_z = r v_\varphi$. Thus, we assume that $f_0 = f_0(E_\perp, E_z, L)$, and the task is to solve the integral equation

$$\rho_0 \theta \left(1 - \frac{r^2}{a^2} - \frac{z^2}{c^2} \right) = \iiint f_0 \, d^3 v. \tag{52}$$

Equation (52) is written in the form:

$$\rho_0 \theta \left(1 - \frac{r^2}{a^2} - \frac{z^2}{c^2} \right) = \iiint \frac{f_0(E_\perp, E_z, L) \, dE_\perp \, dE_z \, dL}{\sqrt{2E_\perp r^2 - L^2 - r^4} \sqrt{2E_z - \omega_0^2 z^2}}. \tag{53}$$

To define the region of integration in (53), consider in more detail the orbit of some particle characterized by the integrals $E_\perp$, E_z, L. In the rotational plane (x, y), the orbit is an ellipse centered on the coordinate origin. The maximum r_{max} and minimum r_{min} radii of the orbit are linked with $E_\perp$ and L by the formulae (let us further assume that $a = 1, \Omega = 1$).

$$r_{max}^2 = E_\perp + \sqrt{E_\perp^2 - L^2}, \tag{54}$$

$$r_{min}^2 = E_\perp - \sqrt{E_\perp^2 - L^2}. \tag{55}$$

It is convenient to assume r_{max}^2 and r_{min}^2 to be new arguments of the distribution function f_0 instead of $E_\perp$ and $L: r_{max}^2 = u$ and $r_{min}^2 = v$. Taking into account the movement along the z-axis, it becomes clear that the orbit of the particle with given u, v, E_z fills up, in the course of time, the surface of the elliptical cylinder, whose height $z_{max} = (2E_z/\omega_0^2)^{1/2}$ cannot be greater than $c\sqrt{1 - r_{max}^2} = c\sqrt{1 - u}$. This condition permits the equation of one of the boundaries of phase region of the system in question

$$u + \frac{2E_z}{\omega_0^2 c^2} = 1. \tag{56}$$

We assume[4] here that the remaining boundaries are as follows:

$$u \geq r^2, \qquad 0 \leq v \leq r^2, \tag{57}$$

$$2E_z \geq \omega_0^2 z^2. \tag{58}$$

Thus, instead of (53) we can write

$$\rho_0 \theta \left(1 - r^2 - \frac{z^2}{c^2} \right) = \iiint \frac{dE_z}{\sqrt{2E_z - \omega_0^2 z^2}} \frac{du \, dv \, f(E_z, u, v)}{\sqrt{(u^2 - r^2)(r^2 - v)}}. \tag{59}$$

Here a new function

$$f \equiv J(u, v) f_0(u, v, E_z) \tag{60}$$

is introduced, while $J(u, v) \equiv (\sqrt{u/v} - \sqrt{v/u})/4$ is the Jacobian of the transform from the variables $E_\perp$ and L to the variables u and v. Integration in (59) expands onto the region defined in (56)–(58) [Fig. 47(a)].

---

[4] Ozernoy and Kondrat'ev noted [16[ad]] that conditions (56)–(58) permit some generalization; in particular, one can write $r_0^2 \leq v \leq k^2 r^2$ instead of $0 \leq v \leq r^2$ (where r_0, k are parameters, $0 < k \leq 1$).

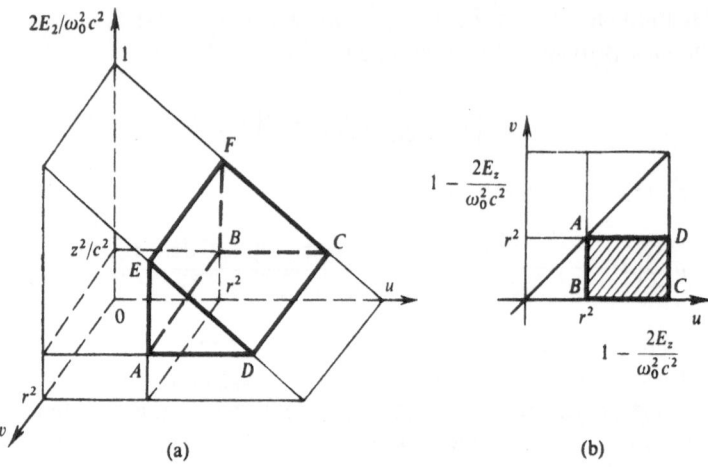

Figure 47. Regions of integration: (a) in formula (59), the prism $ABCDEF$; (b) in formula (61), the rectangle $ABCD$.

The solution for Eq. (59) will be performed in two stages. First we reduce (59) to the equivalent equation, but with a double integral in the right-hand part (eliminate the integral over E_z). Then we seek the solution of the simpler integral equation which ensues.

By denoting the double integral in (59), at a fixed E_z, by $\psi(r^2, E_z)$ it is easy to see that the equation for the ψ function is the standard integral Abel equation (with respect to the argument E_z; and it is essential that ψ be independent of z^2). By solving this equation, we shall have, instead of (59),

$$\frac{\rho_0}{\pi\omega_0 c}\left(1 - r^2 - \frac{2E_z}{\omega_0^2 c^2}\right)^{-1/2} = \iint \frac{f(u, v, E_z)\, du\, dv}{\sqrt{(u - r^2)(r^2 - v)}} \qquad (61)$$

where integration in the plane u, v should be performed over the rectangular region [Fig. 47(b)]

$$r^2 \le u \le 1 - \frac{2E_z}{\omega_0^2 c^2}, \qquad 0 \le v \le r^2. \qquad (62)$$

Note that if we integrate (61) over $dz\, dv_z = (2\pi/\omega_0)\, dE_z$ we shall arrive at the equation for the distribution function of the relevant disk system (which can also be obtained directly)

$$\sigma_0(0)\sqrt{1 - r^2} = \iint \frac{du\, dv}{\sqrt{(u - r^2)(r^2 - v)}} f'_0(u, v), \qquad \sigma_0(0) \equiv 2\rho_0 c. \quad (63)$$

Here f'_0 denotes $\frac{1}{2}\iint f\, dv_z\, dz$ and $\sigma_0(0)\cdot\sqrt{1 - r^2}$ is the surface density of the uniform ellipsoid.

It is easy to indicate the recipe for seeking some classes of formal solutions of Eqs. (61) and (63), for instance, of all multiplicative solutions of (63) or those similar to them for Eq. (61). However, in each specific case it is necessary to test additionally the positiveness of the ensuing distribution functions (this condition is not automatically fulfilled, as can be observed in the simplest examples). Therefore, we shall restrict ourselves here to the indication of one particular solution of (61), which, in this sense, is thought obviously to be "good":

$$f = \frac{2\rho_0}{\pi^2 \omega_0 c} v^{-1/2} \delta\left(1 - u - \frac{2E_z}{\omega_0^2 c^2}\right), \tag{64}$$

or according to (60)

$$f_0 = \frac{8\rho_0 \sqrt{u}}{\pi^2 \omega_0 c(u - v)} \delta\left(1 - u - \frac{2E_z}{\omega_0^2 c^2}\right). \tag{65}$$

In the disk limit $c \to 0$, this function becomes:

$$f_0 = \text{const} \frac{\sqrt{u}}{u - v} = \text{const} \frac{\sqrt{E + \sqrt{E^2 - L^2}}}{\sqrt{E^2 - L^2}} \tag{66}$$

Starting from any distribution function of a nonrotating spheroid, e.g., in (64), one can construct an infinite number of models of rotating uniform spheroids, by reversing the sign of L_z in some particles. However, then the manner of rotation in the general case will differ from a solid body. For instance, if all the particles rotate in the same direction, then the average velocity is

$$\bar{v} = \bar{v}_\varphi = \frac{1}{\pi}\left(1 + \frac{r^2}{\sqrt{1 - r^2}} \ln \frac{1 + \sqrt{1 - r^2}}{r}\right). \tag{67}$$

We find distribution functions of ellipsoids *rotating as a solid body*, similar to (64)–(65). Let $f_0^{(+)}$ and $f_0^{(-)}$ be the distribution functions of particles with $L_z > 0$ and $L_z < 0$, respectively. If one now denotes by γ the angular velocity of the ellipsoid, then we obtain the following equation for the difference $(f_0^{(+)} - f_0^{(-)})$:

$$\int (f_0^{(+)} - f_0^{(-)}) v_\varphi \, d^3v = \rho_0 \cdot \gamma r. \tag{68}$$

To solve this equation, it is convenient again to transform to the variables u and v, and instead of $f_0^{(+)}$ and $f_0^{(-)}$ to consider the $f^{(+)}$ and $f^{(-)}$ functions:

$$f^{(\pm)} = \frac{u - v}{4\sqrt{uv}} f_0^{(\pm)}. \tag{69}$$

The solution similar to (64) will be written in the form

$$f^{(+)} - f^{(-)} = \frac{4\gamma\rho_0}{\pi^2 \omega_0 c} u^{-1/2} \delta\left(1 - u - \frac{2E_z}{\omega_0^2 c^2}\right). \tag{70}$$

Since, on the other hand, according to (64)

$$f = f^{(+)} + f^{(-)} = \frac{2\rho_0}{\pi^2 \omega_0 c} v^{-1/2} \delta\left(1 - u - \frac{2E_z}{\omega_0^2 c^2}\right), \tag{71}$$

then ultimately the distribution function of a solid-body rotating ellipsoid can be represented in the form

$$f(u, v, E_z; \gamma)$$

$$= \frac{2\rho_0}{\pi^2 \omega_0 c} \delta\left(1 - u - \frac{2E_z}{\omega_0^2 c^2}\right) [\varphi^{(+)}(u, v)\theta(L_z) + \varphi^{(-)}(u, v)\theta(-L_z)], \quad (72)$$

where

$$\varphi^{(\pm)}(u, v) = v^{-1/2} \pm 2\gamma u^{-1/2}. \tag{73}$$

From the condition of the positiveness of the $\varphi^{(\pm)}$ functions, we find with due regard for (57) the following restriction to the angular velocity of rotation of the ellipsoid:

$$|\gamma| \le \tfrac{1}{2} \quad \text{(in } \Omega \text{ units).} \tag{74}$$

The question of the stability of the models thus suggested is interesting. The ellipsoids rotating as solid bodies seem to be stable. Among the differentially rotating models there must be unstable ones [for example, the model in (67)]. We shall discuss this point in Section 3.2.

The distribution functions similar to (72) can be written also for uniform circular cylinders:

$$f(u, v) = \frac{\rho_0}{\pi^2 \sqrt{1 - u}} [(v^{-1/2} + 2\gamma u^{-1/2})\theta(L_z) + (v^{-1/2} - 2\gamma u^{-1/2})\theta(-L_z)]. \tag{75}$$

Inhomogeneous "ellipsoidal" systems. If we turn to inhomogeneous systems, then a question arises at once: Which systems are to be considered as "ellipsoidal"? Seemingly, the answer to this question is ambiguous.

It is possible, for instance, to term as "ellipsoidal" those systems whose surfaces of identical density are concentric ellipsoids. A simple example of inhomogeneous spheroids[5] of this kind was found by Kuz'min [60]. The density in these spheroids is

$$\rho_0(r, z) = \frac{\rho_0(0, 0)}{\{1 + [(1 - \varepsilon^2)/z_0^2](r^2 + z^2/\varepsilon^2)\}^2}; \tag{76}$$

it is constant on the surfaces of ellipsoids of revolution with the ratio of semiaxes equal to ε. Another parameter of the models in (76) is z_0. The potential corresponding to such a mass distribution is

$$\Phi_0 = -\frac{2}{\pi} \frac{GM}{\zeta_1^2 - \zeta_2^2} \left[\zeta_1 \arctan \frac{\zeta_1 \sqrt{1 - \varepsilon^2}}{z_0 \varepsilon} - \zeta_2 \arctan \frac{\zeta_2 \sqrt{1 - \varepsilon^2}}{z_0 \varepsilon} \right]. \tag{77}$$

Here $M = \pi^2 \varepsilon z_0^2 (1 - \varepsilon^2)^{-3/2} \rho_0(0, 0)$ is the total mass; (ζ_1, ζ_2) are the elliptical coordinates of the point (r, z). They are expressed by (r, z) as

---

[5] They are distinguished in that they admit a third integral of motion *quadratic* in velocity (see Introduction).

solutions of the quadratic equation with respect to ζ^2

$$\frac{r^2}{\zeta^2 - z_0^2} + \frac{z^2}{\zeta^2} = 1, \tag{78}$$

$|\zeta_1| > z_0$, $|\zeta_2| < z_0$ [the curves ζ_1 = const in the (r, z) plane are ellipses, while the curves ζ_2 = const are hyperboles].

The problem of the construction of self-consistent models of such systems encounters significant difficulties. No such models are known as yet.

Another approach is to seek stationary systems with the density distribution *similar to the observed* ones, for instance, in elliptical galaxies. As is known [34], the surfaces of identical density (more precisely, isophotes) in elliptical galaxies have only roughly the shape of spheroids.

Therefore, it is possible to try to construct suitable models by solving the integral equation for the distribution function of the form $f_0 = f_0(E, L_z)$:

$$\rho_0(r, z) = \int f_0(E, L_z)\, d^3v, \tag{79}$$

by assuming the density $\rho_0(r, z)$ to be known. At the same time, one may consider the potential $\Phi_0(r, z)$ to be known also.

Eliminating z from the equations

$$\rho_0 = \rho_0(r, z), \qquad \Phi_0 = \Phi_0(r, z), \tag{80}$$

we arrive at the $\rho_0 = \rho_0(r, \Phi_0)$ function. It is easy to see that Eq. (79) defines only the part of the distribution function even in L_z, which we will denote by $f_0(E, L_z^2)$.

It is convenient to rewrite Eq. (79) in the form [280]

$$\rho_0(r^2, \Phi_0) = 4\pi \int_{X=\Phi_0}^{0} dX \int_{Y=0}^{-X} (2Y)^{-1/2} f_0(Y + X, 2Yr^2)\, dY, \tag{81}$$

where

$$Y \equiv \tfrac{1}{2}v_\varphi^2, \qquad X = E - Y, \qquad F = \tfrac{1}{2}v^2 + \Phi_0.$$

By differentiating over Φ_0, we shall obtain an apparently simple equation, $\psi = -\Phi_0$:

$$\frac{\partial \rho_0(r^2, \psi)}{\partial \psi} = 4\pi \int_{Y=0}^{\psi} (2Y)^{-1/2} f_0(Y - \psi, 2Yr^2)\, dY. \tag{82}$$

The general method of solving this equation was suggested by Lynden-Bell [280]. Since the right-hand side of (82) has the form of convolution, in which part is taken by the first argument f_0, it is advantageous to perform the Laplace transformation with respect to ψ. Let s be the variable of this transformation; then we have

$$\frac{s}{4\pi} \int_0^\infty e^{-s\psi} \rho_0(r^2, \psi)\, d\psi = \int_0^\infty e^{-s\psi}\, d\psi \int_0^\psi (2Y)^{-1/2} f_0(Y - \psi, 2Yr^2)\, dY$$

$$= \int_0^\infty (2Y)^{-1/2}\, dY \int_{\psi=Y}^\infty e^{-s\psi}(Y - \psi, 2Yr^2)\, d\psi$$

$$= \int_{Y=0}^\infty e^{-sY}\, dY \int_{B=0}^\infty e^{-sB}(2Y)^{-1/2} f_0(-B, 2Yr^2)\, dB,$$

$$(83)$$

where $B \equiv \psi - Y$, and the equation

$$\rho_0(r^2, 0) = 0 \qquad (\psi = 0 \text{ at infinity}) \tag{84}$$

is employed.

Introducing new variables $t = 2Yr^2$, $u = s/2r^2$, Eq. (83) then converts in the following manner:

$$\frac{s}{2\pi}\left(\frac{s}{2u}\right)^{1/2} \int_0^\infty e^{-s\psi} \rho_0\left(\frac{s}{2u}, \psi\right) d\psi = \int_0^\infty e^{-ut}\, dt \int_0^\infty e^{-sB}\frac{f_0(-B, t)}{t^{1/2}}\, dB. \tag{85}$$

Since the $\rho_0(r^2, \psi)$ function is assumed to be a given one, the left-hand side of (85) is known, while the double Laplace transformation is on the right-hand side. Therefore, the twofold inverse Laplace transformation of the left-hand side of Eq. (85) will yield the sought-for distribution function $f_0(E, L_z^2)$.

Thus, the problem is solved *in principle* for any initial mass distribution $\rho_0(r, z)$. However, in practice, the realization of the suggested procedure for the determination of the distribution function in the explicit form encounters difficulties. The first difficulty presents the need to explicitly calculate the $\rho_0(r^2, \Phi_0)$ function. The other, even more serious difficulty is associated with the requirement of the double Laplace inversion (each inversion includes integration in a *complex* plane).

Nonetheless, in the simplest cases one manages to find explicit solutions. Lynden-Bell illustrates his method in the example of systems with the gravitational potential[6]

$$\Phi_0 = -A\lambda^{-1/4}, \qquad \lambda \equiv (r^2 + a^2)^2 - 2b^2 r^2 \sin^2\theta, \tag{86}$$

where r, θ, φ are spherical coordinates, A and a, b are the parameters, and λ is positive, if $b^2 < 2a^2$. The density

$$\rho = \frac{A}{\pi G}\lambda^{-9/4}[(3a^2 - 2b^2)(r^2 + a^2)^2 + (4a^2 - b^2)b^2 r^2 \sin^2\theta] \tag{87}$$

---

[6] It may be noted that the models in (86) are the generalization of the Schuster–Plummer spherical model, for which $\Phi_0 = -A(r^2 + a^2)^{-1/2}$, $\rho_0 \sim (r^2 + a^2)^{-5/2}$.

is positive provided $b^2 < 3a^2/2$. From (86) and (87) we find the $\rho(\psi)$ function, $\psi \equiv -\Phi_0$:

$$\rho = D\psi^5 + CR^2\psi^9, \qquad R = r \sin\theta,$$

$$D = \left(3a^2 - \frac{2b^2}{A^4\pi G}\right), \qquad C = \frac{5b^2(2a^2 - b^2)}{A^8\pi G}. \tag{88}$$

Further, we have

$$\int_0^\infty e^{-s\psi}\left(\frac{s}{u}\right)^{1/2}\frac{\partial\rho(s/u, \psi)}{\partial\psi}\,d\psi = \left(5!\,s^{-5}D + 9!\,\frac{s^{-8}}{u}C\right)\left(\frac{s}{u}\right)^{1/2}. \tag{89}$$

The inverse Laplace transformation of the $u^{-n-1/2}$ function is $t^{(n-1/2)}/\Gamma(n+\frac{1}{2})$. Therefore, the inverse Laplace transformation of (89) in u is

$$\pi^{-1/2}(5!\,s^{-4.5}t^{-1/2}D + 9!\,2s^{-7.5}t^{1/2}C). \tag{90}$$

Performing further the inverse transformation on s, we arrive at the following expression for the distribution function (more exactly, its part symmetrical in L_z):

$$f_0 = F_1(-E)^{3.5} + F_2 L_z^2(-E)^{6.5}, \qquad E = \frac{v^2}{2} - \psi, \tag{91}$$

$$F_1 = \frac{\sqrt{2}}{4\pi^{3/2}}\frac{5!}{\Gamma(5.5)}\cdot D \simeq 0.65D, \qquad F_2 = \frac{\sqrt{2}}{4\pi^{3/2}}\frac{9!}{\Gamma(7.5)}C \simeq 12.3C. \tag{92}$$

The properties of this model are described in detail in the previously mentioned work by Lynden-Bell [280].

Hunter [235] somewhat modified Lynden-Bell's method. Using his method, Hunter found [235] an exact distribution function of the system with the potential

$$\Phi_0 = -\frac{A}{[(r^2 + z^2 + a^2)^2 - 2\lambda a^2 r^2]^{(1-\mu)/4}}, \tag{93}$$

where A and a are the dimensional parameters (defining the scale of the system) and λ and μ are the dimensionless parameters. The parameter λ characterizes the degree of deviation from the spherical symmetry, while the parameter μ defines the behavior of Φ_0 at large distances. At small λ, the surfaces of identical density corresponding to the potential in (93) are approximately ellipsoids of revolution with eccentricities which decrease with moving away from the center. The condition of positiveness of the density restricts the values of the parameters μ and λ by

$$0 < \mu < 1, \qquad -6 < \lambda < 1.5. \tag{94}$$

Since at $(r^2 + z^2) \to \infty$

$$\rho_0 \sim (r^2 + z^2)^{(\mu - 3)/2}, \tag{95}$$

the systems of interest have an infinite mass, which is one of the drawbacks of these models. The distribution function found by Hunter has a very cumbersome appearance; therefore we shall not write it here.

An obvious advantage of the methods of Lynden-Bell and Hunter described above is a (principled) possibility of determining the self-consistent models of elliptical galaxies directly from observational data, which give the density $\rho_0(r, z)$. The drawbacks of these methods may involve, apart from calculational difficulties, also the need for verification of the ensuing (generally very complicated) distribution function being positive. The point is that the positiveness of the solution of initial equation (79) for the arbitrary law of the density distribution $\rho_0(r, z)$ is not ensured by any general theorems. Conversely, one may easily find examples of such functions which would have $f_0 < 0$ in some region of phase space.

It might have been suggested that in selecting the density to be maximally close to that *observed* in elliptical galaxies, the solution of Eq. (79) automatically will be proved to be positively defined. In reality, however, we cannot guarantee even this because, for instance, with the above-made selection of the distribution function in the form $f_0 = f_0(E, L_z)$ the possibility of the existence of the third integral of motion is neglected.

These difficulties can be avoided if one goes from the obviously positive functional forms for $f_0(E, L_z)$, with some number of parameters which would provide the possibility of varying the models. The density distribution can then be found by solving (by numerical methods) the Poisson equation corresponding to $f_0(E, L_z)$.

Such an approach is adopted in papers by Prendergast and Tomer [307] and Wilson [351]. For example, in the former paper, f_0 is imposed in the form

$$f_0 = \alpha \exp\left(-\frac{E}{\sigma^3} + \beta L_z\right),\tag{96}$$

with cutoff at higher velocities (to avoid divergence of mass). With such a method of construction of self-consistent models, it is difficult beforehand to foresee the shape of the ensuing surfaces of constant density. Thus, the surfaces of level on models of (96) are greatly different from ellipsoids.

Hunter [232] considered also spheroids with the density

$$\rho_0 = \rho_0(0, 0)\left(1 - \frac{r^2}{a_1^2} - \frac{z^2}{a_3^2}\right)\theta\left(1 - \frac{r^2}{a_1^2} - \frac{z^2}{a_3^2}\right),\tag{97}$$

This is evidently the simplest (after uniform) type of spheroid. By means of a fairly simple formula, also the gravitational potential inside the system

$$\Phi_0 = \alpha r^2 + \beta z^2 + \gamma r^4 + \delta r^2 z^2 + \varepsilon z^4\tag{98}$$

is expressed (expressions for the constant coefficients $\alpha, \beta, \gamma, \delta, \varepsilon$ can be found, for example, in Chandrasekhar's book [148]). At small[7] values of the

---

[7] At arbitrary e, the attempt to find the distribution function $f_0(E, L_z)$ corresponding to (97) and (98) leads in Lynden-Bell's method to the need for Laplace inversion of the function lacking in standard tables of Laplace transforms.

eccentricities $e = (1 - a_3^2/a_1^2)^{1/2}$ it is easy to find, using Lynden-Bell's method, the distribution function of the form

$$f_0 = \frac{\rho_0(0, 0)[1 - 5\mu L_z/4\lambda]\theta[-E - 6\lambda/5 + \frac{3}{2}\mu L_z]}{6\pi^2\sqrt{2}(\mu L_z - \lambda - E)[-E - 6\lambda/5 + \frac{3}{2}\mu L_z]^{1/2}} + o(e^2), \quad (99)$$

where λ and μ are the constants defined by

$$\lambda = \frac{4\pi G\rho_0 a_1^2}{9}\left(1 - \frac{e^2}{7}\right), \qquad \mu = 4e\left(\frac{\pi G\rho_0(0, 0)}{105}\right)^{1/2}. \quad (100)$$

§ 2 Stability of a Three-Axial Ellipsoid and an Elliptical Disk

2.1 Stability of a Three-Axial Ellipsoid

2.1.1. Preliminary Remarks. This section investigates the model of the Freeman three-axial collisionless ellipsoid (28), §1, for stability with respect to small perturbations. Here we restrict ourselves to the largest-scale types of oscillations transmitting the initial ellipsoid again into the ellipsoid with modified semiaxes. Let the equation of ellipsoid boundary be

$$\frac{x^2}{a^2} + \frac{y^2}{b^2} + \frac{z^2}{c^2} = 1. \quad (1)$$

Let us represent the gravitational potential in the form

$$\Phi = \frac{1}{2}(A^2x^2 + B^2y^2 + \gamma^2z^2) + D, \quad (2)$$

where

$$A^2 = 2\pi G\rho_0\alpha_0, \qquad B^2 = 2\pi G\rho_0\beta_0, \qquad \gamma^2 = 2\pi G\rho_0\gamma_0,$$

$$\alpha_0 = F\left(\frac{a^2}{b^2}, \frac{a^2}{c^2}\right), \qquad \beta_0 = F\left(\frac{b^2}{a^2}, \frac{b^2}{c^2}\right), \qquad \gamma_0 = F\left(\frac{c^2}{a^2}, \frac{c^2}{b^2}\right),$$

$$F(u, v) = \int_0^\infty d\lambda\,(1 + \lambda)^{-3/2}[(1 + \lambda u)(1 + \lambda v)]^{-1/2}.$$

We write the distribution function in the following form [203]:

$$f_0 = \frac{\rho_0}{\pi\beta\gamma\mu bc}\,\delta(v_x + 2\Omega y)\delta\left(\frac{c_y^2}{\beta^2 b^2\mu^2} + \frac{v_z^2}{\gamma^2 c^2} + \frac{x^2}{a^2} + \frac{y^2}{b^2} + \frac{z^2}{c^2} - 1\right). \quad (3)$$

2.1.2. Investigation of Stability. The kinetic equation in the frame of reference rotating with the angular velocity $-\Omega$ is of the form

$$\frac{\partial f}{\partial t} + v_x\frac{\partial f}{\partial x} + v_y\frac{\partial f}{\partial y} + v_z\frac{\partial f}{\partial z} + \left(\Omega^2 x - 2\Omega v_y - \frac{\partial\Phi}{\partial x}\right)\frac{\partial f}{\partial v_x}$$

$$+ \left(\Omega^2 y + 2\Omega v_x - \frac{\partial\Phi}{\partial y}\right)\frac{\partial f}{\partial v_y} - \frac{\partial\Phi}{\partial z}\frac{\partial f}{\partial v_z} = 0. \quad (4)$$

It is convenient in (4) to make the substitution $v_x \rightarrow v_x - 2\Omega y$:

$$\frac{\partial f}{\partial t} + (v_x - 2\Omega y)\frac{\partial f}{\partial x} + v_y\frac{\partial f}{\partial y} + v_z\frac{\partial f}{\partial z} + \left(\Omega^2 x - \frac{\partial \Phi}{\partial x}\right)\frac{\partial f}{\partial v_x}$$

$$+ \left(2\Omega v_x - 3\Omega^2 y - \frac{\partial \Phi}{\partial y}\right)\frac{\partial f}{\partial v_y} - \frac{\partial \Phi}{\partial z}\frac{\partial f}{\partial v_z} = 0. \tag{5}$$

Let us seek the perturbed distribution function f in the form ($\varepsilon \ll 1$)

$$f = c_0[\delta(1 - J - \varepsilon A) + \varepsilon C\delta(1 - J)]\delta(v_x) + \varepsilon c_0\,\delta(1 - J - \varepsilon A)B\delta'(v_x). \tag{6}$$

Here we have introduced the unknown functions

$$A = A(x, y, z, v_y, v_z, t), \quad B = B(x, y, z, v_y, v_z, t) \quad C = C(x, y, z, v_y, v_z, t)$$

and denoted

$$c_0 = \frac{\rho_0}{\pi\gamma\beta\mu bc}, \quad J = \frac{c_y^2}{\beta^2 b^2 \mu^2} + \frac{v_z^2}{\gamma^2 c^2} + \frac{x^2}{a^2} + \frac{y^2}{b^2} + \frac{z^2}{c^2}.$$

The dependence of the perturbed values on time will be chosen in the form $\sim e^{-i\omega t}$.

Linearizing Eq. (5) and substituting in it (6), we get the system of differential equations for the A, B, C functions:

$$\hat{L}A - 2B\left\{\frac{x}{a^2}\left[1 - \frac{4\Omega^2}{\beta^2\mu^2}\left(1 - \frac{b^2}{a^2}\right)\right] + \frac{2\Omega v_y}{b^2\beta^2\mu^2}\left(1 - \frac{b^2}{a^2}\right)\right\}$$

$$= 2\frac{\partial \Phi_1}{\partial y}\frac{v_y - 2\Omega b^2 x/a}{b^2\mu^2\beta^2} + 2\frac{\partial \Phi_1}{\partial z}\frac{v_z}{\gamma^2 c^2}, \tag{7}$$

$$\hat{L}C = \frac{\partial B}{\partial x} + 2\Omega\frac{\partial B}{\partial v_y}, \tag{8}$$

$$\hat{L}B = \frac{\partial \Phi_1}{\partial x}, \tag{9}$$

where we have introduced the notation $\hat{L}$ for the following operator:

$$L = \frac{\partial}{\partial t} - 2\Omega y\frac{\partial}{\partial x} + \left(v_y\frac{\partial}{\partial y} - B^2 y\frac{\partial}{\partial v_y}\right) + \left(v_z\frac{\partial}{\partial z} - \gamma^2 z\frac{\partial}{\partial v_z}\right). \tag{10}$$

The equations in (7)–(9) are completed to form a closed system by means of the Poisson equation. The perturbed potential Φ_1 can in this case be represented (as for all other systems with the quadratic potential) in the form of finite polynomials in degrees of the Cartesian coordinates x, y, z. Then the general scheme of the construction of the solution is as follows: by imposing the polynomial Φ_1 in x, y, z with the coefficients indefinite at the beginning, from Eqs. (7)–(9) we find, by solving them in the sequence (9)–(8)–(7), the A, B, C functions. Equations (7)–(9) are the most convenient to solve by the method of integration over unperturbed trajectories, which, as can be shown, have the form

$$x' = \frac{2\Omega y}{\beta} \sin \beta t + \frac{2\Omega v_y}{\beta^2} \cos \beta t + x - \frac{2\Omega v_y}{\beta^2}, \qquad y' = y \cos \beta t + \frac{v_y}{\beta} \sin \beta t$$

$$z' = z \cos \gamma t + \frac{v_z}{\gamma} \sin \gamma t, \qquad v'_y = -\beta y \sin \beta t + v_y \cos \beta t, \tag{11}$$

$$v'_z = -\gamma z \sin \gamma t + v_z \cos \gamma t.$$

Then the perturbed density is determined by the formula

$$\varepsilon \rho_1 = -\varepsilon \frac{c_0}{2} \iint dc_y \, dv_z \, \delta(1 - J)\left(b^2\beta^2\mu^2 \frac{\partial}{\partial c_y}\frac{\chi_3}{c_y} + c^2\gamma^2 \frac{\partial}{\partial v_z}\frac{\chi_2}{v_z}\right)$$

$$+ \varepsilon c_0 \int C\delta(1 - J)\, dc_y \, dv_z, \tag{12}$$

where $A = \chi_0 + \chi_1$, $\chi_0 = A(c_y = 0, v_z = 0)$, $\chi_2 = \chi_1(c_y = 0)$, $\chi_3 = \chi_1 - \chi_2$. The contribution to the perturbed potential Φ_1 corresponding to (12) is determined from the solution of the Poisson equation; one has to add to it also the potential of a simple layer produced by the perturbation of the ellipsoid boundary. The equation of the new boundary is

$$\frac{x^2}{a^2} + \frac{y^2}{b^2} + \frac{z^2}{c^2} = 1 - \varepsilon\chi_0. \tag{13}$$

Finally, in comparing the expressions for Φ_1 thus obtained with the initial one, we find the dispersion equation and the coefficients in the polynomial $\Phi_1(x, y, z)$ not defined at the beginning.

We shall restrict ourselves further to the largest-scale oscillations of the ellipsoid–ellipsoid type retaining their direction of the main z-axis. In this case, the calculations are the simplest; however, the ensuing expressions, including the dispersion equation, are nevertheless rather cumbersome, and therefore are not discussed here (for details see [91]).

The results are given in Fig. 48, which shows the regions of stable and unstable solutions. The boundary separating these regions is defined approximately; it has, as is seen from the figure, a rather complicated character. The band of unstable solutions near $b = 1$ corresponds to the instability of Freeman's ellipsoid of revolution. This ellipsoid is investigated in detail below (§3).

In [91] the table of values of instability growth rates was shown (in β units). The greatest values of growth rates are obtained for circular disks.

2.1.3. The Limit of Strongly Elongated Ellipsoids. The form of spiral galactic bars corresponds evidently to ellipsoids strongly elongated in the plane of rotation (of the "needle" or "stick" type). In Fig. 48 they are represented by the bottom left-hand side; they, consequently, are stable with respect to the above largest-scale perturbations.

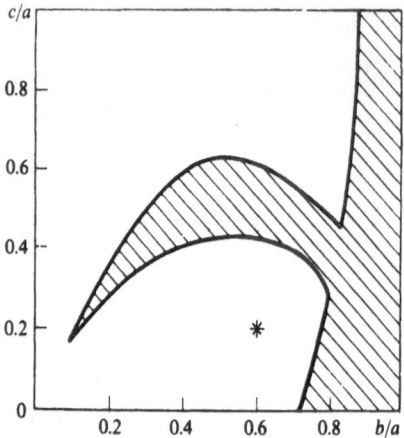

Figure 48. The regions of stable and unstable solutions (the latter is shaded) of the Freeman triaxial ellipsoid. The asterisk marks the approximate location of NGC 4027 according to [344].

However, with respect to small-scale (on the x-axis) perturbations, this model having no dispersion of velocities along the direction of the "bar" (axis of the "spoke") will, of course, be unstable according to Jeans.

This instability of the "cold" system in the x-axis (evident from the physical considerations) can roughly be described if one notes that a very elongated ellipsoid for perturbations with the wavelength in x small compared to a must behave as a cylinder. The cross section of this cylinder has generally an elliptical form; however, if it is assumed that $b \simeq c$, so one may apply the formulae for a circular cylinder.[8]

The dispersion equation for $k_x c \ll 1, k_\perp c \ll 1$ must therefore be of the form:

$$\omega^2 \approx -\text{const} \cdot \beta \cdot k_x^2 v_0^2, \qquad (14)$$

where $v_0 \sim \omega_0 c$ is the characteristic velocity of movement of the particles in the (y, z) plane, $\beta = \ln(1/k_x c)$, const ~ 1.

The maximum growth rate of this, Jeans-like in essence, instability is of the order of ω_0. Thus, strictly speaking, we have at present no *quite acceptable* model of the bar of a spiral galaxy (see, however, Problem 9 in this chapter).

But here one must note the following. According to the results of §2, Chapter II, the cylinder (and, consequently, a fairly strongly elongated ellipsoid) becomes practically stable if the velocity dispersion of the particles along its axis of rotation (the major axis of the ellipsoid, x) exceeds v_0; in our case at

$$v_{xT} > \omega_0 c. \qquad (15)$$

---

[8] By the way, for fairly long-wave perturbations, at $\lambda \gg \max(b, c)$, the shape of the cross section must evidently not play any role.

For a strongly elongated system, this condition is very "soft" (at $c \to 0$, $v_0 \equiv \omega_0 c \to 0$): a relatively small velocity dispersion on the x-axis is enough to make the bar stable.

The introduction of a small $v_{xT} \neq 0$ somewhat modifies Freeman's model; in particular, this will lead to small deviations from an accurately ellipsoidal shape. However, these deviations evidently are inessential to consider the largest-scale perturbations which in this case become most "dangerous." The investigation carried out above in Freeman's *exact* model, show that these perturbations do not lead to instability. Thus, the weak modification mentioned of the ellipsoidal model of Freeman must evidently lead to a stable (and in this sense to a satisfactory) model of the bar of spiral galaxies.

In conclusion note one interesting fact. The dispersion equation for perturbations of an ellipsoid very elongated in the x-axis can, with logarithmic accuracy, be obtained from the considerations of dimensionality. Indeed, from the values G and k_x and the linear density of the "spoke" $\rho \equiv \pi \rho_0 bc$, one may make a single combination of dimensionality of the frequency square, $G k_x^2 \rho$. Therefore, for a system "cold" on the x-axis, there must be

$$\omega^2 \approx -G k_x^2 \rho, \tag{16}$$

which just yields, within an accuracy to the factor $\beta = \ln(1/k_x c)$, Eq. (14) describing the perturbations of a "cold" system. The same considerations of dimensionality show that with due regard for the finite-velocity dispersion v_{xT} we must have instead of (16) the following approximate dispersion equation:

$$\omega^2 \approx k_x^2 (\alpha_1 v_{xT}^2 - \alpha_2 G \rho_0) \qquad (\alpha_1, \alpha_2 \sim 1) \tag{17}$$

from which follows the universal[9] condition (15) of stabilization of the instability [this criterion does not involve k_x since the dependence of the two terms in (17) on k_x^2 is identical].

Allowance for rotation will lead, evidently, to a positive contribution to the right-hand side of (17) $\sim \Omega^2$, which is, however, nonessential at $k_x \to \infty$.

Note that, in the case of the disk system, similar considerations of dimensionality lead, instead of (16), to the well-known Toomre dispersion equation

$$\omega^2 \approx -\text{const} \cdot G \sigma_0 k_x, \tag{18}$$

where σ_0 is the surface density of the disk, an accurate value of const $= 2\pi$. Consideration of perturbations with the wavelength greatly exceeding the disk thickness led Toomre to the model of an *infinitely thin* disk, incomparably simpler than in the three-dimensional model. Similarly, it should be expected that the consideration of the quasi-one-dimensional "spokes" as the models of real strongly elongated bars of spiral galaxies also will lead to simplifications and a possibility of some generalizations.

---

[9] We neglect the possibility of instabilities with exponentially small growth rates.

Note finally that apart from perturbations of the Jeans type (14) there must exist perturbations which would bend the "spoke" perpendicular to the plane of rotation. For thin Freeman ellipsoids which are "cold," such disturbances, of course, seem to be stable, while the dispersion equation corresponding to them differs from (14) in the sign of the right-hand side. This is similar to the case of an infinitely thin layer (or disk) where, apart from oscillation in the plane, described by the equation

$$\omega^2 = -2\pi G\sigma_0 k \tag{19}$$

there are also "membrane" oscillations with the dispersion equation

$$\omega^2 = +2\pi G\sigma_0 k \tag{20}$$

(see §2, of Chapter I). We consider in more detail the stability of prolated ellipsoids in Problem 9 in this chapter.

2.1.4. On the "Polynomial" Form of Eigenfunctions. We recall first of all why finite polynomials in degrees of the Cartesian coordinates may serve as eigenfunctions of the problem of small oscillations of uniform incompressible ellipsoids.

The system of equations and boundary conditions in this case is thus (in the rotating frame of reference):

$$\operatorname{div} \mathbf{v}_1 = 0, \tag{21}$$

$$\frac{\partial \mathbf{v}_1}{\partial t} + 2[\Omega \mathbf{v}_1] = -\nabla \frac{P_1}{\rho_0} - \nabla \Phi_1, \tag{22}$$

$$\Delta \Phi_1 = 0, \tag{23}$$

$$\mathbf{v}_1 = \frac{\partial \xi}{\partial t} \quad (\xi \text{ is Lagrange displacement}), \tag{24}$$

$$(P_1 + \xi \nabla P_0)\Big|_{S_0} = 0. \tag{25}$$

The first equation is the condition of incompressibility, the second, the Euler equation, and the third equation means that perturbation of the potential Φ_1 is created only due to curvature of the ellipsoid boundary. The condition in (25) is the fact of constant pressure on the perturbed boundary; it is expressed in terms of the values which are to be calculated on the initial boundary S_0. Five (scalar) equations in (21)–(23) must define five unknown functions:

$$\xi(\xi_1, \xi_2, \xi_3), P_1, \Phi_1.$$

Assume that the potential Φ_1 is imposed in the form of the polynomial of the power n with unknown coefficients that are functions of time. Then, by representing P_1 and the components ξ by similar polynomials (respectively, of the degree n and $n - 1$) we express from Eqs. (21)–(25) the coefficients of these polynomials in the form of linear combinations of coefficients of polynomial expansion of the potential Φ_1. Here it is important that P_0 be expressed as is known, by the *second-order polynomial*.

Now one has only to employ the Laplace equation (23). For this purpose, we calculate a normal displacement on the unperturbed boundary: $\xi_n = (\boldsymbol{\xi} \mathbf{n})|_{S_0}$, where $\mathbf{n}$ is the vector of the normal to the S_0 surface.

Since it is easy to see that

$$\mathbf{n} = l_0 \left(\frac{x}{a^2}, \frac{y}{b^2}, \frac{z}{c^2} \right), \qquad l_0 \equiv \frac{1}{\sqrt{x^2/a^4 + y^2/b^4 + z^2/c^4}}, \qquad (26)$$

and $\boldsymbol{\xi}$ is the polynomial of the degree $(n-1)$, then ξ_n ensues in the form

$$\xi_n = l_0 P_n(x, y, z), \qquad (27)$$

where P_n is the polynomial of the degree n. However, such an ξ_n corresponds to the potential Φ_1, which inside the ellipsoid is again expressed by the polynomial of degree n. This state is proved in the potential theory [127, 15], where, in particular, the following relations (see, e.g., the "fundamental formulas" in Appel's book [15], §65):

$$\xi_n = l_0 \sum_k \frac{2n+1}{4\pi} \alpha_k M_k N_k, \qquad \Phi_1 \big|_{S_0} = \sum_k \alpha_k R_k^{(0)} S_k^{(0)} M_k N_k,$$

$$\Phi_{1<} = \sum_k \alpha_k S_k^{(0)} R_k M_k N_k, \qquad \Phi_{1>} = \sum_k \alpha_k R_k^{(0)} S_k M_k N_k \qquad (28)$$

are established, where $\Phi_{1<}$ and $\Phi_{1>}$ are, respectively, the potentials inside and outside the ellipsoid S_0, $\Phi_1|_{S_0}$ is the potential on the surface S_0; $R(\rho)$, $M(\mu)$, $N(\nu)$ are Lamé functions of the first kind; ρ, μ, ν are the elliptical coordinates; RMN are Lamé products, $S_k(\rho)$ is a Lamé function of the second kind; the index (0) denotes that the function is calculated on the surface S_0 (which corresponds to $\rho = \rho_0$).

It is essential for us that Lamé products RMN, through which, according to (28), the internal potential $\Phi_{1<}$ is expressed, are reduced in Cartesian coordinates to harmonic polynomials [15].

Thus, the calculations are non-contradictorially closed and result in a uniform system of ordinary differential equations (with constant coefficients) with respect to the coefficients of the initial polynomial expansion Φ_1.

The property of the "polynomiality" of solutions remains also for those models of uniform *collisionless* ellipsoidal systems[10] whose stability is under investigation.

If one attempts not to specify the form of distribution function, one will face a number of problems. First of all, the systems with different symmetry

---

[10] In particular, one first of all bears in mind confined systems. For the case of unlimited systems (uniform flat layer, cylinder) this property remains only for perturbations of a particular kind independent of those coordinates which may take on infinite values. For the sake of example, one may adduce flutelike oscillations of the cylinder ($k_z = 0$).

have in the general case a different number of unambiguous integrals of motion which are arguments of f_0. Therefore, one has nevertheless to separately consider the three-axis ellipsoids, two-axis ellipsoids (spheroids), cylinders, and spheres. Another uncertainty is due to the fact that the distribution function may depend, apart from energy and angular moment, also on some number of, in a certain sense, "occasional" integrals of motion, the existence of which is due to the symmetry of motion in this or that specific model (see Introduction). An example of such a situation[11] may be the above-considered model of the three-axis Freeman's ellipsoid just having a very special symmetry of movement of particles due to the suggestion about the accurate balance of the gravitational attraction and the centrifugal force along the major semiaxis x. Accordingly, the distribution function found by Freeman is dependent on four integrals, two of which (E_β and E_γ) are standard, but the other two (x_0, I_2) are specific just for this model. Also, it should be noticed that this distribution function yields, possibly, a single physically acceptable solution for a uniform *three-axis* collisionless ellipsoid. In any event, attempts to find distribution functions dependent only on "standard" integrals of motion, did not lead to reasonable solutions (see, e.g., Problem 5).

In this connection, the possibility of a "general" proof for the statement about the polynomiality of solutions appears to be questionable. It is also possible that in the general case this statement is not valid, although it may be proved for some classes of distribution functions of homogeneous ellipsoids. For all homogeneous systems investigated for stability in this book, "polynomiality" is easily established in each specific case separately. And, essentially, only one additional circumstance, with respect to the case of incompressible ellipsoidal systems, is used; namely, polynomial perturbations of the local density also lead to polynomial perturbations of the potential.

2.2 Stability of Freeman Elliptical Disks

The stability of disk models (48), §1, with respect to the largest scale perturbations (of the "disk"–"disk" type) was studied by Tremaine [335]. The investigation is performed by a standard method, which is described in detail above. True, Tremaine uses the apparatus of Lamé functions [15], but, as we saw in the previous section (2.1), for the simplest modes considered by him, one may not use these functions.

Therefore, we at once begin to discuss the results obtained in [335], omitting uncomplicated (though cumbersome) mathematical calculations.

---

[11] This also concerns the systems with circular trajectories, where the role of such "occasional" integrals of motion is played by radii of orbits of the particles. However, the distribution functions of *uniform* systems with circular orbits may be expressed through "ordinary" integrals [in the form $\sim \delta(E - L)$].

Classification of perturbations can, as usual, be performed using the highest degrees n of polynomial expansions of the perturbed potential

$$\Phi_1 = a_{01} x^n + a_{02} x^{n-1} y + \cdots + a_{0,n+1} y^n + \cdots. \tag{1}$$

For $n = 1$, the frequencies

$$\omega = \pm \Omega \tag{2}$$

ensue, which signifies stability of all Freeman's disks with respect to perturbations of the dipole type.

For "quadrupole" perturbations ($n = 2$), nontrivial solutions are determined from the equation

$$C(\omega^2) = 0, \tag{3}$$

where $C(\omega^2)$ is the polynomial of the third degree with respect to ω^2. The coefficients of this polynomial for the general Freeman's elliptical disk are calculated numerically. The analytical expressions for $C(\omega^2)$ are given by Tremaine in two limiting cases.

1. If the disk is at rest, then

$$C(x) = \left\{ \frac{1}{a^2 - b^2} [\alpha^2(3a^2 - b^2) + \beta^2(a^2 - 3b^2)] - x \right\}$$

$$\times \left(x^2 + 2 \frac{b^2\alpha^2 - a^2\beta^2}{a^2 - b^2} x + 2\alpha^2\beta^2 + \frac{b^2\beta^4 - a^2\alpha^4}{a^2 - b^2} \right). \tag{4}$$

The roots of Eq. (4) are real and positive, so that the nonrotating Freeman's disks are stable relative to perturbations with $n = 2$. For a circular disk, this result is a particular case of the general theorem about the stability of arbitrary uniform ellipsoids of revolution at rest (see §3).

2. In the case of a circular disk ($b/a = 1$)

$$C(x) = (1 - x)\left[x^2 - \left(12 \frac{\Omega^2}{A^2} + 5 \right)x + \left(\frac{12\Omega^2}{A^2} - \frac{5}{2} \right)^2 \right]. \tag{5}$$

From there we obtain the stability criterion $\Omega^2 < \frac{5}{6}A^2$, which is coincident with that derived earlier [252, 77] directly for annular disks (see Chapter V, Section 4.4).[12]

The numerical calculations at arbitrary b/a and Ω^2/A^2 ($0 < b/a \le 1$, $0 \le \Omega^2/A^2 \le 1$) show that the elliptical disks are unstable in the region, having an approximately triangular shape limited by the points: $b/a = 1$, $\Omega^2/A^2 = \frac{5}{6}$; $b/a = 1$, $\Omega^2/A^2 = 1$; $b/a = 0.7296$; $\Omega^2/A^2 = 1$ (Fig. 49). Note that the last point was also approximately defined earlier [76, 96]. From Table 1, given in [76], it follows that the disks with $\Omega/A = 1$ must become

---

[12] In comparison, it should be taken into account that the parameter γ used in Section 4.4 is linked with Ω by the equation

$$\gamma = \frac{1}{3} \frac{\Omega}{A} \left(\frac{8\Omega^2}{A^2} - 5 \right).$$

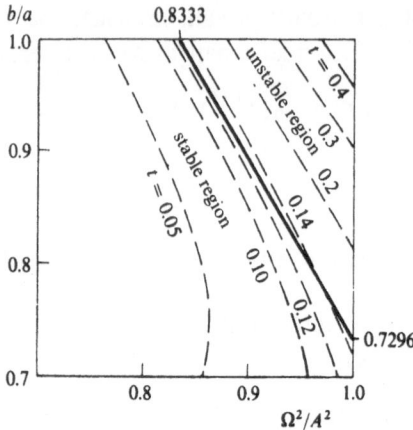

Figure 49. The results of the stability investigation of the Freeman elliptical disk (48) in Section 1 toward "barlike" perturbations. The solid line marks the boundary between the stable models and unstable models. The dashed lines mark the boundary of the constant value of the Peebles–Ostriker parameter: $t = \text{const}$ (see [234]).

unstable at b/a lying within the interval of 0.69 and 0.75. Tremaine [335] partly specified this boundary.[13]

It should be emphasized that the stability boundary given in Fig. 49 refers only to the modes which are expressed by the second-order polynomials (in x and y). Apparently, many models must be unstable with respect to perturbations with $n \geq 3$. We shall see (for example, in the next chapter, Section 4.3) that all annular disks are unstable with respect to excitation of at least one mode with $3 \leq n \leq 5$ [76, 96]. In addition, one should expect the Jeans instability of "too cold" disks with respect to short-wave perturbations ($n \gg 1$) in accordance with the familiar energy considerations of Toomre [333] (see the beginning of Chapter V.)

Let us consider in more detail very flattened disk models ($b/a \ll 1$) rotating relatively slowly ($A^2 \gg \Omega^2$, $B^2 \gg \Omega^2$). For that purpose, perform expansion of all values in (48), §1, in b/a. The frequencies α and β of movements prove to be thus:

$$\alpha \approx A_1 \left(1 - \frac{2\Omega^2}{B_1^2}\right), \qquad \beta = B_1 \left(1 + \frac{2\Omega^2}{B_1^2} + \frac{2\Omega^2 A_1^2}{B_1^4} - \frac{2\Omega^4}{B_1^4}\right), \qquad (6)$$

where $A_1^2 = A^2 - \Omega^2$, $B_1^2 = B^2 - \Omega^2$. The values k_α and k_β entering into f_0 are roughly equal to

$$k_\alpha \approx \frac{2\Omega A_1}{B_1^2}, \qquad k_\beta = \frac{B_1}{2\Omega}\left(1 + \frac{2\Omega^2 - A_1^2}{B_1^2}\right). \qquad (7)$$

---

[13] Chandrasekhar (see [148]) was the first to determine $(b/a)_{cr} = 0.7296$ when he investigated the problem of the stability of incompressible Riemann's ellipsoids; in the disk limit $c/a \to 0$ they prove to be identical to collisionless elliptical Freeman's disks at $\Omega^2/A^2 = 1$.

For $A_1 \approx A$ and $B_1 \approx B$ (i.e., $A \gg \Omega$, $B \gg \Omega$), we get

$$A_1^2 \approx \frac{3MG}{a^3} \ln \frac{4a}{b}, \qquad B_1^2 \approx \frac{3MG}{ab^2}, \tag{8}$$

so that

$$\Lambda^2 \approx \frac{1}{a^2 b^2 B_1^2} \approx \frac{1}{3aMG},$$

$$\frac{b^2}{k_\alpha^2 k_\beta^2} \approx \frac{a^2}{\ln(4a/b)},$$

$$I \approx \frac{1}{3aMG} \left[a^2 \left(v_y^2 + 2\Omega x \frac{b^2}{a^2} \ln \frac{4a}{b} \right)^2 + \frac{a^2}{\ln(4a/b)} \left(v_x - 2\Omega y \ln \frac{4a}{b} \right)^2 \right].$$

$$\tag{9}$$

Therefore, the distribution function in (48), §1, takes the form

$$f \simeq f(I) \approx \frac{1}{4\pi^2 G \sqrt{\ln(a/b)}} \left[1 - \frac{x^2}{a^2} - \frac{y^2}{b^2} - \frac{a}{3GM} \left(v_y^2 + \frac{v_x^2}{\ln(a/b)} \right) \right]^{-1/2}, \tag{10}$$

while the one-dimensional distribution function

$$f(v_x) = \int f(J) \, dc_y \simeq \frac{1}{4\pi} \sqrt{\frac{3M}{Ga \ln(a/b)}} \, \theta \left[1 - \frac{x^2}{a^2} - \frac{y^2}{b^2} - \frac{a v_x^2}{\ln(a/b)} \right]. \tag{11}$$

The velocity dispersions corresponding to (10) are

$$\sigma_{xx} \approx \frac{GM}{a} \ln \frac{4a}{b} \left(1 - \frac{x^2}{a^2} - \frac{y^2}{b^2} \right), \qquad \sigma_{yy} \approx \frac{GM}{a} \left(1 - \frac{x^2}{a^2} - \frac{y^2}{b^2} \right),$$

and the mean velocities in the x- and y-axes are

$$\overline{v_x} \approx 2\Omega y \ln \frac{4a}{b}, \qquad \overline{v_y} \approx -2\Omega x \frac{b^2}{a^2} \ln \frac{4a}{b}.$$

The two last values are small as $b/a \to 0$, so that the system "is maintained" mainly due to large velocity dispersion in the x-axis.

This picture of the equilibrium states of very compressed elliptical Freeman's disks with $A \gg \Omega$, $B \gg \Omega$ is likely to suggest stability of such systems (see discussion of stability of elongated ellipsoids at the end of subsection 2.1.3, as well as Problem 1 of Chapter II).

Conversely, in the case of *rapidly rotating* ($A \approx \Omega$) very flattened disks one should expect instabilities with respect to perturbations with $n \geq 3$. In conclusion, we note that *nonlinear* evolution of Freeman's circular and elliptical disks is considered in the problems and the large-scale stability of Freeman's elliptical cylinders is also investigated in the Problem 10.

§3 Stability of Two-Axial Collisionless Ellipsoidal Systems

This section deals with the stability of uniform stellar spheroids with a
density ρ_0 and the boundary equation

$$\frac{x^2 + y^2}{a^2} + \frac{z^2}{c^2} = 1.$$

Let us write the gravitational potential in the form

$$\Phi_0 = \frac{\Omega^2(x^2 + y^2)}{2} + \frac{\omega_0^2 z^2}{2} + \text{const},$$

where the constants Ω and ω_0, having the meaning of the oscillation frequencies of stars respectively in the (x, y) plane and along the z-axis, are
linked with the density ρ_0 and the semiaxes a, c by the relations

$$\Omega^2 = 2\pi G\rho_0[(\zeta^2 + 1)\zeta \operatorname{arccot} \zeta - \zeta^2],$$

$$\omega_0^2 = 4\pi G\rho_0(1 - \zeta \operatorname{arccot} \zeta)(1 + \zeta^2), \qquad a = \frac{(\zeta^2 + 1)^{1/2}}{\zeta} c.$$

3.1 Stability of Freeman's Spheroids

1. Assume that all particles rotate in one direction with an angular velocity
Ω. The distribution function under investigation, in the rotating frame of
reference, has the form (see §1)

$$f_0 = \frac{\rho_0}{\pi\omega_0 c} \delta(v_x)\delta(v_y) \bigg/ \sqrt{1 - \frac{x^2 + y^2}{a^2} - \frac{z^2}{c^2} - \frac{v_z^2}{\omega_0^2 c^2}}. \tag{1}$$

In the limit $c \to 0$, distribution function (1) becomes a δ-like one:

$$f_0 \sim \delta(v_x)\delta(v_y)\delta(v_z), \tag{2}$$

corresponding to the "Maclaurin" disk with circular orbits of particles.
They were thoroughly investigated for stability by Hunter [226] (see
Section 2.2 of Chapter V).

The elliptical configurations in (1) are unstable for there is no local velocity
dispersion of stars in the (x, y) plane. The growth rates of some unstable modes
can increase with a decrease in the scale of perturbations in this plane. If,
however, one bears in mind that real systems (for example, flat galaxies)
possess some (small) velocity dispersion of stars in the rotational plane, then
it is clear that just the shortest-wave perturbations will be stabilized. At the
same time, larger-scale oscillations, representing in the sense of loss of
stability the greatest "danger", are correctly described by distribution
function (1) which does not contain velocity dispersion.

2. The perturbation theory is constructed in the following way. First, the distribution function f is represented in the form (similarly to Section 3.1 of Chapter III)

$$f = A\delta(v_x)\delta(v_y) + \varepsilon B\delta'(v_x)\delta(v_y) + \varepsilon C\delta(v_z)\delta'(v_y), \tag{3}$$

where A, B, C are the x, y, z, v_z, t functions; ε is the parameter of expansion of perturbation theory. Substitution of (3) into the kinetic equation in Cartesian coordinates yields within an accuracy of up to the first order terms in ε:

$$\frac{\partial A}{\partial t} + v_z \frac{\partial A}{\partial z} - \varepsilon \frac{\partial B}{\partial x} - \varepsilon \frac{\partial C}{\partial y} - \omega_0^2 z \frac{\partial A}{\partial v_z} = \varepsilon \frac{\partial \Phi}{\partial z} \frac{\partial A_0}{\partial v_z}, \tag{4}$$

$$\frac{\partial B}{\partial t} + v_z \frac{\partial B}{\partial z} - 2C - \omega_0^2 z \frac{\partial B}{\partial v_z} = \frac{\partial \Phi_1}{\partial x} A_0, \tag{5}$$

$$\frac{\partial C}{\partial t} + v_z \frac{\partial C}{\partial z} + 2B - \omega_0^2 z \frac{\partial C}{\partial v_z} = \frac{\partial \Phi_1}{\partial y} A_0, \tag{6}$$

where $A_0 = A(\varepsilon = 0)$ and it is assumed, as usual, that the potential

$$\Phi = \Phi_0 + \varepsilon \Phi_1.$$

Further, we substitute the A function in the form

$$A = \frac{\rho_0}{\pi \omega_0 c} \left(1 - x^2 - y^2 - \frac{z^2}{c^2} - \frac{v_z^2}{\omega_0^2 c^2} - \varepsilon \chi \right)^{-1/2}, \tag{7}$$

where $\varepsilon \chi$ is perturbation. Substituting (7) into (4) and separating the first-order terms, we shall find

$$\frac{\partial \chi}{\partial t} + v_z \frac{\partial \chi}{\partial z} - \omega_0^2 z \frac{\partial \chi}{\partial v_z} - 2\left(\frac{\partial B}{\partial x} + \frac{\partial C}{\partial y} \right)\left(1 - x^2 - y^2 - \frac{z^2}{c^2} - \frac{v_z^2}{\omega_0^2 c^2} \right)^{3/2}$$

$$= \frac{2v_z}{\omega_0^2 c^2} \frac{\partial \Phi_1}{\partial z}, \tag{8}$$

$$\frac{\partial B}{\partial t} + v_z \frac{\partial B}{\partial z} - \omega_0^2 z \frac{\partial B}{\partial v_z} - 2C = \frac{\partial \Phi_1}{\partial x}\left(1 - x^2 - y^2 - \frac{z^2}{c^2} - \frac{v_z^2}{\omega_0^2 c^2} \right)^{-1/2}, \tag{9}$$

$$\frac{\partial C}{\partial t} + v_z \frac{\partial C}{\partial z} - \omega_0^2 z \frac{\partial C}{\partial v_z} + 2B = \frac{\partial \Phi_1}{\partial y}\left(1 - x^2 - y^2 - \frac{z^2}{c^2} - \frac{v_z^2}{\omega_0^2 c^2} \right)^{-1/2}. \tag{10}$$

Finally, introducing instead of B and C the functions

$$B' = B\left(1 - x^2 - y^2 - \frac{z^2}{c^2} - \frac{v_z^2}{\omega_0^2 c^2} \right)^{1/2}, \tag{11}$$

$$C' = C\left(1 - x^2 - y^2 - \frac{z^2}{c^2} - \frac{v_z^2}{\omega_0^2 c^2} \right)^{1/2}, \tag{12}$$

and taking into account that the operator[14]

$$\hat{L} \equiv \frac{\partial}{\partial t} + v_z \frac{\partial}{\partial z} - \omega_0^2 z \frac{\partial}{\partial v_z}$$

[on the left-hand side in eqs. (9) and (10)] acting on the function from $(1 - x^2 - y^2 - z^2/c^2 - v_z^2/\omega_0^2 c^2)$ yields zero, we obtain finally the following system of equations:

$$\hat{L}\chi - 2\left(1 - x^2 - y^2 - \frac{z^2}{c^2} - \frac{v_z^2}{\omega_0^2 c^2}\right)\left(\frac{\partial B'}{\partial x} + \frac{\partial C'}{\partial y}\right) - 2(xB' + yC')$$

$$= \frac{2v_z}{\omega_0^2 c^2} \frac{\partial \Phi_1}{\partial z}, \tag{13}$$

$$\hat{L}B' - 2C' = \frac{\partial \Phi_1}{\partial x}, \tag{14}$$

$$\hat{L}C' + 2B' = \frac{\partial \Phi_1}{\partial y}. \tag{15}$$

This system of equations with respect to χ, B', C', Φ_1 must be solved together with the Poisson equation, and in the latter, apart from the change of the local density ρ_1, one should also take into account the variation of the ellipsoid boundary.

Split the χ function into two parts:

$$\chi = \chi_0 + \chi_1, \qquad \chi_0 = \chi(v_z = 0), \qquad \chi_1 = \chi - \chi_0, \tag{16}$$

by singling out the part of χ_0 independent of velocities. Then, as will easily be shown, the perturbation of the density ρ_1 is determined using the known χ_1 function according to the formula

$$\varepsilon \rho_1 = -\frac{\rho_0}{2\pi\omega_0 c} \varepsilon \int dv_z \frac{\partial}{\partial v_z}\left(\frac{\chi_1}{v_z}\right)\left(1 - x^2 - y^2 - \frac{z^2}{c^2} - \frac{v_z^2}{\omega_0^2 c^2}\right)^{-1/2}, \tag{17}$$

and the equation of the ellipsoid boundary is defined from the relation

$$1 - x^2 - y^2 - \frac{z^2}{c^2} - \varepsilon\chi_0 = 0. \tag{18}$$

As for other systems with the quadratic potential, the perturbations Φ_1 of the ellipsoid may be represented in the form of polynomials in degrees of Cartesian coordinates. Imposing some polynomial with undetermined coefficients and the time dependence $\sim \exp(-i\omega t)$, the general scheme of calculation of eigenfrequencies and eigenfunctions can be described in the following way. To begin with, from (14) and (15) we determine the B' and C' functions. Then B' and C' are substituted into Eq. (13), from which the function χ is determined.

---

[14] It is easy to see that the operator $\hat{L}$ is equal to the full derivative over unperturbed trajectory: $\hat{L} = d/dt$.

Equations (13)–(15) are readily solved both by the method of integration over trajectories and by the method of integration over angle. Thereafter, from (17), we find the perturbation of the density ρ_1, and, from (18), the perturbation of the boundary. Then we restore the potential Φ_1, by solving the corresponding problem of the potential theory.[15] And, finally, comparing the expression for the potential Φ_1 thus obtained with the initial one, we can calculate the oscillation frequencies and eigenfunctions corresponding to them.

3. Consider as an example the perturbation of the simplest form [91, 96]

$$\Phi_1 = (x + iy)[\alpha z^2 + \beta(x^2 + y^2) + \gamma], \tag{19}$$

where α, β, γ are the coefficients not defined at the beginning.

The calculations using the general scheme described above lead to a rather cumbersome dispersion equation which must be solved numerically.

All ellipsoids apart from those that lie within a narrow interval $0.405 < c/a < 0.455$ are unstable (with respect to a given mode). For each c/a (in the region of unstable c/a) there is only one unstable root.

The maximum of the instability growth rate for the mode in question is reached at $c = 0$, i.e., for a disk. This is quite understandable: with an increase in the ellipsoid thickness, the (destabilizing) gravitational force in the plane of rotation decreases.

It is interesting that during the transition over the stability region, the real part of the frequency Re ω changes sign: at $c < 0.405$ the increasing waves of stellar density trail from rotating stars, and at $c > 0.455$ lead them.

Numerically were also calculated the perturbed potential in the plane $z = 0$ and the surface density σ. The equipotential lines[16] for a typical case are presented in Fig. 51. It is seen that they possess insignificant asymmetry with respect to the x- and y-axes (this asymmetry is of the order of the ratio Im γ/β).

Note that, in the limit case of the disk ($c \to 0$), eigenfunction (19) transforms to the following (Chapter V, Section 2.2):

$$\Phi_1 \to e^{i\varphi}P_3^1(\sqrt{1 - x^2 - y^2}) = (x + iy)[5(x^2 + y^2) - 4], \tag{20}$$

where $P_n^m(\eta)$ are the associated Legendre functions. In this limit $\gamma/\beta = -\frac{4}{5}$, and the frequencies of oscillations are defined by the equation

$$1 + \frac{\frac{3}{8}}{\omega^2 - 4}\left(11 - \frac{2}{\omega}\right) = 0. \tag{21}$$

4. The calculation of the eigenfrequencies of even the simplest oscillation modes changing the local density of the ellipsoid are rather cumbersome. The calculations simplify very much (and can easily be performed in the general form) for oscillations of the ellipsoid boundary at which the local

---

[15] Contributions to Φ_1 from the perturbations in the ellipsoid boundary and in the density are readily calculated separately, by transforming temporarily to spheroidal coordinates.

[16] The lines of equisurface density are similar.

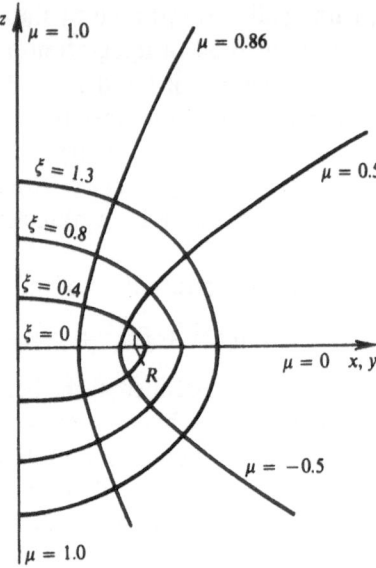

Figure 50. Spheroidal coordinates.

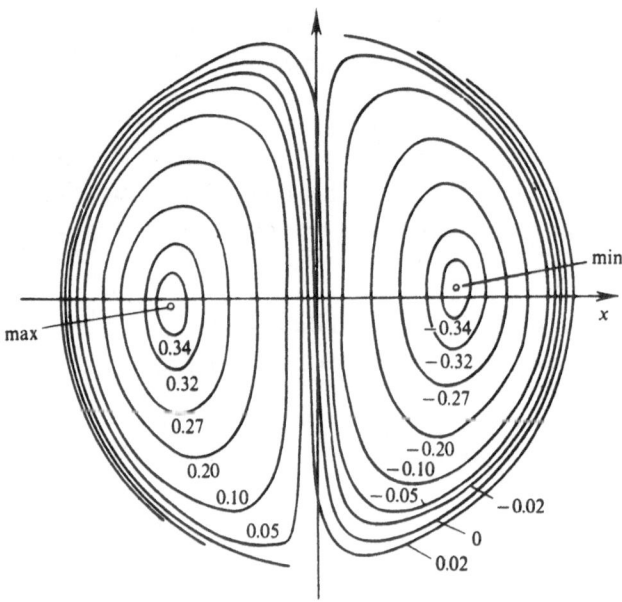

Figure 51. Isolines of the perturbed potential (or surface density) for $c/a = 0.5$ ($m = 1$).

density is unaltered. The potential perturbation in the case of interest is independent of z and is of the form:

$$\Phi_{(\zeta < \zeta_0)} = p_m^m(\zeta) P_m^m(\mu) e^{im\varphi} \equiv A(x + iy)^m, \qquad p_m^m(x) = P_m^m(ix). \qquad (22)$$

Here ζ, μ, φ are the spheroidal coordinates (Fig. 50)

$$x = R_0[(\zeta^2 + 1)(1 - \mu^2)]^{1/2} \cos \varphi,$$
$$y = R_0[(\zeta^2 + 1)(1 - \mu^2)]^{1/2} \sin \varphi, \qquad (23)$$
$$z = R_0 \zeta \mu,$$

such that the equation of the unperturbed ellipsoid boundary has the form

$$\zeta = \zeta_0 \qquad \left[R_0^2 = \frac{c^2}{\zeta_0^2}, a^2 \equiv 1 = R_0^2(1 + \zeta_0^2) \right].$$

The system of (13)–(15) is solved simply and yields

$$\chi = -\frac{2mA}{\omega(\omega + 2)}(x + iy)^m. \qquad (24)$$

We switch again to the spheroidal coordinates of (23):

$$\chi = -\frac{2m}{\omega(\omega + R)} R_0^m p_m^m(\zeta) P_m^m(\mu) e^{im\varphi} \equiv B p_m^m P_m^m e^{im\varphi}. \qquad (25)$$

Find the normal boundary displacement on the basis of the equation for the perturbed boundary

$$x^2 + y^2 + \frac{z^2}{c^2} + \varepsilon\chi = 1.$$

It turns out to be

$$h = (\zeta - \zeta_0)h = -\frac{\varepsilon\zeta_0}{R_0(\zeta_0^2 + \mu^2)^{1/2}} \frac{\chi}{(1 + \zeta_0^2)^{1/2}}. \qquad (26)$$

We have the perturbed potential as a potential of a simple layer with the surface density $\sigma = \rho_0 h$:

$$\Phi_{(\zeta > \zeta_0)} = c q_m^m(\zeta) P_m^m(\mu) e^{im\varphi}, \qquad \Phi_{(\zeta < \zeta_0)} = b p_m^m(\zeta) P_m^m(\mu) e^{im\varphi},$$

$$q_m^m(x) = (-1)^m (2m)! \, p_m^m(x) \int_x^\infty (1 + y^2)^{-1} [p_m^m(y)]^{-2} \, dy. \qquad (27)$$

From the condition of matching on the boundary, the coefficients c and b are determined; the coefficient b necessary further on is

$$b = -\frac{4\pi G \rho_0 \zeta_0 B p_m^m}{W_m(1 + \zeta_0^2)}, \qquad (28)$$

where W_m denotes a Wronskian

$$W_m = p_m^{m'} q_m^m - p_m^m q_m^{m'} = (-1)^m \frac{(2m)!}{1 + \zeta^2}, \qquad (29)$$

and all the functions in (28) are calculated at $\zeta = \zeta_0$. Comparing the calculated potential with the initial one, we get the following equation for frequencies of eigenoscillations:

$$\omega^2 + 2\omega + a = 0, \tag{30}$$

where

$$a = mp\zeta_0(1 + \zeta_0^2)^m \int_{\zeta_0}^{\infty} \frac{dx}{(1 + x^2)^{m+1}}, \qquad \rho = \frac{2}{(1 + \zeta^2)\zeta \arctan(1/\zeta) - \zeta^2}. \tag{31}$$

Determine from (30) the eigenfrequencies of oscillations:

$$\omega = -1 \pm \sqrt{1 - a}. \tag{32}$$

The instability condition apparently is $a > 1$.

Equation (30) was investigated in detail numerically at different values of oblateness, the calculated growth rates of instability depending on the ratios of semiaxes c/a are to be found in [91].

In the cold disk ($c = 0$), the growth rate of instability of the models in question increases infinitely with the increase in the number m, proportional to $m^{1/4}$ (see Chapter V, Section 2.2).

The tables in [91] show that with increasing c, the increase in the growth rate at larger m slows down. In the case of the sphere ($c = 1$) the expression for the growth rate can be obtained analytically in the form

$$\gamma = \left(\frac{3m}{2m + 1} - 1\right)^{1/2}; \tag{33}$$

γ asymptotically tends to the value $\gamma_\infty = 1/\sqrt{2}$.

Since $\omega = \omega' - m$, where ω' is the frequency in a fixed coordinate system, then the azimuthal angular velocity of the unstable wave in this system is

$$\Omega_w = \frac{\omega'}{m} = 1 - \frac{1}{m}. \tag{34}$$

It is seen that the density wave always lags behind the stellar disk rotating with the velocity Ω, and the drift velocity $|\Omega_w - \Omega| = 1/m$ (measured in Ω units) is independent of the degree of oblateness of the ellipsoid and decreases in inverse proportion to the number of the mode m.

5. Calculation of the oscillation frequencies of Freeman's ellipsoid for arbitrary modes presents some technical problems. In any event, it must lead to some poorly foreseeable characteristic equation. Calculations for ellipsoids with *small thickness* are comparatively simple. In this case we deal in essence with the allowance for the stabilizing influence of the finite thickness on the Jeans instability of a cold disk (see the beginning of Chapter V). For the case of *axially-symmetrical* perturbations, the corresponding

dispersion equation has the form[17]

$$\omega^2 = 4[1 - 2n(2n + 1)\gamma_{2n}] + [\tfrac{8}{9}n(n - 4) + 8n(n + 1) - 16\gamma_{2n}$$

$$\times\, 2n(2n + 1)]\frac{c}{\pi}, \qquad \gamma_{2n} = \frac{[(2n)!]^2}{2^{4n+1}[n!]^4}, \tag{35}$$

where n is any positive integer.

At $c = 0$, this yields the equation which describes the radial oscillations of a cold disk [see formula (63) in Section 2.2, Chapter V]:

$$\omega^2 = 4[1 - 2n(2n + 1)\gamma_{2n}]. \tag{36}$$

The term in (35) proportional to c, describes this effect of the Jeans instability stabilization of the disk with finite thickness.

Take for example the largest-scale unstable mode; it corresponds in (35) to $n = 2$. At small c we have

$$\frac{\gamma^2}{\Omega^2} = -\frac{\omega^2}{\Omega^2} \approx \mathrm{const}\left(1 - 14\,\frac{c}{\pi}\right), \tag{37}$$

so that evidently the stability occurs (a rough estimate) at

$$c = c_{\mathrm{cr}} \simeq \frac{\pi}{14} \simeq 0.23.$$

The remaining modes with $m = 0$ stabilize at smaller c, although at $c = 0$ they have larger growth rates of instability. Thus, at $n = 3$, c_{cr} (calculated in the same way) is of the order of 0.22, while at larger n, taking into account that

$$\gamma_{2n} \approx \frac{1}{2\pi n}\ (n \gg 1),$$

we have from (35)

$$c_{\mathrm{cr}}(n) \approx \frac{1}{n}. \tag{38}$$

Of course, the accuracy of such estimates is small.

Note that the "incompressible" modes and, in particular, the "barlike" mode ($m = 2$) is stabilized by the thickness more slowly: We have already seen this from the accurate expression for eigenfrequencies of the corresponding perturbations, but we could have derived it from asymptotic formulae similar to (37). For $m = 2$ at $c \to 0$, one can obtain

$$\gamma \approx \mathrm{const}\sqrt{1 - 4c/\pi}. \tag{39}$$

If in (39) it is formally assumed that $\gamma = 0$, then for the critical c we shall find $c_{\mathrm{cr}} \approx \pi/4 \approx 1$ (in reality, in this model, $c_{\mathrm{cr}} = \infty$).

---

[17] Derived with I. G. Shukhman's participation. It is easy to obtain the equation similar to (35) also for *arbitrary* perturbations.

6. The model of the two-axis Freeman's ellipsoid in (1) in which all the particles rotate in the same direction, admit the evident generalization [21]

$$f_0 = \frac{\rho_0}{\pi} \frac{\alpha \delta(v_r)\delta(v_\varphi - \Omega r) + \beta \delta(v_r)\delta(v_\varphi + \Omega r)}{\sqrt{1 - r^2 - z^2/c^2 - v_z^2/\omega_0^2 c^2}} \qquad (\alpha + \beta = 1). \quad (40)$$

This distribution function corresponds to the superposition of two Freeman's ellipsoids (with weights α and β) rotating in the opposite directions.

Investigation of the stability of such systems, in principle, presents no problems. However, the dispersion equations become somewhat more cumbersome (in particular, the order of these equations increases), and we do not consider them here.

Then model (40) is convenient for the investigation of the stabilization effect with a decrease in angular velocity of rotation of the system $|\alpha - \beta|\Omega$. Of most interest are the "barlike" perturbations (ellipsoid of rotation–three-axis ellipsoid) since they generally are the most unstable (for more detail, see next section).

It may be shown (Problem 7) that the ellipsoids (as a whole!) at rest are stable with respect to such perturbations. In particular, this is true also for the model in question, (40), at $\alpha = \beta$. Consequently, there must exist the boundary velocity of rotation (or the angular moment) such that the more slowly rotating systems are stable with respect to their conversion into the three-axis ellipsoid (see the next subsection).

On the other hand, the thickness exerts a stabilizing effect on Jeans instability (cutting into rings with axial-symmetrical perturbations of a strongly flattened "cold" system). Such perturbations (at $m = 0$) should be independent of the velocity of the system revolution $|\alpha - \beta|\Omega$. Thus, in this vein it is possible to obtain estimates (from above) of the critical oblateness c/a separating the stable and unstable ellipsoidal systems. In particular, from an asymptotic formula (35) such an estimate yields, as we see, $c \approx 0.23$. It is interesting that the exact calculation [45^{ad}] for the Jeans mode with one radial node leads to the same value of c_{cr}.

3.2 Peebles–Ostriker Stability Criterion. Stability of Uniform Ellipsoids, "Hot" in the Plane of Rotation

Rather probable is the hypothesis of Peebles and Ostriker [301], according to which stability or instability of an isolated axial-symmetrical system is defined by the ratio $t = T_{rot}/|W|$, where T_{rot} is the kinetic energy of rotation (in the inertial system of reference) and W is the potential energy. From the virial theorem it follows that $0 \leq t \leq \frac{1}{2}$.

The critical value of the parameter t for strongly flattened systems is defined by the authors of the hypothesis as

$$t_{cr} \approx 0.14 \pm 0.03. \quad (1)$$

If the system is so "hot" that $t < t_{cr}$, then, according to this hypothesis, *strong* instabilities in it must be suppressed. We call "strong" those instabilities which produce essential rearrangement of the initial state for the time comparable to the period of system revolution. Weaker instabilities which do not lead to such dramatic consequences are possible also at $t < t_{cr}$.[18]

Isolated systems with $t > t_{cr}$ must be strongly unstable. Peebles and Ostriker tested their hypothesis in computer experiments by simulating the macroscopic gravitating system by the combination of several hundreds of points attracting according to Newton's law. Even in that case if *locally* the systems with $t > t_{cr}$ are stable, nevertheless, they rapidly reconstruct their shape with apparent contribution of the large-scale *barlike* mode and are "heated" (on account of the rotational energy) so that the parameter t decreases to the value $\sim t_{cr}$.

The beauty of such a simple and general hypothesis is evident. The hypothesis is formulated for any sufficiently flattened axial-symmetrical systems, both liquid (individual stars) and collisionless (stellar systems). Such generality is ultimately due to the importance of barlike perturbations both in the former and latter cases (and with approximately identical conditions of their excitation).

For the sequence of Maclaurin's ellipsoids

$$t = \tfrac{1}{2}[(3e^{-2} - 2) - 3(e^{-2} - 1)^{1/2}(\arcsin e)^{-1}], \qquad (2)$$

where e is the eccentricity of the meridional cross section. Along this sequence, t varies from 0 (at $e = 1$, disk) to $\tfrac{1}{2}$ (at $e = 0$, sphere), i.e., t takes on all possible values. Spheroids with $t > 0.1376$ are secularly[19] unstable (and at $t > 0.2738$ also dynamically unstable) with respect to the conversion to a three-axis ellipsoid. The point $t_{cr} = 0.1376$ ($e_{cr} = 0.81267$) is a known point of bifurcation of the sequence of Maclaurin spheroids with the three-axis ("barlike") Jacobi ellipsoids [15, 64] having a lower total energy (at the same angular moment and mass).

It is intriguing that this result proves not to be associated with the selection of a special artificial model (of Maclaurin's spheroids). A thorough investigation [167, 300] of liquid "stars" within a large range both of angular velocities and the degree of matter concentration toward the center showed that they become secularly unstable at nearly the same value of the parameter $t_{cr} = 0.137 \pm 0.002$. In all cases, instability is associated (as for Maclaurin's spheroids) with the presence beyond the critical point t_{cr} of barlike equilibria with lower total energy but identical angular moment, mass, and central density.

How much motivated is the hypothesis of Peebles and Ostriker (from now on, POH) in the case of collisionless systems? Numerical experiments performed in an original work [301] (cf. also [220]), though they do make a strong impression, nevertheless need a proof by the results of a strict

---

[18] See in this connection discussion in Section 2, Chapter XI.

[19] The concept of secular as well as dynamic instabilities will be discussed in Section 3.4.

theoretical consideration of the stability of rotating stellar systems (which is not a trivial problem at all). In this connection, it should be noted that the studies carried out so far (mainly by our group as well as by Kalnajs) *do not contradict* POH. For the use of disk systems we shall turn to this issue in the next chapter (Sections 4.4 and 4.5). Now let us recall some results already described earlier.

Above we have proved the instability of Freeman's rotational ellipsoids at any value of the ratio of the semiaxes c/a. The instability of Freeman's flattened ellipsoids is natural in terms of POH. Indeed, in this case

$$T_{rot} = \frac{M\Omega^2 a^2}{5},$$ (3)

and the potential energy for oblate ($c/a \leq 1$) spheroids

$$|W| = \frac{3GM^2}{5\sqrt{a^2 - c^2}} \arccos \frac{c}{a}.$$

Therefore,

$$t = \frac{\Omega^2 a^2 \sqrt{a^2 - c^2}}{3GM \arccos c/a}.$$ (4)

At $0 < c/a < 1$, this quantity varies from $\frac{1}{2}$ to $\frac{1}{3}$, which is much more than t_{cr}.

The stability of rotating uniform spheres with circular orbits proved in Chapter III (Section 3.3) is not surprising from this point of view as well, since in this case $t = \mu^2/12$ and the maximum value of the parameter t is $\frac{1}{12}$; it explicitly lies in the stable region.

Further, in Fig. 52 we present [45^{ad}] critical values of parameter t (corresponding to the stability boundary with respect to the barlike mode) for the simplest model in the form of superposition of two counter rotating Freeman ellipsoids. The latter is described in the item 6 of the preceding section. It is seen that the criterion (1) in this case holds even for sufficiently "thick" ellipsoids, $c/a \lesssim 0.5$.

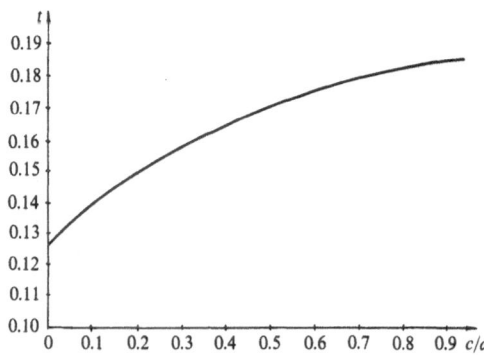

Figure 52. Dependence of the critical value of t for the superposition of two counter rotating Freeman's spheroids on the ratio of semi-axes c/a.

The criterion similar to (1) for a thin ellipsoid immersed into a "halo" was formulated in previous work [40^{ad}]. It turns out that criterion (1) is valid only in disk limit. Dependence on halo mass is rather strong, so that the error is not included within the limits given by (1). One is, however, able to formulate the criterion of excitations of the barlike mode (for systems of arbitrary thickness), which, however, does not admit universality with respect to the mass of the halo:

$$\tfrac{1}{2}u_\perp + \beta(\mu)u_z < \alpha_{cr}(\mu), \tag{5}$$

where $u_r = U_r/W, u_z = U_z/W, U_{r,z}$ is the kinetic energy of chaotic motion of particles on axes r, z, $W = |W_d| + 2U_0$ is the potential energy, and W_d is the energy of interaction of galaxy stars with each other, U_0 is the energy of interaction of stars with the halo, $\alpha_{cr} = \tfrac{1}{4}[1 - (4 - a)^3/27a^2]$, $a = \tfrac{3}{2}/(1 + 4\mu/3\pi)$. Here β depends only on the halo mass. The values of $\beta(\mu)$ for several values of μ are as follows: $\beta(0) = 0.69 \pm 0.01$, $\beta(0.2) = 0.93 \pm 0.03$, $\beta(0.4) = 1.26 \pm 0.07$. For $c = 0$, $u_z = 0$, $\mu = 0$ the criterion (5) is coincident with the criterion of Ostriker and Peebles. From (5), in particular it follows that the model becomes stable (even for a disk of zero thickness) for $\mu \approx 1.1$. For $c/a \approx \tfrac{1}{20}$ the model is stable if $\mu > 1.0$ (see Fig. 53).

The Peebles–Ostriker criterion specifies the natural expectation that the behavior of the simplest, the largest-scale perturbations involving the whole system is defined by a small number of "*global*" characteristics of the system, for example, of the first momenta of its stationary distribution function averaged over the phase volume occupied by the system. For homogeneous ellipsoids, even a more general statement is likely to be true: the character of the evolution of any perturbation of a given spatial type is completely described by the set of momenta of the distribution function of orders which are smaller than given ones, i.e. by the quantities

$$\alpha_{ij} = \iint d\mathbf{r}\, d\mathbf{v}\, (r_i)^{p_i}(v_j)^{p_j} f_0(\mathbf{r}, \mathbf{v}),$$

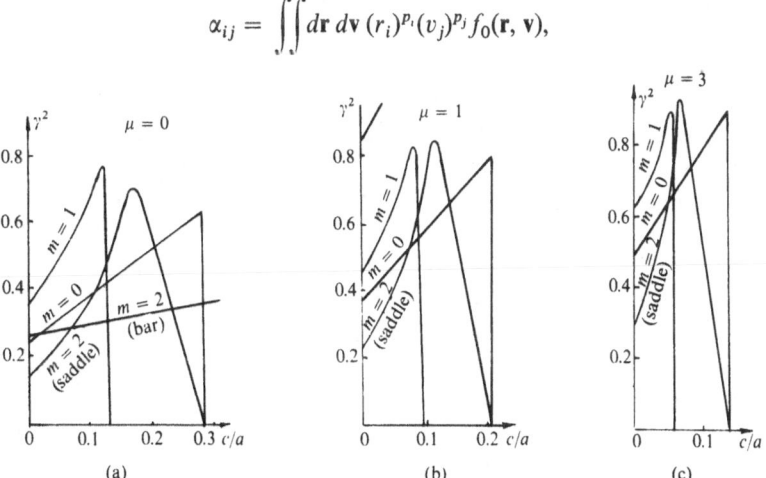

Figure 53. Regions of instability for the homogeneous spheroids immersed into a "halo" (for the mode "bar"—above the relevant curves, for the membrane modes—under the curves); (a) a halo mass $M_h = 0$, (b) $M_h = M_d$, (c) $M_h = 3M_d$.

at $p_i + p_j \leq N$. In concrete examples considered by us of ellipsoidal systems (two- and three-axis Freeman's ellipsoids) this follows already from the very possibility of the representation of the potential perturbations in the form of finite polynomials in degrees of Cartesian coordinates:

$$\Phi_1(x, y, z) = ax^n + by^n + cz^n + \cdots, \tag{6}$$

which determines, for each number n, a finite number of "degrees of freedom."

In a more direct way we become convinced of this in the example of disks rotating as a solid body (which are a limit form of uniform ellipsoids), for which, in Chapter V, equations for frequencies of *all* possible eigenoscillations will be obtained.

One may note also some "elementary" manifestations of the above-mentioned property of solutions. For example, the uniform spherical systems with arbitrary distribution functions (see Chapter III) at expansions–compressions always oscillate with the same frequency $\omega^2 = \Omega^2$, while the simplest nonaxial mode of oscillations (uniform sphere–uniform ellipsoid) has a frequency $\omega^2 = \frac{14}{5}\Omega_0^2$. Further, it may be shown (Problem 7) that, for arbitrary uniform spheroids being at rest in the equilibrium state, the frequencies of oscillations, at which the initial ellipsoid transforms to the uniform ellipsoid with other semiaxes, are completely determined only by its density and geometric dimensions.

In connection with the above, for example, the investigation of the large-scale stability of the model of bars of spiral galaxies of Freeman in the form of three-axis collisionless ellipsoids (§1), carried out in §2, acquires more significance. Indeed, the results must not change very much at some modification of the distribution function considered. since then we are to preserve the basic property of such systems probably supported by observations (see §3, Chapter XI): the presence in them of "counterflows" (see §1) which is determined by the first moments of the distribution function and on which depends the character of stability of large-scale modes in this case.

It is possible that the behavior of the largest-scale modes in the systems not possessing the axial symmetry (and not necessarily very flattened) is also defined by a certain simple criterion that generalizes the criterion of Peebles and Ostriker (POH).

The applicability of this criterion (in its literal formulation) to the case of Freeman's elliptical disks was considered by Tremaine [335]. It turned out (cf. Fig. 49) that the instability region boundary is predicted by this criterion surprisingly exactly. The value of the parameter t increases monotonically along the boundary from $(125/972) = 0.1286$ at $b/a = 1, \Omega^2/A^2 = \frac{5}{6}$ to 0.1446 at $b/a = 0.7296$, $\Omega^2/A^2 = 1$ remaining all the time within the limits given by Peebles and Ostriker for axially symmetrical systems. However, on the other hand, as follows, for example, from the contour map in Figure 54 (taken by us from [234, Hunter]), at $b/a \to 0$ and $\Omega^2/A^2 \to 1$ the parameter $t \to 0.5$, although these systems are stable with respect to modes considered here. A similar result is known also in the stability theory of liquid equilibrium figures where it is proved [148] that the Jacobi ellip-

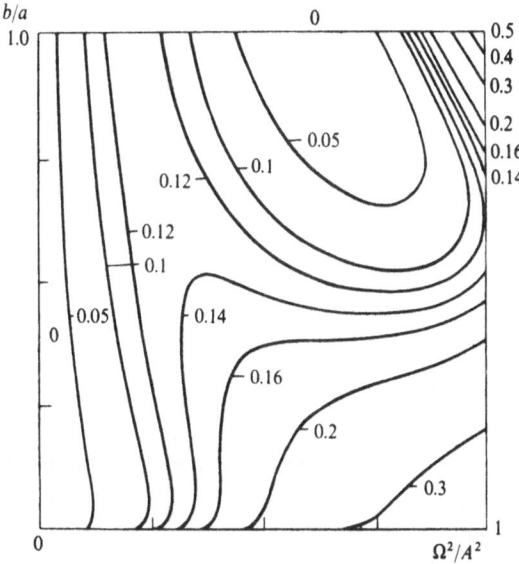

Figure 54. The lines of the constant value of the Peebles–Ostriker parameter t for the Freeman elliptical disks [234].

soids are stable for perturbations with $n = 2$, although the parameter t leaves for the limits of the region indicated by POH ($0.1376 \le t \le 0.5$). Therefore, it may be argued [335] that this criterion seems to be applicable to systems close to the axially symmetrical ones but is apparently violated for very elongated systems.

To conclude this section, consider the problem of the stability of uniform ellipsoids of revolution "hot" in the plane of rotation (§1).

For models rotating as a solid body [(72), §1] this question must be solved generally by accurate methods: the Peebles–Ostriker instability criterion $t = T_{\text{rot}}/|W| > 0.14 \pm 0.03$ here is too rough. For example, for the disklike systems $t = \gamma^2/2$, so for $\gamma = \gamma_{\max} = \frac{1}{2}$ we have $t_{\max} = 0.125$. The uniformly-rotating ellipsoids in all likelihood are stable. This is in any event true for "barlike" modes (which generally are more unstable). Indeed, the stability of uniformly rotating disks may be investigated in the general form without specifying the form of the distribution function (cf. Section 4.4, Chapter V). Either stability or instability of any such model is defined by the value of the single parameter: the angular velocity of rotation γ. If $\gamma < 0.507\ldots$, the model is stable. Since for the disk models under consideration $\gamma \le 0.500$, they must be stable. Moreover, all systems (72), §1, with *finite thickness*, $c \ne 0$, must be stable. If one, for example, calculates the parameter t for the maximally possible velocity of rotation $\gamma = \frac{1}{2}\Omega$, then it happens that

$$t_{\max}(e) = \frac{1}{8}\left(\frac{1}{e^2} - \frac{\sqrt{1 - e^2}}{e \arcsin e}\right),\qquad (7)$$

where $e = \sqrt{a^2 - c^2}/a$ is the eccentricity of the meridional cross section of the ellipsoid. The $t_{max}(e)$ function decreases monotonically from the value $t = \frac{1}{8}$ at $e = 1$ (disk limit) to $t = \frac{1}{12}$ at $e = 0$ (sphere).

Thus, the uniformly rotating models are likely to be stable; however, the "reserve of stability" decreasing with increasing degree of system flattening, becomes quite insignificant in the disk limit.

The differentially rotating model (67), §1, according to the Peebles–Ostriker criterion, must be unstable.

3.3 The Fire-Hose Instability of Ellipsoidal Stellar Systems [37^{ad}]

3.3.1. Linear Theory of Stability. Below we consider the stability of the stellar ellipsoid of revolution[20] with respect to bendings of its equatorial plane, taking into account the possible influence of the nonflat system's components ("halo"). The latter, due to increasing of the velocity dispersion at the plane of equator, act on the system as a destabilizing factor.

Thus, we shall investigate the stability of the system in the form of a spheroid with homogeneous density (with the semiaxes a and c) of stars, immersed into an extended halo of a spherical shape which is also homogeneous. The gravitational potential inside the spheroid is

$$\Phi_0 = \tfrac{1}{2}\Omega^2(x^2 + y^2) + \tfrac{1}{2}\omega_0^2 z^2.$$

It involves the halo potential and the potential of the spheroid itself: $\Omega^2 = GM_h/a^3 + \Omega_d^2$; $\omega_0^2 = GM_h/a^3 + \omega_d^2$; G is the gravitational constant, and Ω_d^2 and ω_d^2 are dependent on the ellipsoid density and on the values of its semiaxes (explicit expressions for Ω_d^2 and ω_d^2 may be found, for example, in §1, Chapter IV). The momenta of the distribution functions, which will be needed later, are of the form

$$\overline{v_r^2} = \tfrac{1}{2}(1 - \gamma^2)\left(1 - \frac{r^2}{a^2} - \frac{z^2}{c^2}\right), \qquad \overline{(v_\varphi - \overline{v}_\varphi)^2} = \overline{v_r^2},$$

$$\overline{v}_\varphi = \Omega\gamma r, \qquad \overline{v_z^2} = \tfrac{1}{2}\omega_0^2 c^2\left(1 - \frac{r^2}{a^2} - \frac{z^2}{c^2}\right). \tag{1}$$

Here the parameter γ is the ratio of the angular velocity of the centroid to the circular velocity, $\gamma^2 < 1$; r, φ, z are the cylindrical coordinates; the overbar means an average over the distribution function of the system.

Let us restrict ourselves to the consideration of the following modes (Φ_1 is the potential perturbation):

(1) $m = 0$ mode ("bell"):

$$\Phi_1 = z(az^2 + br^2 + d) \tag{2}$$

---

[20] The fire-hose instability of the *prolate* spheroids is considered in Problem 9 of this chapter.

(2) $m = 1$ mode:

$$\Phi_1 = z(x + iy)(ar^2 + bz^2). \tag{3}$$

(3) $m = 2$ mode ("saddle"):

$$\Phi_1 = az(x + iy)^2. \tag{4}$$

The terms "bell" and "saddle" are connected with the perturbation shape of the ellipsoid symmetry plane, the coefficients a, b, d are dependent only on a time $(\sim e^{-i\omega t})$.

These modes are interesting, first of all, as the largest-scale ones and, therefore, as the most "dangerous" from the point of view of stability loss. In addition, one is to expect that the behavior of similar, largest-scale modes involving the whole system is defined by a few of its "global" characteristics. Therefore, the results ensuing from the consideration of the simplest homogeneous models should not alter very much as the other more complicated models are considered. On the other hand, the investigation of the stability with respect to the modes listed above does not require specification of the distribution function of the system: it is quite enough to know the momenta given above, (1). For the homogeneous ellipsoids under consideration, the expressions for the momenta in (1) are determined unambiguously, if only one assumes the isotropy of velocity dispersion in the rotation plane. This last assumption is in any event valid for systems with orbits close to circular ones (cf. §1, Chapter V).

For the anisotropic model of the ellipsoid "hot" in the rotational plane (considered in §1, Chapter IV) very similar results are obtained.

The technique of investigation of stability is standard. First of all, from the equations for Lagrange displacements X, Y, Z, these values are expressed through the perturbed potential Φ_1. Then the density perturbation ρ_1 is calculated as well as the normal boundary displacement. Solving further the Poisson equation and comparing the potential obtained with the initial one (2)–(4), we arrive at the system of linear equations with respect to coefficients that enter the potential. The zero-equality of the determinant of this system yields the sought-for dispersion equation.

The dispersion equations for the above-listed modes (2)–(4)[21] were solved numerically. The results are given in Fig. 53 (a–c).[22]

In Fig. 53(a–c), on the abscissa axis, the oblateness of the spheroid c/a, on the ordinate axis, the value of γ^2 (which characterizes the contribution of the centrifugal force to the equilibrium) are marked. The value $(1 - \gamma^2)$ is proportional to the stellar velocity dispersion in the rotational plane. The different curves in Fig. 53 correspond to different values of the parameter $\mu = M_h/M_d$, the ratios of the halo mass to the spheroid mass. From Figure 53(a–c), it is easy to see that, in the disk limit, each mode is unstable for

---

[21] These equations are quite complex and are not given here.

[22] In Fig. 53, small regions of instability near $\gamma^2 = 1$ and of resonance frequencies, which are not essential for further considerations, are omitted.

sufficient large stellar velocity dispersion. With increasing thickness of the spheroid, the instability region first increases, i.e., still "colder" systems become unstable, and then vanishes quite rapidly. The most essential effect is the dependence of the instability region boundaries on the halo mass: for a sufficiently large halo mass, systems with a small velocity dispersion in the plane may become unstable. This implies just the destabilizing role of the halo.

Note that the sharp decrease of the curves occurs at such a thickness, for which in the homogeneous model used there are resonances between the stellar oscillation frequencies with respect to z and the frequencies in the rotational plane (x, y). For the modes $m = 0$ and $m = 2$, there is resonance $\omega_0 = 2\Omega$; for the $m = 1$ mode, the resonance is $\omega_0 = 3\Omega$. Such a strong influence of these resonances is due, of course, to the idealization of the model. For the real nonhomogeneous systems, the curves will have a smoother form.

It is important to note that for $(c/a) > (c/a)_{cr}$, including that sufficiently far from the boundary (where the influence of the resonances must be weak), instability no longer appears.

For elliptical galaxies, in which rotation is weak, one may, from the condition of stability with respect to the bendings, determine the value of maximum oblateness of c/a. For a mass of the halo equal to zero, this value [cf. Fig. 53(a)] is of the order of $c/a \simeq 0.3$. With increasing value of M_h, the lower boundary of c/a decreases.

Putting off further discussion of possible applications of the stability theory of stellar ellipsoids (including numerical estimates of halo or latent mass influence) till §2, Chapter X, we now turn to the investigation of non-linear evolution of initial disturbances by "the method of particles."

3.3.2. Simulation of Nonlinear Evolution of Perturbations in the Homogeneous Non-rotating Ellipsoids. Here we give a short description of the stability investigation of anisotropic models (64), §1, performed by "the method of particles" (an account of the simulation technique is given in §3, Appendix).

In Fig. 55 we represent numbers of particles in cylindrical layers for one of the typical realizations of the model (with $N = 150$, the ratio of semiaxes $b/a = 0.5$, the cutoff radius $c = 0.05$) and, in addition, for the exact equilibrium.

Evolution of this model (with an artificially introduced large-scale bending disturbance of the "bell" type) is demonstrated in Fig. 56. For the ellipsoid with the ratio of semiaxes $b/a = 0.06$ (cutoff radius was assumed as $c = 0.02$) the evolution of perturbations of the same kind as above is depicted in Fig. 57.

It is seen that, in the first case, the system changes insignificantly during a counting time (which was, in the given example, equal to one period of particle oscillations at the equatorial plane, $T = 2\pi/\Omega_0$) while, in the second case, the fast development of the instability occurs. We emphasize that it is just the instability (i.e., a collective phenomenon) but not a consequence of

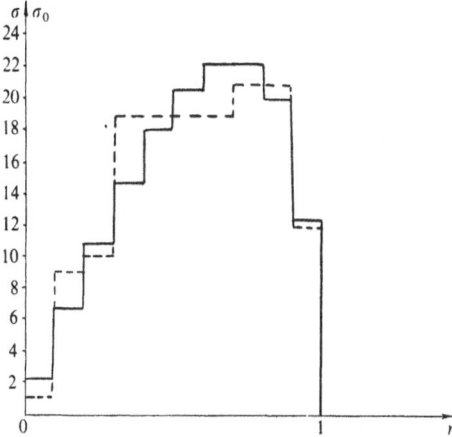

Figure 55. Particle numbers in the cylindrical layers for one of the model realizations (σ, the solid line) and for the exact equilibrium (σ_0, the dotted line) of the "hot" spheroid.

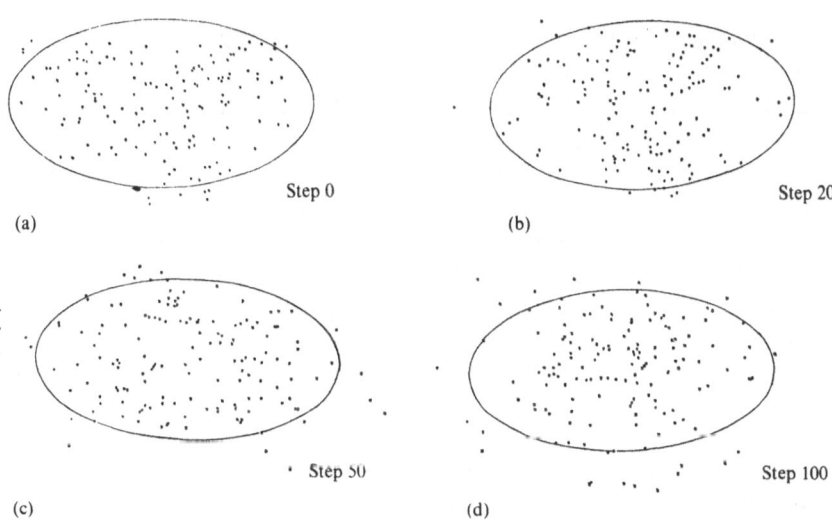

Figure 56. Projections of points-stars' locations at different times, for the computer model of the homogeneous spheroid with $b/a = 0.5$.

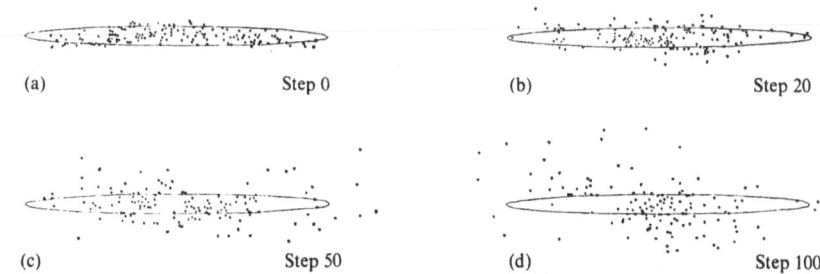

Figure 57. Projections of points-stars' locations at different times, for the computer model of the homogeneous spheroid with $b/a = 0.06$.

collisions since the latter doesn't manifest for such short time. Detailed description of the relevant computer experiments can be found in [36^{ad}]. There, the evolution both models obtained by pure-statistical playing and models with additional artificial membrane-like perturbations of various types ("bell," "saddle") are considered. The counting time was within the limits $t = 1 \div 3$ of the period T (outside these limits, $t > 3T$, the collision effects become to act). Seemingly, the most interesting result of these considerations consists of that it turns out that the evolution of perturbations, in the strongly flattened systems (which are, consequently, strongly unstable), drive these systems in the quasi-stationary state corresponding to the boundary of the fire-hose instability.

3.4 Secular and Dynamical Instability. Characteristic Equation for Eigenfrequencies of Oscillations of Maclaurin Ellipsoids

In dealing with collisionless systems, we consider only the special kind of instability which, according to the classification of the classical theory of stability of liquid equilibrium figures is called the *dynamical* or *ordinary* instability [15]. These terms distinguish this kind of instability (in the case of "purely" collisionless systems it is unique from the so-called *secular* instability due to dissipative mechanisms, above all viscosity.

The concepts of dynamical (based on analysis of equations of motion without friction) and secular instability in the general case, of course, do not coincide.[23] For the secular stability, it is necessary that the equilibrium configuration in question possess a minimum complete (mechanical) energy as compared to closed states of the system. The classical example is the behavior of Maclaurin's ellipsoids with their gradual compression: the secular instability occurs significantly earlier than the dynamical one.[24]

This issue was clarified by Poincaré (cf. [15]), who had shown that at the point of bifurcation with the sequence of Jacobi ellipsoids, Maclaurin's spheroid may be deformed to the body with a lower total mechanical energy (preserving, however, its full angular moment).

Dissipative mechanisms manifest, evidently, for the times of the order (or more) of the time between collisions. For most stellar systems which are practically collisionless, these mechanisms and the possibility of secular instability connected with them may be neglected.[25]

The dynamical stability may remain in spite of the loss of secular stability.

---

[23] Although the cases of coincidence are not rare (rather, vice versa).

[24] Accordingly, at eccentricities of the meridional cross section equal to 0.813 and 0.953. Then the secular stability comes over to three-axis ellipsoids of Jacobi.

[25] They may possibly play some role in the case of galactic stellar clusters. Note, however, that for the nonrotating systems such as, evidently, most of these clusters, the criteria of ordinary and secular stabilities coincide.

The dispersion equation for eigenfrequencies of oscillations of nonviscous Maclaurin's ellipsoids was derived by Bryan [172]. It gives the oscillation frequencies ω under the perturbations of the potential of the form

$$\Phi_1 \sim \begin{cases} p_n^m(\zeta)P_n^m(\mu)e^{im\varphi}, & \zeta \leq \zeta_0, \\ q_n^m(\zeta)P_n^m(\mu)e^{im\varphi}, & \zeta \geq \zeta_0, \end{cases} \tag{1}$$

where ζ, μ, φ are oblate spheroidal coordinates, the inequality $\zeta < \zeta_0$ corresponds to the internal region of the system. Perturbations of the type of (1) arise evidently as a result of distortions of the boundary surface of the ellipsoid.

In a system rotating together with the ellipsoid (angular velocity Ω) the dispersion equation can be reduced to the form

$$p_1^0(\zeta)q_1^0(\zeta) - p_n^m(\zeta)q_n^m(\zeta) \tag{2}$$

$$= \frac{\omega^2(\omega + 2\Omega)}{\Omega^2} \frac{q_2^0(\zeta)\hat{D}^m P_n(v_0)}{m\omega\hat{D}^m P_n(v_0) + (\omega + 2\Omega)v_0(1 + 1/\zeta^2)\hat{D}^{m+1}P_n(v_0)}.$$

The following notations are introduced here: $P_n(z)$ are the Legendre polynomials, $p_n^m(z) = i^{m-n}P_n^m(iz)$; $P_n^m(z)$ are the associated Legendre functions,

$$q_n^m(z) = p_n^m(z) \int_z^\infty \frac{dt}{(1 + t^2)[p_n^m(t)]^2}, \qquad v_0 = -\frac{\omega\zeta}{\sqrt{4\Omega^2(1 + \zeta^2) - \omega^2}};$$

$\hat{D}$ is the operator of differentiation.

From this equation, one may define the minimum value of the ratio c/a (or the corresponding eccentricity of the meridional cross section $e = \sqrt{a^2 - c^2}/a$), when stability is lost.

The most unstable mode ($n = m = 2$) corresponds to the compression of the spheroid in the plane of rotation with its conversion to the three-axis ellipsoid; for this mode $e_{cr} = 0.95289$.

In the case of collisionless ellipsoids of revolution (for example, of Freeman's model) there are also the modes of oscillations of the "incompressible" type, at which the local density of matter remains unchanged. They are described by dispersion equation (30) of Section 3.1. The latter is, however, essentially different from the corresponding equation for Maclaurin spheroids [i.e., equation (2) at $n = m$]. For example, as we saw, Freeman's model proves to be unstable with respect to the barlike mode ($n = m = 2$) for all the values of c/a. This is not surprising: even equilibrium states in the two cases in question are quite different. Indeed, for Maclaurin spheroids with finite thickness ($c \neq 0$), an essential role is played by the pressure which equals zero in "cold" collisionless ellipsoids. Equilibria are identical only in the disk limit ($c = 0$). In this limit also dispersion equations are coincident (see Section 2.2, next chapter).

In conclusion of the present section devoted to stability investigation of collisionless ellipsoids of revolution, we indicate a gap which exists, up to now, in the total picture of stability of the such systems. In the sufficiently oblate ellipsoids with primarily *radial* stellar trajectories there must obviously develop (apart from fire-hose) Jeans instability for non-axial-symmetrical large-scale perturbations. This instability is analogous to the instability of anisotropic spherical systems considered in more detail in §§5 and 6 of Chapter III. It is also necessary to derive relevant *quantitative stability criteria* for highly oblate systems. Since the rotation of the elliptical galaxies is slow (see §2, Chapter X), "usual" Jeans instability (which leads to ring splitting of the system) cannot occur; hence only the fire-hose instability and "slanting" Jeans instability mentioned above are competing. These two instabilities alone lead to certain restrictions on the possible parameters of E-type systems.

The investigation of real models of *inhomogeneous* elliptical systems is more interesting. However, the construction of corresponding exact models is very difficult. Therefore, one can use as a first step the approximate method similar to that which we applied for the examination of "thin" systems' instability in Chapters I and II. In this way a complicated problem of the investigation of perturbed motions in the three-dimensional system may be essentially reduced to a much simpler, two-dimensional problem. Moreover, one may use all numerous phase models of the disklike systems known at present.

Problems

1. Derive distribution function (47), §1, of the uniform elliptical cylinder (Freeman [202] and Hunter [234]).

Solution. The movement of particles in the rotating frame of reference (in which the equation of the system boundary remains unaltered) occurs in accordance with formula (44), §1. The orbits are (cf. Fig. 58) elliptical epicycles, and the β-movement occurs always counterclockwise, while the α-movement is clockwise (assume that $\Omega^2 < A^2$). The orbits must evidently lie in the limits of the boundary ellipse. Accordingly, let us calculate [234] the maximum of the value

$$e(x, y) = \frac{x^2}{a^2} + \frac{y^2}{b^2} \tag{1}$$

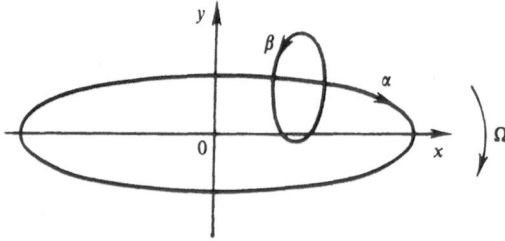

Figure 58. A particle orbit in the Freeman elliptical cylinder.

under additional conditions

$$g_1(x, y, \xi, \eta) = (x - \xi)^2 + \frac{(y - \eta)^2}{k_\alpha^2} - A_\alpha^2 = 0, \tag{2}$$

$$g_2(x, y, \xi, \eta) = \xi^2 + \eta^2/k_\beta^2 - A_\beta^2 = 0. \tag{3}$$

Here, the notations ξ and η for the β-movement are introduced:

$$\xi = A_\beta \sin(\beta t + \varepsilon_\beta), \qquad \eta = -k_\beta A_\beta \cos(\beta t + \varepsilon_\beta).$$

Introducing, as usual, the indefinite factors of Lagrange, λ_1 and λ_2, we seek extreme values of the $f \equiv (e + \lambda_1 g_1 + \lambda_2 g_2)$ function. Equating to zero the derivatives in x, y, ξ, η, we obtain

$$\frac{x}{a^2} = \lambda_1(\xi - x) = -\lambda_2 \xi, \tag{4}$$

$$\frac{y}{b^2} = \lambda_1(\eta - y) = -\frac{\lambda_2 \eta}{k_0^2}. \tag{5}$$

The possible solutions of system (4), (5) are:

(1)
$$x = y = 0, \qquad \lambda_1 = \lambda_2 = 0, \tag{6}$$

$$\xi^2 = \frac{k_\beta^2 A_\beta^2 - k_\alpha^2 A_\alpha^2}{k_\beta^2 - k_\alpha^2}, \qquad \eta^2 = \frac{k_\alpha^2 k_\beta^2 (A_\alpha^2 - A_\beta^2)}{k_\beta^2 - k_\alpha^2},$$

(2)
$$x = \xi = 0, \qquad \eta = \pm k_\beta A_\beta, \qquad y = \eta \pm k_\alpha A_\alpha, \tag{7}$$

(3)
$$y = \eta = 0, \qquad \xi = \pm A_\beta, \qquad x = \xi \pm A_\alpha, \tag{8}$$

(4)
$$x = \frac{a^2(k_\beta^2 - k_\alpha^2)\xi}{b^2 - a^2 k_\alpha^2}, \qquad y = \frac{b^2(k_\beta^2 - k_\alpha^2)\eta}{k_\beta^2(b^2 - a^2 k_\alpha^2)}, \tag{9}$$

$$\xi^2 = \frac{1}{k_\beta^2 - k_\alpha^2}\left[\frac{k_\beta^2(b^2 - a^2 k_\alpha^2)^2 A_\alpha^2}{(a^2 k_\beta^2 - b^2)^2} - k_\alpha^2 A_\beta^2\right],$$

$$\eta^2 = \frac{k_\beta^4}{k_\beta^2 - k_\alpha^2}\left[A_\beta^2 - \frac{(b^2 - a^2 k_\alpha^2)^2 A_\alpha^2}{(a^2 k_\beta^2 - b^2)^2}\right].$$

In the case of (6), we have minimum, $e = 0$, in the other cases, maximum. The solution of (7) yields a minimum of e, which is reached on the y-axis:

$$e_{\max} = \frac{(k_\beta A_\beta + |k_\alpha| A_\alpha)^2}{b^2}; \tag{10}$$

and the solution of (8), on the x-axis:

$$e_{\max} = \frac{(A_\alpha + A_\beta)^2}{a^2}. \tag{11}$$

Finally, the solution of (9) considered by Freeman [202] yields a maximum, which may occur at a more general point and is

$$e_{\max} = J(A_\alpha, A_\beta) = (k_\beta^2 - k_\alpha^2)\left[\frac{A_\alpha^2}{a^2 k_\beta^2 - b^2} + \frac{A_\beta^2}{b^2 - a^2 k_\alpha^2}\right]. \tag{12}$$

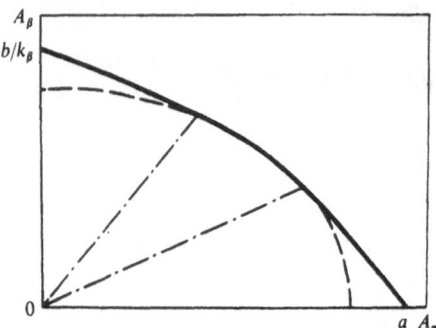

Figure 59. The region of the admissible values of oscillation amplitudes (A_α, A_β) for orbits that stay within the ellipse with $b/a = 0.7$ and $\Omega^2/A^2 = 0.8$ [234]. The solid line marks the region's boundary. The dashed lines mark the continuation of the ellipse $J = 1$, the dotted lines mark the boundary separating the regions in the inequalities (13)–(15) [234].

Since $k_\beta > 1$, always $a^2 k_\beta^2 > b^2$; moreover, it is necessary that the inequality $b^2 > a^2 k_\alpha^2$ be satisfied. In fact, otherwise (10) is the largest from (10)–(12), but it is impossible to fill up the ellipse by the orbits of the particles if the orbits which may reach the boundary do it necessarily at the ends of the minor axis.

Under these conditions, the maximum value of e_{max} can be found by comparing (10)–(12); it is equal to the expressions

(10),   for $k_\alpha A_\beta (a^2 k_\beta^2 - b^2)/k_\beta > (b^2 - a^2 k_\alpha^2) A_\alpha,$ (13)

(11),   for $(b^2 - a^2 k_\alpha^2) A_\alpha > (a^2 k_\beta^2 - b^2) A_\beta,$ (14)

(12),   for $(a^2 k_\beta^2 - b^2) A_\beta > (b^2 - a^2 k_\alpha^2) A_\alpha > \dfrac{k_\alpha A_\beta (a^2 k_\beta^2 - b^2)}{k_\beta}.$ (15)

The range of values of the two amplitudes A_α and A_β admissible at $x^2/a^2 + y^2/b^2 \leq 1$ is represented in Fig. 59; it is confined by the segments of the lines

$$k_\alpha A_\alpha + k_\beta A_\beta = b, \qquad A_\alpha + A_\beta = a,$$

and by part of the ellipse $J = 1$; this boundary represents orbits which reach the external boundary of the elliptical cylinder.

Construct now the self-consistent solution of the collisionless kinetic equation. Let us seek the distribution function of the form

$$f_0 = f_0(A_\alpha, A_\beta). \tag{16}$$

The density must be equal to the integral

$$\iint f_0(A_\alpha, A_\beta)\, dv_x\, dv_y = \rho_0 \theta \left(1 - \frac{x^2}{a^2} - \frac{y^2}{b^2} \right). \tag{17}$$

The latter equality is the two-dimensional integral equation of the first kind for f. We shall seek, after Freeman [202], the solution[26] in the form of $f = f(J)$, where J, defined

---

[26] In a more general case (16), the range of integration in the (A_α, A_β) plane (after respective change of the integration variables) has a very complicated form.

by Eq. (12), can be reduced to the form

$$J = \frac{x^2}{a^2} + \frac{y^2}{b^2} + \Lambda^2(a^2\tilde{c}_y^2 + b^2\tilde{c}_x^2/k_\alpha^2 k_\beta^2), \tag{18}$$

$$\tilde{c}_x = \dot{x} + \frac{\theta y}{b^2}, \qquad \tilde{c}_y = \dot{y} - \frac{\theta x}{a^2}, \tag{19}$$

$$\theta = \frac{\beta k_\beta(b^2 - a^2 k_\alpha^2) - \alpha k_\alpha(a^2 k_\beta^2 - b^2)}{k_\beta^2 - k_\alpha^2}, \tag{20}$$

$$\Lambda^2 = \frac{4\Omega^2(k_\beta^2 - k_\alpha^2)^2}{(a^2 k_\beta^2 - b^2)(b^2 - a^2 k_\alpha^2)(\beta^2 - \alpha^2)^2}. \tag{21}$$

Then, although, as follows from Fig. 59, $J > 1$ for some orbits which do not leave the limits of the boundary ellipse as yet, all orbits are properly confined if the requirement

$$f(J) = 0 \qquad \text{for } J > 1 \tag{22}$$

is imposed.

This requirement confines the points (x, y) by the values

$$\frac{x^2}{a^2} + \frac{y^2}{b^2} \le J \le 1. \tag{23}$$

The integral in (17) is now readily calculated by integrating over $\tilde{c}_x$, $\tilde{c}_y$; then the range of integration is confined by the ellipse:

$$0 \le a^2\tilde{c}_y^2 + \frac{b^2\tilde{c}_x^2}{k_\alpha^2 k_\beta^2} \le \frac{1 - x^2/a^2 - y^2/b^2}{\Lambda^2}. \tag{24}$$

The integral of (17) over velocities can, for $f = f(J)$, be presented in the form

$$\frac{\pi k_\alpha k_\beta}{ab\Lambda^2} \int_{x^2/a^2 + y^2/b^2}^{1} f(J)\, dJ. \tag{25}$$

Hence for the cylinder with $\rho = $ const we get the sought-for solution

$$f = \frac{\rho ab\Lambda^2 \delta(1 - J)}{\pi k_\alpha k_\beta}. \tag{26}$$

2. Derive the distribution function of (48), §1, of the elliptical disk with the surface density

$$\sigma_0 = \frac{3M}{2\pi ab}\left(1 - \frac{x^2}{a^2} - \frac{y^2}{b^2}\right)^{1/2}$$

(Freeman [204]).

Solution. The properties of particle orbits in this case do not differ formally in anything from the previous problem; therefore, the integral of (25), of Problem 1, must be equated to the surface density,

$$\frac{\pi}{k_\alpha k_\beta ab\Lambda^2} \int_{x^2/a^2 + y^2/b^2}^{1} f(J)\, dJ = \frac{3M}{2\pi ab}\left[1 - \left(\frac{x^2}{a^2} + \frac{y^2}{b^2}\right)\right]^{1/2}. \tag{1}$$

By differentiating this equation, we arrive at (48).

3. Prove the uniqueness of Freeman's solution (48) for the case of the elliptical disk at rest among the distribution functions of the form $f_0 = f_0(A_\alpha^2, B_\beta^2)$ (Hunter [234]).

Solution. The movements of an individual particle along the x, y-axes in the case of the disk being at rest are split

$$x = A_\alpha \sin(At + \varepsilon_\alpha), \qquad y = -B_\beta \cos(Bt + \varepsilon_\beta). \tag{1}$$

If the frequencies α and β are incommensurable, the orbit of (1) fills up the "box" with edges in $(x = \pm A_\alpha, y = \pm B_\beta)$. This "box" lies within the limits of the ellipse boundaries if

$$J = \frac{A_\alpha^2}{a^2} + \frac{B_\beta^2}{b^2} \leq 1. \tag{2}$$

Further, it is easy to see that

$$\dot{x}^2 = A^2(A_\alpha^2 - x^2), \qquad \dot{y}^2 = B^2(B_\beta^2 - y^2), \tag{3}$$

and, finally, for the distribution function of the form of $f = f(A_\alpha^2, B_\beta^2)$ we have

$$\sigma = AB \int_{x^2}^{a^2(1 - y^2/b^2)} dA_\alpha^2 \int_{y^2}^{b^2(1 - A_\alpha^2/a^2)} \frac{f(A_\alpha^2, B_\beta^2)\, dB_\beta^2}{[(A_\alpha^2 - x^2)(B_\beta^2 - y^2)]^{1/2}}. \tag{4}$$

This integral equation for $f(A_\alpha^2, B_\beta^2)$ is the two-dimensional generalization of the Abel equation. The solution for this equation can be obtained by the method similar to that applied for solution of the ordinary Abel equation. By multiplying (4) by $[(x^2 - \lambda)(y^2 - \mu)]^{-1/2}$ (λ, μ are the parameters), integrating over x^2, y^2 in the range $x^2 \geq \lambda$, $y^2 \geq \mu$, $x^2/a^2 + y^2/b^2 \leq 1$, then, interchanging the order of integration, we obtain the following expression on the right-hand side of (4):

$$AB \int_\lambda^{a^2(1 - \mu/b^2)} dA_\alpha^2 \int_\mu^{b^2(1 - A_\alpha^2/a^2)} f\, dB_\beta^2$$

$$\times \int_\lambda^{A_\alpha^2} \frac{dx^2}{[(A_\alpha^2 - x^2)(x^2 - \lambda)]^{1/2}} \int_\mu^{B_\beta^2} \frac{dy^2}{[(B_\beta^2 - y^2)(y^2 - \mu)]^{1/2}}. \tag{5}$$

This expression simplifies very much, taking into account that the two internal integrals are equal to π, and we find instead of (5)

$$\pi^2 AB \int_\lambda^{a^2(1 - \mu/b^2)} dA_\alpha^2 \int_\mu^{b^2(1 - A_\alpha^2/a^2)} f\, dB_\beta^2 = \frac{3M}{2\pi ab} \int_\lambda^{a^2(1 - \mu/b^2)} \frac{dx^2}{(x^2 - \lambda)^{1/2}}$$

$$\times \int_\mu^{b^2(1 - x^2/a^2)} \frac{(1 - x^2/a^2 - y^2/b^2)^{1/2}}{(y^2 - \mu)^{1/2}}\, dy^2 = M\left(1 - \frac{\lambda}{a^2} - \frac{\mu}{b^2}\right)^{3/2} \theta\left(1 - \frac{\lambda}{a^2} - \frac{\mu}{b^2}\right). \tag{6}$$

Finally, differentiating (6) sequentially over λ and μ we shall obtain the sought-for solution in the form

$$f(\lambda, \mu) = 3M\left(1 - \frac{\lambda}{a^2} - \frac{\mu}{b^2}\right)^{-1/2} \theta\left(1 - \frac{\lambda}{a^2} - \frac{\mu}{b^2}\right) \frac{1}{4\pi^2 a^2 b^2 AB}, \tag{7}$$

or, otherwise,

$$f(A_\alpha^2, A_\beta^2) = 3M(1 - J)^{-1/2} \frac{\theta(1 - J)}{4\pi^2 a^2 b^2 AB}. \tag{8}$$

This solution coincides with that given by Freeman (48).

Note here that in a similar way one may prove the uniqueness of the solution of (47) for an elliptical cylinder.

4. Prove the lack of physically acceptable solutions of the form $f_0 = f_0(E_x, E_y, E_z)$ for ellipsoids at rest; E_x, E_y, E_z are the energies of movement of an individual particle, respectively along the x, y, z-axes (Hunter [234]).

Solution. The equations of movement of a star along the axes are split:

$$x = A_\alpha \sin(At + \varepsilon_\alpha), \qquad y = -B_\beta \cos(Bt + \varepsilon_\beta), \qquad z = C_\gamma \sin(Ct + \varepsilon_\gamma). \tag{1}$$

For the distribution function of the form $f = f(A_\alpha^2, B_\beta^2, C_\gamma^2)$

$$A_\alpha^2 \sim E_x, \qquad B_\beta^2 \sim E_y, \qquad C_\gamma^2 \sim E_z, \tag{2}$$

instead of Eq. (4) of Problem 3 we obviously have the following equation:

$$\rho_0 = ABC \int_{x^2}^{u^2} dA_\alpha^2 \int_{y^2}^{v^2} dB_\beta^2 \int_{z^2}^{w^2} \frac{f(A_\alpha^2, B_\beta^2, C_\gamma^2) \, dC_\gamma^2}{[(A_\alpha^2 - x^2)(B_\beta^2 - y^2)(C_\gamma^2 - z^2)]^{1/2}}, \tag{3}$$

where

$$u^2 = a^2\left(1 - \frac{y^2}{b^2} - \frac{z^2}{c^2}\right), \qquad v^2 = b^2\left(1 - \frac{A_\alpha^2}{a^2} - \frac{z^2}{c^2}\right), \qquad w^2 = c^2\left(1 - \frac{A_\alpha^2}{a^2} - \frac{B_\beta^2}{b^2}\right),$$

We multiply (3) by $[(x^2 - \lambda)(y^2 - \mu)(z^2 - \nu)]^{-1/2}$ and integrate over x^2, y^2, z^2 over the range $x^2 \geq \lambda$, $y^2 \geq \mu$, $z^2 \geq \nu$; $x^2/a^2 + y^2/b^2 + z^2/c^2 \leq 1$.

As in the previous problem, by changing the order of integration, etc., we obtain

$$\left(\frac{4\pi}{3}\right)\rho abc\left[1 - \frac{\lambda}{a^2} - \frac{\mu}{b^2} - \frac{\nu}{c^2}\right]^{3/2} \theta\left(1 - \frac{\lambda}{a^2} - \frac{\mu}{b^2} - \frac{\nu}{c^2}\right)$$

$$= \pi^3 ABC \int_\lambda^{a^2(1 - \mu/b^2 - \nu/c^2)} dA_\alpha^2 \int_\mu^{b^2(1 - A_\alpha^2/a^2 - \nu/c^2)} dB_\beta^2 \int_\nu^{c^2(1 - A_\alpha^2/a^2 - B_\beta^2/b^2)} f(A_\alpha^2, B_\beta^Y, C_\gamma^2) \, dC_\gamma^2. \tag{4}$$

Differentiating sequentially over λ and μ, we have

$$\frac{\rho c}{\pi^2 ab ABC}\left(1 - \frac{\lambda}{a^2} - \frac{\mu}{b^2} - \frac{\nu}{c^2}\right)\theta\left(1 - \frac{\lambda}{a^2} - \frac{\mu}{b^2} - \frac{\nu}{c^2}\right) = \int_\nu^{c^2(1 - \lambda/a^2 - \mu/b^2)} f(\lambda, \mu, C_\gamma^2) \, dC_\gamma^2. \tag{5}$$

Finally, differentiating over ν and renotating the variables, we get the solution in the form

$$f(A_\alpha^2, B_\beta^2, C_\gamma^2) = -\frac{\rho}{\pi^2 abc \, ABC} \frac{d}{dI}[(1 - I)^{-1/2}\theta(1 - I)]$$

$$= \frac{\rho}{2\pi^2 abc \, ABC}[-(1 - I)^{-3/2}\theta(1 - I) + 2(1 - I)^{-1/2}\delta(1 - I)], \tag{6}$$

where

$$I \equiv \frac{A_\alpha^2}{a^2} + \frac{B_\beta^2}{b^2} + \frac{C_\gamma^2}{c^2} = \frac{x^2}{a^2} + \frac{y^2}{b^2} + \frac{z^2}{c^2} + \frac{\dot{x}^2}{a^2 A^2} + \frac{\dot{y}^2}{b^2 B^2} + \frac{\dot{z}^2}{c^2 C^2}. \tag{7}$$

This solution is not acceptable physically since $f < 0$ everywhere at $I < 1$.

The cause lies in the excessive number of high-energy orbits necessary for the establishment of uniform density up to the system boundary.

The singularity of (6) at $I = 1$ is a sufficiently strong positive one to compensate negative values of f and to yield a positive full density at each point of space.

The above may be formulated in a somewhat different way. As is well known, the bounded gravitating systems satisfy the virial theorem

$$2T + U = 0, \tag{8}$$

where T and U is the total kinetic and potential energy of the system, respectively.

For the density of kinetic energy we have

$$\varepsilon = \frac{\rho v^2}{2} = \frac{3}{4}\rho\mu^2\left(1 - \frac{r^2}{a^2}\right), \tag{9}$$

where $\mu^2 = 4\pi G\rho a^2/3$, $r^2 \equiv x^2 + y^2 + z^2$. Therefore, the total kinetic energy is

$$T = \frac{3}{10}\frac{GM^2}{a}, \tag{10}$$

and the potential energy

$$U = -\frac{3}{5}\frac{M^2 G}{a}, \tag{11}$$

where G is the gravitational constant and M is the mass of the system. Thus, (8) holds true. However, the following paradox arises. As is seen from (9), the mean square of the velocity is $\frac{3}{2}\mu^2(1 - r^2/a^2)$; therefore, in the system, there must be particles with velocities exceeding the velocity of "escape" $v_0 = \mu$. Hence a conclusion may be drawn that there are no real ellipsoidal systems at rest of the form thus considered.

One of the essential suggestions [cf. (2)] is the dependence of the distribution function only on three integrals of motion A_α^2, B_β^2, C_γ^2. This assumption unambiguously defines the solution, and it appears not to be physical. If (as, for example, in [21]) one analyzes models depending on four integrals, we immediately obtain a larger variety of solutions, among which there are also ellipsoids at rest.[27]

Besides, the question remains as to the models of ellipsoidal systems with distribution functions depending on any other integrals of motion reflecting the specifics of the gravitational potential of such systems (for instance, models of the types of the Einstein models of spherically symmetrical systems).

5. Find distribution functions of the form (6) of Problem 4 for rotating ellipsoids. In which cases do the physically acceptable solutions correspond to them?

Solution. The equations of movement of a particle in the frame of reference rotating about the z-axis with the angular speed $(-\Omega)$ is

$$x = A_\alpha \sin(\alpha t + \varepsilon_\alpha) + A_\beta \sin(\beta t + \varepsilon_\beta),$$
$$y = k_\alpha A_\alpha \cos(\alpha t + \varepsilon_\alpha) - k_\beta A_\beta \cos(\beta t + \varepsilon_\beta). \tag{1}$$

All orbits will be confined by the given ellipsoid, if $I \le 1$, where now by I we denote the expression

$$I \equiv J + \frac{C_\gamma^2}{c^2} = \frac{x^2}{a^2} + \frac{y^2}{b^2} + \frac{z^2}{c^2} + \frac{z^2}{c^2 C^2} + \Lambda^2\left(a^2\widetilde{c_y^2} + \frac{b^2 \widetilde{c_x^2}}{k_\alpha^2 k_\beta^2}\right). \tag{2}$$

---

[27] In [21], this is simply the superposition of two Freeman's ellipsoids of (39), §1, rotating in opposite directions.

The solution of a special form $f = f(I)$, vanishing for $I > 1$, is possible if again

$$\rho = \frac{2\pi k_\alpha k_\beta cC}{ab\Lambda^2} \int_\zeta^1 (I - \zeta)^{1/2} f(I)\, dI \qquad (0 \le \zeta \le 1), \tag{3}$$

$\zeta \equiv x^2/a^2 + y^2/b^2 + z^2/c^2$. For the solution of the generalized Abel equation let us define the subsidiary function

$$h(I) = \int_I^1 f(I')\, dI'; \tag{4}$$

then (3) will reduce to the ordinary Abel equation

$$\int_\zeta^1 \frac{h(I)\, dI}{(I - \zeta)^{1/2}} = \frac{ab\Lambda^2\rho}{\pi k_\alpha k_\beta cC}. \tag{5}$$

The solution for this equation:

$$f(I) = -\frac{ab\Lambda^2\rho}{\pi^2 k_\alpha k_\beta cC} \frac{d}{dI} [(1 - I)^{-1/2}\theta(1 - I)]. \tag{6}$$

Since, according to (6), the orbits with $I < 1$ have negative "occupation numbers," it may be concluded that there are generally no collisionless uniform ellipsoids of the kind mentioned.

Physically reasonable solutions ensue from (6) only in two particular cases considered by Freeman: in the disk limit $c \to 0$ and in the case of the ellipsoid rotating with a maximum possible angular velocity $\Omega = A$. In the disk limit $c \to 0$ from (6) one can again obtain the Freeman solution (48), §1. Indeed, integrating (6) over all $\dot{z} > 0$ and z, we have

$$4 \int_0^{c(1-J)^{1/2}} dz \int_0^{cC(1-J-z^2/c^2)^{1/2}} dz\, f(I) = 2cC \int_0^{c(1-J)^{1/2}} dz \int_{J+z^2/c^2}^1 \frac{f(I)\, dI}{(I - J - z^2/c^2)^{1/2}}$$

$$= 2cC \int_J^1 f(I)\, dI \int_0^{c(I-J)^{1/2}} \frac{dz}{(I - J - z^2/c^2)^{1/2}}$$

$$= \pi c^2 C \int_J^1 f(I)\, dI = 3M\Lambda^2(1 - J)^{-1/2} \frac{\theta(1 - J)}{4\pi^2 k_\alpha k_\beta}, \tag{7}$$

which coincides with (48), §1.

The ellipsoidal solution of Freeman (28), §1, also can be attained from (6) at $k_\alpha \to 0$, $k_\beta \to \beta/2\Omega$,

$$\beta \to (B^2 + 3A^2)^{1/2}, \qquad \theta \to 2\Omega b^2, \qquad \Lambda \to \frac{1}{ab\beta\mu}, \qquad \mu^2 = 1 - \frac{4\Omega^2 b^2}{\beta^2 a^2}.$$

Let us define

$$I = J' + \frac{\tilde{c}_x^2}{\varepsilon^2}, \qquad \varepsilon = \frac{k_\alpha k_\beta}{b\Lambda}, \qquad \Lambda' = \frac{x^2}{a^2} + \frac{y^2}{b^2} + \frac{z^2}{c^2} + \frac{\tilde{c}_y^2}{b^2\beta^2\mu^2} + \frac{z^2}{c^2C^2}. \tag{8}$$

Then the sought-for distribution function is equal to the limit

$$f = -\frac{\rho}{\pi^2 b\beta cC\mu} \lim_{\varepsilon \to 0} \frac{1}{\varepsilon} \frac{d}{dI} [(1 - I)^{-1/2}\theta(1 - I)]. \tag{9}$$

It is clear that f will have singularity of δ-like type depending on $\tilde{c}_x$ at $\varepsilon \to 0$ for the finite range of variation of I restricts $\tilde{c}_x$ by the range $-\varepsilon \leq \tilde{c}_x \leq \varepsilon$. A simple way of calculating the limit is to multiply (9) by a "good" function $F(J', \tilde{c}_x)$ and to integrate over $\tilde{c}_x$, J' to give

$$\lim_{\varepsilon \to 0} \int_{-\varepsilon}^{\varepsilon} \frac{d\tilde{c}_x}{\varepsilon} \int_{0}^{1 - \tilde{c}_x^2/\varepsilon^2} dJ' \, F(J', \tilde{c}_x) \frac{\partial}{\partial J'} \left[\left(1 - J' - \frac{\tilde{c}_x^2}{\varepsilon^2}\right) \theta\left(1 - J' - \frac{\tilde{c}_x^2}{\varepsilon^2}\right) \right]. \tag{10}$$

Take the internal integral in (10) and assume that $\tilde{c}_x = \varepsilon s$; then instead of (10) we have

$$\lim_{\varepsilon \to 0} \left\{ - \int_{-1}^{1} \frac{ds \, F(0, \varepsilon s)}{(1 - s^2)^{1/2}} - \int_{-1}^{1} ds \int_{0}^{1 - s^2} \frac{\partial F(J', \varepsilon s)}{\partial J'} \frac{dJ'}{(1 - J' - s^2)^{1/2}} \right.$$

In the second integral, we invert the order of the integrations:

$$-\pi F(0, 0) - \pi \int_{0}^{1} \frac{\partial F \, (J', 0)}{\partial J'} dJ' = -\pi F(1, 0). \tag{11}$$

Hence it follows that (9) in the limit must be the product of two δ-functions:

$$-\pi \delta(1 - J') \delta(\tilde{c}_x). \tag{12}$$

This is the familiar Freeman distribution function of (28), §1.

Quite similar difficulties are encountered also in the endeavour to construct collisionless analogs of the Maclaurin and Jacobi ellipsoids if one seeks respective phase distributions depending only on the energy E of a particle in a rotating system:

$$E = \frac{v^2}{2} + (A^2 - \Omega^2) \frac{x^2}{2} + (B^2 - \Omega^2) \frac{y^2}{2} + \frac{C^2 z^2}{2}. \tag{13}$$

For these ellipsoids, the condition

$$(A^2 - \Omega^2)a^2 = (B^2 - \Omega^2)b^2 = C^2 c^2 \equiv \mu^2 \tag{14}$$

is fulfilled, so that

$$E = \frac{v^2}{2} + \frac{\mu^2}{2} \left(\frac{x^2}{a^2} + \frac{y^2}{b^2} + \frac{z^2}{c^2} \right). \tag{15}$$

Then, for the distribution function of the form $f_0 = f_0(J)$, where

$$J = \mu^2 - 2E, \tag{16}$$

we obtain the Abel integral equation

$$\frac{\rho_0 \theta(\varkappa)}{2\pi} = \int_{0}^{\varkappa} f(J)\sqrt{\varkappa - J} \, dJ, \qquad \varkappa \equiv \mu^2 \left[1 - \left(\frac{x^2}{a^2} + \frac{y^2}{b^2} + \frac{z^2}{c^2} \right) \right]. \tag{17}$$

The solution of this equation

$$f_0(J) = \frac{\rho_0}{\pi^2} \frac{d}{dJ} [J^{-1/2} \theta(J)] \tag{18}$$

involves all the above-noted difficulties characteristic of the distribution function in (6).

6. Investigate the conditions of the existence of ellipsoidal solutions with the simplest picture of movements of particles in the plane of rotation (x, y)—in elliptical orbits similar to and concentric with the boundary ellipse; along the z-axis, the particles perform harmonic oscillations.

Solution. Denote the frequency of internal circulation by v and the angular velocity of revolution of the ellipsoid itself, by Ω. In a rotating system, the equations of motion formally admit the solution of the form

$$x = ae^{ivt}, \qquad y = -ibe^{ivt}. \tag{1}$$

Accordingly, the distribution function

$$f_0 \sim \delta\left(v_x + \frac{av}{b}y\right)\delta\left(v_y - \frac{bv}{a}x\right)F(x, y, z, v_z). \tag{2}$$

Substituting (2) into the kinetic equation written in a rotating system (in Cartesian coordinates),

$$v_x\frac{\partial f}{\partial x} + v_y\frac{\partial f}{\partial y} + v_z\frac{\partial f}{\partial z} + (\Omega^2 x + 2\Omega v_y - A^2 x)\frac{\partial f}{\partial v_x}$$

$$+ (\Omega^2 y - 2\Omega v_x - B^2 y)\frac{\partial f}{\partial v_y} - C^2 z\frac{\partial f}{\partial v_z} = 0, \tag{3}$$

we obtain the following conditions for Ω and v:

$$a(\Omega^2 - A^2 + v^2) + 2\Omega vb = 0, \qquad b(\Omega^2 - B^2 + v^2) + 2\Omega va = 0, \tag{4}$$

as well as the equation for the F function

$$v_z\frac{\partial F}{\partial z} - C^2 z\frac{\partial F}{\partial v_z} = 0. \tag{5}$$

Note that the conditions in (4) could, of course, have been obtained also from the equations of motion

$$\ddot{x} = -\frac{\partial \Phi}{\partial x} + \Omega^2 x + 2\Omega\dot{y}, \qquad \ddot{y} = -\frac{\partial \Phi}{\partial y} + \Omega^2 y - 2\Omega\dot{x}, \tag{6}$$

if (1) is substituted into them. The solution of Eq. (5) is

$$F \sim \left(1 - \frac{x^2}{a^2} - \frac{y^2}{b^2} - \frac{z^2}{c^2} - \frac{v_z^2}{c^2 C^2}\right)^{-1/2},$$

so that the complete distribution function of the system thus described ensues formally in the following form:

$$f_0 = \frac{\rho_0}{\pi c C}\delta\left(v_x + \frac{av}{b}y\right)\delta\left(v_y - \frac{bv}{a}x\right)\theta\left(1 - \frac{x^2}{a^2} - \frac{y^2}{b^2} - \frac{z^2}{c^2} - \frac{v_z^2}{c^2 C^2}\right)$$

$$\times \left(1 - \frac{x^2}{a^2} - \frac{y^2}{b^2} - \frac{z^2}{c^2} - \frac{v_z^2}{c^2 C^2}\right)^{-1/2}. \tag{7}$$

It can, however, be shown that Eqs. (4) have reasonable solutions only in two cases: either for $a = b$, or as $c \to \infty$.

For the case of the ellipsoid of revolution $a = b$ from (6) follows the already familiar Freeman distribution function (39), §1.

Eliminating v from (4) and solving the equation obtained with respect to Ω^2, we find

$$\Omega^2 = \frac{(a^2 A^2 - b^2 B^2) \pm \sqrt{(a^2 - b^2)(a^2 A^4 - b^2 B^4)}}{2(a^2 - b^2)}. \tag{8}$$

Hence it follows that

$$aA^2 \geq bB^2 \tag{9}$$

ought to be.

Prove that at $c \neq \infty$ this condition is violated for all b/a apart from the trivial case $b/a = 1$ (ellipsoid of revolution).

Condition (9) may be rewritten in the following way ($\beta \equiv b/a$, $\gamma \equiv c/a$):

$$f(\beta, \gamma) \equiv \int_0^\infty \frac{(1 - \beta)(\beta - \lambda)\, d\lambda}{(1 + \lambda)^{3/2}(\beta^2 + \lambda)^{3/2}(\gamma^2 + \lambda)^{1/2}} \leq 0.$$

In reality, it happens that $f \geq 0$. This can easily be shown by writing

$$f = \int_0^\beta + \int_\beta^\infty$$

and transforming the second integral to the interval $(0, \beta)$ by means of a substitution $\lambda \to \lambda' = \beta^2/\lambda$. We obtain

$$f = \int_0^\beta \frac{d\lambda\, (\beta - \lambda)}{(1 + \lambda)^{3/2}(\beta^2 + \lambda)^{3/2}} \left[\frac{1}{(\gamma^2 + \lambda)^{1/2}} - \frac{\lambda^{1/2}}{(\beta^2 + \gamma^2\lambda)^{1/2}} \right].$$

But in the interval $(0, \beta)$

$$\frac{1}{\gamma^2 + \lambda} \geq \frac{\lambda}{\beta^2 + \gamma^2\lambda},$$

so that indeed $f \geq 0$. It is proved that the simplest conceivable model of the *three-axis* collisionless ellipsoid which is characterized by hydrodynamical movements of all particles in the plane of rotation (with trajectories, in the rotating system, concentric and similar to the boundary ellipse) does not exist.

In concluding, let us notice that for the elliptical cylinder the distribution function F in "longitudinal" velocities v_z, can be arbitrary, and

$$\Omega^2 = \frac{2\pi G\rho_0\, ab}{(a + b)^2}. \tag{10}$$

7. Prove the stability of a uniform spheroid at rest with respect to its conversion to the uniform ellipsoid with other values of semiaxes.

Solution.[28] The problem can be solved in the general form, without specifying the stationary distribution function f_0.[29] By denoting by (x, y, z) and (x_0, y_0, z_0), respectively, the current and initial Lagrange coordinates of the particles, we consider the perturbations of the form (cf., for instance, Problem 5, Chapter II)

$$x = x_0 a_1 + \dot{x}_0 b_1, \tag{1}$$

$$y = y_0 a_2 + \dot{y}_0 b_2, \tag{2}$$

$$z = z_0 a_3 + \dot{z}_0 b_3, \tag{3}$$

where $a_1, a_2, a_3, b_1, b_2, b_3$ are the unknown functions of time.

---

[28] Carried out with participation of I. G. Shukhman.

[29] Here, however, it is supposed that the substitution below [(1)–(3)] again yields the uniform ellipsoid.

The value of the perturbed semiaxis c is defined as a maximum of (3) on the z-axis provided that

$$z_0^2 + \left(\frac{\dot z_0}{\omega_0}\right)^2 = c_0^2,\tag{4}$$

which links z_0 and $\dot z_0$ (the integral of energy for the particles reaching the boundary of the ellipsoid). Here c_0 is the unperturbed semiaxis, and ω_0 is the frequency of oscillations of particles in the z-direction in the unperturbed potential.

As a result, we have

$$c^2 = (a_3^2 + b_3^2\omega_0^2)c_0^2.\tag{5}$$

In a similar way, other semiaxes are also determined:

$$a^2 = a_1^2 + b_1^2, \qquad b^2 = a_2^2 + b_2^2 \qquad (\Omega = 1).\tag{6}$$

The total kinetic energy of the system is

$$T = \tfrac12 \int d\mathbf{r}\, d\mathbf{v}\, f_0(\dot x_0^2 a_1^2 + \dot x_0^2 b_1^2 + \dot y_0^2 a_2^2 + \dot y_0^2 b_2^2 + \dot z_0^2 a_3^2 + \dot z_0^2 b_3^2).$$

We calculate first of all the integral

$$A_1 = \tfrac12 \int d\mathbf{r}\, d\mathbf{v}\, f_0 \dot x_0^2 = \tfrac12 \int d\mathbf{r}\, d\mathbf{v}\, f_0 \dot y_0^2 = \frac{2\pi\rho_0}{15} c_0.$$

In a similar way, the integral

$$\tfrac12 \int d\mathbf{r}\, d\mathbf{v}\, f_0 \dot z_0^2 = \frac{4\pi\rho_0}{15} c_0^3 \equiv 2A_2$$

is calculated, where

$$A_2 \equiv \frac{2\pi\rho_0}{15} c_0^3 = A_1 c_0^2.$$

If we introduce denotations for the integrals

$$\tfrac12 \int d\mathbf{r}\, d\mathbf{v}\, f_0 \dot x_0^2 = \tfrac12 \int d\mathbf{r}\, d\mathbf{v}\, f_0 \dot y_0^2 \equiv p_1, \qquad \tfrac12 \int d\mathbf{r}\, d\mathbf{v}\, f_0 \dot z_0^2 \equiv p_2,$$

then the expression for the kinetic energy will be written as

$$T = A_1(\dot a_1^2 + \dot a_2^2) + A_2 \dot a_3^2 + p_1(\dot b_1^2 + \dot b_2^2) + \frac{p_2}{\omega_0^2}(\omega_0^2 \dot b_3^2).\tag{7}$$

We perform the transform

$$a_1 = a\cos\varphi_1, \qquad b_1 = a\sin\varphi_1,$$

$$a_2 = b\cos\varphi_2, \qquad b_2 = b\sin\varphi_2,$$

$$a_3 = \frac{c}{c_0}\cos\varphi_3, \qquad \omega_0 b_3 = \frac{c}{c_0}\sin\varphi_3;$$

then

$$T = A_1(\dot{a}^2 \cos^2 \varphi_1 + a^2 \sin^2 \varphi_1 \dot{\varphi}_1^2 - a\dot{a} \sin 2\varphi_1 \dot{\varphi}_1 + \dot{b}^2 \cos^2 \varphi_2$$

$$+ b^2 \sin^2 \varphi_2 \dot{\varphi}_2^2 - b\dot{b} \sin 2\varphi_2 \dot{\varphi}_2)$$

$$+ \frac{A_2}{c_0^2}(\dot{c}^2 \cos^2 \varphi_3 + c^2 \sin^2 \varphi_3 \dot{\varphi}_3^2 - c\dot{c} \sin 2\varphi_3 \dot{\varphi}_3)$$

$$+ p_1(\dot{a}^2 \sin^2 \varphi_1 + a^2 \cos^2 \varphi_1 \dot{\varphi}_1^2 + a\dot{a} \sin 2\varphi_1 \dot{\varphi}_1 + \dot{b}^2 \sin^2 \varphi_2$$

$$+ b^2 \cos^2 \varphi_2 \dot{\varphi}_2^2 + b\dot{b} \sin 2\varphi_2 \dot{\varphi}_2) + \frac{p_2}{\omega_0^2 c_0^2}(\dot{c}^2 \sin^2 \varphi_3 + c^2 \cos^2 \varphi_3 \dot{\varphi}_3^2$$

$$+ c\dot{c} \sin 2\varphi_3 \dot{\varphi}_3). \tag{8}$$

Since $A_1 = p_1$, $A_2 = p_2/\omega_0^2$ (i.e., $\langle x_0^2 \rangle = \langle y_0^2 \rangle = \langle \dot{x}_0^2 \rangle = \langle \dot{y}_0^2 \rangle$; $\omega_0^2 \langle z_0^2 \rangle = \langle \dot{z}_0^2 \rangle$), then

$$T = A_1(\dot{a}^2 + a^2 \dot{\varphi}_1^2 + \dot{b}^2 + b^2 \dot{\varphi}_2^2 + \dot{c}^2 + c^2 \dot{\varphi}_3^2). \tag{9}$$

The potential energy of the homogeneous three-axis ellipsoid is given by the expression

$$U = -\frac{3GM^2}{10}\Phi, \qquad \Phi = \int_0^\infty \frac{ds}{\sqrt{(a^2 + s)(b^2 + s)(c^2 + s)}} \equiv \Phi(a, b, c).$$

The Lagrangian is $L = T - U$, and the Lagrange equations are accordingly the following:

$$\frac{d}{dt}\frac{\partial L}{\partial \dot{q}} = \frac{\partial L}{\partial q} \qquad [q = (a, b, c, \varphi_1, \varphi_2, \varphi_3)]. \tag{10}$$

They give

$$A_1 2\ddot{a} = 2A_1 a\dot{\varphi}_1^2 - \frac{\partial U}{\partial a}, \qquad A_1 2\ddot{b} = 2A_1 b\dot{\varphi}_2^2 - \frac{\partial U}{\partial b}, \qquad A_1 2\ddot{c} = 2A_1 c\dot{\varphi}_3^2 - \frac{\partial U}{\partial c}, \tag{11}$$

$$\frac{d}{dt}(a^2\dot{\varphi}_1) = 0, \qquad \frac{d}{dt}(b^2\dot{\varphi}_2) = 0, \qquad \frac{d}{dt}(c^2\dot{\varphi}_3) = 0. \tag{12}$$

From Eqs. (12) follow the laws of conservation

$$a^2\dot{\varphi}_1 = \text{const}, \qquad b^2\dot{\varphi}_2 = \text{const}, \qquad c^2\dot{\varphi}_3 = \text{const}. \tag{13}$$

We define the constants by equating the right-hand sides of (13) to their equilibrium values:

$$a^2\dot{\varphi}_1 = 1, \qquad b^2\dot{\varphi}_2 = 1, \qquad c^2\dot{\varphi}_3 = c_0^2\omega_0. \tag{14}$$

By means of these expressions, one can eliminate $\dot{\varphi}_1, \dot{\varphi}_2, \dot{\varphi}_3$ from Eqs. (11):

$$A_1 2\ddot{a} = -\frac{\partial U}{\partial a} + A_1\frac{2}{a^3}, \qquad A_1 2\ddot{b} = -\frac{\partial U}{\partial b} + A_1\frac{2}{b^3}, \qquad A_1 2\ddot{c} = -\frac{\partial U}{\partial c} + A_1\frac{2\omega_0^2 c_0^4}{c^3}. \tag{15}$$

We rewrite (15) in such a form:

$$\ddot{a} = A\frac{\partial \Phi}{\partial a} + \frac{1}{a^3}, \qquad \ddot{b} = A\frac{\partial \Phi}{\partial b} + \frac{1}{b^3}, \qquad \ddot{c} = A\frac{\partial \Phi}{\partial c} + \frac{\omega_0^2 c_0^4}{c^3}, \tag{16}$$

or, in a more compact form,

$$\ddot{R}_i = -\frac{\partial \psi}{\partial R_i}, \qquad \mathbf{R} = (a, b, c), \tag{17}$$

where the effective potential ψ is

$$\psi = -A\Phi + \frac{1}{2a^2} + \frac{1}{2b^2} + \frac{\omega_0^2 c_0^4}{2c^2}. \tag{18}$$

In formulas (16)–(18), the quantity A is defined by the formula

$$A = 2\pi G \rho_0 c_0 = \frac{1}{\sqrt{1 + \zeta^2}} \frac{1}{(\zeta^2 + 1)\arctan(1/\zeta) - \zeta}, \tag{19}$$

where

$$\zeta = \frac{c_0}{\sqrt{1 - c_0^2}}.$$

The equations derived describe the *nonlinear* evolution of the system. As particular cases, they may give similar equations for nonlinear perturbations of spheres, disks, and cylinders at rest.

The stability for not too strong initial perturbations is fairly evident from (18). Find the frequencies of small oscillations near the equilibrium state (where $\partial\psi/\partial a = \partial\psi/\partial b = \partial\psi/\partial c = 0$). Linearizing Eqs. (16) by the substitution

$$a = 1 + \varepsilon_1, \qquad b = 1 + \varepsilon_2, \qquad c = c_0(1 + \varepsilon_3), \qquad \varepsilon_1, \varepsilon_2, \varepsilon_3 \ll 1, \tag{20}$$

we shall obtain the following system:

$$\ddot{\varepsilon}_1 = A(\Phi_{aa}\varepsilon_1 + \Phi_{ab}\varepsilon_2 + \Phi_{ac}c_0\varepsilon_3) - 3\varepsilon_1,$$

$$\ddot{\varepsilon}_2 = A(\Phi_{bb}\varepsilon_2 + \Phi_{bc}c_0\varepsilon_3 + \Phi_{ab}\varepsilon_1) - 3\varepsilon_2, \tag{21}$$

$$\ddot{\varepsilon}_3 = A\left(\Phi_{cc}\varepsilon_3 + \Phi_{ac}\frac{1}{c_0}\varepsilon_1 + \Phi_{bc}\frac{1}{c_0}\varepsilon_2\right) - 3\omega_0^2\varepsilon_3,$$

where

$$\Phi_{bb} = \Phi_{aa} \equiv \frac{\partial^2\Phi}{\partial a^2} = \frac{\partial^2\Phi}{\partial b^2} = \int_0^\infty \frac{(2-s)\,ds}{(1+s)^3\sqrt{c_0^2 + s}},$$

$$\Phi_{cc} = \frac{\partial^2\Phi}{\partial c^2} = \int_0^\infty \frac{(2c_0^2 - s)\,ds}{(1+s)(c_0^2 + s)^{3/2}}, \qquad \Phi_{ab} = \frac{\partial^2\Phi}{\partial a\,\partial b} = \int_0^\infty \frac{ds}{(1+s)^3\sqrt{c_0^2 + s}},$$

$$\Phi_{ac} = \Phi_{bc} = \frac{\partial^2\Phi}{\partial a\,\partial c} = \frac{\partial^2\Phi}{\partial b\,\partial c} = \int_0^\infty \frac{c_0\,ds}{(1+s)^2(c_0^2 + s)^{3/2}}.$$

If in (20) we substitute $\partial/\partial t$ for $(-i\omega)$ and make the determinant of the system equal zero, one obtains the characteristic equation of third order with respect to ω^2. It is possible to utilize at once the limitations imposed by the symmetry of individual modes $(\sim e^{im\varphi})$. Let first $m = 0, a = b, \varepsilon_1 = \varepsilon_2$, then instead of (20) we have

$$[A(\Phi_{aa} + \Phi_{ab}) - 3 + \omega^2]\varepsilon_1 + A\Phi_{ac}c_0\varepsilon_3 = 0,$$

$$2\frac{A}{c_0}\Phi_{ac}\varepsilon_1 + [A\Phi_{cc} - 3\omega_0^2 + \omega^2]\varepsilon_3 = 0. \tag{22}$$

This gives the characteristic equation

$$\omega^4 - [3(1 + \omega_0^2) - A(\Phi_{aa} + \Phi_{cc} + \Phi_{ab})]\omega^2 + [A(\Phi_{aa} + \Phi_{ab}) - 3]$$
$$\times [A\Phi_{cc} - 3\omega_0^2] - 2A^2\Phi_{ac}^2 = 0. \tag{23}$$

A test for (23) may be the calculation of the roots in some known limiting cases (disk, sphere). It provides correct results:

$$\omega^2 = 1 \quad \text{as } c \to 0, \qquad \omega^2 = 1, \tfrac{14}{5} \quad \text{as } c \to 1.$$

Prove that the roots of Eq. (23) are positive. Represent this equation as

$$(\omega^2 - \omega_1^2)(\omega^2 - \omega_2^2) = C,$$

where

$$\omega_1^2 = -A(\Phi_{aa} + \Phi_{ab}) + 3 = 2\pi G\rho c$$

$$\times \left[3 \int \frac{ds}{(1 + s)^2\sqrt{c^2 + s}} - \int \frac{(2 - s)\, ds}{(1 + s)^3\sqrt{c^2 + s}} - \int \frac{ds}{(1 + s)^3\sqrt{c^2 + s}} \right]$$

$$= 8\pi G\rho c \int \frac{s\, ds}{(1 + s)^3\sqrt{c^2 + s}} > 0,$$

$$\omega_2^2 = 3\omega_0^2 - A\Phi_{cc} = 2\pi G\rho c \left[\int \frac{ds}{(1 + s)(c^2 + s)^{3/2}} + 3 \int \frac{s\, ds}{(1 + s)(c^2 + s)^{5/2}} \right] > 0,$$

$$C = 4\pi G\rho \int \frac{c^2\, ds}{(1 + s)^2(c^2 + s)^{3/2}}.$$

It is easy to see that it is sufficient to prove the inequality

$$\omega_1^2\omega_2^2 > C,$$

i.e.,

$$2 \int \frac{s\, ds}{(1 + s)^3(c^2 + s)^{1/2}} \int \frac{(c^2 + 4s)\, ds}{(1 + s)(c^2 + s)^{5/2}} > \left[\int \frac{c\, ds}{(1 + s)^2(c^2 + s)^{3/2}} \right]^2. \tag{24}$$

Apply the Cauchy inequality

$$\int \varphi_1\varphi_2 < \sqrt{\int \varphi_1^2}\sqrt{\int \varphi_2^2}, \tag{25}$$

assuming that

$$\varphi_1 = \frac{c}{(1 + s)^{3/2}(c^2 + s)^{1/4}\sqrt{c^2 + 4s}}, \qquad \varphi_2 = \frac{\sqrt{c^2 + 4s}}{(1 + s)^{1/2}(c^2 + s)^{5/4}}.$$

Then inequality (24) will only become stronger:

$$\int \frac{2s\, ds}{(1 + s)^3(c^2 + s)^{1/2}} > \int \frac{c^2\, ds}{(1 + s)^3(c^2 + s)^{1/2}(c^2 + 4s)}.$$

It is possible to prove this inequality by using once more the Cauchy inequality (25), this time with

$$\varphi_1 = \frac{1}{(1 + s)^3(c^2 + s)^{1/2}}, \qquad \varphi_2 = \frac{c^2}{c^2 + 4s},$$

i.e.,

$$\int\frac{ds\,c^2}{(1+s)^3(c^2+s)^{1/2}(c^2+4s)} < \sqrt{\int\frac{ds}{(1+s)^6(c^2+s)}}\sqrt{\int\frac{ds\,c^4}{(c^2+4s)^2}}$$

$$= \frac{c}{2}\sqrt{\int\frac{ds}{(1+s)^6(c^2+s)}}.$$

Thus, one has only to prove that

$$J_1 \equiv \int\frac{2s\,ds}{(1+s)^3(c^2+s)^{1/2}} > J_2 \equiv \frac{c}{2}\sqrt{\int\frac{ds}{(1+s)^6(c^2+s)}}.$$

By a direct calculation we find that $J_{1\,\text{min}} = J_1(c=1) = 8/15$, and, since $c^2/(c^2+s) = 1 - s/(c^2+s) < 1$, then

$$J_{2\,\text{max}} < \tfrac{1}{2}\sqrt{\int\frac{ds}{(1+s)^6}} = \tfrac{1}{2}\sqrt{\tfrac{1}{5}} < \tfrac{1}{4},$$

so that $J_1 > J_2$, and the statement is proved.

Consider now the case $m = 2$, $\varepsilon_1 = -\varepsilon_2$:

$$[A(\Phi_{aa} - \Phi_{ab}) - 3 + \omega^2]\varepsilon_1 + A\Phi_{ac}C_0\varepsilon_3 = 0, \qquad [A\Phi_{cc} - 3\omega_0^2 + \omega^2]\varepsilon_3 = 0. \quad (26)$$

Therefore, the possible frequencies in this case are defined elementarily:

$$\omega_1^2 = 3 + A(\Phi_{ab} - \Phi_{aa}), \tag{27}$$

$$\omega_2^2 = 3\omega_0^2 - A\Phi_{cc}. \tag{28}$$

Substituting into (27), (28) the explicit expressions for ω_0^2, A, Φ_{aa}, Φ_{ab}, Φ_{cc}, we get

$$\omega_1^2 = 4\pi G\rho c_0 \int_0^\infty\frac{(1+2s)\,ds}{(1+s)^3\sqrt{c_0^2+s}}, \tag{29}$$

$$\omega_2^2 = 2\pi G\rho c_0\left(\int_0^\infty\frac{ds}{(1+s)(c_0^2+s)^{3/2}} + 3\int_0^\infty\frac{s\,ds}{(1+s)(c_0^2+s)^{5/2}}\right). \tag{30}$$

It is easy to see that $\omega_1^2 > 0$ and $\omega_2^2 > 0$, which implies stability of the oscillation modes in question.

8. Derive equations describing nonlinear perturbations (of the ellipsoid–ellipsoid type) of the rotating Freeman spheroid.

Solution. The current coordinates of the particles will be expressed through the initial ones in the form

$$x = \alpha x_0 + \beta y_0, \qquad y = \gamma x_0 + \delta y_0, \qquad z = mz_0 + n\dot z_0, \tag{1}$$

where α, β, γ, δ, m, n are functions of time, and the system in (1) is written in the inertial system. Under such perturbations the system, as can easily be proved, remains a uniform ellipsoid, and the z-axis remains the principal axis.

The equation of projection of the modified surface is of the form

$$\frac{(\delta x - \beta y)^2 + (\alpha y - \gamma x)^2}{a_0^2} = (\alpha\delta - \beta\gamma)^2, \qquad \frac{1}{a^2b^2} = \frac{1}{(\alpha\delta - \beta\gamma)^2 4a_0}, $$

$$\frac{1}{a^2} + \frac{1}{b^2} = \frac{1}{(\alpha\delta - \beta\gamma)^2}\left(\frac{\beta^2 + \delta^2 + \alpha^2 + \gamma^2}{a_0^2}\right), \tag{2}$$

where a_0 and c_0 are the unperturbed semiaxes. Determine the value of the third semi-axis from the condition of maximum at $x = y = 0$:

$$c = \max(mz_0 + n\dot{z}_0), \tag{3}$$

under the additional condition

$$z_0^2 + \left(\frac{\dot{z}_0}{\omega_0}\right)^2 = c_0^2. \tag{4}$$

This yields

$$c = c_0\sqrt{m^2 + n^2\omega_0^2}. \tag{5}$$

We calculate the kinetic energy

$$T = \tfrac{1}{2}\int (\dot{x}^2 + \dot{y}^2 + \dot{z}^2)\,dM = \tfrac{1}{10}M[(\dot{\alpha}^2 + \dot{\gamma}^2 + \dot{\beta}^2 + \dot{\delta}^2)a_0^2 + (\dot{m}^2 + \dot{n}^2\omega_0^2)c_0^2]. \tag{6}$$

The potential energy of the uniform ellipsoid is

$$W = -\tfrac{3}{10}GM^2 \int_0^\infty \frac{ds}{\sqrt{(a^2 + s)(b^2 + s)(c^2 + s)}}. \tag{7}$$

Then, varying the action integral, we get the "equations of motion," first in the variables $\alpha, \beta, \gamma, \delta, m, n$. However, the number of unknown functions can be decreased if we take into account the laws of conservation:

$$j_1 = m\dot{n} - \dot{m}n = \text{const}, \qquad j_2 = a_0^2(\dot{\alpha}\gamma - \alpha\dot{\gamma} + \dot{\beta}\delta - \beta\dot{\delta}) = \text{const},$$

$$j_3 = \alpha\dot{\beta} - \delta\dot{\gamma} - \dot{\alpha}\beta + \dot{\delta}\gamma = \text{const}.$$

These quantities are as follows (within an accuracy up to constant factors): j_1 is the phase density in the projection onto the $(z, \dot{z})$ plane, j_2 is the z-component of the total angular moment of the system, j_3 is the density of the vortex in the projection onto (xy). The conservation of the latter (Thomson theorem) is due to the hydrodynamical character of movement in the plane of rotation.

The equations of motion are readily formulated in terms of the following geometrical parameters:

$$\tau_1 = a - b, \qquad \tau_2 = a + b, \qquad \tau_3 = c\sqrt{2}.$$

It turns out that

$$\ddot{\tau}_i = -\frac{\partial\psi}{\partial\tau_i} \qquad (i = 1, 2, 3), \tag{8}$$

where the effective potential

$$\psi = -3GM \int_0^\infty \frac{ds}{\sqrt{\{s + [(\tau_1 + \tau_2)/2]^2\}\{s + [(\tau_1 - \tau_2)/2]^2\}(s + \tau_3^2/2)}}$$

$$+ \frac{8a_0^4\Omega^2}{2\tau_2^2} + \frac{2c_0^4\omega_0^2}{\tau_3^2}. \tag{9}$$

Equations (8) were derived by V. A. Antonov.

We are already aware that the system in question is unstable. Make sure about this once more by using a new formulation of the problem contained in Eqs. (8) and (9).

The stationary state is defined by equations

$$\frac{\partial \psi}{\partial \tau_1} = \frac{\partial \psi}{\partial \tau_2} = \frac{\partial \psi}{\partial \tau_3} = 0. \tag{10}$$

which give

$$\tau_1^{(0)} = 0, \qquad \tau_2^{(0)} = 2, \qquad \tau_3^{(0)} = c_0\sqrt{2}. \tag{11}$$

Put

$$\tau_1 = 2\varepsilon_1, \qquad \tau_2 = 2(1 + \varepsilon_2), \qquad \tau_3 = c_0\sqrt{2}(1 + \varepsilon_3), \qquad \varepsilon_1, \varepsilon_2, \varepsilon_3 \ll 1, \tag{12}$$

and, to begin with, let us consider perturbations with $\varepsilon_1 \neq 0, \varepsilon_2 = \varepsilon_3 = 0$ (they are just unstable). For the effective potential, we get

$$\psi \approx \text{const} + 3GM \cdot \varepsilon_1^2 \int_0^\infty \frac{(s-1)\,ds}{\sqrt{s + c_0^2(s+1)^3}}, \tag{13}$$

so that either stability or instability of the spheroid with a given c_0 is dependent on the sign of the integral

$$I(c_0) \equiv \int_0^\infty \frac{(s-1)\,ds}{\sqrt{s + c_0^2(s+1)^2}}. \tag{14}$$

Since, as is easily shown, $I(c_0) < 0$ at all finite c_0, then there occurs the (exponential) instability. The vanishing of this integral for $c_0 = \infty$ (i.e., for the cylindrical limit) corresponds to the familiar fact of the *power* instability of a cylinder (Chapter II).

For perturbations with $\varepsilon_1 = 0, \varepsilon_2 \neq 0, \varepsilon_3 \neq 0$, we can find

$$\psi \simeq \text{const} + \alpha\varepsilon_2^2 - \beta\varepsilon_2\varepsilon_3 + \gamma\varepsilon_3^2, \tag{15}$$

where

$$\alpha = 6 + 3 \int_0^\infty \frac{ds}{(s+1)(s+c_0^2)^{1/2}} \left[\frac{1}{s+1} - \frac{4}{(s+1)^2} \right],$$

$$\gamma = 3c_0^2\omega_0^2 + \frac{3}{2} \int_0^\infty \frac{ds}{(s+1)(s+c_0^2)^{3/2}} \left[c_0^2 - \frac{3}{(s+c_0^2)} \right],$$

$$\beta = 6c_0^2 \int_0^\infty \frac{ds}{(s+1)^2(s+c_0^2)^{3/2}}.$$

For example, for the Freeman *sphere*, this yields

$$\psi \simeq \text{const} + \tfrac{16}{5}\varepsilon_2^2 + \tfrac{11}{5}\varepsilon_3^2 - \tfrac{12}{5}\varepsilon_2\varepsilon_3, \tag{16}$$

and the equations of motion (8) give oscillations with the frequencies $\omega_1^2 = 1, \omega_2^2 = \tfrac{14}{5}$. This implies that the effective potential near the equilibrium state $c_0 = 1$ has a "saddle-shaped" appearance.

9. Determine the eigenfrequencies of bending and Jeansonian perturbations of thin, prolate, homogeneous, nonrotating spheroid (with the volume density ρ_0, major semi-axis a, minor semiaxis c). Consider also the stability of the largest-scale bending mode for such a rotating spheroid.

Solution. We take the x-axis of Cartesian coordinates along the major axis of the spheroid, the axes y, z in the perpendicular plane. Besides that, in this case it is natural to use the prolate spheroidal coordinates ζ, μ, φ:

$$x = k\mu\zeta, \qquad y = k(1 - \mu^2)^{1/2}(\zeta^2 - 1)^{1/2} \cos \varphi, \qquad z = k(1 - \mu^2)^{1/2}(\zeta^2 - 1)^{1/2} \sin \varphi.$$

We begin with the small *bending* oscillations since only they may be unstable in the case under consideration. For these perturbations the volume density ρ does not change, so the potential perturbation Φ_1 is completely due to bendings of the spheroid's boundary. Then the potential obeys the Laplace equation $\Delta\Phi_1 = 0$, which may be written in the coordinates ζ, μ, φ as follows:

$$\frac{\partial}{\partial \mu}\left[(1 - \mu^2)\frac{\partial \Phi}{\partial \mu}\right] + \frac{1}{1 - \mu^2}\frac{\partial^2 \Phi}{\partial \varphi^2} = \frac{\partial}{\partial \zeta}\left[(1 - \zeta^2)\frac{\partial \Phi}{\partial \zeta}\right] + \frac{1}{1 - \zeta^2}\frac{\partial^2 \Phi}{\partial \varphi^2}. \tag{1}$$

The elementary solutions of this equation (with separated variables) have the form

$$\Phi_< = AP_n^m(\mu)B_n^m(\zeta)e^{im\varphi}, \tag{2}$$

$$\Phi_> = BP_n^m(\mu)D_n^m(\zeta)e^{im\varphi}. \tag{3}$$

The conditions of matching of the external solution with the internal solution at the surface of the spheroid are

$$BD_n^m(\zeta_0) = AB_n^m(\zeta_0),$$

$$BD_n^{m'}(\zeta_0) - AB_n^{m'}(\zeta_0) = 4\pi G\rho_0\zeta_n h_\zeta, \tag{4}$$

where $h_\zeta\zeta_n \equiv h_\zeta(\zeta\mathbf{n})$ is the normal displacement of the boundary, $h_\zeta = k\sqrt{\zeta^2 - \mu^2}/\sqrt{\zeta^2 - 1}$ is the Lamé coefficient, $\mathbf{n}$ is the vector of the normal to the spheroid's surface: $\mathbf{n} = \mathbf{c}/|\mathbf{c}|$, $\mathbf{c} = (x/a^2, y/c^2, z/c^2)$. The internal solution of system (4) is the following:

$$A = \frac{4\pi G\rho_0\chi D_n^m(\zeta_0)}{W}, \qquad W \equiv D_n^{m'}B_n^m - D_n^m B_n^{m'}|_{\zeta=\zeta_0}. \tag{5}$$

Consequently, we obtain that the potential

$$\Phi_< = B_n^m(\zeta)P_n^m(\mu)e^{im\varphi} \tag{6}$$

corresponds to the displacement

$$\zeta_n h_\zeta = \frac{W(\zeta_0)}{4\pi G\rho_0 D_n^m(\zeta_0)} P_n^m(\mu)e^{im\varphi}. \tag{7}$$

In the limit of the "needle" $\zeta = a/\sqrt{a^2 - c^2} \to 1$ we have $x = \pm k\mu$ or, for $a - 1$, $x = \pm\mu$. For the "fire-hose modes" ($m = 1$) in the thin "needle," one can write that the displacement

$$\psi(x) = \frac{cW(\zeta_0 \simeq 1)}{4\pi G\rho_0 D_n^1(\zeta_0 \simeq 1)}\frac{P_n^1(\pm x)}{\sqrt{1 - x^2}}$$

$$= \frac{n(n + 1)}{2}\frac{1}{\omega_0^2} \cdot \frac{1 - 1/2\omega_0^2}{1 - n(n + 1)/4\omega_0^2}\frac{dP_n}{dx} \tag{8}$$

corresponds to the potential

$$\Phi_<(\zeta \simeq 1) = B_n^1(\zeta)P_n^1(\pm x) \sin \varphi,$$

or, after some transformations,

$$\Phi_< = -\frac{n(n + 1)}{2}\frac{d}{dx}P_n(x) \cdot z. \tag{9}$$

The equation of motion for the case under consideration is

$$(D^2 + \omega_0^2)\psi = -\frac{\partial \Phi_<}{\partial z} = \frac{n(n+1)}{2}\frac{d}{dx}P_n(x) \equiv T(x). \tag{10}$$

For derivation of the dispersion equation it is sufficient to keep only the highest power of x in the polynomial $T(x)$:

$$\frac{d}{dx}P_n(x) = \alpha x^{n-1} + \cdots, \qquad T(x) = \frac{n(n+1)}{2}\alpha x^{n-1} + \cdots,$$

so that

$$(D^2 + \omega_0^2)\psi = \frac{n(n+1)}{2}\alpha x^{n-1} + \cdots;$$

hence

$$\psi \propto \frac{1}{2i\omega_0}\left[\frac{1}{D - i\omega_0} - \frac{1}{D + i\omega_0}\right]x^{n-1}\frac{n(n+1)}{2}\alpha. \tag{11}$$

For the particles locating on the axis of the needle $z = y = 0$, $v_z = v_y = 0$

$$\hat{D} = -i\omega + v_x\frac{\partial}{\partial x} - A^2 x\frac{\partial}{\partial v_x} = -i\omega + A\frac{\partial}{\partial \varphi},$$

where $v_x/A = \rho\cos\varphi$ and $x = \rho\sin\varphi$. By integrating over the angle φ we find (after some additional transformations and averaging over the equilibrium distribution):

$$\psi_n \propto \frac{n(n+1)}{2}\alpha\frac{x^2 p}{2^{2p}}\sum_{m=-p}^{p}\frac{\mu_m^{(p)}}{\omega_0^2 - (\omega + 2m)^2}, \tag{12}$$

where

$$\mu_m^{(n)} = C_{2p}^{p+m}\frac{2}{\pi}\int_{-\pi/2}^{\pi/2}d\varphi\,(1 + \sin\varphi)^{p+1-m}(1 - \sin\varphi)^{p+1+m}$$

$$= \frac{(2p+1-2m)!!\,(2p+1+2m)!!}{(p+1)(2p+1)(p+m)!(p-m)!}, \tag{13}$$

$n = 2p + 1$, $p = 0, 1, 2, \ldots$. Taking into account (8), we obtain the dispersion equation in the form

$$\frac{1}{2^{2p}}\sum_{m=-p}^{p}\frac{\mu_m^{(p)}}{1 - (\omega + 2m)^2/\omega_0^2} = \frac{1 + \frac{1}{2}c^2\ln(c^2/4)}{1 + \frac{1}{2}(1 + p)(2p + 1)c^2\ln(c^2/4)}. \tag{14}$$

For $p = 0$ we have, from (14), $\omega^2 = 0$ for arbitrary c^2, as it must be for the displacement of the system as a whole (the trivial mode). For the largest-scale nontrivial mode $p = 1$ ("arch") we have the equation

$$1 + \frac{\frac{5}{2}}{\omega_0^2} = \frac{\omega_0^2}{2}\frac{(2\omega_0^4 - 11\omega_0^2 + 12) - (4\omega_0^2 + 1)\omega^2 + 2\omega^4}{[(\omega_0 - 2)^2 - \omega^2][(\omega_0 + 2)^2 - \omega^2](\omega_0^2 - \omega^2)}. \tag{15}$$

As $\omega_0^2 \to \infty$ we have $\omega^2 = 0$, for the small $c/a \neq 0$ ($\omega_0^2 \gg \omega_0^2 \neq \infty$) the system is unstable: $\omega^2 \simeq -10/\omega_0^2$. The stability boundary corresponds to $\omega_0^2 = 4$. Indeed, for $\omega_0 = 2 + \varepsilon$ one can obtain from (15) $\omega^2 = -\frac{4}{7}\varepsilon$, so that $\omega^2 > 0$ (stability) for $\varepsilon < 0$ and $\omega^2 < 0$ (instability) for $\varepsilon > 0$. In Fig. 60(a) the graph of the dependence of ω^2 from ratio

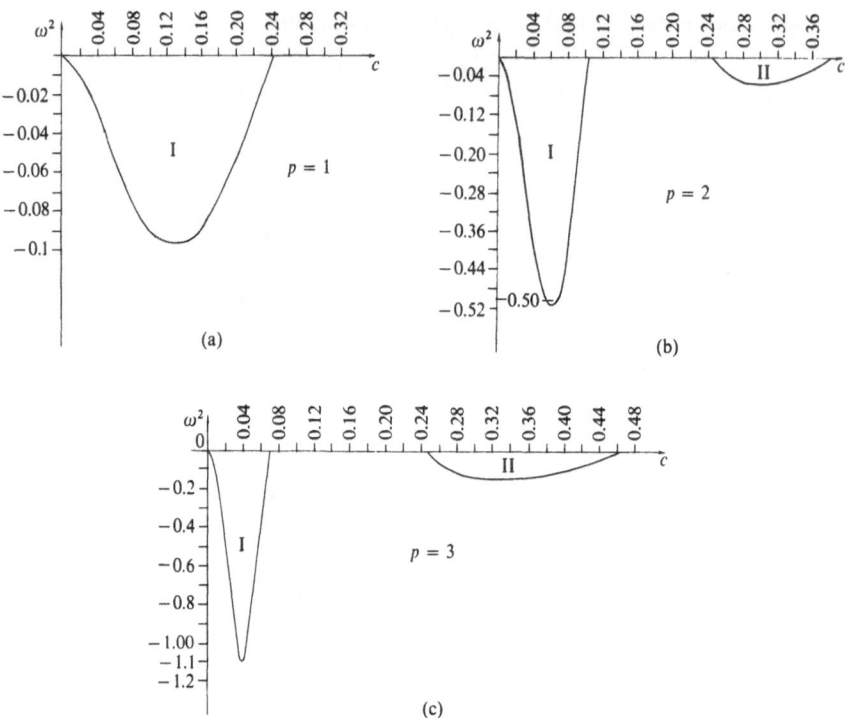

Figure 60. Dependence of the growth rate of the fire-hose instability on the ratio of semi-axes c, for prolate homogeneous spheroids; (a) $p = 1$ (a nodeless mode, "arch"), (b) $p = 2$ (one node), (c) $p = 3$ (two nodes).

c/a for the mode $p = 1$ is presented. It is seen that the instability is stabilized for $c/a \simeq 0.25$. Such value c/a corresponds to $\omega_0^2 = 4$, if we calculate ω_0^2 at the stability boundary by the same approximate formula as for $c/a \to 0$ ($\omega_0^2 \to \infty$). Under calculation by the exact formula $\omega_0^2 = 4$ corresponds to $c/a \simeq 0.34$.

In Figs. 60(b) and 60(c) we represent the graphs $\omega^2(c/a)$ which corresponds to modes $p = 2$ and $p = 3$ [calculated according to (15)]. Apart from the main unstable region (I), lying near small c/a, for each of these modes there is another unstable region (II) which is due to the strong influence of resonance $\omega_0^2 = 4$ in the case of the homogeneous spheroid considered here. The region of the hydrodynamical fire-hose instability (I) becomes narrow with decreasing of the scale of disturbance (i.e., with increasing p); the widest unstable region corresponds as was to be expected to the largest-scale mode $p = 1$. For this mode the instability drops also in the resonance point $\omega_0^2 = 4$ ($c/a \simeq 0.34$). Essentially, however, the instability is absent for all the larger c/a.[30]

The global anisotropy, i.e., the ratio of the kinetic energies along major and minor axes of the homogeneous spheroid, equals

$$\xi = \frac{T_x}{T_\perp/2} = \frac{\omega_x^2 a^2}{\omega_z^2 c^2}, \tag{16}$$

---

[30] For the ellipsoids elongated along x-axis and having the *elliptical* cross section in the (x, y) plane, the stability boundary for the largest-scale fire-hose mode, $\omega_z^2 = 4\omega_x^2$, corresponds to c/a lying in the narrow interval: $0.30 < c/a < 0.34$ (the critical c/a slowly decreases from 0.34 for $b = c$ to 0.30 for $b = 1$).

where $\omega_z = \omega_0$ and ω_x are the frequencies of oscillations in axes Z and X, respectively. At the stability boundary of the largest-scale mode $p = 1$ $(\omega_0^2/\omega_x^2 = 4, c/a \simeq 0.34)$ we have

$$\xi = \xi_c \simeq 2.15, \qquad (17)$$

where the spheroids with $\xi > \xi_c$ are unstable. The stability criterion reformulated in this manner has probably more universal meaning, being approximately valid also for nonhomogeneous "needles" with arbitrary arrangement.

We derived above Eq. (14), which determines the eigenfrequencies of the bending oscillations of a prolate homogeneous spheroid. In this case, however, we can show that the functions

$$\Phi_<^{(n)} = -zC_{n-1}^{3/2}(x) = -z\frac{dP_n(x)}{dx} \qquad (n = 2p + 1, p = 0, 1, 2, \ldots) \qquad (18)$$

are, in the limit of infinitely thin needle, the *exact* eigenfunctions. Let us assume the expression (18) for $\Phi_<$, and, correspondingly, the expression

$$\psi = B_N C_{n-1}^{3/2}(x), \qquad B_N \equiv \frac{1}{\omega_0^2}\frac{1 - 1/2\omega_0^2}{1 - n(n + 1)/4\omega_0^2}. \qquad (19)$$

Solving then the equation of motion (10) and averaging over the one-dimensional distribution function of the "needle"

$$f_0 = \frac{2\rho_l(0)}{\pi}\sqrt{1 - x^2 - v_x^2} \qquad (20)$$

(the linear density $\rho_l(x) = \rho_l(0)\varkappa^2, \varkappa^2 \equiv 1 - x^2/a^2$), we find

$$\bar{\psi} = -\frac{1}{\omega_0}\int_{-\infty}^0 dt\, e^{-i\omega t}\sin \omega_0 t\, \frac{2}{\pi}\int_0^\pi d\varphi\, \sin^2 \varphi C_N^{3/2}(x \cos t + \sqrt{1 - x^2}\, \sin t \cos \varphi)$$

$$(N = n - 1). \qquad (21)$$

By applying the addition theorem to $C_N^{3/2}$ (analogously to that made in Problem 2 of Chapter I) we obtain

$$\bar{\psi} = -C_N^{3/2}(x) \cdot \frac{1}{\omega_0}\int_{-\infty}^0 dt\, e^{-i\omega t}\sin \omega_0 t\, \frac{2}{(N + 1)(N + 2)}\, C_N^{3/2}(\cos t), \qquad (22)$$

which proves the hypothesis assumed at the beginning that Eqs. (18) and (19) are the eigenfunctions of our problem. Using Eq. (19), we obtain also the dispersion equation in the form

$$\frac{1 - 1/(2\omega_0^2)}{1 - (p + 1)(2p + 1)/2\omega_0^2} + \frac{\omega_0}{(p + 1)(2p + 1)}\int_{-\infty}^0 dt\, e^{-i\omega t}\sin \omega_0 t C_{2p}^{3/2}(\cos t) = 0. \qquad (23)$$

Jeans perturbations in the thin prolated nonrotating systems are evidently stable. Therefore, we shall restrict ourselves to derivation of the appropriate characteristic equation for the limiting case of the infinitely thin needle. Assuming that

$$\Phi_1 = -P_n(x) \qquad (n = 2, 4, \ldots), \qquad (24)$$

we then solve the equation

$$(D^2 + 1)\psi_x = \frac{\partial P_n}{\partial x} = C_{n-1}^{3/2}(x). \qquad (25)$$

Hence we obtain

$$\bar{\psi}_x = -\int_{-\infty}^{0} dt\, e^{-i\omega t} \sin t\, \frac{2}{(n+1)n}\, C_{n-1}^{3/2}(\cos t)C_{n-1}^{3/2}(x) \equiv AC_{n-1}^{3/2}(x). \tag{26}$$

The perturbed density equals

$$\rho_1 = -\frac{\partial}{\partial x}\rho_l(x)\bar{\psi} = A\rho_l(0)n(n+1)P_n(x) \equiv CP_n(x). \tag{27}$$

For the thin needle

$$\Phi_1 = \rho_1 \left. \frac{2GD_n(\zeta_0)}{CW(\zeta_0)\sqrt{\zeta_0^2-1}}\, B_n(\zeta)\right|_{\zeta=\zeta_0} \approx -G \ln\frac{1}{c^2}\, CP_n(x). \tag{28}$$

Thus the characteristic equation for the eigenfrequencies will be the following:

$$1 + \int_{-\infty}^{0} dt\, e^{-i\omega t} \sin t\, C_{n-1}^{3/2}(\cos t) = 0. \tag{29}$$

Simultaneously, we proved that functions $\Phi_1 = -P_n(x)$ are the eigenfunctions of the given problem. In particular, for $n = 2$ one can obtain from (29): $\omega^2 = 1$, for $n = 4$ $\omega^2 = 1$ and 9, and so on.

Let us give without derivation the picture of stability for the arch-shaped bending of a thin rotating "needle" (Fig. 61). This picture is very similar to one represented in Fig. 53 (for the bell-shaped fire-hose mode). In the vicinity of $\omega_0^2 = 4$ we can obtain

$$\omega^2 \simeq -\varepsilon\, \frac{26t^2 - 26t + 10}{78t^2 + 34t - 42}, \tag{30}$$

where $t = 1 - \Omega^2$, $\omega_0^2 = 4 + \varepsilon$. As $c^2 \to 0$, the approximate equation of the stability boundary

$$\Omega^2 \simeq 4c^2 \ln\frac{4}{c^2}. \tag{31}$$

In conclusion we write (also without derivation) the simple characteristic equation for the bending oscillations of the infinitely thin needle:

$$\omega^2 = \Omega^2 p(2p+3)/2 \qquad (p = 0, 1, 2, \cdots). \tag{32}$$

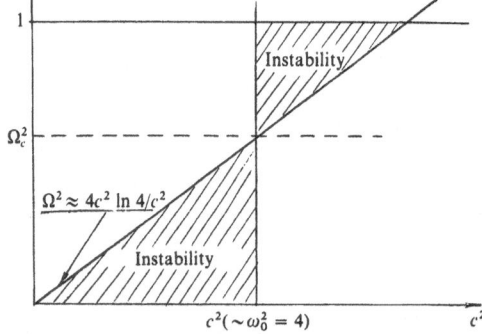

Figure 61. Regions of the fire-hose instability relative to the nodeless mode ("arch"), for the rotating prolate homogeneous spheroids.

10. Investigate the linear stability of collisionless elliptical Freeman cylinders (47), §1, with respect to the largest-scale perturbations [92^{ad}].

Solution. Let the equation of boundary ellipse be

$$\frac{x^2}{a^2} + \frac{y^2}{b^2} = 1 \qquad (a > b);$$

it rotates around the axis z with an angular velocity Ω. The unperturbed potential

$$\Phi_0 = 2\pi G\rho_0 \left[\frac{bx^2}{a+b} + \frac{ay^2}{a+b} + \text{const} \right].$$

The orbits of stars in such a potential are elliptical epicycles. We also write here for the sake of convenience the equilibrium distribution function:

$$f_0(J) = \frac{ab\rho_0}{\pi k_\alpha k_\beta} \Lambda^2 \delta(1 - J), \tag{1}$$

where

$$\begin{aligned}
J &= \Lambda^2 \left[a^2 \tilde{c}_y^2 + \left(\frac{b}{k_\alpha k_\beta} \right)^2 \tilde{c}_x^2 \right] + \frac{x^2}{a^2} + \frac{y^2}{b^2}, \\[2mm]
\tilde{c}_x &= c_x + \frac{y}{b^2}\,\theta, \\[2mm]
\tilde{c}_y &= c_y - \frac{x}{a^2}\,\theta, \\[2mm]
\theta &= \frac{\beta k_\beta (b^2 - a^2 k_\alpha^2) - \alpha k_\alpha (a^2 k_\beta^2 - b^2)}{k_\beta^2 - k_\alpha^2},
\end{aligned} \tag{2}$$

α and β are the frequencies of the elliptical epicycle; k_α, k_β, and Λ^2 are the constants (cf. §1). The structure of the cylinders in question is characterized by two dimensionless parameters b/a and $\Omega^* \equiv (\Omega^2/2\pi G\rho_0)^{1/2}$, where $0 \le \Omega^{*2} \le \Omega_\beta^{*2} \equiv 2b/(a+b)$, $0 \le b/a \le 1$. At $\Omega^* = \Omega_\beta^*$, the centrifugal and gravitational forces compensate for each other along the large axis ("balanced" cylinders). At $\Omega^{*2} = \Omega_j^{*2} \equiv 2ab/(a+b)^2$, the velocity dispersion in the (x, y) plane is isotropic while the mean velocity (of "hydrodynamical" flows) within the cylinder is zero ("Jacobian" cylinders). The "cold" cylinders correspond to $\Omega^{*2} = \Omega_c^2 \equiv ab/(a+b)^2$; then, the velocity dispersion of stars throughout the cylinder disappears, while the mean velocity is given by the formulae

$$\bar{c}_x = -\theta y/b^2, \qquad \bar{c}_y = \theta x/a^2,$$

where θ represents circulation of the system. The mean velocity is direct (with respect to the boundary rotation) for $0 \le \Omega^{*2} \le \Omega_j^{*2}$ and reverse for $\Omega_j^{*2} < \Omega^{*2} \le \Omega_\beta^{*2}$. In the limit $b/a = 1$, the cylinder transforms to a uniformly rotating circular cylinder (2), §1, Chapter II.

Substituting, as, for example, in formula (37), §6, Chapter I, into the kinetic equation $f = f_0(J + \varepsilon\chi)$, where $\chi = \chi(\tilde{c}_x, \tilde{c}_y, x, y, t)$ is the perturbation, $\varepsilon \ll 1$, we obtain

$$\frac{d\chi}{dt} = -\nabla\Phi \frac{\partial J}{\partial c}, \tag{3}$$

where d/dt is the time derivative along the unperturbed orbit of a star and Φ is the potential perturbation. The location of the star (x, y) and its peculiar velocity $(\tilde{c}_x, \tilde{c}_y)$

at any time $t + \tau$ may be expressed with these values at time t:

$$\tilde{c}_x(t + \tau) = Ax(\tau)x(t) + Ay(\tau)y(t) + A\tilde{c}_x(\tau)\tilde{c}_x(t) + A\tilde{c}_y(\tau)\tilde{c}_y(t),$$

$$\tilde{c}_y(t + \tau) = Bx(\tau)x(t) + By(\tau)y(t) + B\tilde{c}_x(\tau)\tilde{c}_x(t) + B\tilde{c}_y(\tau)\tilde{c}_y(t),$$

$$x(t + \tau) = Cx(\tau)x(t) + Cy(\tau)y(t) + C\tilde{c}_x(\tau)\tilde{c}_x(t) + C\tilde{c}_y(\tau)\tilde{c}_y(t),$$

$$y(t + \tau) = Dx(\tau)x(t) + Dy(\tau)y(t) + D\tilde{c}_x(\tau)\tilde{c}_x(t) + D\tilde{c}_y(\tau)\tilde{c}_y(t),$$

$$(4)$$

where the coefficients $A(\tau)$, $B(\tau)$, $C(\tau)$, $D(\tau)$ are expressed as linear combinations of $\sin \alpha\tau$, $\sin \beta\tau$, $\cos \alpha\tau$, $\cos \beta\tau$ (we do not give them here).

Assuming, as usual, that $\Phi(\tau)$ is the increasing function of time τ, $\chi \to 0$ as $\tau \to -\infty$, we integrate Eq. (3) along the unperturbed trajectory (4) from $\tau = -\infty$ to $\tau = 0$; then we obtain

$$\chi(\tilde{c}_x, \tilde{c}_y, x, y, t) = -2\Lambda^2 \int_{-\infty}^{0} \left[\left(\frac{b}{k_\alpha k_\beta} \right)^2 \tilde{c}_x(t + \tau)\Phi_x(x(t + \tau), y(t + \tau), t + \tau) \right]$$

$$+ a^2 c_y(t + \tau)\Phi_y(x(t + \tau), y(t + \tau), t + \tau) \, d\tau, \qquad (5)$$

where $(\Phi_x, \Phi_y) \equiv (\partial\Phi/\partial x, \partial\Phi/\partial y)$. For the polynomial with respect to x, y potential Φ, as is easily seen from Eq. (5), χ is reduced to the polynomial over x, y, $\tilde{c}_x$, $\tilde{c}_y$ of the same degree as Φ.

The local density perturbation may be expressed by the formula

$$\rho_1 = \frac{\rho_0 \Lambda^2}{2\pi} \int_0^{2\pi} d\varphi \int_0^{\infty} \delta(1 - J - \varepsilon\chi) \, du^2 - \rho_0, \qquad (6)$$

where the polar coordinates u, φ [$u \cos \varphi = (b/k_\alpha k_\beta)\tilde{c}_x$; $u \sin \varphi = a\tilde{c}_y$] are introduced (then χ will become a polynomial in u^2, $\cos \varphi$, $\sin \varphi$, x, y). In considering the equation $1 - J - \varepsilon\chi = 0$ as the algebraic equation with respect to u^2, we obtain

$$u_0^2 \simeq \frac{1 - x^2/a^2 - y^2/b^2}{\Lambda^2}. \qquad (7)$$

Using the property of the δ function, Eq. (6) is reduced to

$$\rho_1 = -\varepsilon \frac{\rho_0}{2\pi\Lambda^2} \int_0^{2\pi} \left[\frac{\partial\chi}{\partial u^2} \bigg|_{u^2 = u_0^2} \right] d\varphi. \qquad (8)$$

Consequently, ρ_1 is expressed as the polynomial with respect to x, y.

In the unperturbed state, the peculiar velocities $(\tilde{c}_x, \tilde{c}_y)$ disappear on the boundary (x_0, y_0), where $x_0^2/a^2 + y_0^2/b^2 = 1$. Let the boundary (x_0, y_0) and the velocity $(0, 0)$ in the perturbed state become $(x_0 + \Delta x, y_0 + \Delta y)$ and $(\Delta\tilde{c}_x, \Delta\tilde{c}_y)$. Since $1 - J - \varepsilon\chi = 0$ on the boundary of the phase volume, then

$$1 - J - \varepsilon\chi = -2\left(\frac{x_0 \Delta x}{a^2} + \frac{y_0 \Delta y}{b^2} \right) - \varepsilon\chi(0, 0, x_0, y_0, t) + O(\varepsilon^2). \qquad (9)$$

The displacement normal to the unperturbed boundary is

$$\Delta\xi = abl_0\left(\frac{x_0 \Delta x}{a^2} + \frac{y_0 \Delta y}{b^2} \right),$$

where abl_0 is the distance from the origin of coordinates to the tangent to the ellipse at the point (x_0, y_0). From Eq. (9), we find the connection of $\Delta\xi$ and χ:

$$\Delta\xi = -\frac{\varepsilon}{2} abl_0\,\chi(0, 0, x_0, y_0, t) + O(\varepsilon^2). \tag{10}$$

The potential response to the potential perturbation Φ is composed of the response ψ_d due to the density perturbation ρ_1,

$$\Delta\psi_d = \begin{cases} -4\pi G\rho_1 & \left(\dfrac{x^2}{a^2} + \dfrac{y^2}{b^2} \leq 1\right), \\[2mm] 0 & \left(\dfrac{x^2}{a^2} + \dfrac{y^2}{b^2} > 1\right), \end{cases} \tag{11}$$

and the response ψ_b arising due to the boundary deformation,

$$\Delta\psi_b = \begin{cases} -4\pi G\rho_0\Delta\xi & \left(\dfrac{x^2}{a^2} + \dfrac{y^2}{b^2} = 1\right), \\[2mm] 0 & \left(\dfrac{x^2}{a^2} + \dfrac{y^2}{b^2} \neq 1\right). \end{cases} \tag{12}$$

Equation (11) may be solved in a usual way so that the internal solution is smoothly transformed to the external solution. For the solution of Eq. (12), the Globe–Mikhailenko theorem (cf. [65^{ad}]) is useful: If $\Delta\xi$ is presented in the form of expansion over the two-dimensional Lamé functions of the form

$$\Delta\xi = l_0 \sum_n (\lambda_{2n-1} M_{2n-1} + \lambda_{2n} M_{2n}),$$

the potential of the corresponding simple layer inside the cylinder will be

$$\psi_b = \pi G\rho_0 \sum_n \frac{1}{n} (\lambda_{2n-1} S^0_{2n-1} R_{2n-1} M_{2n-1} + \lambda_{2n} S^0_{2n} R_{2n} M_{2n}), \tag{13}$$

where M are R are the two-dimensional Lamé functions of the first, and S^0 of second kind ("0" denotes the value on the boundary). Expressions for $M, R, \ldots, S$ may be found, for example, in Appel's book [15]. The total response of the potential $\Phi = \Phi_d + \Phi_b$. From the self-consistency condition $\Phi = \Phi_1$ follows both the dispersion equation and the eigenfunctions.

In such a way disturbances in the form of polynomials of second and third orders (for Φ_1) were considered in [92^{ad}]. It was found that elliptical Freeman's cylinders are stable with respect to elliptical deformation (to disturbances of the second order). More cumbersome calculations for disturbances of the third order lead to the dispersion equation so that its numerical investigation gives the stability picture represented in Fig. 62. Regions I are stable. They, in particular, involve "cold" cylinders. Regions II and III (involving the "balanced" and "Jacobian" cylinders) have complex frequencies so that there the instability occurs. Regions III have only purely imaginary frequencies (aperiodical instability). The narrow stability region appears at $b/a \sim 0.5$ and $\Omega^{*2} \sim 0$, while at $b/a > 0.8$ and $\Omega^{*2} > 0.4$ the picture becomes complicated.

The picture of stability shown in Fig. 62 is rather complicated (this especially refers to the above band at $b/a \gtrsim 0.8$; here the cause is likely to be in a strong artificiality of the system under consideration). At a fixed value of b/a, in the larger part of the plane of Fig. 62, instability at small Ω^2 is changed by the region of stable solutions (it involves

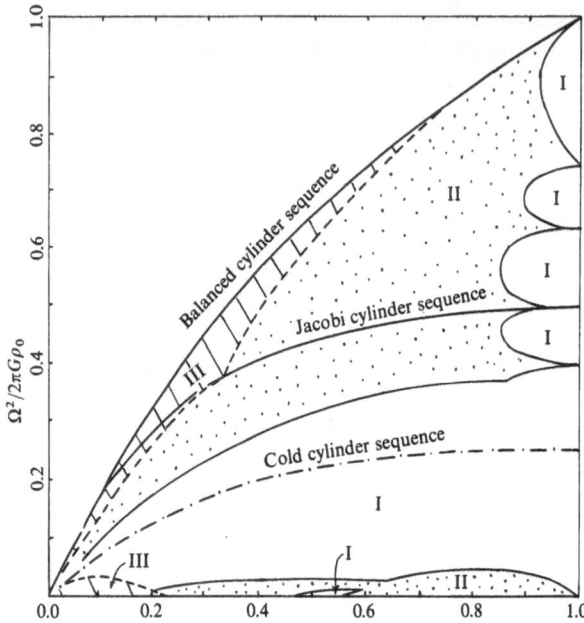

Figure 62. Instability regions for Freeman's elliptical cylinders [92^{dd}].

"cold" cylinders); then, at sufficiently large Ω^2, instability occurs again (the latter involves, in particular, the sequence of Jacobian cylinders and extends up to the maximum possible Ω^2, corresponding to the sequence of "balanced" cylinders). Roughly (especially roughly for large b/a), such a picture may be interpreted as instability of "fire-hose" type in hot weakly rotating (small Ω^2) systems and instability of the Jeans type for a rapid rotation (and, consequently, for a lesser part of the "pressure" of stars in equilibrium).

CHAPTER V

Equilibrium and Stability of Flat Gravitating Systems

Due to a very great flatness of spiral galaxies, they are normally, for the sake of theoretical analysis, represented in the form of infinitely thin disks [34, 355]. A similar representation proves to be useful also for the description of objects of a quite different nature possessing a similar geometry: Saturn's rings, the protoplanetary cloud of the Solar System, and others (see Chapter XII).[1]

In §1 of this chapter, the principles of the theory of stationary states of flat gravitating systems are discussed, while §§2–4 are devoted to the investigation of their small oscillations and stability.

The simplest example of a disk system is a disk with the particles revolving in circular trajectories ("cold" disk). The equilibrium states (Section 1.1) as well as stability (§2) of such systems are investigated most comprehensively.

Instability of a cold disk with respect to perturbation lying in its plane was first demonstrated by Mestel [292] and Hunter [226]. The qualitative picture of the development of this instability, the role of different stabilizing factors (velocity dispersion of stars, rotation, finite thickness of the system) are excellently described by Toomre [333]. He was the first to give also the quantitative analysis of the short-wave instability of cold rotating disks obtaining, in particular, the corresponding dispersion equation [(14), Section 2.2].

---

[1] Here the finiteness of the real system thickness may, if desired, be taken into account as a small correction in the solutions obtained in the approximation of an infinitely thin disk (c.f., e.g., below Sections 1.1 and 4.1.).

Hunter in [229] developed a method similar to the WKB method allowing one, in any order of the perturbation theory, to calculate the eigenfrequencies and eigenfunctions of short-wave oscillations. In particular, it proved that the latter have (for unstable solutions) the form of "leading" spirals (cf. Section 2.2, Fig. 67).

The analytical solution for *large-scale* modes was obtained only for a disk rotating as a solid body [226].

In the case of *nonuniformly rotating* disks, the analysis of large-scale oscillations was performed largely by numerical methods.

A very detailed investigation of spectra of small oscillations of cold rotating disks is given by Hunter [228], who had considered both oscillations lying in the disk plane (Section 2.2) and membrane-type oscillations perpendicular to it, which bend the disk plane (Section 2.1). The oscillation spectra can have both discrete and continuous parts.

Membrane oscillations of cold disks were studied by Hunter and Toomre [230] and Hunter [233] (cf. also [237]) with the aim of explaining the observable bending of the plane of the Galaxy. We discuss the contents of these works in Section 2.1 of this chapter and in Chapter XII. Membrane oscillations of cold disks are normally stable.

Since the gas in galaxies is generally speaking ionized, the study of the influence of the magnetic field on the dynamics of these systems is of interest (cf. §3). The question of the magnetic field effect on the formation of the spiral arm structure in our Galaxy was discussed in detail earlier by Pikelner on the basis of the conditions of plasma equilibrium in the magnetic field [108]. Later, Hosking [221] suggested a model of a rotating, infinitely thin magnetized disk for the investigation of the magnetic field effect on the spiral structure formation in galaxies. He also (and later in [27]) obtained the stabilization criterion for the short-wave gravitational instability by the magnetic field which is already easily derived from qualitative considerations (cf. §3). In §3, it is shown [95, 97] that, in the Hosking model [221], the frequencies of small oscillations, in addition to the discrete spectrum, produce also a continuous spectrum. This section deals with the analysis [95, 97] of not only short-wave but also long-wave perturbations of the disk, both in the self-consistent version (where the magnetic field is produced by currents flowing in the disk itself) and in the version with the external magnetic field. It is shown that, in the case of a fairly strong external field, the spectrum of oscillations consists of a discrete set of real frequencies. If the criterion of the local stability is satisfied only in a part of the disk, then the model possesses a continuous spectrum of frequencies (and, in particular, there is a range of unstable frequencies). In the self-consistent version, only the second case is always realized, so that all the models with their own magnetic field turn out to be unstable [27, 95, 97]. However, the particular character of the models investigated so far for gravitating disks with the magnetic field should be noted.

The influence of chaotic motions of stars on the disk stability is considered in §4. This problem was first investigated by Toomre [333]. To derive the stability criterion, Toomre solved the system involving the time-independent linearized collisionless kinetic equation and the Poisson equation (in the local approximation). It turned out that the radial localized instabilities become completely suppressed if the radial-velocity dispersion of the particles at any point exceeds a certain critical value defined by the equilibrium parameters of the disk (at the same point).

Making use of his stability criterion, Toomre calculated [333] the local values of velocity dispersion minimally necessary for stability in different models of galaxies. It turned out, in particular, that the velocity dispersion available in the solar vicinity of the Galaxy is enough to provide local stability of this region (for more detail cf. §3, Chapter XI).

Lin and Shu generalized [270, 271] the dispersion equation for the cold disk (Toomre) for the case of short-wave oscillations of the disk of stars which possess, apart from a regular velocity, also chaotic velocities (small in comparison with the regular one; Section 4.1). At the same time, they took into account the possibility of the presence of the gaseous component in the system. They used the dispersion equation obtained by them to form the basis of the theory of galactic spiral structure (we consider in detail the relevant issues in §3, Chapter XI).

Later, Mark [290] noticed that some approximations assumed by Lin and Shu in the derivation of the dispersion equation may in fact be valid only for regions of the disk sufficiently separated from resonance circumferences, on which the frequency of intersections by the particles of crests (or valleys) of the spiral wave of perturbations is equal to the integer multiple of the epicyclic frequency. The "improved" dispersion equation inferred by Mark [290] (also for short-wave perturbations) is true everywhere, with the exception of a comparatively narrow region near the so-called ring of corotation, where the particles move with the same angular speed as the wave does. The basic effect following from the Mark theory involves the damping of the short-wave spiral perturbation propagating toward the inside of the disk system (like our Galaxy) from the ring of corotation where it approaches the resonance circumference. The physical cause of damping is the interaction of the wave with resonant particles, i.e., nothing but the "Landau damping" familiar from plasma physics [65].

In the other subsection of §4 (4.4) we consider small perturbations of "hot" disks rotating as a solid body, for which it was possible to obtain (in works by the authors [93, 111, 113] and by Kalnajs [252]) exact characteristic equations for eigenfrequencies and explicit expressions for eigenfunctions. The solutions derived, unlike those considered in the previous section, describe not only short-wave (small-scale) but also large-scale perturbations and not only for "nearly cold" (with orbits close to circular ones) but also for "hot" systems in which the radial and circular velocities of the particles are of the same order.

Investigation of these characteristic equations allows us in particular, to conclude that the most "dangerous," in the sense of loss of stability, for the disk systems in question, are large-scale non-axially symmetrical modes, and, first of all, the "barlike" mode, at which the circular disk becomes elliptical. The instability on this mode develops rapidly, even if the Jeans instability under radial perturbations is suppressed. (The latter is achieved, as noted above, even at the value of radial velocity dispersion of the particles which is given by the Toomre criterion.[2]) Only the "quite hot" disks (for accurate conditions, cf. Section 4.4) turn out to be stable with respect to the formation of a "bar."

The theoretical inferences stated above find confirmation in numerical experiments made by Hohl in [219]. On the other hand, the numerical experiments of Hohl in [220], Miller, Prendergast, and Quirk [294, 308] and finally Ostriker and Peebles [301] show that the barlike mode plays a dominating role also in the case of differentially rotating disks. Because of the undoubted importance of barlike perturbations, they are dealt with in Section 4.4 in more detail (following mainly [254]).

The above discussion about the stability of disk systems may be summed up in the following way. The cold rotating disk is strongly unstable with respect to most of the various types of perturbations. Instabilities of radial perturbations having Jeans nature (cf. beginning of Section 2.2) are suppressed by a relatively small velocity disperion of the particles in accordance with the Toomre criterion. However, then there still remain instabilities of the large-scale nonaxially symmetrical modes linked to the energetic advantage of reconstruction of the circular shape of a rapidly rotating system to produce an elliptical one. For the stabilization of this instability, the radial velocities of the particles must be of the order of azimuthal ones.

Instabilities may be suppressed completely if the system contains a large central mass as is the case of Saturn's rings (cf. §1, Chapter XII). For flat galaxies, the concentration of the density toward the center (equivalent to some central mass) is as a rule insufficient for complete stabilization. For such systems, in the case where the radial velocities of the particles are much less than the circular ones (as, e.g., for the stars in the Galaxy), the conclusion about the existence of a "halo" with a mass of the order of that of a flat subsystem is probably compulsory, owing to the suggestion about the relative stability of the system [301]. Otherwise, for the time of the existence of the Galaxy, there would have developed a noticeable ellipticity of its shape and essential "heating" would have occurred.

Such a conclusion derived directly from the analysis of the disk system stability is one of the most serious arguments for the hypothesis about the existence of a significant "latent mass" (which in part may be concentrated in the galactic "halo," cf., e.g., [301] and references on this point therein).

---

[2] More precisely, close to the value given by this criterion which is, strictly speaking, valid only for short-wave perturbations.

§ 1 Equilibrium States of Flat Gaseous and Collisionless Systems

1.1 Systems with Circular Particle Orbits

The simplest model of a flat gravitating system is an infinitely thin disk with the particles revolving in circular orbits. Equilibrium in this case results from the balance of gravitational attraction of particles and the centrifugal force. The equilibrium condition is therefore expressed by the formula

$$\Omega_0^2(r)r = \frac{\partial \Phi_0}{\partial r}, \tag{1}$$

where Φ_0 is the unperturbed gravitational potential, $\Omega_0(r)$ the local angular velocity of rotation linked with the linear velocity via $v_0(r) = \Omega_0(r) \cdot r$. The surface density $\sigma_0 = \sigma_0(r)$ defines the potential $\Phi_0(r)$ according to the familiar expression for the potential of a simple layer

$$\Phi_0(r) = -G \int_0^R \int_0^{2\pi} \frac{\sigma_0(r')r' \, dr' \, d\varphi'}{\sqrt{r^2 + r'^2 - 2rr' \cos \varphi'}}, \tag{2}$$

R is the radius of the disk.

If the $\sigma_0(r)$ function is assumed to be known, then the respective angular velocity is inferred from (1) and (2) by an elementary calculation. One can easily obtain, in particular, the following examples of the $\sigma_0(r)$ and $\Omega_0(r)$ functions. If $\sigma_0 = \text{const}$, then $\Omega_0^2 = -4G\sigma_0 E'(r/R)/r$, where $E(x)$ is the complete elliptical integral. From the fact of the presence of the (logarithmic) singularity of the $E'(r/R)$ function as $r \to R$, it is necessary that the unperturbed distribution of density $\sigma_0(r)$ vanishes fairly smoothly as $r \to R$. For $\sigma_0(r) = \sigma_0(0)\sqrt{1 - r^2/R^2}$, we have the uniform law of rotation with the angular velocity $\Omega_0^2 = \pi^2 G\sigma_0(0)/2R = 3\pi GM/4R^3$, where M is the mass of the disk; for the density $\sigma_0(r) = \sigma_0(0)(1 - r^2/R^2)^{3/2}$, the square of the rate of revolution: $\Omega_0^2(r) = (3\pi^2/4R)G\sigma_0(0)\sqrt{1 - 3r^2/4R^2}$. The profile of the velocity $v_0(r) = r\Omega_0(r)$, described by this formula, corresponds to the uniform character of the rotation at small r, and reaches its maximum at $r = R\sqrt{\frac{2}{3}} \simeq 0.81R$.

Rewrite (1) and (2) in the form

$$G \int_0^R \int_0^{2\pi} \frac{\sigma_0(r')r' \, dr' \, d\varphi'}{\sqrt{r^2 + r'^2 - 2rr' \cos \varphi'}} = -\Phi_0(0) - f_1(r), \tag{3}$$

where

$$f_1(r) = \Phi_0(r) - \Phi_0(0) = \int_0^r \Omega_0^2(t)t \, dt. \tag{4}$$

If $\Omega_0(r)$ is considered a known function, then (3) is the integral equation with respect to the unperturbed density σ_0. In [355], Eq. (3) was solved for the case where $\sigma_0(r)$ is representable in the form of a power series; the coefficients of this series were chosen so that the best agreement of the $\Omega_0(r)$ function with the observed circular rotational curves of galaxies is produced.

However, integral equation (3) can be solved in the general form [59, 61] for an arbitrary angular velocity $\Omega_0(r)$. A similar problem arises in electrostatics, when it is necessary, using a given potential, to determine the distribution of the surface charge (i.e., in essence, to solve integral equation (3) with the opposite sign on its left-hand side). Following [71], we reproduce briefly the solution of this equation.

Integrating in (3) over φ', we get

$$4G \int_0^R \frac{dr'}{r+r'} K\left(\frac{2\sqrt{rr'}}{r+r'}\right) r'\sigma_0(r') = -\Phi_0(r). \tag{5}$$

Write this equation thus:

$$4G \int_0^r dr' \frac{r'\sigma_0(r')}{r} \frac{1}{1+r'/r} K\left(\frac{2\sqrt{r'/r}}{1+r'/r}\right)$$

$$+ 4G \int_r^R dr' \frac{r'\sigma_0(r')}{r'} \frac{1}{1+r/r'} K\left(\frac{2\sqrt{r/r'}}{1+r/r'}\right) = -\Phi_0(r). \tag{6}$$

Now we can make use of the Landen transformation [133]

$$\frac{1}{1+k} K\left(\frac{2\sqrt{k}}{1+k}\right) = K(k), \tag{7}$$

which holds true at $k < 1$. As a result, we have

$$4G \int_0^r dr' \frac{r'\sigma_0(r')}{r} K\left(\frac{r'}{r}\right) + 4G \int_r^R dr' \frac{r'\sigma_0(r')}{r'} K\left(\frac{r}{r'}\right) = -\Phi_0(r) \tag{8}$$

Taking into account the formulae

$$\frac{1}{r} K\left(\frac{r'}{r}\right) = \int_0^{r'} \frac{ds}{\sqrt{(r'^2 - s^2)(r^2 - s^2)}}, \tag{9}$$

$$\frac{1}{r'} K\left(\frac{r}{r'}\right) = \int_0^{r} \frac{ds}{\sqrt{(r'^2 - s^2)(r^2 - s^2)}}, \tag{10}$$

following directly from the definition of the $K(x)$ function, we represent (8) in the form

$$4G \int_0^r r'\sigma_0(r')\, dr' \int_0^{r'} \frac{ds}{\sqrt{(r'^2 - s^2)(r^2 - s^2)}}$$

$$+ 4G \int_r^R r'\sigma_0(r')\, dr' \int_0^{r} \frac{ds}{\sqrt{(r'^2 - s^2)(r^2 - s^2)}} = -\Phi_0(r). \tag{11}$$

The expression on the left-hand side of (11) can be written in the form of a double integral taken over the area of a trapezium limited by straight lines:

$s = 0$, $s = r$, $s = r'$, $r' = R$. By changing in this double integral the order of integrations [as compared to (11)], we arrive at the following convenient form of the equation:

$$4G \int_0^r \frac{ds}{\sqrt{r^2 - s^2}} \int_s^R \frac{r' \sigma_0(r') \, dr'}{\sqrt{r'^2 - s^2}} = -\Phi_0(r). \qquad (12)$$

If one puts down

$$\int_s^R \frac{r' \sigma_0(r') \, dr'}{\sqrt{r'^2 - s^2}} \equiv \psi(s) \qquad (0 \leq s \leq R), \qquad (13)$$

then Eq. (12) will go over to the Schlömilch integral equation for the $\psi(s)$ function:

$$4G \int_0^r \frac{\psi(s) \, ds}{\sqrt{r^2 - s^2}} = -\Phi_0(r). \qquad (14)$$

The solution for this equation is familiar [132]:

$$\psi(s) = -\frac{d}{ds} \int_0^s \frac{\Phi_0(t) t \, dt}{\sqrt{s^2 - t^2}}. \qquad (15)$$

For the given $\psi(s)$, (13) is also the integral equation with a familiar solution

$$r' \sigma_0(r') = -4G \frac{d}{dr'} \int_{r'}^R \frac{s \psi(s) \, ds}{\sqrt{s^2 - r'^2}}. \qquad (16)$$

Thus, finally the solution of the initial equation of (3) can be written in the form

$$\sigma_0(r) = -\frac{1}{\pi^2 G r} \frac{d}{dr} \int_r^R \frac{s \, ds}{\sqrt{s^2 - r^2}} \frac{d}{ds} \int_0^s \frac{f_1(t) t \, dt}{\sqrt{s^2 - t^2}} - \frac{\Phi_0(0)}{\pi^2 G \sqrt{R^2 - r^2}}. \qquad (17)$$

For each specific $f_1(r)$ function, the constant $\Phi_0(0)$ is defined after calculating the integral in (17) from the condition of the lack in $\sigma_0(r)$ of a singularity at $r = R$. We illustrate this in the example of uniform rotation, when

$$f_1(r) = \frac{\Omega_0^2 r^2}{2}. \qquad (18)$$

The calculation of the iterated integral in (17) leads here to the following expression for the surface density:

$$\sigma_0(r) = \frac{1}{\pi^2 G} \left[\Omega_0^2 \sqrt{R^2 - r^2} - \frac{\Phi_0(0) + \Omega_0^2 r^2}{\sqrt{R^2 - r^2}} \right]. \qquad (19)$$

Hence, it is clear that $\Phi_0(0) = -\Omega_0^2 R^2$, and we again get

$$\sigma_0(r) = \frac{2\Omega_0^2}{\pi^2 G} \sqrt{R^2 - r^2}. \qquad (20)$$

In original work by Kuzmin, the solution for the initial integral equation is given in the following equivalent form:

$$\sigma_0(r') = -\frac{4}{\pi^2 G r'^2} \int_0^\infty \frac{dr^3 \, \Omega_0^2(r)}{dr} \, \varphi\left(\frac{r}{r'}\right) dr, \qquad (21)$$

where

$$\varphi(x) \equiv \begin{cases} xK(x), & x \leq 1, \\ x^{-2}K(x^{-1}), & x > 1. \end{cases} \qquad (22)$$

In theoretical investigations of stability of a cold disk [226, 228, 333], exact solutions in the form of (17) and (21) turn out to be of little convenience, and the representations of solutions in the form of series over various complete systems of the eigenfunctions of the Poisson equation for a flat layer are frequently used. We are aware of several such complete systems [251] whose application depends on the range (confined or infinite) of the change in the disk radius. These systems of functions correspond, as we shall see, to multiplicative solutions of the Laplace equation ensuing in the separation of the variables in this or that frame of reference. For finite disks, the use of the system of Legendre polynomials [226] is most convenient. In [356, 358], for this very case, the method of expansion in Bessel functions was employed. The latter is more convenient, seemingly, for disks not confined in radius $(0 < r < \infty)$. That case was first suggested by Toomre [332]. Note also another complete system of eigenfunctions, which is considered in [251]— logarithmic spirals.

In the method of expansion over Legendre polynomials applied by Hunter [226], the oblate spheroidal coordinates (ζ, μ, φ) (Fig. 50) are introduced.[3] The Laplace equation for the axially symmetrical potential $\Phi_0 = \Phi_0(\zeta, \mu)$ in spheroidal coordinates is written as

$$\frac{\partial}{\partial \mu}\left\{(1 - \mu^2)\frac{\partial \Phi_0}{\partial \mu}\right\} + \frac{\partial}{\partial \zeta}\left\{(\zeta^2 + 1)\frac{\partial \Phi_0}{\partial \zeta}\right\} = 0. \qquad (23)$$

The elementary multiplicative solutions for this equation, as is readily shown, are

$$\Phi_0^{(1)} = p_n(\zeta)P_n(\mu), \qquad (24)$$

$$\Phi_0^{(2)} = q_n(\zeta)P_n(\mu), \qquad (25)$$

where $P_n(\mu)$ are the ordinary Legendre polynomials, and $q_n(\zeta)$ are the Legendre functions of the second kind of the imaginary argument defined by the equality

$$q_n(\zeta) = p_n(\zeta) \int_\zeta^\infty \frac{d\zeta}{[p_n(\zeta)]^2(\zeta^2 + 1)}. \qquad (26)$$

---

[3] Of course, in electrostatics, these expansions had been employed long before Hunter used them.

The general axially symmetrical solution for the Laplace equation outside the disk, regular at infinity, is thus written in the form

$$\Phi_0(\zeta, \mu) = \sum_n A_n q_n(\zeta) P_n(\mu), \tag{27}$$

where A_n are the expansion constants. On the disk surface, $\zeta = 0$ Hunter defines the coordinates $\eta = \sqrt{1 - r^2/R^2} = |\mu|$, $0 \leq \eta \leq 1$. If the surface density σ_0 and the potential Φ_0 on the disk surface are represented in the form

$$2\pi G\sigma_0 = \frac{1}{R\eta} \sum_n \alpha_{2n} P_{2n}(\eta), \tag{28}$$

$$\Phi_0|_{\zeta=0} = \sum_n A_{2n} q_{2n}(0) P_{2n}(\eta), \tag{29}$$

then from the boundary conditions

$$\Phi_0|_{z=-0}^{z=+0} = 0, \tag{30}$$

$$\frac{\partial \Phi_0}{\partial z}\bigg|_{z=+0} - \frac{\partial \Phi_0}{\partial z}\bigg|_{z=-0} = 4\pi G\sigma_0, \tag{31}$$

it is possible to infer the relationship between the coefficients of these expansions: $A_{2n} = \alpha_{2n}/q'_{2n}(0)$. If the density is assumed to be present in the form of (28), then for the potential in the plane of the disk we obtain the expansion

$$\Phi_0(\zeta = 0) = \sum_n \alpha_{2n} \frac{q_{2n}(0)}{q'_{2n}(0)} P_{2n}(\eta) = -\frac{\pi}{2} \sum_n \alpha_{2n} \left[\frac{(2n-1)!!}{(2n)!!} \right]^2 P_{2n}(\eta). \tag{32}$$

Now the linear velocity $v_0(r)$ is defined from the equilibrium condition.

The series obtained in (28) and (32) are evidently expansions over "gravitational multipoles;" for example, the quadrupole field corresponds to the term $n = 2$ in (32).

In [226], specific models of equilibrium disks as calculated in this way are presented. Out of the linked systems of the σ_0, Φ_0, $\Omega_0^2(r)$ functions, let us give the following [226]:

$$\sigma_0 = \frac{3M}{2\pi R^2} [\eta + \alpha(5\eta^3 - 3\eta)] = \frac{3M}{2\pi R^2} \left(1 - \frac{r^2}{R^2}\right)^{1/2} \left(1 + 2\alpha - 5\alpha \frac{r^2}{R^2}\right),$$

$$\Phi_0 = -\frac{3\pi GM}{8R} \left[2 - \frac{3\alpha}{2} - \left(1 + \frac{27}{4}\alpha\right)\left(\frac{r}{R}\right)^2 + \frac{45}{16}\alpha\left(\frac{r}{R}\right)^4\right], \tag{33}$$

$$\Omega_0^2 = \frac{3\pi GM}{4R^3} \left[1 + \frac{9\alpha}{2} - \frac{45}{8}\alpha\left(\frac{r}{R}\right)^2\right] \quad (0 < \alpha < 1)$$

In a later work [227], Hunter, by the same method, obtained an interesting sequence of the models of disks of a finite radius (n is any positive integer):

$$v_0^2(r) = r^2 \Omega_0^2(r) = \frac{2n+1}{4n} \left(\frac{\pi GM}{R} \right)(1 - \zeta^{2n})$$

$$\sigma_{0,n}(r) = \frac{M}{2\pi R^2} \sum_{k=1}^{n} b_{k,n} \zeta^{2k-1}, \qquad \zeta^2 = 1 - \frac{r^2}{R^2}, \tag{34}$$

$$b_{1,n} = \frac{2n+1}{2n-1}, \qquad b_{k,n} = \frac{4(k-1)(n-k+1)}{(2k-1)(2n-2k+1)} b_{k-1,n}.$$

The case $n = 1$ in (34) refers to the model of a uniform rotating disk. With an increase of the number n of the model, rotation becomes more nonuniform, while within the limit $n \to \infty$ there ensues a singular model [292] of a disk with a constant linear velocity

$$v_0^2 = \frac{\pi GM}{2R} = \text{const}, \qquad \sigma_0(r) = \frac{M \arcsin \zeta}{2\pi r R}. \tag{35}$$

Hunter and Toomre in [230], on the basis of the series of (34), constructed a large number of other models, being certain superpositions of the solution in (34), and made an attempt to select a not very complicated model satisfactorily describing the distribution of mass and the curve of galactic rotation.

Since we frequently use below, for the sake of illustration, the Hunter and Toomre models let us briefly describe the method of the derivation of these models (irrespective of the problem of the degree of similarity or difference with the Galaxy [230]).

The initial models (34) are denoted simply by the number n. The second series of models denoted by the symbol n/q is constructed in the following way (take, e.g., $n = 16$, as in [230]):

$$\sigma_{0(16/q)} = \frac{\sigma_{0(16)} - a\sigma_{0(1)}}{1 - a}, \qquad a = \left[1 + \frac{100}{q} \left(\frac{b_{1,1}}{b_{1,16}} - 1 \right) \right]^{-1},$$

where the parameter q can take on any values between 0 and 100. The models of $16/q$ resemble the model of 16; however, with increasing q from 0 to 100, from $\sigma_0(\zeta)$ is subtracted a still larger part of the term proportional to $\zeta((b_{1,16} - ab_{1,1})\zeta)$ and finally, at $q = 100$ ($a = b_{1,16}/b_{1,1}$), this term is subtracted completely. The limiting model 16/100 is denoted in [230] by the symbol $16x$.

One can in a similar way define the models $16//q$ differing from the model $16x$ mainly in a gradual subtraction of the term that is proportional to ζ^3, with the limiting model (when this term is eliminated completely) denoted as $16\,xx$. This procedure can be continued indefinitely; its purpose is to obtain models (in the series n, nx, nxx, $nxxx$, etc.) with a still smoother trend of the surface density to zero at the end of the disk or, as we shall otherwise say, models with a still "thinner edge." (With an increase in the number n, we turn to the systems with the increasing degree of concentration of the mass toward the center.)

Separation of the variables in cylindrical coordinates naturally leads to the Bessel functions. If the surface density σ_0 is represented in the form

$$\sigma_0(r) = \int_0^\infty J_0(kr)kS(k)\,dk, \tag{36}$$

where $S(k)$ is some weight function, then the potential is expressed with the help of a similar formula by the same function $S(k)$:

$$\Phi_0(r) = -2\pi G \int_0^\infty J_0(kr)S(k)\,dk. \tag{37}$$

The transformation inverse to (36) yields $S(k)$ through $\sigma_0(k)$:

$$S(k) = \int_0^\infty J_0(ur)u\sigma_0(u)\,du. \tag{38}$$

The method of expansion in Bessel functions is well illustrated by the practically important example of "exponential disks," i.e., disks with the surface density $\sigma = \sigma_0 e^{-\alpha r}$ [205]. In this case, the Toomre method [332] allows one to obtain a simple analytical expression for the equilibrium rotation velocity. Substituting (38) into (37) and calculating [for the case $\sigma_0(r) = \sigma_0 e^{-\alpha r}$] the ensuing integrals, it is possible to find $\Phi_0(r)$ and, further, the velocity of rotation $v_0(r)$ in the form

$$v_0^2(r) = \pi G \sigma_0 \alpha r^2 \left[I_0\!\left(\frac{\alpha r}{2}\right) K_0\!\left(\frac{\alpha r}{2}\right) - I_1\!\left(\frac{\alpha r}{2}\right) K_1\!\left(\frac{\alpha r}{2}\right) \right], \tag{39}$$

where I and K are the known cylindrical functions. The law of rotation (39), as it appears [205], coincides satisfactorily with the curves of rotation of some galaxies.

The problem of restoration of the mass distribution from the observed rotational curve arises constantly in galactic astronomy, and, in this connection, some rather fine methods of calculation were created. In particular, they take into account the finite thickness of the disks of spiral galaxies and their structure in the vertical direction. The simplest and most widely used model of such a type is the model of the nonuniform spheroid (surfaces of identical density—similar spheroids), suggested first by Kuz'min [59, 61].

In [59, 61] it is shown that, in this case, the velocity of rotation in the equatorial plane is expressed by the formula

$$v_0^2(r) = G \int_0^r \frac{\mu(a)\,da}{\sqrt{r^2 - e^2 a^2}}, \tag{40}$$

where the so-called "function of mass" $\mu(a) = 4\pi a^2 \varepsilon \rho(a)$, with $\rho(a)$ the volume density depending on the value of the equatorial semiaxis of the spheroid of identical density, $a = (r^2 + \varepsilon^{-2}z^2)^{1/2}$, ε is the ratio of the polar and equatorial semiaxis of the elliptical meridional cross section, and e is its eccentricity. Equation (40) was later [59, 61] derived in [173], where it is solved by means of representation of the $\rho(a)$ function in the form of a power polynomial; then $v_0^2(r)$ is also obtained in the form of polynomial.

In the cases of small thickness of the disk, an analytical expression [59] may be derived for corrections to the formulae of the model of an infinitely thin disk. Equation (40) in this case ($\varepsilon^2 \ll 1$) is reduced to the following:

$$v_0^2(r) = G\left[\int_0^r \frac{\mu(a)\,da}{\sqrt{r^2 - a^2}} - \varepsilon\mu(r)\right]. \tag{41}$$

By solving it according to perturbation theory, it is possible to obtain [59]

$$\mu(a) = \mu_0(a) + \frac{2\varepsilon}{\pi}\int_0^a \frac{d\mu_0(r)r}{dr}\frac{r\,dr}{a\sqrt{a^2 - r^2}}, \tag{42}$$

where $\mu_0(a)$ is the "function of mass" for an infinitely thin disk. In [59], other effective methods of investigation of such equilibrium systems are also developed.

More complicated, heterogeneous models of spiral galaxies consisting of spherical and flat subsystems are considered in [94]. In this work, the volume density of the spherical subsystem and the angular frequency of rotation of the flat subsystem are imposed (within rather wide limits). Also, observed data on our and neighboring spiral galaxies was essentially used. In all the cases considered, the surface density decreased exponentially outside the boundaries of the nucleus of the spherical subsystem, and the dependence of the value of the exponential index on the relative size of the nucleus of the spherical subsystem agrees qualitatively with observed data. Also the dependence (a power character) between the total mass and total momentum of the flat subsystem is established. It turns out that the power index ν of this dependence lies [78] within narrow limits $1.6 < \nu < 1.8$.

1.2 Plasma Systems with a Magnetic Field

The technique of obtaining exact models of infinitely thin disks with the poloidal magnetic field similar to that in [226] is developed by Bisnovaty-Kogan and Blinnikov in [27].

The equilibrium of such a disk is described by the equation

$$\Omega_0^2 r = -\frac{\partial\Phi_0}{\partial r} + \frac{1}{c\sigma_0}J_{0\varphi}H_{0z}, \tag{1}$$

where H_{0z} and $J_{0\varphi}$ are respectively the vertical component of the magnetic field and the density of the azimuthal current on the disk surface. The **H** and **J** vectors are linked by the Maxwellian equation

$$\text{rot }\mathbf{H} = \frac{4\pi}{c}\mathbf{J}\,\delta(z). \tag{2}$$

Outside the disk rot $\mathbf{H} = 0$, so that one can introduce the scalar potential Φ_H of the magnetic field

$$\mathbf{H} = -\operatorname{grad}\Phi_H. \tag{3}$$

This potential Φ_H satisfies at $z \neq 0$ the Laplace equation

$$\Delta\Phi_H = 0. \tag{4}$$

The normal component H_z of the magnetic field is continuous on the disk, whereas the tangential one undergoes discontinuity with the value $(4\pi/c)J_\varphi$; therefore, the corresponding boundary conditions for the potential Φ_H are written in the form

$$\frac{\partial\Phi_H}{\partial z}\bigg|_{z=-0}^{z=+0} = 0, \tag{5}$$

$$\frac{\partial\Phi_H}{\partial r}\bigg|_{z=-0}^{z=+0} = -\frac{4\pi}{c}J_\varphi. \tag{6}$$

The general expansion of Φ_H at $z \neq 0$ is similar to (27) of Section 1.1:

$$\Phi_H = \sum_n B_n q_n(\zeta)P_n(\mu), \tag{7}$$

where B_n are the new constants. However, owing to the condition in (5) there now remain only odd terms in expansion (7). The second boundary condition in (6) in spheroidal coordinates, with due regard for symmetry of the problem, will be written in the form

$$\frac{1}{h_\mu}\frac{\partial\Phi_H}{\partial\mu}\bigg|_{\zeta=0} = \frac{2\pi}{c}J_{0\varphi}, \tag{8}$$

where the Lamé parameter

$$h_\mu = R\left(\frac{\zeta^2 + \mu^2}{1 - \mu^2}\right)^{1/2}. \tag{9}$$

It is convenient to take for the current $J_{0\varphi}$ a certain expansion in derivatives from the Legendre polynomials

$$\frac{2\pi}{c}J_{0\varphi} = \frac{(1 - \eta^2)^{1/2}}{R\eta}\sum_n \beta_{2n+1}P'_{2n+1}(\eta), \tag{10}$$

where β_{2n+1} are the constants of expansion. Then from (8) we obtain the following link between β_{2n+1} and B_{2n+1}:

$$B_{2n+1} = \frac{\beta_{2n+1}}{q_{2n+1}(0)}. \tag{11}$$

The magnetic field H_z in the disk plane is

$$H_z|_{\zeta=0} = -\frac{1}{R\eta}\frac{\partial\Phi_H}{\partial\zeta}\bigg|_{\zeta=0} = \frac{1}{R\eta}\sum_n \beta_{2n+1}\frac{q'_{2n+1}(0)}{q_{2n+1}(0)}P_{2n+1}(\eta)$$

$$= \frac{\pi}{2R\eta}\sum_n \beta_{2n+1}\left[\frac{(2n+1)!!}{(2n)!!}\right]^2 P_{2n+1}(\eta). \tag{12}$$

Equations (10) and (12) do represent the sought-for pair of the linked expansions of $J_{0\varphi}$ and $H_{0z}|_{z=0}$. Thereafter, from equilibrium equation (1) one can also find the angular velocity Ω_0.

To illustrate, let us give an example that is the simplest generalization of the model of the cold uniformly rotating disk for the case of a nonzero magnetic field [27]. For the disk with the surface density

$$\sigma_0 = \frac{3M}{2\pi R^2}\left(1 - \frac{r^2}{R^2}\right)^{1/2}$$

and current $J_{0\varphi} = J_0(r/R)(1 - r^2/R^2)^{1/2}$, it is possible to calculate the magnetic field $H_z = (\pi^2 J_0/2c)[1 - \frac{3}{2}r^2/R^2]$ (c is the light velocity) and the angular velocity

$$\frac{v_0^2}{r^2} = \Omega_0^2 = \frac{3\pi GM}{4R^3} - \frac{4H_0^2 R}{3\pi M}\left[1 - \frac{3}{2}\frac{r^2}{R^2}\right], \qquad H_0 = \frac{\pi}{2c}J_0. \qquad (13)$$

The maximally possible field is equal to $H_0 = (3\pi/4)(\sqrt{GM}/R^2)$, then

$$\Omega^2 = \frac{9\pi GM}{2R^3}\frac{r^2}{R^2}. \qquad (14)$$

At $H > H_0$, we would have $\Omega^2 < 0$ at the center of $r = 0$. Note that the field H_z changes sign inside the disk at the point $r = \sqrt{\frac{2}{3}}R$. This circumstance is not occasional: it may be shown [27, 95, 97] that, at any regular current distribution, the field H_z necessarily changes its sign at $r \leq R$. This fact happens to be important in the analysis of the stability of such systems [22, 95, 97] (see §3, this chapter). It is very easy to prove [95, 97] this by making use, instead of (10) and (12), of a somewhat different representation of H_{z0} and $J_{0\varphi}$, which can be inferred in a similar way:

$$J_{0\varphi} = \frac{1}{\pi}\sqrt{1 - \eta^2}\sum_{n=1}^{\infty}\frac{g_{2n}P'_{2n}(\eta)}{\gamma'_{2n}}, \qquad (15)$$

$$H_{z0} = \sum_{n=1}^{\infty}g_{2n}P_{2n}(\eta), \qquad (16)$$

where

$$\gamma'_{2n} \equiv n(2n - 1)\left[\frac{(2n - 1)!!}{(2n)!!}\right]^2. \qquad (17)$$

In the summations of (15) and (16) there are no terms with $n = 0$, since in the expression for the current this would have corresponded to the singular term. Let us prove this by the rule of contraries, i.e., assume that H_{z0} does not change sign on the disk, for example everywhere $H_{z0} \gtrless 0$.

Then, integrating (16), we arrive at the contradiction

$$0 < \int_0^1 H_{z0}\,d\eta = \sum_{n=1}^{\infty}g_{2n}\int_0^1 P_{2n}\,d\eta = -\sum_{n=1}^{\infty}g_{2n}\int_0^1\frac{[(1 - \eta^2)P'_{2n}]'\,d\eta}{2n(2n + 1)}$$

$$= -\sum_{n=1}^{\infty}\frac{g_{2n}(1 - \eta^2)P'_{2n}(\eta)}{2n(2n + 1)}\Bigg|_0^1 = 0. \qquad (18)$$

Thus, the component H_{z0} of a self-consistent magnetic field ought to change sign on the disk.

In [27], other examples of exact models of the magnetized gravitating disks are given.

1.3 Gaseous Systems

Consider briefly some flat gaseous systems with a pressure. Since the gaseous pressure P_0 is isotropic and the equilibrium in the vertical direction (along the z-axis) requires identity of the pressure force $(1/\rho_0)(dP_0/dz)$ and the gravitational force $\partial\Phi_0/\partial z$, then it is clear that in the case of an infinitely thin disk there must be $P_0 = 0$. The disk thickness h seems to have to be of the order of the Jeans wavelength, i.e., $h^2 \sim D^2/G\rho_0$ (D is the sound velocity) or $h \sim D^2/G\sigma_0$. At the same time, the characteristic scale of length in the disk plane $R \sim v_0^2/G\sigma_0$; hence the natural condition of a strong flatness of the disk follows $(h \ll R): D \ll v_0$. The equation of disk equilibrium that takes into account the force of gaseous pressure P_0', calculated for length unit in the disk plane, may be written in the form

$$r\Omega_0^2 = \frac{\partial\Phi_0}{\partial r} + \frac{1}{\sigma_0}\frac{\partial P_0'}{\partial r}. \tag{1}$$

If one assumes the polytropic equation of state to be $P_0 = k\rho_0^\gamma$, then the corresponding relationship between P_0' and σ_0 is thus [233]

$$P_0' = \text{const } \sigma_0^{3-2/\gamma} \equiv \text{const } \sigma_0^{\gamma^*}. \tag{2}$$

This formula can be inferred already from ordinary dimensional analysis if one writes

$$P_0' = ck^\alpha G^\beta \sigma_0^{\gamma^*} \tag{3}$$

(c is the dimensionless constant) and determines the degrees α, β, γ^* by equaling the dimensions of the values standing in both parts of the equality (note that $[P_0'] = MT^{-2}$, $[k] = M^{1-\gamma}L^{3\gamma-1}T^{-3}$, $[G] = M^{-1}L^3T^{-2}$). Therefore, it is easy to obtain $\alpha = 1/\gamma$, $\beta = 1 - 1/\gamma$, $\gamma^* = 3 - 2/\gamma$. The effective adiabatic index γ^* is, consequently, $\gamma^* = 3 - 2/\gamma$. It is coincident with γ, if $\gamma = 1$ or 2, and for $1 < \gamma < 2: \gamma^* > \gamma$. The ordinary case of $\gamma = \frac{5}{3}$ refers to $\gamma^* = \frac{9}{5}$. The constant c can also be calculated [233]; it turns out to be

$$c = \frac{\pi^{3/2-1/\gamma}\Gamma(2 - 1/\gamma)}{2^{2-1/\gamma}\Gamma(\frac{5}{2} - 1/\gamma)}. \tag{4}$$

In investigations of the stability of flat gaseous systems adopted normally there was [209, 210] the model of a thin layer homogeneous in the plane (x, y), so that the presence of pressure did not influence the equilibrium structure of the layer. Allowance for pressure in nonhomogeneous gaseous disks is not associated with any principled difficulties.

1.4 "Hot" Collisionless Systems

We turn finally to the consideration of the stationary states of flat collisionless systems. Here one usually bears in mind spiral galaxies. It is well known that galaxies have a spiral and a circular structure only in the case of a sufficiently strong flatness. But then the gravitational potential of the real ellipsoid of a galaxy may be approximated by a simpler potential of an infinitely thin disk. In theoretical investigations of spiral galaxies, the above-considered equilibrium models are used most frequently because the contribution of the random-velocity dispersion to the equilibrium equation is normally small. However, when the residual velocities of the particles "are switched on," then differences of collisionless systems from gaseous ones (collisional) are manifested at once. One of them implies that the former may in principle have an anisotropic distribution in velocities. Therefore, in particular, in this case one can quite strictly consider the models of infinitely thin but nevertheless "hot" disks, i.e., those having a finite random-velocity dispersion in its plane.

The distribution function $f_0 \sim \delta(v_r)\delta(v_\varphi - \Omega_0 r)$ for a cold disk, where $r\Omega_0^2 = d\Phi_0/dr$ and v_r and v_φ are respectively the radial and azimuthal velocities of particles. The distribution function of any quasistationary disk system with orbits of stars close to circular ones can be represented in a similar way as in item 3, §1, Chapter III, in the form of a formal series in the δ function from the arguments v_r and $v_\varphi - \Omega r$ and their derivatives:

$$f_0 = a_1 \delta(v_r)\delta(v_\varphi - \Omega r) + b_1 \delta'(v_r)\delta(v_\varphi - \Omega r) + b_2 \delta(v_r)\delta'(v_\varphi - \Omega r)$$
$$+ c_1 \delta''(v_r)\delta(v_\varphi - \Omega r) + c_2 \delta'(v_r)\delta'(v_\varphi - \Omega r) + c_3 \delta(v_r)\delta''(v_\varphi - \Omega r). \quad (1)$$

Substituting expansion (1) into the collisionless kinetic equation

$$v_r \frac{\partial f}{\partial r} - \frac{\partial \Phi_0}{\partial r}\frac{\partial f}{\partial v_r} + \frac{v_\varphi^2}{r}\frac{\partial f}{\partial v_r} - \frac{v_r v_\varphi}{r}\frac{\partial f}{\partial v_p} = 0, \quad (2)$$

and setting equal to zero the coefficients in different combinations of derivatives of the δ functions, we obtain[4] the system of equations

$$\frac{1}{r}\frac{\partial}{\partial r}(rb_1) = 0, \qquad -2\Omega b_2 - \frac{2}{r}\frac{\partial}{\partial r}(rc_1) + \frac{2c_3}{r} = 0,$$

$$\frac{\varkappa^2}{2\Omega}b_1 - \frac{1}{r^2}\frac{\partial}{\partial r}(r^2 c_2) = 0, \qquad -2\Omega c_2 = 0, \quad (3)$$

$$-4\Omega c_3 + \frac{\varkappa^2}{2\Omega}2c_1 = 0, \qquad \frac{\varkappa^2}{2\Omega}c_2 = 0,$$

where $\varkappa^2 \equiv 4\Omega^2(1 + r\Omega'/2\Omega)$ is the square of the epicyclic frequency. Hence at $\Omega \neq 0$ equalities follow (being, as we see, simply a consequence of sug-

---

[4] It is clear that zero equalities imply here, in fact, that the corresponding combinations of the values must be small, such as $O(c_{r,\varphi}^{3/2}/v_0^{3/2})$.

gestions about an infinitely thin disk and about the closeness of the trajectories to the circular ones): $c_2 = 0$ ($\overline{v_r v_\varphi} = 0$); $b_1 = 0$ (i.e. the lack of a radial flux: $\overline{v}_r = 0$); $c_3 = \varkappa^2 c_1/4\Omega^2$ (i.e., the familiar Lindblad link between the radial c_r and transversal c_φ velocity dispersions of stars: $c_\varphi = \varkappa c_r/2\Omega$); $b_2 = [c_3 - (rc_1)']/\Omega r$ (relation between the transversal flux $\overline{v}_\varphi \neq 0$, the velocity dispersion $c_{r,\varphi}$ and density [101]).

Thus, there ensues the general expression for the distribution functions of such systems [93] as follows:

$$
f_0 = \sigma_0(r)\left\{\delta(v_r)\delta(v_\varphi - \Omega r) + \frac{1}{2\Omega r}\left[\frac{\varkappa^2}{4\Omega^2}c_r^2 - \frac{\partial}{\partial r}(rc_r^2)\right]\right.
$$
$$
\times \delta(v_r)\delta'(v_\varphi - \Omega r) + \tfrac{1}{2}c_r^2(r)\delta''(v_r)\delta(v_\varphi - \Omega r)
$$
$$
\left. + \frac{1}{2}\frac{\varkappa^2}{4\Omega^2}c_r^2(r)\delta(v_r)\delta''(v_\varphi - \Omega r) + \cdots\right\}
\tag{4}
$$

The coefficients for different combinations of derivatives of the δ functions in (4) are evidently proportional to the corresponding moments of lower orders of the distribution function. In the general case one can arbitrarily impose the surface density $\sigma_0(r)$ [or the law of rotation of $\Omega_0(r)$; the relation between $\sigma_0(r)$ and $\Omega_0(r)$ is defined from the Poisson equation] and the velocity dispersion $c_\varphi(r)$ [or $c_r(r)$] which ought to be small. Expansions of the form in (4) are convenient also for taking into account the influence of the small velocity dispersion on the hydrodynamical Jeans instability.

The exact distribution functions for disk systems with a finite (and not necessarily small) velocity dispersion are known for a small number of cases, mainly for the surface density of the form $\sigma_0(r) = \sigma_0(0)\sqrt{1 - r^2/R^2}$ (R is the radius of the boundary).

The distribution function for the elliptical disk at rest with the surface density $\sigma_0(x, y) - \sigma_0(1 - x^2/a^2 - y^2/b^2)^{1/2}$ and the potential $\Phi_0(x, y) - \tfrac{1}{2}A^2x^2 + \tfrac{1}{2}B^2y^2 + \text{const}$ was derived by Freeman [204]. It has the form (see §1, Chapter IV).

$$
f_0 = \frac{\sigma_0}{2\pi AB}\left(1 - \frac{v_x^2}{a^2 A^2} - \frac{v_y^2}{b^2 B^2} - \frac{x^2}{a^2} - \frac{y^2}{b^2}\right)^{-1/2}.
$$

At $a = b$, this distribution function goes over to the following:

$$
f_0 = \frac{\sigma_0}{2\pi}\left(1 - \frac{r^2}{a^2} - \frac{v^2}{2a^2 A^2}\right)^{-1/2}.
\tag{5}
$$

It describes the nonrotating circular disk with an isotropic velocity distribution. The trajectories of stars in it are ellipses with arbitrary eccentricities. In the work of Bisnovatyi-Kogan and Zel'dovich [23], there is the generalization of (5) for the case of uniformly rotating circular disks with the surface density $\sigma = \sigma_0(1 - r^2)^{1/2}$ and the potential $\Phi_0 = \tfrac{1}{2}\Omega^2 r^2 + \text{const}$. In the frame of reference associated with the disk, this function is of the form

$$f_0 = \frac{\sigma_0}{2\pi\sqrt{1-\gamma^2}}[(1-\gamma^2)(1-r^2) - v_r^2 - v_\varphi^2]^{-1/2}, \qquad |\gamma| \le 1, \qquad (6)$$

where γ is the angular velocity of rotation with respect to the z-axis in Ω_0 units (it is assumed that $\Omega_0 = 1$) and the radius of the disk is assumed to be unity. At $|\gamma| \to 0$, (6) describes the nonrotating disk (5), while at $|\gamma| \to 1$, a disk with circular orbits (all stars rotate in the same direction with the angular velocity $\gamma = \Omega_0 = 1$). The distribution function in (6) can, of course, be represented in the form which would explicitly show its dependence on the integrals of motion E, L_z:

$$f_0 \sim (E_0 - E - \gamma L_z)^{-1/2}.$$

The stability of disks of (6) is investigated in [93, 111, 113, 252] (cf. Section 4.3 below). In [23], it is noticed that the distribution function of the uniformly rotating disks of the type of (6) permits an essential generalization

$$f_{0\gamma} \to f_0 = \sigma_0(0) \int_{-1}^{1} A(\gamma) f'_{0\gamma} \, d\gamma, \qquad (7)$$

where $A(\gamma)$ is the function of the parameter γ, satisfying the condition of normalization of $\int_{-1}^{1} A(\gamma) \, d\gamma = 1$, and arbitrary otherwise; $f'_{0\gamma}$ is the original distribution function of (6) but written, for all γ, in the same frame of reference (inertial). The integral in (7) is equivalent to the following [93] $[A(\gamma) \equiv a(\gamma)\sqrt{1-\gamma^2}]$:

$$f_0 = \frac{\sigma_0(0)}{2\pi} \int_{-\alpha}^{\alpha} \frac{a(x + rv_\varphi) \, dx}{\sqrt{\alpha^2 - x^2}}, \qquad \alpha^2 = (1-r^2)(1 - v_\varphi^2) - v_r^2. \qquad (8)$$

Among this continuum of distribution functions, one can, of course, find distributions with a quite realistic dependence on velocity, for example, close to the Maxwellian one (of course, in the range of sufficiently small peculiar velocities in comparison with the velocity of escape). Even among the simplest examples given below, there are functions corresponding to everywhere monotonically decreasing one-dimensional distribution functions. It is remarkable that the dispersion equations for frequencies of small oscillations of all these disks are quite automatically obtained from the dispersion equation corresponding to the system with the initial distribution function $f_{0\gamma}$ (cf. Section 4.4). Let us give some particular examples of distribution functions of the type of (8) for some $a(\gamma)$. For $a(\gamma) = \text{const}$ we have [23]

$$f_0 = \frac{\sigma_0}{\pi}\theta[(1-r^2)(1-v_\varphi^2) - v_r^2] \equiv \frac{\sigma_0}{\pi}\theta(\alpha^2). \qquad (9)$$

The corresponding one-dimensional distributions

$$f_0(v_\varphi) \sim (1-r^2)(1-v_\varphi^2), \qquad f_0(v_r) \sim 1 - r^2 - v_r^2$$

are everywhere decreasing velocity functions with a "Maxwellian" behaviour near $v = 0$, the velocity dispersion in this case is of the order of the circular velocity (~ 1).

Let now $a(\gamma) = \text{const} \cdot \gamma^2$; then

$$f_0 = \text{const} \cdot \theta(\alpha^2)(\alpha^2 + 2r^2 v_\varphi^2) \equiv \text{const} \cdot \theta[(1 - r^2)(1 - v_\varphi^2) - v_r^2]$$
$$\times [1 - r^2 - v_r^2 - (1 - 3r^2)v_\varphi^2].$$

The one-dimensional distribution function in v_φ thus becomes the following:

$$f_0(v_\varphi) \sim \sqrt{(1 - r^2)(1 - v_\varphi^2)}[1 - r^2 - (1 - 4r^2)v_\varphi^2].$$

This distribution is a monotonically decreasing one for $r < 1/\sqrt{3}$, while for $r > 1/\sqrt{3}$, f_0 increases with an increase in v_φ from 0 to

$$v_\varphi = \sqrt{(1 - 3r^2)/(1 - 4r^2)},$$

then it decreases up to zero at $r = 1$. Similarly, the distribution function of radial velocities is

$$f_0(v_r) \sim (1 - r^2 - v_r^2)^{3/2};$$

it everywhere decreases monotonically.

For $a(\gamma) = \text{const}(1 - \gamma^2)$, we shall obtain

$$f_0 = \text{const} \, \theta(\alpha^2)(2 - 2r^2 v_\varphi^2 - \alpha^2)$$
$$\equiv \text{const} \, \theta[(1 - r^2)(1 - v_\varphi^2) - v_r^2][(1 + r^2) + v_r^2 + (1 - 3r^2)v_\varphi^2].$$

The corresponding one-dimensional distribution functions are written in the form

$$f_0(v_\varphi) \sim \sqrt{1 - v_\varphi^2}[2 + r^2 + (1 - 4r^2)v_\varphi^2],$$
$$f_0(v_r) \sim \sqrt{1 - r^2 - v_r^2}(2 - 2r^2 + v_r^2).$$

It can be shown that these two functions are monotonically decreasing ones (within the velocity range admissible for the particles).

In [29], the distribution functions of a customary form

$$f_0 = AL_z^m(-E)^n \tag{10}$$

are obtained for unlimited disks with the power dependence of the surface density and the potential (in the plane of the disk) on the radius

$$\sigma_0 = \varkappa r^p, \tag{11}$$

$$\Phi_0 = -kr^{p+1}. \tag{12}$$

The constants $\varkappa$ and k are connected by the easily provable relation

$$k = -\pi G\varkappa \frac{\Gamma[-(p + 1)/2]\Gamma[(p + 2)/2]}{\Gamma(-p/2)\Gamma[(p + 3)/2]}. \tag{13}$$

On the other hand, integrating (10) over velocities, we find the surface density

$$\sigma_0 = Aa_{mn}k^{1 + n + m/2}r^{m + (p + 1)(1 + n + m/2)} \equiv \varkappa r^p, \tag{14}$$

where

$$a_{mn} = 2^{1+m/2} \frac{\sqrt{\pi}}{4} \frac{\Gamma(n+1)\Gamma[(m+1)/2]}{\Gamma(n+m/2+2)}. \tag{15}$$

Comparing (14) and (11), we find the following relations between the above derived constant parameters:

$$p = -\frac{3m+2n+2}{n+2m}, \qquad n = -\frac{2+m(p+3)}{2(p+1)}, \tag{16}$$

$$A = \frac{\varkappa}{a_{mn}k^{1+n+m/2}} \qquad (m, n > -1).$$

It is evident that, as always, with the power law of density variation, it is impossible to achieve finiteness of the total mass in the models under consideration. It must diverge either at zero or at infinity. Requiring [29] that the mass

$$M(r) = 2\pi \int_0^r \sigma_0 r\, dr = \frac{2\pi\varkappa}{p+2} r^{p+2} \tag{17}$$

converges at zero as well as the potential $\Phi_0(r)$ vanishes at infinity, we obtain restrictions on admissible values of p:

$$-2 < p < -1. \tag{18}$$

At $m = 0$, $n = -1/(p+1)$, the distribution function is isotropic while, with an increase in m, n, it becomes still more anisotropic and at $n, m \to \infty$ there ensues a disk with circular orbits, where

$$p \to -\frac{2n+3m}{2n+m}, \qquad \frac{1}{2} < \lim_{n,m\to\infty} \left(\frac{n}{m}\right) < \infty.$$

In conclusion, let us consider the approximate models of collisionless disks with a finite (but small) thickness, as studied by Vandervoort [336–338]. The author indicates that, in a strongly flattened galaxy, the stars oscillate in the direction perpendicular to the galactic plane, at a frequency high in comparison with other frequencies that characterize motions of stars. A corresponding regular procedure for the solution of the kinetic equation and the Poisson equation is developed, also taking into account the closeness of stellar trajectories to circular ones.

In [336], the same author constructed a family of equilibrium configurations for a strongly flattened, rapidly rotating galaxy that is a superposition of the subsystems. The method is taken from the earlier work [337]. The solutions obtained are employed for constructing the model of the solar neighborhood of the Galaxy.

§ 2 Stability of a "Cold" Rotating Disk

2.1 Membrane Oscillations of the Disk

We shall begin to discuss the problems that refer to small perturbations and stability of flat gravitating systems with the simplest case: membrane oscillations which bend the disk plane.

The mathematical theory describing small bending oscillations as well as the response of an infinitely thin rotating disk to an external force (primarily of tidal nature) is developed by Hunter and Toomre [330]. Their aim was to explain the observed bending of the "plane" of the Galaxy (for more detail see §3, Chapter XII).

Note immediately that for an infinitely thin disk the membrane oscillations, in linear approximation, "become split away" from the oscillations in the disk plane (considered in Section 2.2).

2.1.1. Equation of Motion. Let the plane of the disk undergo deformation characterized by the vertical displacements of $h(r, \varphi)$. The effects of perturbed movements in the disk plane may be neglected as values of the second order of smallness. The vertical acceleration of the particle at the point (r, φ) will evidently be

$$\left[\frac{\partial}{\partial t} + \Omega(r)\frac{\partial}{\partial \varphi}\right]^2 h(r, \varphi, t), \tag{1}$$

where the operator in square brackets is the Lagrange derivative d/dt. Then the equation of motion will be written as

$$\left[\frac{\partial}{\partial t} + \Omega(r)\frac{\partial}{\partial \varphi}\right]^2 h = F + F^*, \tag{2}$$

where F^* is the external force perpendicular to the plane, per unit mass, and F is the perturbed force due to the deformed disk itself. We calculate the force acting on the element of the disk being at the point **r**. Another specific element with an area of ds', located at the point **r**′, acts on the former with a force $G\sigma_0(r')\,ds'/(r - r')^2$. The vertical component of this force is

$$\frac{G\sigma_0(r')\,ds'}{(r - r')^2} \cdot \frac{[h(r', t) - h(r, t)]}{|r - r'|} .$$

To obtain the total force, it is apparently necessary to sum all such contributions. In addition, in order to exclude possible singularity, it is necessary to eliminate previously the small ring of the radius ε centered on the point **r**. Therefore, the sought-for force F should be understood as the principal value of the corresponding integral

$$F(r, t) = G \lim_{\varepsilon \to 0} \iint \frac{\sigma_0(r')[h(r', t) - h(r, t)]\,ds'}{|r - r'|^3} . \tag{3}$$

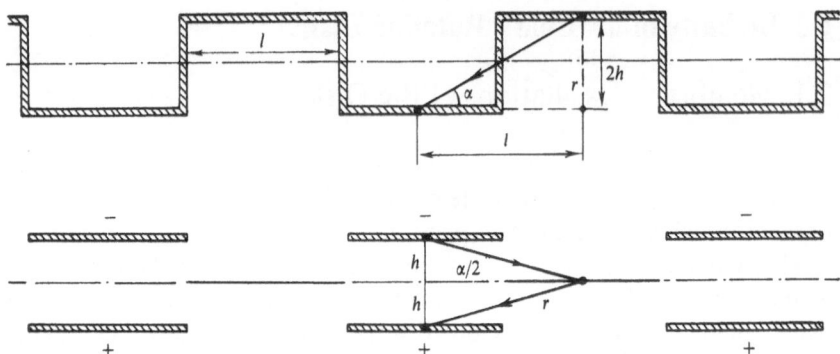

Figure 63. Perturbations of the membrane type as a double layer.

The right-hand side of (3) is useful to interpret as a force acting from the side of the vertically positioned gravitational dipoles with the surface density $\sigma(r')[h(r', t) - h(r, t)]$, which are located in the unperturbed galactic plane. This equivalent dipole density is dependent on the point $\mathbf{r}$, at which the force is to be calculated.

It is natural to represent the total force F, which is given by the integral in (3), in the form of the sum of the forces F_1 and F_2 due to the dipole distributions $\sigma(r')h(r')$ and $-\sigma(r')h(r)$, respectively:

$$F = F_1 + F_2. \tag{4}$$

Thus, we arrive at the problem of seeking the gravitational potential (and force) produced by dipoles of a double layer distributed on the disk[5] (Fig. 63).

2.1.2. Elementary Solutions of the Poisson Equation. Since the disk is confined, it is convenient to make use of computations by oblate spheroidal coordinates (cf. the previous section). Elementary solutions of the Laplace equation outside the disk of the radius R are, as easily shown by the method of separation of variables [64, 226], the following products (which generalize the axially symmetrical solutions of §1):

$$\Phi_{\text{ext}} = Aq_n^m(\zeta)P_n^m(\mu)e^{im\varphi}, \tag{5}$$

where n and m are integers, $n \geq m$, A is constant,

$$p_n^m(\zeta) = i^{m-n}P_n^m(i\zeta),$$

$$q_n^m(\zeta) = (-1)^m \frac{(n+m)!}{(n-m)!} p_n^m(\zeta) \int_\zeta^\infty \frac{dy}{(1+y^2)[p_n^m(y)]^2}, \tag{6}$$

and $P_n^m(x)$ are the associated Legendre functions. The gravitational potential in (5) must be produced by a double layer with a corresponding density $\tilde{\mu}$.

---

[5] Each of the forces, F_1 and F_2, will separately be discontinuous in the disk plane, but the total force F (and the total potential Φ), as is seen from (3), is continuous as it intersects the plane of the double layer.

As is known [130], the potential of the double layer undergoes a jump of the value $4\pi G\tilde{\mu}$ while passing the layer:

$$\Phi_+ - \Phi_- = Aq_n^m(0)P_n^m(+\mu)e^{im\varphi} - Aq_n^m(0)P_n^m(-\mu)e^{im\varphi} = -4\pi G\tilde{\mu}; \quad (7)$$

hence, first of all, it follows that the difference $(n - m)$ in this case must be odd[6] since, otherwise, a zero is on the left-hand side of Eq. (7). Let $n = 2k + m - 1, k$ being an integer; then from (7) we obtain that the dipole distribution

$$\tilde{\mu} = P_{2k+m-1}^m(\xi)e^{im\varphi} \quad \left(\xi = \sqrt{1 - \frac{r^2}{R^2}}\right) \quad (8)$$

implies the potential

$$-\Phi_{ext} = \frac{2\pi G}{q_{2k+m-1}^m(0)} q_{2k+m-1}^m(\zeta)P_{2k+m-1}^m(\mu)e^{im\varphi}. \quad (9)$$

The vertical force is obtained from (9) by differentiating:

$$-F = \frac{\partial\Phi_{ext}}{\partial z}\bigg|_{z=0} = \left(\frac{1}{h_\zeta}\frac{\partial\Phi_{ext}}{\partial\zeta}\right)_{\zeta=0}. \quad (10)$$

Since $h_{\zeta=0} = R\xi$, then

$$F = -\frac{2\pi^2 G}{R\xi}P_{2k+m-1}^m(\xi)e^{im\varphi}\Gamma_k^m, \quad (11)$$

where the notation

$$\Gamma_k^m = \frac{-q_{2k+m-1}^{m'}(0)}{\pi q_{2k+m-1}^m(0)} \quad (12)$$

is introduced. Γ_k^m are the positive numbers defined by the recurrent relations [230]

$$\Gamma_1^0 = \frac{1}{2}, \qquad \Gamma_1^m = \frac{2m+1}{2m}\Gamma_1^{m-1},$$

$$\Gamma_k^m = \frac{(2k-1)(2k+2m-1)}{(2k-2)(2k+2m-2)}\Gamma_{k-1}^m. \quad (13)$$

From (13), it is possible to derive

$$\Gamma_j^m = \frac{(2j-1)!(2j+2m-1)!}{2^{2m+4j-3}[(j-1)!(j+m-1)!]^2}. \quad (13')$$

In the general case of a disk with some arbitrary $\sigma_0(r)$ and $\Omega_0(r)$, the solution of Eq. (1) can be found [230] by means of expansion of all the values over a suitable system of functions of the type of (11), which leads to a matrix

---

[6] The solutions with even $(n - m)$ describe evidently the potentials produced by the mass redistribution in the disk plane $z = 0$. Such disturbances, which do not lead to bending of the disk plane, are considered in Section 2.2.

formulation of the problem (cf. below), and the further solution is performed, as a rule, by numerical methods. However, in one case, namely for the uniformly rotating disk, Ω = const, one can derive a simple analytical solution.[7]

2.1.3. Uniformly Rotating "Cold" Disks (Maclaurin's Disks). The force F_1 is produced by a double gravitational layer with a density $\tilde{\mu}_1 = \sigma_0(0)\xi'h(\xi', t)$, for $\sigma_0(\xi') = \sigma_0(0)\xi'$. Let the vertical displacement $h(\xi)$ be

$$h(\xi) = \frac{1}{\xi} P^m_{2n+m-1}(\xi).$$

(14)

Then the density $\tilde{\mu}_1$ is

$$\tilde{\mu}_1(\xi') = \frac{3M}{2\pi R^2} P^m_{2n+m-1}(\xi'),$$

(15)

and the force F_1 will be, according to (11),

$$F_1(\xi) = \frac{3\pi GM}{R^3} \Gamma^m_n \frac{1}{\xi} P^m_{2n+m-1}(\xi).$$

(16)

The density of the second dipole distribution

$$\tilde{\mu}_2 = -\sigma_0(0)\xi'h(\xi, t) = -\frac{3M}{2\pi R^2} \frac{P^m_{2n+m-1}(\xi)}{\xi} P_1(\xi')$$

(17)

so that the relevant vertical component of the force is

$$F_2 = \frac{3\pi GM}{2R^3} \frac{P^m_{2n+m-1}(\xi)}{\xi}.$$

(18)

The total force $F_1 + F_2$ is therefore

$$F(\xi) = F_1 + F_2 = -\frac{3\pi GM}{R^3} (\Gamma^m_n - \tfrac{1}{2})h(\xi).$$

(19)

The left-hand side of (1) in this case (Ω = const) is also proportional to the displacement h:

$$\left(\frac{\partial}{\partial t} + \Omega \frac{\partial}{\partial \varphi}\right)^2 h = -(\omega - m\Omega)^2 h.$$

(20)

As a result, by comparing (20) and (19), we get the dispersion equation[8]

$$(\omega - m\Omega)^2 = \frac{3\pi GM}{R^3} (\Gamma^m_n - \tfrac{1}{2}).$$

(21)

---

[7] As we shall see below, the analytical solutions in the case of uniformly rotating disks is derived also for oscillations lying in their plane, both for "cold" (Section 2.2) and for "hot" systems (Section 4.4).

[8] Note [180] that the frequencies of (21) as well as the frequencies for perturbations lying in the disk plane (Section 2.2) could have been obtained as limiting (at vanishing thickness) for the general spheroid of incompressible liquid of Maclaurin [172]. The frequencies of most modes of the spheroid vanish in flattening to form a disk; there remain only those modes considered here, as well as those in Section 2.2.

All frequencies defined from (21) are real, since, according to (13') all $\Gamma_n^m \geq \frac{1}{2}$. The frequency $\omega = 0$ ensues in two cases: at $m = 0$, $n = 1$ and at $m = 1$, $n = 1$ (and in the former case, the root is multiple). All the remaining modes are purely oscillating. The former of the modes mentioned with $\omega = 0$ corresponds to the disk displacement as a whole, while the latter is a stationary rotation about a new axis slightly declined with respect to the initial one. Thus, these roots are also quite "harmless" from the standpoint of stability loss, and we may infer that the uniformly rotating cold disk is stable with respect to membrane oscillations bending its plane.

All eigenmodes occur in pairs. Each nonaxial symmetrical pair consists of a mode which rotates more rapidly than the disk, and a mode which rotates more slowly, with the same magnitude.

Asymptotically at $n \to \infty$, the equations in (21) and (13) yield

$$\omega \sim \left[\frac{m}{2} \pm \left(\frac{2n}{\pi}\right)^{1/2}\right] \cdot 2\Omega, \tag{22}$$

for any fixed m.

Of particular interest for applications are modes with $m = 1$ since both the tidal forces and the observed deformed plane of the Galaxy have roughly just this dependence on the azimuthal angle (cf. §3, Chapter XII). The pair of the modes $n = 1$ involves, apart from the trivial mode already mentioned, another mode with $\omega = 2\Omega$, which also corresponds to displacements $h(r, \varphi, t)$ proportional to the radius r. The disk oscillates near an unperturbed plane as a solid body. From the viewpoint of an observer at rest, deformation travels in the direction of rotation with a doubled velocity of disk rotation.

The remaining modes with $m = 1$ propagate around the disk either in the direction of rotation (we shall say in the "direct" direction) at a velocity higher than 2Ω, or in the opposite direction.

We shall denote [230] all "direct" modes as $D1, D2, \cdots$, and all "inverse" modes as $R1, R2, \cdots$ (the latter begin with a trivial $\omega = 0$). The first non-trivial mode $R2$ corresponds to $n = 2$:

$$h^{(R2)} = \text{Re}\left\{Cr\left[\frac{4}{7} - \left(\frac{r}{R}\right)^2\right] \exp[i(\varphi - \omega^{(R2)}t)]\right\}, \tag{23}$$

where the frequency $\omega^{(R2)} = -0.783(2/\sqrt{3})\Omega$. The $D2$ mode has already a significantly higher frequency: $\omega^{(D2)} = 2.515(2/\sqrt{3})\Omega$.

2.1.4. Arbitrary Disks. The Matrix Formulation. The other models, as we have already mentioned, can be calculated only numerically. Here we bear in mind the possibility of calculating eigenfrequencies and eigenfunctions of different types of disk oscillations (or the evolution of initial perturbations). If one concerns stability, then it can in some cases be established for arbitrary disks. Hunter and Toomre [230] analytically proved the stability with respect to perturbations with $m = 0$ and $m = 1$.

Any disk deformation perpendicular to its plane increases, of course, its potential energy. But this does not yet mean stability [230] since, in principle, a part of rotational kinetic energy of the disk may compensate for this excess of potential energy. And indeed, for example, for superposition of two identical cold Maclaurin disks rotating in the opposite directions and deformed in the vertical direction, one can prove instability of all the modes with $m \geq 2$.[9]

Hunter and Toomre [230] further give the matrix formulation of the problem of membrane oscillations. It is found in the following way. Assume that the product $\sigma_0 h$ permits the expansion

$$\sigma_0(r)h(r, \varphi, t) = \frac{M}{2\pi R} \text{Re}\left\{\exp(im\varphi) \sum_{k=1}^{\infty} X_k^m(t)P_{2k+m-1}^m(\xi)\right\}, \qquad (24)$$

where $X_n^m(t)$ are the dimensionless functions of time (generally, complex functions), and

$$p_{2k+m-1}^m(\xi) \equiv \left[\frac{(4k + 2m - 1)(2k - 1)!}{(2k + 2m - 1)!}\right]^{1/2} P_{2k+m-1}^m(\xi) \qquad (25)$$

are the associated Legendre functions renormalized so that the integrals from their squares over the interval $(0, 1)$ are unity. This expansion is general since the system of the functions p_{2k+m-1}^m is orthonormal and complete on the interval $(0, 1)$.

Expand also the surface density:

$$\sigma_0(r) = \sum_{k=1}^{\infty} \left[\int_0^1 \sigma_0(r')P_{2k-1}(\xi')\, d\xi'\right]P_{2k-1}(\xi). \qquad (26)$$

The vertical force F_1 due to the observer-independent part of the dipole density $\sigma(r')h(r', t)$ is given by the expression

$$F_1(r, \varphi, t) = -\frac{\pi GM}{R^2\xi} \text{Re}\left[\exp(im\varphi) \sum_{k=1}^{\infty} \Gamma_k^m X_k^m(t)p_{2k+m-1}^m(\xi)\right]. \qquad (27)$$

The force F_2 corresponding to the dipole density $[-\sigma(r')h(r, t)]$ is equal to the value $[-h(r, t)]$ multiplied by the force in the position r, produced because of the axially symmetrical distribution of dipoles with a density $\sigma_0(r')$:

$$F_2(r, \varphi, t) = \frac{2\pi^2 G}{R\xi} h(r, \varphi, t) \sum_{k=1}^{\infty} \Gamma_k^0\left(\int_0^1 \sigma_0 P_{2k-1}(\xi')\, d\xi'\right)P_{2k-1}(\xi). \qquad (28)$$

The last expression can be transformed to give a simpler form

$$F_2(r, \varphi, t) = \left[2\Omega^2(r) + r\frac{d\Omega^2}{dr}\right]h(r, \varphi, t). \qquad (29)$$

---

[9] This directly follows from the corresponding dispersion equation which is easily obtained similarly to (21):

$$\omega^2 - 2(m^2 + 2\Gamma_n^m - 2)\omega^2 + (m^2 + 2)(m^2 - 4\Gamma_n^m + 2) = 0,$$

where we assumed that $\Omega = 1$; for more detail cf. Section 4.2, Chapter V.

This result can be inferred also in another way, and just in the form of (29). The double layer with a density $\delta \cdot \sigma_0(r')$ is equivalent to two ordinary disks (with densities $\pm \sigma_0$) separated by a small distance δ. Then it is easy to see that the vertical force acting on the particle located just above the disk is produced due to the fact that the particle at a distance δ is closer to the upper (positive, for certainty) disk than to the lower (negative) one. Let $(-\partial\Phi_0/\partial z)$ denote the vertical component of the force near the disk with a density $\sigma_0(r)$. Then the mentioned excess of force will apparently be $\partial^2\Phi_0/\partial z^2|_{z=+0}$. On the other hand, however, the Lagrange equation allows one to turn from differentiation over z to differentiation over r:

$$\frac{\partial^2\Phi}{\partial z^2}\bigg|_{0+} = -\frac{\partial^2\Phi}{\partial r^2}\bigg|_{0+} - r^{-1}\frac{\partial\Phi}{\partial r}\bigg|_{0+}. \tag{30}$$

By using now the condition of the balance of the centrifugal and gravitational forces, we get

$$\frac{\partial^2\Phi}{\partial z^2}\bigg|_{0+} = -\frac{d}{dr}[r^2\Omega^2(r)] - \Omega^2(r). \tag{31}$$

If (31) is multiplied by h, we shall arrive at (29) for F_2.

It is convenient to define the dimensionless time τ, the velocity of rotation $\omega(\xi)$, and the surface density $s(\xi)$:

$$t = \left(\frac{R^3}{\pi GM}\right)^{1/2}\tau, \quad \Omega(r) = \left(\frac{\pi GM}{R^3}\right)^{1/2}\omega(\xi), \quad \sigma(r) = \left(\frac{M}{2\pi R^2}\right)s(\xi). \tag{32}$$

Then from (27), (29), (2) it is easy to infer the following infinite system of linear equations:

$$\frac{d^2 X_n^m}{d\tau^2} + i\sum_{j=1}^{\infty} A_{nj}^m \frac{dX_n^m}{d\tau} + \sum_{j=1}^{\infty} B_{nj}^m X_j^m = g_n^m(\tau). \tag{33}$$

Matrix elements A^m and B^m are constants defined by the integrals

$$A_{nj}^m = 2m \int_0^1 \omega(\xi) p_{2n+m-1}^m(\xi) p_{2j+m-1}^m(\xi)\, d\xi,$$

$$B_{nj}^m = \int_0^1 d\xi \left\{ \Gamma_j^m \left[\frac{s(\xi)}{\xi}\right] - (2+m^2)\omega^2(\xi) \right. \tag{34}$$

$$\left. + \left[\frac{(1-\xi^2)}{\xi}\right]\frac{d\omega^2}{d\xi} \right\} p_{2j+m-1}^m(\xi) p_{2n+m-1}^m(\xi),$$

and the vector $g^m(\tau)$ is linked to the external force by the equation

$$g_j^m(\tau) = \frac{R^2}{\pi GM} \int_0^1 \xi F^m(\tau) p_{2j+m-1}^m(\xi)\, d\xi. \tag{35}$$

The system of equations (33) is the basic one for numerical computation in [230]. By using it, Hunter and Toomre [230] calculated oscillations of a large number of models. We employ Eqs. (33) in the description in subsection

2.1.6 of Hunter's [228] method of a strict investigation of the conditions of the appearance of a *continuous* spectrum of frequencies of membrane oscillations of the disk. Nevertheless, let us first consider this issue qualitatively.

2.1.5. Continuous Spectrum. Qualitative Treatment [230]. Consider the membrane oscillations of a homogeneous nonrotating thin layer with infinite extension. Introduce the horizontal wave number $k = 2\pi/\lambda$, which characterizes the bending perturbation in the form of a plane wave in the direction of the x-axis.

The potential of the double layer with the density

$$\mu = e^{ikx} \qquad (k > 0) \tag{36}$$

is equal, as can be readily shown, to the expression

$$\Phi = -2\pi G \, \mathrm{sgn} \, z e^{-k|z|} e^{ikx}, \tag{37}$$

and the force, accordingly, is

$$F|_{z=+0} = -2\pi G k e^{ikx}. \tag{38}$$

Taking into account that in this case there is no rotation $dh/dt = \partial h/\partial t = -i\omega h$, we obtain the following formula for the oscillation frequency:

$$\omega^2 = 2\pi G \sigma_0 k. \tag{39}$$

For nonhomogeneous systems, $\sigma_0 = \sigma_0(r)$, (39) is a local dispersion relation; it provides correct understanding of the WKB type oscillations with a sufficiently large number of nodes in the radius. The latter is approximately equal to the integral

$$\frac{\omega^2}{2\pi^2 G} \int_0^R \frac{dr}{\sigma_0(r)}. \tag{40}$$

If we set Eq. (40) equal to a large positive integer N, we obtain the asymptotic formula for discrete eigenfrequencies

$$\omega^{(N)} \simeq \pi (2GN)^{1/2} \left[\int_0^R \frac{dr}{\sigma_0(r)} \right]^{-1/2}. \tag{41}$$

This formula describes correctly, in particular, the asymptotic frequencies of membrane oscillations of a disk rotating as a solid body:

$$\omega^{(N)} \approx \left(\frac{\pi GM}{R^3} \right) \left(\frac{6N}{\pi} \right)^{1/2} \qquad (N \to \infty). \tag{42}$$

It is more important that (41) shows that the frequency interval $\omega^{(N+1)} - \omega^{(N)}$, separating the neighboring modes N and $(N + 1)$, vanishes if one decreases the density $\sigma_0(r)$ on the edge of the disk. Thus, Eq. (41) predicts that the oscillation frequency spectrum of the disk ceases to be discrete but becomes *continuous* if the value inverse to the density, $[\sigma_0(r)]^{-1}$, becomes nonintegrable[10] at the edge, at $r = R$.

The models with continuous spectra possess quite new properties.[11]

Ultimately, we are always interested in how in a given system the initial perturbations will evolve. From this viewpoint, the difference between the systems with a discrete and a continuous spectrum are cardinal. In the latter case (unlike the first one), *individual* modes lose their value since simultaneously there must be an excited continuum of such modes with approximately equal amplitudes.

The system with strongly separated discrete modes, once excited, tends to oscillate without any radical change of its form. The system, however, having a continuous spectrum, will evolve continually due to the fact that in the course of time in a still greater degree the initial synchronism of oscillations of "neighboring" modes will be violated. In essence, the very appearance of the continuous spectrum is a consequence of the weakness of the mutual influence of oscillations of the particles moving in different orbits (each orbit practically precesses with its own angular velocity).

This fundamental difference is illustrated in [230] in many examples. Thus, for "smooth" initial excitations, in the models with a discrete spectrum there is a strong predominance of the basic mode (of the largest scale) which receives a significant part of the excitation energy. The models with a continuous spectrum respond to the same initial excitations in a quite different way. Their behavior is characterized by a constantly increasing number of nodes, each moving toward the edge of the disk. At the same time, the amplitude of each ridge increases, while the central flat region of the disk expands. This corresponds to "manifestation" with time of a still larger number of individual modes. Perturbations are "drifting" toward the edge, where they "are accumulated."

Physically, such a characteristic of the evolution of the initial perturbation may be understood in analogy with more customary examples (oscillations of the reservoir with a slowly varying depth, the exponential atmosphere). In all these cases, the perturbation energy has the same trend to travel from deeper or denser parts of the system to the opposite ones, even for a nearly uniform initial perturbation. Practically, such propagation always leads to a more rapid damping of disturbances. It may be expected that bending oscillations of disks with a continuous spectrum also will be damping if one takes into account even a small viscosity in the region of the edge.

---

[10] A strict proof of the existence of regions of a continuous spectrum of eigenfrequencies requires a more thorough analysis, which also was performed by Hunter in [228]. We shall return to this issue in subsection 2.1.6. This analysis, in addition, indicates the boundaries of regions of the continuous spectrum.

[11] At the same time, the models having only a discrete spectrum (many of them are calculated by Hunter and Toomre in [230]) have properties similar to those of the Maclaurin disk (subsection 2.1.3). Again, all frequencies are divided into "direct" and "inverse" families which may be classified also in a similar way ($\cdots R2, R1, D1, D2 \cdots$); the first "inverse" modes have frequencies ω, significantly less in value than for any "direct" mode, etc.

We can analytically describe the evolution of initial perturbations in the system with the help of the following approximate method (in the work of Hunter and Toomre, this problem is calculated numerically).

We write the differential equation corresponding to the dispersion relation in (39), which must describe the function $h(x, t)$, varying fairly rapidly in space and time

$$\frac{\partial^2 h}{\partial t^2} = 2\pi i G \sigma_0 \frac{\partial h}{\partial x} \qquad (x \equiv 1 - r). \tag{43}$$

Let us solve this equation for perturbations of the disk whose surface density, for the sake of simplicity, is presented in the form $\sigma_0 = \alpha_0 x^n$ (then the parameter $n > 0$). The ensuing equation

$$\frac{\partial^2 h}{\partial t^2} = 2\pi i G \alpha_0 x^n \frac{\partial h}{\partial x} \tag{44}$$

can be reduced to the equation with constant coefficients, by making the substitution

$$y = \frac{x^{1-n}}{n-1}; \tag{45}$$

then we have[12]

$$\frac{\partial^2 h}{\partial t^2} + ia \frac{\partial h}{\partial y} = 0 \qquad (a \equiv 2\pi G \alpha_0). \tag{46}$$

This equation has eigensolutions in the form of plane waves over y: $h \sim e^{-i\mu y}$ with $\mu = $ const; they have eigenfrequencies obeying the dispersion equation

$$\omega^2 = |\mu| a. \tag{47}$$

Of course, these eigenfunctions could have been found also from (39) if one writes

$$h \sim e^{-i\int^x k(x)\,dx}, \qquad k(x) = \frac{\omega^2}{2\pi G \sigma_0(x)}. \tag{48}$$

It is seen, in particular, that the behavior of eigenfunctions near the disk edge $x = 0$ changes radically during transition from the case $n < 1$ to $n > 1$ [cf. (45)]: then the "good" behavior is changed by infinite oscillations as one approaches $x = 0$. The oscillation spectrum in this case becomes continuous, and individual eigenmodes do not provide direct information

---

[12] It is, of course, possible to derive the equation similar to (46) also in the general case [for any $\sigma_0 = \sigma_0(x)$] if the substitution

$$x \to y = \text{const} \int^x \frac{dx}{\sigma_0(x)}$$

is made.

regarding the character of the evolution of the perturbation imposed at the initial time. Equation (46), however, is in principle easily solved for the initial perturbation of any form:

$$h(t = 0) = f(x) = \varphi(y),\tag{49}$$

$$\frac{\partial h}{\partial t}(t = 0) = f_1(x) = \varphi_1(y).\tag{50}$$

Then, it is easy to see that

$$h(y, t) = \frac{1}{2\pi} \int d\mu e^{i\mu y}[A(\mu)e^{i\omega(\mu)t} + B(\mu)e^{-i\omega(\mu)t}],\tag{51}$$

where $\omega(\mu) = \sqrt{a|\mu|}$, and the $A(\mu)$ and $B(\mu)$ functions are expressed through the Fourier transforms of $\varphi(y)$ and $\varphi_1(y)$:

$$A(\mu) + B(\mu) = \varphi_\mu \equiv \int_{-\infty}^{\infty} \varphi(y)e^{-i\mu y}\, dy,\tag{52}$$

$$i\omega(\mu)[A(\mu) - B(\mu)] = \varphi_{1\mu} \equiv \int_{-\infty}^{\infty} \varphi_1(y)e^{-i\mu y}\, dy.\tag{53}$$

By using the method of the stationary phase, it is easy to find asymptotic (at large t) expressions for $h(y, t)$. Thus, for perturbations with $A(\mu) = 0$, we obtain

$$h \approx \sqrt{\frac{2}{\pi|\omega''(\mu_0)t|}}\, B(\mu_0) \exp\{i[\mu_0 y - \omega(\mu_0)t - \tfrac{1}{4}\pi \operatorname{sgn} \omega''(\mu_0)]\},\tag{54}$$

where μ_0 is the positive root of the equation

$$\frac{d\omega}{d\mu} \equiv \omega' = \frac{y}{t}.\tag{55}$$

The formulae of the type of (54) are valid when for large t we have a rather rapidly oscillating solution (requirements of approximation of geometrical optics are fulfilled). The obtained solution is a wave packet with a variable wave number $\mu_0(x, t)$, the frequency $\omega(x, t) = \sqrt{a\mu_0(x, t)}$, and the amplitude proportional to $|\omega''(\mu_0)t|^{-1/2} \cdot |B(\mu_0)|$. The amplitude must be defined for each particular initial perturbation separately. As far as the perturbation phase ψ [which, in particular, defines the motion of zeros of the $h(x, t)$ function] is concerned, however, then it is dependent only, according to (54), on the form of the dispersion equation. In the give case

$$\Psi \sim \mu_0 y - \omega(\mu_0)t \sim \text{const}\, \frac{t^2}{y} \sim \text{const}\, t^2 x^{n-1};\tag{56}$$

therefore,

$$h \sim \exp(i\Psi) \sim \exp(i \cdot \text{const}\, t^2 x^{n-1}).\tag{57}$$

It is seen that the number of zeros in the course of time increases, and they appear to move in the direction toward the disk edge according to the law

$$x \sim \text{const} \cdot t^{-2/(n-1)}. \tag{58}$$

This agrees qualitatively with the above-described picture of the evolution of perturbations in disks with a "thin" edge numerically calculated by Hunter and Toomre [230].

In the framework of geometrical optics, it is particularly simple to calculate the evolution of an individual wave packet characterized by a mean wave number $\mathbf{k} = \mathbf{k}(t)$ and localized near the point $\mathbf{r} = \mathbf{r}(t)$. As is well known [70], the $\mathbf{k}(t)$ and $\mathbf{r}(t)$ functions obey the equations of the type of Hamilton equations of mechanics (where the role of the Hamiltonian is played by the frequency ω):

$$\frac{d\mathbf{r}}{dt} = \frac{\partial \omega}{\partial \mathbf{k}}, \tag{59}$$

$$\frac{d\mathbf{k}}{dt} = -\frac{\partial \omega}{\partial \mathbf{r}}. \tag{60}$$

Since in this case $\omega^2 = 2\pi G \sigma_0(r)k$, so for the radially propagating wave packet we have

$$\dot{x} = -\sqrt{\tfrac{1}{2}\pi G}\, k^{-1/2} x^{n/2}, \tag{61}$$

$$\dot{k} = n\sqrt{\tfrac{1}{2}\pi G}\, k^{1/2} x^{n/2-1}, \tag{62}$$

where we have again assumed that $x = 1 - r$, $\sigma_0 = x^n$. Equations (61) and (62) are readily solved. In particular, one can find the integral of motion $A \equiv x^{-n^2/2+n} \cdot k^{-n/4+1/2} = \text{const}$; hence it follows that $k \sim x^{-2n}$. This relation shows that the wave number k is infinitely increasing as the packet approaches the edge of the disk.

2.1.6. Continuous Spectrum. Strict Treatment [228].

To begin with, consider a simple partial (but typical) case of axially symmetrical oscillations of the following *special one-parametric family of disks*:

$$s = \frac{2\pi R^2 \sigma_0}{M} = 3(1 - \beta)\xi + 5\beta\xi^3, \tag{63}$$

$$\Omega^2 = \frac{R^3 \Omega_0^2}{\pi G M} = \frac{3}{4}\left(1 - \frac{3\beta}{8}\right) + \frac{45\beta\xi^2}{32}, \tag{64}$$

where β is the parameter, $0 \le \beta \le 1$. In the limit $\beta = 0$, (63) gives a uniformly rotating disk. Of most interest is the opposite limit $\beta = 1$, when in (63) there is no term $\sim \xi$, the density vanishes at the edge more smoothly than in all other cases, and one may expect the presence of a continuous spectrum since the integral in (41), subsection 2.1.5, is diverging.

For axially symmetrical modes, the system of equations in (33) is reduced to the following ($A = 0$, $m = 0$, $X \sim e^{-i\lambda t}$):

$$\sum_{j=1}^{\infty} B_{jn}^0 X_j^0 = \lambda^2 X_n^0, \tag{65}$$

where the matrix B for (63) and (64) is

$$B_{jn}^0 = \int_0^1 \{[3(1 - \beta) + 5\beta\xi^2]\Gamma_j^0 - \tfrac{3}{8}[4 - 9\beta + 15\beta\xi^2]\} p_{2j-1}(\xi) p_{2n-1}(\xi)\, d\xi,$$

and

$$\Gamma_j^0 = \frac{[(2j - 1)!]^2}{2^{4j-3}[(j - 1)!]^4}.$$

Recurrence relations for Legendre polynomials show that $\xi^2 p_{2j-1}(\xi)$ may be expressed by p_{2j-3}, p_{2j-1}, p_{2j+1}. Hence it follows (taking into account the properties of the orthogonality of the polynomials used) that the matrix B^0 is tridiagonal and Eq. (65) is reduced to the following three-term linear difference equation:

$$-\frac{10\beta n(2n + 1)(\Gamma_{n+1}^0 - \tfrac{9}{8}) X_{n+1}^0}{(4n + 1)[(4n - 1)(4n + 3)]^{1/2}}$$

$$+\left[3(1 - \beta)\Gamma_n^0 - \frac{3}{2} + \frac{27\beta}{8} - \lambda^2 + \frac{5\beta(8n^2 + 4n - 1)(\Gamma_n^0 - \tfrac{9}{8})}{(4n - 3)(4n + 1)}\right] X_n^0$$

$$+\frac{10\beta(n - 1)(2n - 1)(\Gamma_{n-1}^0 - \tfrac{9}{8}) X_{n-1}^0}{(4n - 3)[(4n - 1)(4n - 5)]^{1/2}} = 0, \tag{66}$$

where n runs through the values from 1 to ∞. At $\beta = 0$, hence, we have, of course, the familiar spectrum of oscillations of the uniformly rotating disk:

$$\lambda^2 = \tfrac{3}{2} - 3\Gamma_n^0. \tag{67}$$

The displacement of the disk as a whole obeys

$$\lambda^2 = 0, \quad X_1^0 = 1, \quad X_2^0 = \frac{2\beta}{\sqrt{21}}, \quad X_n^0 = 0 \text{ for } n > 2, \quad \frac{h}{R} = \frac{1}{\sqrt{3}}. \tag{68}$$

The frequency universal for all the disks of (63) is also inherent to another oscillation

$$\lambda^2 = \tfrac{15}{8}, \quad X_2^0 = 1, \quad X_n^0 = 0 \quad \text{for } n \neq 2,$$

$$\frac{h}{R} = \frac{\sqrt{7(5\xi^2 - 3)}}{2[3(1 - \beta) + 5\beta\xi^2]} \cos \lambda\tau. \tag{69}$$

This is the nontrivial mode of the lower order for disks with any β.

It is impossible to derive other exact solutions, but the character of the spectrum λ^2 may be defined already from the general consideration of (66).

This equation is a linear difference equation of second order with a free parameter λ^2 and with two boundary conditions: at $n = 0$ and $n = \infty$. The condition at $n = 0$ arises from Eq. (66) for $n = 1$. Hence it follows that only the coefficient X_1^0 may be chosen arbitrarily; all subsequent coefficients X_n^0 are then defined for any given λ^2. Thus, one of two independent solutions of (66) is eliminated. On the other hand, analysis of the behavior of the solutions of (66) as $n \to \infty$ (see below) shows that in the general case one independent solution becomes infinitely large, while the other vanishes. Only under the condition when $X_n \to 0$ as $n \to \infty$ do we have a satisfactory solution. This condition can be satisfied, generally speaking, only for some values λ^2 (for the spectrum).[13] By using the asymptotic representation (m is fixed, while $n \to \infty$)

$$\Gamma_n^m \simeq \frac{2}{\pi}\left[n + \frac{2m-1}{4} + O\!\left(\frac{1}{n}\right)\right],\tag{70}$$

we reduce Eq. (66) to the form

$$5\beta\left[1 + \frac{3(4-3n)}{16n} + O\!\left(\frac{1}{n^2}\right)\right]X_{n+1}^0$$

$$+ \left[2(6-\beta) + \frac{1}{n}(\tfrac{1}{2}\beta - 3 - 3n + \tfrac{9}{8}\pi\beta - 2\pi\lambda^2) + O\!\left(\frac{1}{n^2}\right)\right]X_n^0$$

$$+ 5\beta\left[1 - \frac{(20+9\pi)}{16n} + O\!\left(\frac{1}{n^2}\right)\right]X_{n-1}^0 = 0.\tag{71}$$

If, however, one neglects in (71) all the coefficients vanishing at $n = \infty$, then we get

$$5\beta X_{n+1}^0 + 2(6-\beta)X_n^0 + 5\beta X_{n-1}^0 = 0.\tag{72}$$

The general solution of this difference equation with constant coefficient is the following:

$$X_n^0 = c_1 t_1^n + c_2 t_2^n,\tag{73}$$

where c_1 and c_2 are the constants and t_1 and t_2 are the roots of the characteristic equation

$$5\beta t^2 + 2(6-\beta)t + 5\beta = 0.\tag{74}$$

Both roots are real and negative. Since the product $t_1 t_2 = 1$, one of them, at $0 \le \beta < 1$, is more than 1 in magnitude, while the other is less. The latter root gives a converging, and the former a diverging, solution. Such a situation corresponds to a discrete spectrum, which thus must take place at all $0 \le \beta < 1$.

The case $\beta = 1$ is distinguished in that according to (74) it has the multiple root $t = -1$, which requires a more thorough investigation of the solution

---

[13] It must be admitted that there is a close analogy of the described procedure of "quantization" of the spectrum with a more customary procedure arising, for example, in solving the differential equations with the boundary conditions in zero and at infinity.

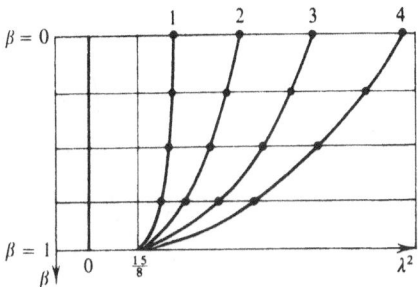

Figure 64. "Trajectories" of eigenfrequencies of membrane oscillations for the models (43) [228]. Curves 1, 2, 3, ... correspond to various types of oscillations.

behavior at larger n that takes into account the terms of the order of $O(1/n)$. The solution is sought for in the form

$$X_n^0 = (-1)^n \exp(\gamma n^{1/2})n^{-3/4}\left[1 + \sum_{k=1}^{\infty} d_k n^{-k/2}\right], \tag{75}$$

where d_k are the constants which can be defined successively. The value γ is readily calculated:

$$\gamma = \gamma_\pm = \pm\sqrt{\pi\left(3 - \frac{8\lambda^2}{5}\right)}. \tag{76}$$

γ_+ and γ_- correspond to two independent solutions of the original equation. At $\lambda^2 < \frac{15}{8}$, we again have the familiar situation with a discrete spectrum, but at $\lambda^2 > \frac{15}{8}$ *both* values of γ give possible solutions of the type of (75): factor $\exp(\gamma n^{1/2})$ then becomes oscillative, while $n^{-3/4}$ ensures[14] the required decreasing of X_n^0 as $n \to \infty$. Therefore, for $\lambda^2 > \frac{15}{8}$, we have no condition of "quantization" as $n \to \infty$, and the frequency spectrum is continuous.

The eigenfrequencies for the disk (63) can be determined numerically. Figure 64 shows a *qualitative* picture of "trajectories" of frequencies as the parameter of the models varies. All frequencies, apart from one, $\lambda^2 = 0$, tend, in the limit $\beta \to 1$, toward the edge of continuum $\lambda^2 = \frac{15}{8}$. At $\beta = 1$, in this example there is only one discrete eigenfrequency $\lambda^2 = 0$, and the continuous spectrum occupies the entire semiaxis $\lambda^2 > \frac{15}{8}$.

Establish now the correspondence between the above analysis of the asymptotic behavior of the coefficient X_n^0 as $n \to \infty$ and the analysis of the WKB type in subsection 2.1.5. The latter predicts the following behavior characteristic of eigenfrequencies near the edge ($\xi \to 0$) (Fig. 65):

$$s \cdot h \sim \sin\left[-\pi\lambda^2 \int^\xi \frac{\xi \, d\xi}{s(1 - \xi^2)^{1/2}} + \text{limited function}\right]. \tag{77}$$

The asymptotic behavior of the coefficients of the Fourier type is defined, as

---

[14] The analysis made is inapplicable at the marginal point $\lambda^2 = \frac{15}{8}$; however, in this case we have an exact solution (69).

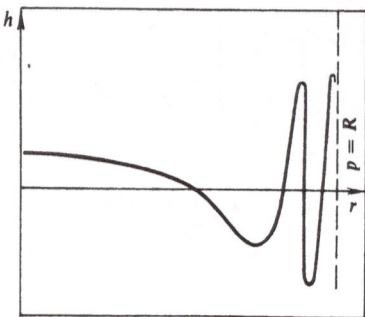

Figure 65. Behavior of the perturbations near the "thin" edge of the disk.

is known, by the most singular part of the function; in this case, this is the infinitely speeding-up oscillations at $\xi \to 0$. The coefficients of the expansion that we use are calculated by the formula

$$R \cdot \text{Re}\{X_n^0 e^{-i\lambda t}\} = \int_0^1 s \cdot h \, p_{2n-1}(\xi) \, d\xi. \tag{78}$$

A correct asymptotic behavior of X_n^0 at $n \to \infty$ will be obtained by using (70) and the asymptotic representation of the Legendre polynomials

$$
\begin{aligned}
X_n^0 &\sim \frac{2(-1)^n}{\pi^{1/2}} \int_0^1 \sin\left(\frac{\pi\lambda^2}{5\xi} + c_0\right) \cos(2\pi\xi + c_1) \, d\xi \\
&= \frac{(-1)^n}{(\pi n)^{1/2}} \text{Im}\left\{ e^{i(c_0+c_1)} \int_0^{\sqrt{n}} \exp\left[i\sqrt{n}\left(2y + \frac{\pi\lambda^2}{5y}\right)\right] dy \right. \\
&\quad \left. + e^{i(c_0-c_1)} \int_0^{\sqrt{n}} \exp\left[i\sqrt{n}\left(\frac{\pi\lambda^2}{5y} - 2y\right)\right] dy \right\},
\end{aligned}
\tag{79}
$$

where c_0 and c_1 are the constants. Integrals in (79) are estimated by the stationary-phase method [130], where the first integral is dominant, whose calculation yields the first approximation for X_n^0:

$$X_n^0 \sim (-1)^n n^{-3/4} \text{Im}\left\{ \text{const} \cdot \exp\left[i\lambda\left(\frac{8\pi n}{5}\right)^{1/2}\right]\right\}, \tag{80}$$

which coincides with (75). This coincidence does serve as verification of the correctness of the results obtained earlier by the WKB method.

Consider now the general case of membrane oscillations of disks, whose structure in the equilibrium state can be described by the following *poly-nomial expansions* in powers ξ^2:

$$\frac{s}{\xi} = \sum_{j=0}^N s_j \xi^{2j}, \qquad \Omega = \sum_{j=0}^N \omega_j \xi^{2j}, \qquad \Omega^2 = \sum_{j=0}^N W_j \xi^{2j}. \tag{81}$$

It is clear that for fairly large N (and below, N is arbitrary) using such expansions, it is possible to represent nearly any reasonable model with sufficient accuracy. The oscillation frequencies are defined by the equations of (33), 2.1.4, in which the A and B matrices are now $(2N + 1)$-diagonal (above we have dealt with the case $N = 1$). Respectively, the ensuing difference equation resulting for the coefficients X_n^m also has a higher order as compared with the previous case. However, the investigation of this equation is performed using the same recipe. The coefficients $X_1^m, X_2^m, \ldots, X_N^m$ may be chosen arbitrarily, while all the remaining coefficients $X_{N+1}^m, X_{N+2}^m, \cdots$ are then defined successively from the equation for $n = 1, 2, \cdots$. Altogether there are $2N$ independent solutions for the difference equation. The condition of generation of the continuous spectrum is again obtained from the consideration of the asymptotic behavior of solutions at $n \to \infty$. If N solutions are converging, then the spectrum must be discrete, since it is determined from the vanishing condition of the remaining N solutions expressed through $X_1^m, X_2^m, \ldots, X_N^m$ (in the case of an *arbitrary* λ they are diverging). If, however, more than N independent solutions are convergent, then possibly the appearance of the regions of a continuous spectrum occurs: the choice of λ becomes less restricted.

For the investigation of asymptotics as $n \to \infty$, it is enough to calculate the matrix elements within an accuracy of the terms of the order of magnitude $O(1/n^2)$.

Using (70) and (81) as well as the relation

$$\xi^2 p_{2n+m-1}^m(\xi) = \left[\frac{1}{4} + O\left(\frac{1}{n^2}\right)\right][p_{2n+m+1}^m(\xi) + 2p_{2n+m-1}^m(\xi) + p_{2n+m-3}^m(\xi)],$$
(82)

one can derive the following difference equation:

$$\frac{2}{\pi}\sum_{j=0}^{N}4^{-j}s_j(E + 2 + E^{-1})^j X_n^m + \frac{1}{n}\left\{-\lambda^2 X_n^m + \sum_{j=0}^{N}4^{-j}\left[\left(m - \frac{1}{2}\right)\frac{s_j}{\pi}\right.\right.$$

$$+ 2m\lambda\omega_j - (2 + 2j + m^2)W_j\bigg](E + 2 + E^{-1})^j X_n^m$$

$$+ \frac{2}{\pi}\sum_{j=1}^{N}js_j 4^{-j}(E - E^{-1})(E + 2 + E^{-1})^{j-1}X_n^m$$

$$+ 8\sum_{j=1}^{N}jW_j 4^{-j}(E + 2 + E^{-1})^{j-1}X_n^m\bigg\} + O\left(\frac{1}{n^2}\right) = 0,$$
(83)

where E is the shift operator

$$EX_n^m = X_{n+1}^m, \qquad E^{-1}X_n^m = X_{n-1}^m.$$
(84)

We neglect first of all in (83) all terms of the order of $O(1/n)$, $O(1/n^2)$, etc. Then, assuming that $X_n^m \sim t^n$, we arrive at the characteristic equation for t of the order of $2N$:

$$\sum_{j=0}^{N}s_j\left[\frac{(t + 1)^2}{4t}\right]^j = 0.$$
(85)

Let ζ be the root of this polynomial for $(t + 1)^2/4t$, i.e.,

$$t^2 + (2 - 4\zeta)t + 1 = 0. \tag{86}$$

The product of two roots of Eq. (86) is unity. If one of them is more than unity in magnitude, then the other is less. Accordingly, the second root yields a converging solution, and the first, a diverging root. But there is another possibility: the roots of (86) can be a complexly conjugate pair (with the magnitude equal to unity). In this case, the sum of the roots $(4\zeta - 2)$ must be real and lie in the range $(-2, 2)$; therefore, ζ must be within the interval $0 \leq \zeta \leq 1$. Now note that (85) can be rewritten in the form

$$\sum_{j=0}^{N} s_j \zeta^j = 0, \tag{87}$$

i.e., in the form of the condition of vanishing of the polynomial in (81) for s/ξ (with the substitution $\xi \to \zeta^{1/2}$. Therefore, the occurrence of the root ζ of Eq. (85) within this interval means that the surface density $s(\xi)$ must be zero at the point $\xi = \zeta^{1/2}$. If one considers only such distributions of density $s(\xi)$ which do not vanish inside the disk (which will just be assumed), then there remains only one possibility: $\zeta = 0$, which corresponds to vanishing on the disk edge of the value s/ξ. Thus, only in the case $(s/\xi)_{\xi=0} = 0$, $s_0 = 0$ can there arise a continuous spectrum; otherwise the difference equation in (83) will have N converging solutions, so that the spectrum is discrete.

Assume that k of the first coefficients s_j vanish: $s_0 = \cdots = s_{k-1} = 0$, $s_k \neq 0$. In this case, the root $t = -1$ of Eq. (85) has multiplicity $2k$. Such is the number of independent solutions of (83) having the form different from the simple power $X_n^m \sim t^n$ form; more precisely, the following [similarly to (75)]:

$$X_n^m = (-1)^n \left[\exp\left(\sum_{s=1}^{2k-1} \gamma_s n^{s/2k} \right) \right] n^{-1/2 - 1/4k} \left(1 + \sum_{l=1}^{\infty} d_l n^{-k/2l} \right). \tag{88}$$

The asymptotic behavior of the solutions of (88) defines the coefficient γ_{2k-1}, for which it is easy to obtain the equation

$$(\gamma_{2k-1})^{2k} = \frac{\pi}{2 s_k} \left[\frac{4ik}{(2k-1)} \right]^{2k} [2W_0 - 2W_1 + (\lambda - m\omega_0)^2]. \tag{89}$$

This equation has $2k$ roots corresponding to $2k$ independent solutions. If the expression in square brackets in (89) is negative, then k roots of γ_{2k-1} have positive real parts, and k, negative. In this case, the frequency spectrum is discrete. But if this expression is real and positive, then $(k - 1)$ solutions are exponentially converging, $(k - 1)$ are diverging (also exponentially), and for two remaining solutions, γ_{2k-1} turns out to be purely imaginary. These two solutions must also be investigated, and such an investigation [228] shows that both are converging. Thus, we have $(N + 1)$ converging and $(N - 1)$ diverging solutions when

$$\left[(\lambda - m\Omega)^2 + 2\Omega^2 + r \frac{d\Omega^2}{dr} \right]_{r=R} > 0. \tag{90}$$

The condition in (90) just determines the region of the continuous spectrum. Note that the presence of a continuous spectrum is defined only by the lack in the expansion of the surface density $s(\xi)$ of the term proportional to ξ.

Here one can again, as in the previous case, establish the correspondence with the WKB method. Additionally, the analysis made above establishes an accurate position of the boundary of the continuous spectrum, giving $\lambda^2 = \frac{15}{8}$. In reality, the WKB method also allows these boundaries to be determined, but for this purpose it is necessary, of course, in the calculation of oscillation frequencies to take into account not only the main order of perturbation theory but also the subsequent one [228]. Since for membrane oscillations these calculations in themselves are of no special interest, we omit them, the more so that in the next section 2.2 a similar problem will be considered for oscillations in the disk plane.[15] Let us give only the final result, the equation for eigenfrequencies of membrane oscillations:

$$\lambda^2 I_1 - 2m\lambda I_2 + (m^2 + 2)I_3 - I_4 + O\left(\frac{1}{\lambda^2}\right) = N + \frac{m}{2} - \frac{1}{4}, \quad (91)$$

where the following notations for the integrals are introduced:

$$I_1 = \int_0^1 \frac{\xi \, d\xi}{s(1 - \xi^2)^{1/2}}, \qquad I_2 = \int_0^1 \frac{\xi\Omega \, ds}{s(1 - \xi^2)^{1/2}}, \qquad (92)$$

$$I_3 = \int_0^1 \frac{\xi\Omega^2 \, d\xi}{s(1 - \xi^2)^{1/2}}, \qquad I_4 = \int_0^1 \frac{(1 - \xi^2)^{1/2}}{s} \frac{d\Omega^2}{d\xi} d\xi \qquad (93)$$

The continuous spectrum is obtained when the integral I_1 is divergent at $\xi \to 0$. The rule for the determination of the boundaries of the continuous spectrum from (91) is as follows. We introduce instead of I_1, I_2, I_3, I_4 the integrals with a variable lower limit (ξ instead of 0), for example,

$$I_1(\xi) = \int_\xi^1 \frac{\xi \, d\xi}{s(1 - \xi^2)^{1/2}}, \qquad (94)$$

and similarly for the remaining integrals. We divide (91) by $I_1(\xi)$ and proceed to the limit as $\xi \to 0$, assuming that

$$\lim_{\xi \to 0} \left(\frac{N}{I_1}\right) = 0. \qquad (95)$$

As a result, we get the sought-for equation for the boundaries of the continuous spectrum

$$\lambda^2 - 2m\lambda\Omega(\xi = 0) + (m^2 + 2)\Omega^2(\xi = 0) - \left[\frac{1}{\xi}\frac{d\Omega^2}{d\xi}\right]_{\xi=0} = 0. \quad (96)$$

It is easy to test that the roots of this equation coincide with the roots of (90). Note that the same procedure remains true also in other cases (e.g.,

---

[15] It is interesting in that such calculations define the form of the increasing spiral perturbation in the cold disk ("leading" spiral).

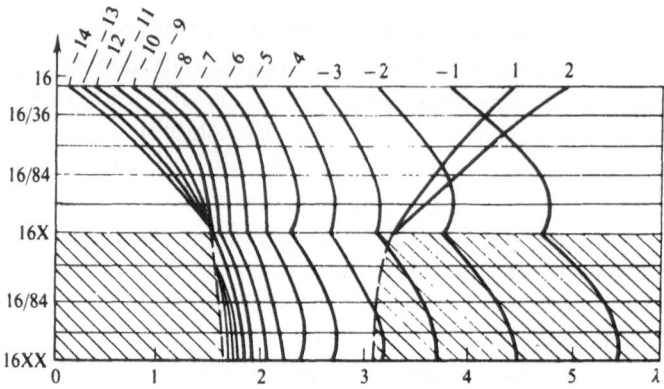

Figure 66. Spectra of $(m = 4)$ eigenfrequencies of membrane oscillations for model sequences with increasingly "thin" edge [228] (the degree of "thinness" of the edge of the disk increases from the upper to the lower models, see Section 1). Curves correspond to various modes. The cross-hatched area is the continuum spectrum.

for oscillation in the disk plane, Section 2.2, or for oscillations of the gravitating plasma disk in the magnetic field, §3). The boundary frequency in (90) is a "vertical" analog of the epicyclic frequency on the edge of the disk.

To conclude this section, let us note that Hunter in [228] calculated for a number of models also a large number of discrete modes, which, in particular, can coexist with the regions of the continuous spectrum. In Fig. 66, taken from [228], "tracks" of eigenfrequencies are shown as one gradually changes (corresponding to the up–down movement on the ordinate axis depicted in Fig. 66) the equilibrium state which makes the edge of the disk still thinner.

Applications of this theory to the explanation of the observed bending of the plane of the Galaxy is discussed in §4, Chapter VIII.

2.2 Oscillations in the Plane of the Disk

2.2.1. Qualitative Consideration. Let $\sigma_0(r)$ be an equilibrium density of the disk, $\Omega_0(r)$ being the angular velocity of particles. Rotation of the disk, counteracting the gravity, stabilizes it only with respect to general compression. If one considers a small region of the disk with the linear dimension l and assumes that it is compressed by the value εl ($\varepsilon \ll 1$), then the change in the force of gravitational attraction with such compression, equal to (in the order of magnitude)

$$\Delta F_{\text{grav}} \sim G\sigma_0 l^2 [(l - \varepsilon l)^{-2} - l^{-2}] \sim \varepsilon G\sigma_0,$$

promotes further increase in density. The corresponding increase of centrifugal force

$$\Delta F_c \sim \Delta(\Omega_0^2 r) = \Delta\left(\frac{\Omega_0^2 r^4}{r^3}\right) = \Omega_0^2 r^4 \Delta\left(\frac{1}{r^3}\right) \simeq \varepsilon l \Omega_0^2$$

is the consequence of the conservation of angular momentum $L = r^2\Omega_0$, thereby hindering compression.

The above estimates (Toomre [333]) show that a sufficiently small region of perturbation l of the disk must be gravitationally unstable: $|\Delta F_{grav}| > |\Delta F_c|$, if $l < G\sigma_0/\Omega_0^2$. From the equilibrium condition it follows that the critical scale $l_{cr} \sim G\sigma_0/\Omega_0^2$ of the order of the characteristic size of the disk $R: l_{cr} \sim R$; i.e., according to the estimate made, the disk must be unstable with respect to practically all wavelengths.

2.2.2. Short-Wave Oscillations. Local Dispersion Equation. Let us consider perturbations $\sim e^{i(m\varphi - \omega t)}$ (ω is the frequency; m is the number of the azimuthal mode) with the wavelength in the radial direction $\lambda = 2\pi/k$, small in comparison with the radius R. More exactly, we shall assume that $kr \gg m$.

The linearized Poisson equation for a flat, infinitely thin gravitating layer in cylindrical coordinates,

$$\Delta\Phi_1 = \frac{\partial^2\Phi_1}{\partial r^2} + \frac{1}{r}\frac{\partial\Phi_1}{\partial r} + \frac{1}{r^2}\frac{\partial^2\Phi_1}{\partial\varphi^2} + \frac{\partial^2\Phi_1}{\partial z^2} = 4\pi G\sigma_1\delta(z), \tag{1}$$

in this approximation is reduced to the following:

$$\frac{\partial^2\Phi_1}{\partial r^2} + \frac{\partial^2\Phi_1}{\partial z^2} = 4\pi G\sigma_1\delta(z). \tag{2}$$

As was to be expected, this equation coincides (if we substitute $r \to x$) with the Poisson equation for the case of an infinitely thin layer, not confined in the (x, y)-plane (cf. §2, Chapter I). Therefore, we can immediately write, similarly to the corresponding formulae in §2, Chapter I,

$$\Phi_1(r, z) = e^{ikr}e^{-k|z|}, \tag{3}$$

$$\sigma_1(r) = -\frac{k}{2\pi G}e^{ikr}, \tag{4}$$

or, denoting through $\Phi_1(r)$ the potential in the disk plane (Toomre [333]),

$$\Phi_1(r) = -\frac{2\pi G\sigma_1(r)}{k}. \tag{5}$$

The presence of a simple local coupling (5) between Φ_1 and σ_1 is the main simplification of the short-wave approximation.

To obtain the dispersion equation, we must derive also another independent relation between Φ_1 and σ_1. In the case under consideration of the "cold" disk, one may utilize for that purpose the linearized hydrodynamical equations with a pressure equal to zero:

$$\frac{\partial\sigma_1}{\partial t} + \Omega\frac{\partial\sigma_1}{\partial\varphi} + \frac{1}{r}\frac{\partial}{\partial r}(r\sigma_0 v_{r1}) + \frac{\sigma_0}{r}\frac{\partial v_{\varphi 1}}{\partial\varphi} = 0, \tag{6}$$

$$\frac{\partial v_{r1}}{\partial t} + \Omega\frac{\partial v_{r1}}{\partial\varphi} - 2\Omega v_{\varphi 1} = -\frac{\partial\Phi_1}{\partial r}, \tag{7}$$

$$\frac{\partial v_{\varphi 1}}{\partial t} + \Omega\frac{\partial v_{\varphi 1}}{\partial\varphi} + \frac{\varkappa^2}{2\Omega}v_{r1} = -\frac{\partial\Phi_1}{r\,\partial\varphi}, \tag{8}$$

where v_{r1} and $v_{\varphi 1}$ are the perturbed velocities and $\varkappa$ is the epicyclic frequency. From the last two equations we find

$$v_{r1} = \frac{1}{\omega_*^2 - \varkappa^2} \left(-i\omega_* \frac{\partial \Phi_1}{\partial r} + \frac{2im\Omega}{r} \Phi_1 \right), \tag{9}$$

$$v_{\varphi 1} = \frac{1}{\omega_*^2 - \varkappa^2} \left(\frac{m\omega_*}{r} \Phi_1 - \frac{\varkappa^2}{2\Omega} \frac{\partial \Phi_1}{\partial r} \right), \tag{10}$$

where the notation $\omega_* = \omega - m\Omega$ is introduced. Substituting (9) and (10) into (6), we find a simple relation between Φ_1 and σ_1:

$$-\sigma_1 = \frac{1}{r} \frac{\partial}{\partial r} \left(r\varepsilon \frac{\partial \Phi_1}{\partial r} \right) - \varepsilon \frac{m^2}{r^2} \Phi_1 - \frac{2m}{r\omega_*} \frac{\partial}{\partial r} (\varepsilon\Omega)\Phi_1, \tag{11}$$

where

$$\varepsilon = \frac{\sigma_0(r)}{\omega_*^2 - \varkappa^2}. \tag{12}$$

In the derivation of this relation we have not yet made any assumption as to the perturbation scale; further, (11) will be used repeatedly.

In the short-wave approximation, formula (11) becomes strongly simplified:

$$\sigma_1 = -\frac{\sigma_0}{\omega_*^2 - \varkappa^2} \frac{\partial^2 \Phi_1}{\partial r^2} = \frac{k^2\sigma_0}{\omega_*^2 - \varkappa^2} \Phi_1. \tag{13}$$

This relation, along with (5), does give the sought-for dispersion equation of Toomre for short-wave perturbations of a cold rotating disk [333]

$$\omega_*^2 = \varkappa^2 - 2\pi G\sigma_0 k. \tag{14}$$

From dispersion equation (14) it follows that the disk is unstable for perturbations with the wavelength less than the critical Toomre wavelength $\lambda_T = 4\pi^2 G\sigma_0/\varkappa^2$. This instability criterion specifies the similar criterion given above, obtained from qualitative considerations.

The first term of Eq. (14) describes the stabilizing (if $\varkappa^2 > 0$!) effect of rotation, while the second term describes the Jeans compression of small volumes of a cold infinitely thin layer [it, of course, coincides with the right-hand side of dispersion equation (11), §2, Chapter I].

Strictly speaking, (14) holds true only for $\lambda \ll \lambda_T$,[16] which corresponds to $\lambda \ll R$; therefore,

$$-(\omega - m\Omega)^2 = \gamma^2 \simeq 2\pi G\sigma_0 k,$$

i.e., as $k \to \infty$, the instability growth rate $\gamma \to \infty$ according to the law $\gamma \sim \sqrt{k}$.

---

[16] Provided that the disk does not contain a great central mass (cf. §1, Chapter XII).

2.2.3. Short-Wave Oscillations. The WKB Method. "Leading" Spirals in a Cold Disk. Hunter [229] developed a method similar to the WKB method, which in principle allows one to calculate the eigenfrequencies and eigenfunctions of short-wave oscillations in any order of perturbation theory.

The starting point of Hunter's analysis is the establishment of the asymptotic coupling between the surface density and the potential that corresponds to it. The calculation is carried out in spheroidal coordinates (ζ, μ, φ), which are convenient in the case of confined disks. We consider the solutions rapidly oscillating in the plane of the disk,

$$\Phi_1 = \frac{\pi GM}{R} \exp\{i[m\varphi + \psi(\mu, \zeta)]\}, \tag{15}$$

where the large phase ψ is represented in the form of a series over inverse powers of the dimensionless frequency $\lambda = \omega(R^3/\pi GM)^{1/2}$:

$$\psi(\mu, \zeta) = \lambda^2 \psi_0 + \lambda \psi_1 + \psi_2 + \frac{\psi_3}{\lambda} + \cdots. \tag{16}$$

To begin with, we consider the behavior of the solution outside the disk $\zeta \neq 0$. We substitute (15) and (16) into the Laplace equation having in the ζ, μ, φ coordinate the form

$$\frac{\partial}{\partial \mu}\left[(1 - \mu^2)\frac{\partial \Phi_1}{\partial \mu}\right] + \frac{\partial}{\partial \zeta}\left[(1 + \zeta^2)\frac{\partial \Phi_1}{\partial \zeta}\right] + \left[\frac{1}{1 - \mu^2} - \frac{1}{1 + \zeta^2}\right]\frac{\partial^2 \Phi_1}{\partial \varphi^2} = 0. \tag{17}$$

In the highest order (λ^4), we obtain the "eikonal" equation for ψ_0:

$$(1 - \mu^2)\left(\frac{\partial \psi_0}{\partial \mu}\right)^2 + (1 + \zeta^2)\left(\frac{\partial \psi_0}{\partial \zeta}\right)^2 = 0. \tag{18}$$

Equation (18) can be factorized:

$$(1 - \mu^2)^{1/2}\left(\frac{\partial \psi_0}{\partial \mu}\right) = \pm i(1 + \zeta^2)^{1/2}\frac{\partial \psi_0}{\partial \zeta}. \tag{19}$$

The general solution for the pair of equations of first order thus obtained is the following:

$$\psi_0 = h_0(\arcsin \mu \pm i \operatorname{arcsinh} \zeta), \tag{20}$$

where h_0 (as well as below $h_1, h_2, \cdots$) denotes the arbitrary functions of argument standing in parenthesis.

In the following order (λ^3), we have the equation for the ψ_1 function:

$$(1 + \zeta^2)\frac{\partial \psi_0}{\partial \zeta}\frac{\partial \psi_1}{\partial \zeta} + (1 - \mu^2)\frac{\partial \psi_0}{\partial \mu}\frac{\partial \psi_1}{\partial \mu} = 0, \tag{21}$$

which, taking into account (20), has a general solution of the form

$$\psi_1 = h_1(\arcsin \mu \pm i \operatorname{arcsinh} \zeta). \tag{22}$$

In a similar way we find equations which must be satisfied by the ψ_2 and ψ_3 functions:

$$2(1 - \mu^2)\frac{\partial\psi_0}{\partial\mu}\frac{\partial\psi_2}{\partial\mu} + 2(1 + \zeta^2)\frac{\partial\psi_0}{\partial\zeta}\frac{\partial\psi_2}{\partial\zeta} + 2i\left(\mu\frac{\partial\psi_0}{\partial\mu} - \zeta\frac{\partial\psi_0}{\partial\zeta}\right) = 0, \quad (23)$$

$$2(1 - \mu^2)\left(\frac{\partial\psi_0}{\partial\mu}\frac{\partial\psi_3}{\partial\mu} + \frac{\partial\psi_1}{\partial\mu}\frac{\partial\psi_2}{\partial\mu}\right)$$

$$+ 2(1 + \zeta^2)\left(\frac{\partial\psi_0}{\partial\zeta}\frac{\partial\psi_3}{\partial\zeta} + \frac{\partial\psi_1}{\partial\zeta}\frac{\partial\psi_2}{\partial\zeta}\right) + 2i\left(\mu\frac{\partial\psi_1}{\partial\mu} - \zeta\frac{\partial\psi_1}{\partial\zeta}\right) = 0. \quad (24)$$

The solutions of these equations are written thus:

$$\psi_2 = h_2(\arcsin\mu \pm i\,\text{arcsinh}\,\zeta) + \frac{i}{4}\ln[(1 - \mu^2)(1 + \zeta^2)], \quad (25)$$

$$\psi_3 = h_3(\arcsin\mu \pm i\,\text{arcsinh}\,\zeta). \quad (26)$$

The perturbed surface density is proportional to the derivative discontinuity $\partial\Phi_1/\partial z|_{-0}^{+0}$; therefore, for the potential in the disk plane and for the corresponding surface density Σ we have such expansions:

$$\Phi_1 = \exp[i\psi(\zeta = 0, |\mu| = \xi)],$$

$$\psi = \lambda^2 h_0 + \lambda h_1 + h_2 + \tfrac{1}{4}i\ln(1 - \zeta^2) + \lambda^{-1}h_3 + O(\lambda^{-2}),$$

$$\Sigma = \pm\frac{\pi(1 - \xi^2)^{1/2}}{\xi}\left[\lambda^2\frac{dh_0}{d\xi} + \lambda\frac{dh_1}{d\xi} + \frac{dh_2}{d\xi} + \lambda^{-1}\frac{dh_3}{d\xi} + O(\lambda^{-2})\right]e^{i\psi},$$

$$(27)$$

$$h_j = h_j(\arcsin\xi),$$

where $\xi = |\mu| = \sqrt{1 - r^2/R^2}$, and the h_j functions are to be determined.

We write now the linearized system of equations of hydrodynamics in spheroidal coordinates, first proceeding to new dimensionless variables:

$$\xi = \sqrt{1 - r^2/R^2}, \quad \tau = (\pi GM/R^3)^{1/2}t$$

is the dimensionless time; $s = (2\pi R^2/M)\sigma_0$, $\Sigma = (2\pi R^2/M)\sigma_1$ are, respectively, the unperturbed and perturbed surface densities;

$$\Omega = \Omega_0\left(\frac{R^3}{\pi GM}\right)^{1/2}$$

is the frequency of revolution; u, $v = v_r$, $v_\varphi(R/\pi GM)^{1/2}$ are the radial and azimuthal perturbed velocities; $\Phi = (R/\pi GM)\Phi_1$ is the gravitational potential. The dependence of the perturbations on the dimensionless time τ is assumed to be of the form $e^{-i\lambda\tau}$. In these notations, we have

$$-i\lambda_* u - 2\Omega v = \frac{(1 - \xi^2)^{1/2}}{\xi} \frac{d\Phi}{d\xi}, \tag{28}$$

$$-i\lambda_* v + \left[2\Omega + \frac{\xi^2 - 1}{\xi} \frac{d\Omega}{d\xi}\right] u = -\frac{im\Phi}{(1 - \xi^2)^{1/2}}, \tag{29}$$

$$-i\lambda_*(1 - \xi^2)^{1/2}\xi\Sigma + \xi s(u + imv) = (1 - \xi^2)\frac{d(su)}{d\xi}, \tag{30}$$

where $\lambda_* = \lambda - m\Omega$. From this system of equations it is easy to establish that $u = O(\lambda), v = O(1)$, so that the components of the perturbed velocity must be represented by such expansions:

$$u = [\lambda u_0 + u_1 + \lambda^{-1}u_2 + \lambda^{-2}u_3 + O(\lambda^{-3})]e^{i\psi}, \tag{31}$$

$$v = [v_0 + \lambda^{-1}v_1 + \lambda^{-2}v_2 + O(\lambda^{-3})]e^{i\psi}. \tag{32}$$

Substituting expansions (31), (32), and (27) into (28)–(30), we find

$$\frac{dh_0}{d\xi} = \mp \frac{\pi\xi}{s(1 - \xi^2)^{1/2}}, \tag{33}$$

$$\frac{dh_1}{d\xi} = \mp \frac{2\pi m\Omega\xi}{s(1 - \xi^2)^{1/2}}, \tag{34}$$

$$\frac{dh_2}{d\xi} = \pm \frac{\pi\xi}{s(1 - \xi^2)^{1/2}} \left[(4 - m^2)\Omega^2 + \frac{\xi^2 - 1}{\xi} \frac{d\Omega^2}{d\xi}\right], \tag{35}$$

$$\frac{dh_3}{d\xi} = 0, \cdots \tag{36}$$

The natural requirement for a rapid decreasing of the potential produced by a rapidly oscillating surface-density distribution, in moving away from the disk, is met for the solution of both signs,

$$\frac{1}{\Phi_1}\left(\frac{\partial\Phi_1}{\partial\zeta}\right)_{\zeta=0} \sim \mp \lambda^2(1 - \xi^2)^{1/2}\frac{dh_0}{d\xi} = \frac{\lambda^2\pi\xi}{s} < 0, \tag{37}$$

since $\lambda^2 < 0$ (and large in absolute value).

The general solution is written in the form of a linear combination of two independent solutions

$$\Phi = \frac{1}{(1 - \xi^2)^{1/4}}\{C_1 \exp[-i\pi\psi(\xi, \lambda)] + C_2 \exp[i\pi\psi(\xi, \lambda)]\}, \tag{38}$$

where

$$\psi(\xi, \lambda) = -\int_\xi^1 \frac{\xi}{s(1 - \xi^2)^{1/2}} \left[(\lambda - m\Omega)^2 - 4\Omega^2 - \frac{(\xi^2 - 1)}{\xi} \frac{d\Omega^2}{d\xi}\right]d\xi, \tag{39}$$

where C_1 and C_2 are arbitrary constants. To determine them as well as to find the frequency spectrum, it is necessary to impose suitable boundary conditions at the disk center $\xi = 1$ and on the edge $\xi = 0$. Hunter [229] considers solutions regular at the center and on the disk edge. Near the center, $\xi \simeq 1$ ($\mu \simeq \pm 1$, $\zeta = 0$), the asymptotic solution of the form of (38) is not exact. At $\mu > 0$, $\mu = 1 - \eta$ (η is small), the Laplace equation takes the form

$$2\eta \frac{\partial^2 \Phi}{\partial \eta^2} + 2 \frac{\partial \Phi}{\partial \eta} + \frac{\partial^2 \Phi}{\partial \zeta^2} + \frac{1}{2\eta} \frac{\partial^2 \Phi}{\partial \varphi^2} = 0, \tag{40}$$

where we have neglected the terms $O(\eta)$ and $O(\zeta^2)$. The solution of this equation, regular at the center ($\eta = 0$):

$$\Phi = C \exp(im\varphi - k\zeta) J_m(k(2\eta)^{1/2}), \tag{41}$$

where the separation constant k must be such that the solution decreases moving away from the center, i.e., must be $\mathrm{Re}(k) > 0$. At sufficiently large η, the solution of (41) ought to transform into (38). By using the asymptotic representation of the Bessel function, one may readily make sure that such "matching" of the solution occurs [for $k \sim -\pi\lambda^2/s(\xi = 1)$] under the condition

$$C_1 \exp\left[-i\left(\frac{m\pi}{2} + \frac{\pi}{4}\right)\right] = C_2 \exp\left[i\left(\frac{m\pi}{2} + \frac{\pi}{4}\right)\right]. \tag{42}$$

The second boundary condition is placed on the disk edge (at $\xi = 0$). Here the Φ and $\xi\Sigma$ functions behave "well"[17] but the velocities u and v are generally singular because they involve the terms $\sim 1/s$:

$$u = \frac{C_1 e^{-i\pi\psi}}{(1 - \xi^2)^{1/4}} \left[\frac{\pi(\lambda - m\Omega)}{s} + \frac{i}{2\lambda^2} \frac{(\lambda - 5m\Omega)}{(1 - \xi^2)^{1/2}}\right]$$

$$- \frac{C_2 e^{i\pi\psi}}{(1 - \xi^2)^{1/4}} \left[\frac{\pi(\lambda - m\Omega)}{s} - \frac{i}{2\lambda^2} \frac{(\lambda - 5m\Omega)}{(1 - \xi^2)^{1/2}}\right] + O(\lambda^{-3}), \tag{43}$$

$$v = \frac{C_1 e^{-i\pi\psi}}{(1 - \xi^2)^{1/4}} \left[\frac{i\pi}{s}\left(2\Omega + \frac{\xi^2 - 1}{\xi} \frac{d\Omega}{d\xi}\right) - \frac{m}{\lambda(1 - \xi^2)^{1/2}}\right] + O(\lambda^{-2}). \tag{44}$$

In order to avoid these singularities, it is necessary to assume that

$$C_1 \exp[-i\pi\psi(0, \lambda)] = C_2 \exp[i\pi\psi(0, \lambda)]. \tag{45}$$

We obtain the dispersion equation from (42) and (45),

$$\sin\left\{\pi\left[\psi(0, \lambda) - \frac{m}{2} - \frac{1}{4}\right]\right\} = 0,$$

---

[17] Singularity of Σ at $\xi = 0$ is fictitious: it is a consequence of the radial motion of the disk edge.

so that

$$\psi(0, \lambda) = -\lambda^2 I_1 - 2m\lambda I_2 + (4 - m^2)I_3 - I_4 + O(\lambda^{-2}) = N + \frac{m}{2} + \frac{1}{4},$$
(46)

where N is the integer (which should be large and positive), while I_1, I_2, I_3, I_4 are denotations of the already familiar integrals of (92) and (93), Section 2.1. For the dimensionless frequency λ, from (46), we obtain such an

$$\lambda = -m\frac{I_2}{I_1} \pm i\left(\frac{N}{I_2}\right)^{1/2}\left\{1 + \frac{1}{2N}\left[\frac{m}{2} + \frac{1}{4} + I_4 + (m^2 - 4)I_3 - m^2\frac{I_2^2}{I_1}\right]\right\}$$
$$+ O(N^{-3/2}).$$
(47)

In the main order of perturbation theory, (47), of course, corresponds to the result earlier obtained in (14).

Corrections for (14) contained in (47) are themselves, of course, of little interest. Of much more interest is the fact that the same analysis provides the form of eigenfunctions of nonaxially symmetrical unstable modes, which, as we shall see, turn out to be the leading spirals.

The density $\xi\Sigma$ is determined by the expression

$$\xi\Sigma = \frac{k\pi^2}{(1 - \xi^2)^{1/4}}\left(\frac{\xi}{s}\right)\left[-(\lambda - m\Omega)^2 + 4\Omega^2 + \frac{\xi^2 - 1}{\xi}\frac{d\Omega^2}{d\xi}\right]$$
$$\times (\exp\{i\pi[\psi(\xi, \lambda) - \psi(0, \lambda)]\} + \exp\{i\pi[\psi(0, \lambda) - \psi(\xi, \lambda)]\}),$$
(48)

where

$$k = \text{const} = C_1 \exp[-i\pi\psi(0, \lambda)] = C_2 \exp[i\pi\psi(0, \lambda)].$$

For large complex numbers λ, one of the exponents in (48) dominates, and for the unstable root it is the second exponent:

$$\exp\left\{\pi\left[\frac{iN}{I_1}\int_0^\xi \frac{\xi\, d\xi}{s(1 - \xi^2)^{1/2}} + 2m\left(\frac{N}{I_1}\right)^{1/2}\int_0^\xi \frac{\xi(I_2/I_1 - \Omega)\, d\xi}{s(1 - \xi^2)^{1/2}}\right] + O(1)\right\}.$$
(49)

The amplitude of oscillations is given by the term $O(N^{1/2})$. It vanishes at the center $\xi = 1$ and on the disk edge, $\xi = 0$, but is positive at all the remaining points if one assumes that the angular velocity Ω decreases monotonically from the center. Then Ω increases monotonically together with ξ, and since I_2/I_1 is nothing but a weighted average of Ω throughout the disk with everywhere positive weight $\xi/[s(1 - \xi^2)^{1/2}]$, $(I_2/I_1 - \Omega)$ must monotonically decrease from positive values at $\xi = 0$ to negative values at $\xi = 1$. Therefore, the integral $O(N^{1/2})$ that defines the amplitude in (49) grows from zero in $\xi = 0$ to some positive maximum, and then decreases to zero in $\xi = 1$.

The form of perturbation is determined in the main order by the expression

$$R(\xi) \exp\left[\tau\left(\frac{N}{I_1}\right)^{1/2}\right]\cos\left[m\varphi - m\frac{I_2}{I_1}\tau + \frac{N\pi}{I_1}\int_0^\xi \frac{\xi\, d\xi}{s(1 - \xi^2)^{1/2}} + O(N^{1/2})\right],$$
(50)

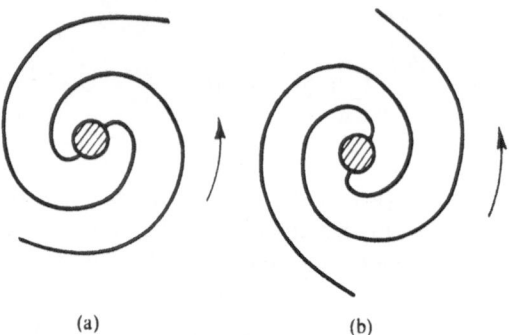

Figure 67. Trailing (a) and leading (b) spirals.

where $R(\xi)$ is some smooth real function. The integral $O(N)$ of the positive function $\xi/s(1 - \xi^2)^{1/2}$ grows monotonically from the disk edge inside, so that the lines of constant value of the perturbed density Σ, which in this approximation coincide with the lines of constant phase of cos in (50), define the leading spirals (Fig. 67).

Rohm [309] (cf. in [229]) also showed theoretically that the leading spirals arise in the particular case of nearly uniformly rotating disks.[18] Hunter [229] noticed that the leading spirals had also been obtained in comparatively earlier numerical experiments of Lindblad [275] in the earlier stages of the computation.

At the present time, however, most of the astronomers consider the spiral branches as trailing. Therefore, the cold disk considered by Hunter is difficult to use as a model of a spiral galaxy. In §3, Chapter XI, we shall continue the discussion of this issue.

2.2.4. The Uniformly Rotating Disk. Stability of a "cold" disk rotating as a solid body has been investigated in detail comparatively recently by Hunter [226], who obtained a dispersion equation exactly coincident with the dispersion equation for the Maclaurin ellipsoid in the disk limit. Remember that the latter was derived in [172] by Bryan as long as 100 years ago.

Such a coincidence is, of course, not unusual. We have already faced a similar situation in the analysis of perturbations of a flat layer in §2, Chapter I. In the zero-thickness limit, oscillations in the plane of the layer and in the perpendicular direction become independent, and the dispersion equations for the incompressible and for a "cold" collisionless layer coincide:

$$\omega^2 = -2\pi G\sigma_0 k \qquad (k \perp z), \qquad (51)$$

$$\omega^2 = +2\pi G\sigma_0 k \qquad (k \parallel z). \qquad (52)$$

In the latter case (52), the identity is obvious and does not require any clarifications. More unexpected is the coincidence of the spectra in the former

---

[18] This is true for solutions regular at the center. If, however, it is assumed that there is singularity at the center, then, as shown in [309] (cf. [229]), there forms a stationary picture of trailing spiral. This singularity corresponds to the presence at the center of a source of the quadrupole type (bar at the center of the Galaxy?).

case.[19] We recall that the reason lies in the fact that these perturbations become, in essence, two-dimensional when the wavelength λ is large as compared with the layer thickness c. They can then be investigated in terms of σ_1, v_x, Φ_1 which are nonsusceptible to the manner of perturbation production (bending of the boundary or a change in local density of matter).

The same causes evidently also explain the identity of the spectra of a "cold" collisionless and incompressible Maclaurin disk, if one also takes into account the identity of the equilibrium state of motion in both cases.

Therefore, to obtain the sought-for dispersion equation, it is sufficient to proceed to the limit $c \to 0$, $\zeta \to 0$ in the Bryan formula in [(4), Section 3.4, Chapter IV)] for the spectrum of frequencies of the Maclaurin ellipsoids.

Here it is necessary to consider separately the cases of even and odd values of the difference $(n - m)$.

1. $(n - m)$ is odd. These perturbations are antisymmetrical with respect to the symmetry plane $z = 0$ of the ellipsoid. Within the disk limit, they correspond evidently to the "membrane" oscillations bending the disk plane. From the mentioned formula at $\zeta \to 0$ one can obtain[20] [233]

$$\omega_*^2 = \frac{(2n - 1)!(2n + 2m - 1)!}{2^{4n + 2m - 5}[(n - 1)!]^2[(n + m - 1)!]^2} - 2. \qquad (53)$$

This equation is written in the rotating system and it is assumed that $\Omega = 1$. Since $\omega_*^2 \geq 0$, these oscillations are stable. The vertical displacements in this limit are expressed by the formula

$$h(\xi) = \frac{1}{\xi} P_{2n+m-1}^m(\xi)e^{im\varphi} \qquad \left(\xi \equiv \sqrt{1 - \frac{r^2}{R^2}}\right). \qquad (54)$$

The potential Φ_1 converts to the potential of a double layer; in the disk plane $z = 0$, it undergoes discontinuity:

$$\Phi_+ - \Phi_- = 4\pi\mu, \qquad (55)$$

corresponding to the dipole mass distribution with the density of dipoles, μ,

$$\mu(\xi) = \frac{3M}{2\pi R^2} \cdot P_{2n+m-1}^m(\xi)e^{im\varphi}. \qquad (56)$$

This question was treated in the previous section in more detail.

---

[19] Since the physical pictures of perturbations in the plane of the incompressible and collisionless layer are quite different, the latter is explained (cf. Chapter I, §2) if one considers a thin (in comparison with the perturbation wavelength) but not infinitely thin layer. This allows one to see that, in the collisionless case, oscillations occur due to variations of the local density, and one may neglect the effect of boundary bending (which is a single cause of perturbations of the incompressible layer).

[20] Note that, from the Bryan formula as $\zeta \to 0$,

$$\omega_*^2 = \frac{P_n^{m+1}(0)}{q_2^0(0)} \lim_{\zeta \to 0} \frac{p_1^0(\zeta)q_1^0(\zeta) - p_n^m(\zeta)q_n^m(\zeta)}{\zeta^2} \lim_{v_0 \to 0} \frac{v_0}{P_n^m(v_0)}$$

is directly obtained. This expression can be reduced to the form of (53).

2. $(n - m)$ is even. These perturbations were first studied by Hunter [226] (and then in connection with the problem of galactic spiral structure also by many other authors). They are symmetrical with respect to the $z = 0$ plane of the ellipsoid, and in the limit of an infinitely thin disk they yield perturbed motions lying in its plane.

The formula ensuing from the Bryan equation as $\zeta \to 0$, can be reduced, in the rotating system, to the following form (Hunter [226])[21]:

$$1 + \frac{4\gamma_n^m}{\omega_*^2 - 4}\left(n^2 + n - m^2 - \frac{2m}{\omega_*}\right) = 0, \tag{57}$$

where

$$\gamma_n^m = \frac{(n + m)!(n - m)!}{2^{2n+1}\{[(n + m)/2]![(n - m)/2]!\}^2}. \tag{58}$$

The perturbed potential and the surface density inside the disk have the form

$$\Phi_1 = -4\gamma_n^m P_n^m(\zeta)e^{im\varphi}, \tag{61}$$

$$\sigma_1 = \sigma_0(0)\frac{1}{\zeta}P_n^m(\zeta)e^{im\varphi}. \tag{62}$$

The singularity in σ_1 at $\zeta = 0$ corresponds evidently to displacements of the disk boundary and does not lead to any difficulties.

The dispersion equation in the form (57) can be obtained also if (59) and (60) are substituted into formula (11) derived by us earlier (from the equation of cold hydrodynamics) linking σ_1 and φ_1.

Let us consider now the properties of solutions of dispersion equation (57) in some detail [226]. Equation (57) is in the general case cubic with respect to ω_*, and for axially symmetrical perturbations it gives immediately the squares of eigenfrequencies

$$\omega^2 = 4\left\{1 - \frac{n(n + 1)(n!)^2}{2^{2n+1}[(n/2)!]^4}\right\} \qquad (n \neq 0, \text{ even}). \tag{63}$$

We have the radial dependence of the perturbation, e.g., of the velocity $\sim[(1 - \eta^2)^{1/2}/\eta]\,dP_n(\eta)/d\eta$; hence it is easy to see that perturbation has on the disk (at $0 \leq \eta < 1$) $(n - 2)/2$ zeros. A single nodeless mode corresponds to $n = 2$. This is simultaneously also the only stable type of oscillation of the cold disk (uniform compressions and expansions), for it, $\omega^2 = \Omega^2$.

---

[21] In the derivation of (57), the relations

$$\gamma_n^m = \frac{q_n^m(0)p_n^m(0)}{4q_2^0(0)} = \frac{-q_n^m(0)}{\pi q_n^{m'}(0)} = \frac{(n + m)!(n - m)!}{2^{2n+1}\{[(n + m)/2]![(n - m)/2]!\}^2}, \tag{59}$$

$$\frac{1}{P_n^m(0)}\lim_{x \to 0}\frac{1}{x}P_n^{m+1}(x) = (m - n)(m + n + 1) \tag{60}$$

are used.

All the remaining modes prove to be unstable. The instability growth rate for radial oscillations, according to (63), increases monotonically with the number of the mode n, i.e., with decreasing wavelength λ of the perturbation. At larger n, $\lambda \sim 2R_0/n$, while the frequencies ω can be expressed from (63) by using the Stirling formula.

As a result,

$$\omega^2 \approx -\frac{4n\Omega^2}{\pi} \tag{64}$$

ensues, which, of course, agrees with the Toomre formula in (14).

Dispersion equation (57) simplifies also in another specific case, for perturbations with $n = m$ (sectorial harmonics):

$$\left(\frac{\omega_*}{2\Omega}\right)^2 + \left(\frac{\omega_*}{2\Omega}\right) + m\gamma_m^m = 0, \tag{65}$$

$$\omega_* = -\Omega \pm \Omega(1 - 4m\gamma_m^m)^{1/2}, \tag{66}$$

$$\gamma_m^m = \frac{(2m)!}{2^{2m+1}(m!)^2}. \tag{67}$$

It may be shown that, for the modes in question, individual liquid elements of the disk do not change their density under perturbations. At $m = 1$, $m\gamma_m^m = \frac{1}{4}$, $\omega_* = -\Omega$, $\omega = 0$, which corresponds simply to the disk displacement as a whole (trivial solution). For $m > 1$, all perturbations are unstable, and the growth rate increases monotonically with m. At larger m, the growth rate of instability is $\gamma \simeq \Omega(4m/\pi)^{1/4}$. Such short-wave perturbations, consequently, increase more slowly than the radial perturbations, in which there is compression of individual liquid elements.

In the general case $m < n$, Eq. (57) can be rewritten in the form

$$y^3 - ay + b = 0, \qquad y = \frac{\omega_*}{2\Omega},$$

$$a = 1 - \gamma_n^m(n^2 + n - m^2), \qquad b = m\gamma_n^m. \tag{68}$$

It has three real roots, if and only if

$$a > 0, \qquad \frac{4a^3}{27} \geq b^2. \tag{69}$$

But the value $\gamma_n^m(n^2 + n - m^2)$ is always positive and increases monotonically with $(n + m)$ for fixed value of $(n - m) \equiv M$. The minimum value therefore is reached at $m = 1$, $n = M + 1$:

$$(n^2 + n - m^2)\gamma_n^m = \frac{(M^2 + 3M + 1)(M + 2)!(M)!}{2^{2M+3}[(M/2)!]^2\{[(M + 2)/2]!\}^2}. \tag{70}$$

This value increases monotonically with M. Consequently, the minimum is reached at $M = 2$:

$$[(n^2 + n - m^2)\gamma_n^m]_{\min} = \tfrac{33}{32}; \tag{71}$$

therefore $a_{max} = -\frac{1}{32} < 0$, and all the remaining a are more negative. Hence it follows that one of the roots of (57) is real, while the other two produce a complex-conjugate pair. Such a situation corresponds to instability of all modes.

Hunter [226] gives further asymptotical expressions for oscillation frequencies at $n \gg 1$, $n - m \gg 1$ when $a \simeq -(n^2 - m^2)^{1/2}/\pi$, $b \simeq m/\pi(n^2 - m^2)^{1/2}$. The real frequency is

$$-\frac{\omega_*}{2\Omega} \simeq \frac{b}{a}, \qquad \omega \simeq m\Omega\left(1 + \frac{2}{n^2 + n - m^2}\right); \tag{72}$$

it corresponds to oscillations propagating in the azimuth with angular velocity

$$\Omega_p = \frac{\omega}{m} = \Omega\left(1 + \frac{2}{n^2 + n - m^2}\right), \tag{73}$$

which is always more than the angular velocity Ω of the disk itself. The point $\omega = m\Omega$ at $n \to \infty$ is the point of accumulation for oscillation frequencies.

The pair of complex-conjugate roots in this case is

$$\frac{\omega_*}{2\Omega} \simeq +\frac{b}{2a} \pm i(-a)^{1/2}; \tag{74}$$

they imply a wave, rotating more slowly than the disk:

$$\Omega_p = \frac{\omega}{m} = \Omega\left(1 - \frac{1}{n^2 + n - m}\right), \tag{75}$$

tending at $n \to \infty$ to Ω. The instability growth rate, according to the expression for a, decreases with an increase of the number of the azimuthal mode ($n = $ const).

2.2.5. Arbitrary Disks. Matrix Formulations. Continuous Spectra.

In the case of *nonuniformly rotating* disks, the analysis of *large-scale* oscillations was performed mainly by numerical methods. Such investigations were pioneered by Toomre [333], who considered the long-wave radial oscillations by changing the real continuous distribution of matter in the disk by the system of axially symmetrical rings.

In a series of works by Yabashita [356–358], a method is developed for expansion of perturbed quantities over Bessel functions which leads to some matrix formulation of the problem of eigenvalues. The works [357, 358] are devoted to the application of this method to the problem of stability of Saturn's rings, while, in [356], the galactic spiral structure is considered. One of the main results of the latter work is the proof for the leading form of spiral two-arm perturbations in a cold nonuniformly rotating disk repeating the above result of Hunter [229] (but for large-scale modes).

The application of the technique of expansion in Legendre polynomials leads to another matrix formulation of the problem of eigenvalues [227].

In [227], this technique is employed to study the behavior of some longest-wave modes in models of disk galaxies with a different mass concentration toward the center. The lowest mode corresponds, as in the case of uniformly rotating disk, to distributions of perturbed values that have no nodes; it is always stable. The increase in central condensation causes stabilization of the instability of the first modes. It turns out that the subsequent mode (with one node) stabilizes already for insignificant deviation of the rotation from the uniform one (for instance, as compared with the inhomogeneity of rotation of the Galaxy).

A detailed study of spectra of small perturbations of cold disks is given by Hunter in [228], where he studied both perturbations lying in the disk plane and the membrane oscillations which occur in the perpendicular direction. Oscillations of the latter type have already been considered by us in the previous section; let us consider now the part of Hunter's investigation that refers to perturbations in the disk plane.

Of most interest to us is the method developed by Hunter of a strict investigation of the conditions of origin and location of the regions of the continuous spectrum.

Hunter [228] conventionally divides the continuous spectrum corresponding to nonradial perturbations of the disk into two classes.

The continuous spectrum of class I arises at sufficiently smooth vanishing of the surface density on the disk edge. The condition of the origin of this class of the continuous spectrum may be seen from the "quantization rule" (46) which in the lowest order of WKB is as $-\lambda^2 I_1 = N$. The distance between the "neighboring" frequencies λ_N^2 and λ_{N+1}^2, corresponding to N and $N + 1$, is $1/I_1$: it vanishes at $I_1 \to \infty$. The condition $I_1 \to \infty$ corresponds to the appearance of the continuous spectrum. (In the previous section we have seen that the condition of the appearance of the continuous spectrum of *membrane* oscillations is just the same.)

The integral I_1 is divergent on the lower limit when the surface density vanishes sufficiently slowly as $\zeta \to 0$ ($r \to R$). Then, the "thin edge" of the disk is similar to an unlimited medium; near $r = R$, there may be an infinite number of oscillations: from the above "quantization rule" it is easy to see that, as $I_1 \to \infty$, the number of nodes $N \to \infty$. We shall see below that the region of the continuous spectrum of class I includes all axially symmetrical unstable modes and most of the nonaxially symmetrical unstable modes, although the latter occur also outside the continuous spectrum.

The continuous spectrum of class II exists only for nonaxially symmetrical oscillations in the plane of the nonuniformly rotating disk. When the angular velocity $\Omega(r)$ drops monotonically toward the disk edge, this spectrum occupies the frequency band $m\Omega(0) > \omega > m\Omega(R)$. The cause of the appearance of this continuous spectrum is the resonance of the orbital movement of particles with a wave velocity ω/m.

The task of the determination of generation conditions and localization of the continuous spectrum for the case in question of oscillations in the disk plane can be investigated similarly to the case of membrane oscillations in

Section 2.1. From the analysis of the main approximation of difference equations it follows that the continuous spectrum can arise only when the function $f = \omega_*(\omega_*^2 - \varkappa^2)\sigma_0/\xi$ vanishes at some point on the interval $0 \leq \xi \leq 1$ (i.e., inside the disk or on its edge).

In principle, there are three possibilities: (1) $\sigma_0/\xi = 0$, (2) $\omega_* = 0$, (3) $\omega_*^2 - \varkappa^2 = 0$. The continuous spectrum of class I corresponds to the first of these possibilities; the frequencies lie in the region [228]

$$\omega_*^2|_{r=R} < \varkappa^2|_{r=R}. \tag{76}$$

On the left-hand side of (76) is the square of the epicyclic frequency on the disk edge. Such a result is similar to that achieved in the previous section [formula (70)]. However, now, the boundary frequencies may be quite simply imaginary (unlike the case of membrane oscillations, where they are always real): This is determined by the degree of central condensation in the mass distribution in the disk. It is evident that the boundary frequency must become real for a sufficiently developed central condensation since, in the limiting case of the Kepler problem, the square of the epicyclic frequency is obviously positive, $\varkappa^2 = \Omega^2$.

The continuous spectrum of class II corresponds evidently to the second of the above possibilities of vanishing of the f function. We have already faced similar continuous spectra (cf. §7, Chapter II, and Section 3.4, Chapter III).

The classes I and II complete all possible types of continuous spectra of collective oscillations of cold disks. The latter of the above alternatives of the vanishing of the f function, linked with the equation $\omega_*^2 - \varkappa^2 = 0$, corresponds evidently to oscillation of individual particles. We have already mentioned similar cases in §7, Chapter II and Section 3.4, Chapter III, where oscillations of inhomogeneous cylinders and spheres with circular orbits are dealt with. In disk systems, however, these oscillations (i.e., the circular orbits of the particles themselves) may be unstable ($\varkappa^2 < 0$, cf. also §3, Chapter XI).

Hunter's investigation of the spectrum of oscillations of a cold rotating disk, described in detail above, may likely be considered as exhaustive. The approach adopted by Hunter is, however, purely formal; consideration is carried out in the language of eigenmodes of oscillations, which, as we have already repeatedly mentioned, in case of a continuous spectrum, are singular and do not provide a direct answer to the question of the evolution of the initial smooth perturbation. On the other hand, it is the answer to this question that is of most interest.

Let us show how the problem of the evolution of initial perturbations located near a certain preliminarily chosen point of the disk is to be solved.

For such perturbations, one can write [333]

$$\sigma_1 = \frac{i\Phi_1'}{2\pi G} \quad \left(\Phi_1' \equiv \frac{d\Phi_1}{dr}\right), \tag{77}$$

which is equivalent to (5). Then the system of two connected equations

[Eq. (11) and the Poisson equation] is reduced to one equation for the perturbed potential

$$\Phi_1'' + \left[\frac{1}{r} + \frac{\sigma_0'}{\sigma_0} + \frac{2\varkappa\varkappa' + 2m\omega_*\Omega'}{\omega_*^2 - \varkappa^2} + \frac{i(\omega_*^2 - \varkappa^2)}{2\pi G\sigma_0} \right] \Phi_1'$$

$$- \left[\frac{m^2}{r^2} + \frac{2m\Omega'}{r\omega_*} + \frac{2m\Omega}{r\omega_*} \frac{[(\omega_*^2 - \varkappa^2)\sigma_0'/\sigma_0 + 2\varkappa\varkappa' + 2m\Omega'\omega_*]}{\omega_*^2 - \varkappa^2} \right] \Phi_1 = 0. \quad (78)$$

Oscillations corresponding to the continuous spectrum $\omega_* = 0$ may be considered by starting from Eq. (78) in exactly the same way as for similar oscillations in a cylinder with circular orbits (see §7, Chapter II). The results of calculations in both cases in essence coincide.

The main features of the wave packet evolution are more readily seen from Hamiltonian equations [cf. (59) and (60) in Section 2.1], which for $\omega = m\Omega(r)$ yield

$$\frac{dr}{dt} = \frac{\partial\omega}{\partial k} = 0, \quad (79)$$

$$\frac{dk}{dt} = -m\Omega'(r). \quad (80)$$

The former equation implies that in this case the perturbation has not naturally drifted. The latter equation shows that in the course of time there occurs size reduction of the spatial scale of oscillations due to different velocities of revolution of the particles located at different radii (cf. §7, Chapter II):

$$k(t) = k_0 - m\Omega'(r)t. \quad (81)$$

Another possibility of the continuous spectrum is associated, as we are aware, with a smooth vanishing of the surface density $\sigma_0(r)$ on the disk edge. In this case also, one is able to find an approximate solution to Eq. (78). In the absence of a significant halo or a large central mass, all short-wave perturbations of the disk (which are contained among the oscillations belonging to the spectrum in question) are unstable. The manner of the evolution of such perturbations is fairly obvious: an aperiodical increase with a gradual singling out of the smallest-scale constituents of the initial perturbation.

2.2.6. Variational Method. To conclude this section, let us consider the alternative approach to the investigation of the stability of flat gravitating systems which use the variational principle [140]. If one introduces instead of the velocities v_r and v_φ the displacements ξ_r and ξ_φ ($v_r = \partial\xi_r/\partial t, v_\varphi = \partial\xi_\varphi/\partial t$) and restricts oneself only to axially symmetrical perturbations, then the employed linearized equation of hydrodynamics of a cold disk and the Poisson equations yield the following integral equation for radial displacement:

$$\frac{\partial^2\xi_r}{\partial t^2} = -\varkappa^2\xi_r - \int_0^R \int_0^{2\pi} \frac{(1/r')(d/dr')[r'\sigma^{(0)}(r')\xi_r(r')]r' \, dr' \, d\varphi'}{\sqrt{r^2 + r'^2 - 2rr' \cos\varphi'}}. \quad (82)$$

This equation may be symbolically represented in the form

$$\sigma^{(0)} \frac{\partial^2 \xi_r}{\partial t^2} = \hat{K}\xi_r, \tag{83}$$

where $\hat{K}$ is the operator, the explicit expression of which is given by the right-hand side of (82).

We prove that the operator $\hat{K}$ is self-conjugate,[22] i.e., for

$$\iint \xi_1 \hat{K} \xi_2 \, ds = \iint \xi_2 \hat{K} \xi_1 \, ds \equiv -2W_{12}. \tag{84}$$

For that purpose, it seems to be enough to show that the form of (84) may be presented in the form explicitly symmetrical with respect to the displacements ξ_1 and ξ_2. Its part linked to $\varkappa^2$ is already symmetrical, while the "gravitational" part ($\sim G$)

$$W_{1,2}^{(G)} = \pi G \int_0^R dr\, r\sigma_0(r)\xi_1 \frac{\partial}{\partial r} \int_0^R \int_0^{2\pi} \frac{(1/r')(\partial/\partial r')[r'\sigma_0(r')\xi_2(r')]r'\, dr'\, d\varphi'}{\sqrt{r^2 + r'^2 - 2rr' \cos \varphi'}} \tag{85}$$

by integrating by parts is reduced to the sought-for symmetrical form

$$W_{1,2}^{(G)} = -\pi G \int_0^R \int_0^R dr\, dr'\, \Phi(r, r') \frac{\partial}{\partial r} [r\sigma_0(r)\xi_1(r)] \frac{\partial}{\partial r'} [r'\sigma_0(r')\xi_2(r')], \tag{86}$$

where the notation

$$\Phi(r, r') = \Phi(r', r) = \int_0^{2\pi} \frac{d\varphi'}{\sqrt{r^2 + r'^2 - 2rr' \cos \varphi'}} = \frac{4}{r + r'} K\left(\frac{2\sqrt{rr'}}{r + r'}\right) \tag{87}$$

is introduced and $K(x)$ is the complete elliptical integral.

The self-conjugateness implies that Eq. (82) may be found from the variational principle of the least action [53]

$$\delta \left\{ \int L \, dt \right\} = 0. \tag{88}$$

The Lagrange function L equals the difference of the kinetic T and potential W energy of small radial perturbations:

$$L = T - W, \qquad T = \tfrac{1}{2} \int ds\, \sigma_0 \left(\frac{\partial \xi_r}{\partial t}\right)^2,$$

$$W = -\tfrac{1}{2} \int \xi_r \hat{K} \xi_r \, ds = W_v + W_G, \tag{89}$$

---

[22] In case of nonaxially symmetrical perturbations, the initial system of hydrodynamic equations is not self-conjugate.

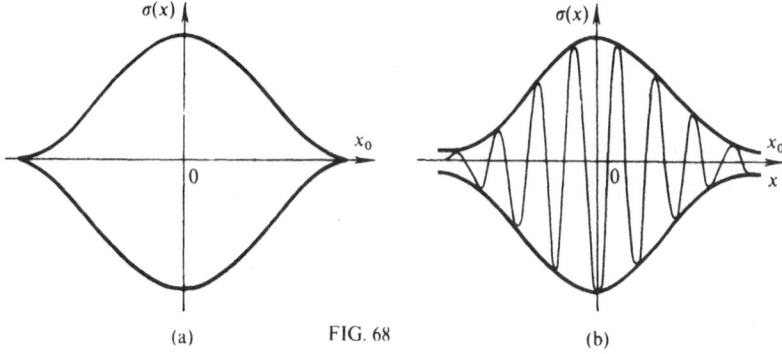

Figure 68. Localized perturbations of various types: (a) $k \approx 0$; (b) $kx_0 \gg 1$.

where

$$W_v = \pi \int_0^R dr\, \sigma_0 \varkappa^2 \xi_r^2,$$

$$W_G = \pi G \int_0^R dr\, \xi_r r \sigma_0 \frac{\partial}{\partial r} \int_0^R \int_0^{2\pi} \frac{(1/r')(\partial/\partial r')(r'\sigma_0(r')\xi_r(r'))r'\, dr'\, d\varphi'}{\sqrt{r^2 + r'^2 - 2rr'\cos\varphi'}}.$$

To analyze the stability of the system, the energy principle is utilized. It states that for stability it is necessary and sufficient that for any displacement ξ_r, $W > 0$. Otherwise, the system is unstable.

The sign of each summand of expression (89) characterizes its contribution to stability of the system. The increment sign of gravitational energy can be negative, i.e., the gravitational attraction exerts a destabilizing action on the system. The sign of the first summand in (89) is defined by the manner of the dependence of the rotation velocity $v_\varphi^{(0)}(r)$ on the radius. If $\varkappa^2 > 0$, i.e.,[23] the derivative $d(rv_\psi^{(0)})/dr$ is positive everywhere: $d(rv_\varphi^{(0)})/dr > 0$, then $W_v > 0$, and in this case rotation is a stabilizing factor.

The energy principle thus formulated allows one to investigate stability of the system in question for any radial perturbations. Regrettably, analytical computation of the gravitational contribution to the perturbed gravitational energy in the general case happens to be difficult. Calculations are comparatively readily performed only for narrowly located perturbations of the type $\sigma_1 = \tilde{\sigma}_1 e^{ikr}$, where $\tilde{\sigma}_1(r) \neq 0$ only within a narrow range of size Δr (Fig. 68). The calculation of the perturbed potential in this case yields

$$\Phi_1 = \frac{4G\sigma_1}{k}\left[\sin(k\Delta r)\ln\frac{\Delta r}{8r} - \int_0^{k\Delta r}\frac{\sin y}{y}\,dy\right].$$

But the first term in this expression vanishes due to mass conservation

$$\int \sigma_1(r)r\,dr = 0;$$

[23] The condition $\varkappa^2 > 0$ is generally not the consequence of the equilibrium equation.

therefore, we get

$$\Phi_1 = -\frac{4G\sigma_1}{k} \int_0^{k\Delta r} \frac{\sin y}{y} \, dy = -\frac{4G\sigma_1}{k} \, \text{Si}(k\Delta r). \tag{90}$$

When $k\Delta r \to \infty$, which corresponds to the short-wave (WKB) approximation, formula (90) transforms to the Toomre expression of (5) linking Φ_1 and σ_1 within this limit: $\Phi_1 = -(2\pi G/k)\sigma_1$. It should be noted that the difference of (90) and (5) is maximally equal to $\text{Si}(\pi)/\text{Si}(\infty) \simeq 1.85/1.57 \simeq 1.2$, and consequently, it is not so essential for estimates. On the other hand, the WKB method, leading to (5), allows one to investigate stability not only of radial but also of azimuthal perturbations (necessarily short-wave ones). A thorough investigation of stability with respect to radial perturbations, on the basis of the variational principle, must be made by numerical methods.

Generalization of the variational principle in (88) for the case of cold disks with a magnetic field are treated below (§3).

§ 3 Stability of a Plasma Disk with a Magnetic Field

This section deals with an investigation of small oscillations of disk gravitating systems in a poloidal magnetic field. The corresponding equilibrium states were considered in Section 1.2.

In Section 3.1, we give a qualitative derivation of the stability condition. In Section 3.2, we prove and discuss the variational principle for the case of radial oscillations of the disk in a magnetic field. Section 3.3 deals with the analysis of short-wave oscillations, while Section 3.4 gives some results of the numerical calculation of oscillation frequencies of a specific model with a uniform external field $B_z = B_0 = \text{const}$ and uniform rotation (in this example, one may demonstrate the appearance of a continuous spectrum of frequencies with decreasing field below some critical limit).

3.1 Qualitative Derivation of the Stability Condition

We derive the stabilizing criterion of gravitational instability of a disk by the magnetic field by making use of energy estimates. The gravitational energy ε_g of the region with a linear size l is $\varepsilon_g \backsim Gm^2/l$, while the magnetic energy $\varepsilon_H \sim (1/c) \int IA \, ds \sim IAl^2/c$, where I is the density of the surface current, rot $\mathbf{A} = \mathbf{B}$, $I \backsim cB$, $\varepsilon_H \backsim B^2 l^3 \backsim \Phi^2 l^{-1}$, Φ is the magnetic flux through S, that is, the value conserving under perturbations. Thus, the gravitational and magnetic energy turn out to be identically dependent on the size of the perturbed region, and we get the stabilization criterion $\varepsilon_H > \varepsilon_g$, i.e., $B_{0z}^2 > (2\pi\sigma_0)^2 G$ universal with respect to the wavelength (at least, in the range of short and "intermediate" wavelengths).

The universality (with respect to any wavelengths) of the stability criterion could have been foreseen after the work of Hunter [229], who showed a good coincidence of the eigenfrequencies obtained by a numerical method, with their values in the WKB approximation, up to the longest-wave modes.

3.2 Variational Principle

We consider now small oscillations of an infinitely thin cold disk in a poloidal magnetic field (Fig. 69). We shall begin with the formulation and proof of the variational principle for small oscillations of the gravitating disk in a proper magnetic field.

We write the linearized set of initial equations in dimensionless variables [95]:

$$i(\lambda - m\Omega)u - 2\Omega v = - \frac{(1 - \xi^2)^{1/2}}{\xi} \frac{\partial \psi}{\partial \xi}$$

$$+ \frac{1}{s}(J_{\varphi_0} H_z + J_\varphi H_{z_0}) - \frac{\Sigma}{s^2} J_{\varphi_0} H_{z_0}, \qquad (1)$$

$$i(\lambda - m\Omega)v + \left(2\Omega + \frac{\xi^2 - 1}{\xi} \frac{d\Omega}{d\xi}\right)u = \frac{im\psi}{(1 - \xi^2)^{1/2}} - \frac{1}{s} J_r H_{z_0}, \qquad (2)$$

$$i(\lambda - m\Omega)(1 - \xi^2)^{1/2}\xi\Sigma + \xi s(u + imv) = (1 - \xi^2)\frac{\partial}{\partial \xi}(su), \qquad (3)$$

$$i(\lambda - m\Omega)\left(\frac{H_z}{s} - \frac{H_{z_0}\Sigma}{s^2}\right) - \frac{\mu}{\xi}(1 - \xi^2)^{1/2}\frac{\partial}{\partial \xi}\left(\frac{H_{z_0}}{s}\right) = 0. \qquad (4)$$

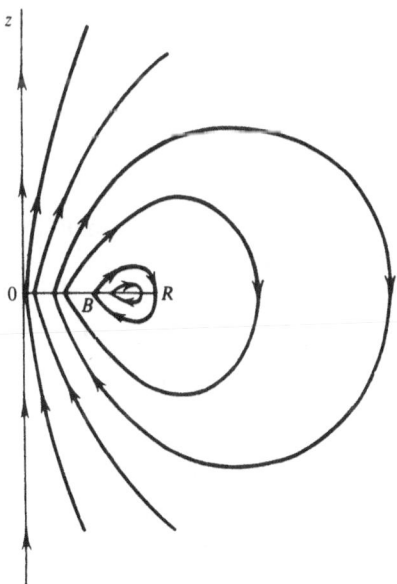

Figure 69. The magnetic field created by currents in the disk [27].

Here

$$H = \left(\frac{R^4}{\pi GM^2}\right)^{1/2} B, \qquad J = \frac{2\pi}{c}\left(\frac{R^4}{\pi GM^2}\right)^{1/2} j,$$

and the remaining notations coincide with those used in Section 2.2. The dependence on the dimensionless time τ is chosen in the form $e^{i\lambda\tau}$. The first three equations are the Euler equations and continuity, while the fourth expresses the condition of freezing. Equations (1)–(4) must be supplemented by the Poisson equation and Maxwellian equations:

$$\Delta\Phi = 4\pi G\sigma\delta(z), \tag{5}$$

$$\text{div }\mathbf{B} = 0, \tag{6}$$

$$\text{rot }\mathbf{B} = \frac{4\pi\mathbf{j}}{c}\delta(z). \tag{7}$$

Let us show that, for radial perturbations, the initial equations may be derived from a certain variational principle of the least action.

Introduce the radial displacement $\eta: u = \partial\eta/\partial\tau$. Then we obtain

$$s\frac{\partial^2\eta}{\partial\tau^2} = s\varkappa_H^2\eta - s\frac{(1-\xi^2)^{1/2}}{\xi}\frac{\partial\psi(\eta)}{\partial\xi}$$

$$+ H_{z0}J_\varphi(\eta) = \hat{K}_1\eta + \hat{K}_2\eta + \hat{K}_3\eta \equiv -\hat{K}\eta. \tag{8}$$

Here we have denoted

$$\varkappa_H^2 = 2\omega\left(2\omega + \frac{\xi^2-1}{\xi}\frac{d\omega}{d\xi}\right) + \frac{J_{\varphi0}(1-\xi^2)^{1/2}}{s^2\xi}\left(H_0\frac{ds}{d\xi} - s\frac{dH_0}{d\xi}\right)$$

and the operators

$$\hat{K}_1\eta = s\varkappa_H^2\eta, \qquad \hat{K}_2\eta = -s\frac{(1-\xi^2)^{1/2}}{\xi}\frac{\partial\psi(\eta(\xi))}{\partial\xi},$$

$$\hat{K}_3\eta = H_{z0}J_\varphi(\eta(\xi)) \tag{9}$$

introduced. It turns out that the equation of small oscillations in (8) may be derived from the following variational principle:

$$\delta\left\{\int L\,d\tau\right\} = 0, \qquad L = T - W, \qquad T = \frac{1}{2}\int s\left(\frac{\partial\eta}{\partial\tau}\right)^2 dS,$$

$$W = \frac{1}{2}\int \eta\hat{K}\eta\,dS, \qquad dS = 2\pi r\,dr = -2\pi\xi\,d\xi. \tag{10}$$

To prove the variational principle in (10), it is necessary to show the self-conjugateness of the operators $\hat{K}_1, \hat{K}_2, \hat{K}_3$. The self-conjugateness of the operators $\hat{K}_1$ and $\hat{K}_2$ was proved at the end of Section 2.2. Show that the operator $\hat{K}_3$ is also self-conjugate, i.e.,

$$\int \eta_1\hat{K}_3\eta_2\,dS = \int \eta_2\hat{K}_3\eta_1\,dS. \tag{11}$$

Make use of the expansions in Legendre polynomials. Let the displacement η_2 be represented in the form of the series.

$$\eta_2 = (1 - \xi^2)^{1/2} \sum_{m=1}^{\infty} \chi_{2m-1} P'_{2m-1}(\xi). \tag{12}$$

The system of the $(1 - \xi^2)^{1/2} P'_{2m-1}$ functions is complete and orthogonal on the interval $(0, 1)$. We shall seek the perturbation of the magnetic field H_z in the form

$$H_z = \frac{1}{\xi} \sum_{n=1}^{\infty} g_{2n-1} \beta_{2n-1} P_{2n-1}(\xi), \qquad \beta_{2n-1} = \frac{\pi}{2} \left[\frac{(2n-1)!!}{(2n-2)!!} \right]^2. \tag{13}$$

Since from (3) and (4) it follows that

$$H_z = -\frac{1}{\xi} \frac{\partial}{\partial \xi} (\sqrt{1 - \xi^2} H_{z0} \eta),$$

then

$$\sqrt{1 - \xi^2} H_{z0} \eta_2 = \int_1^{\xi} \sum_{n=1}^{\infty} g_{2n-1} \beta_{2n-1} P_{2n-1} \, d\xi$$

$$= -\sum_{n=1}^{\infty} \frac{g_{2n-1}}{2n(2n-1)} \beta_{2n-1} (1 - \xi^2) P'_{2n-1}(\xi), \tag{14}$$

where the coefficients g_{2n-1} are readily found from (14):

$$g_{2n-1} = -\frac{1}{\beta_{2n-1}} \sum_{m} \chi_{2m-1} \int_0^1 (1 - \xi^2) P'_{2m-1}(\xi) P'_{2n-1}(\xi) H_0(\xi) \, d\xi. \tag{15}$$

Perturbation J_φ of the current corresponding to (13), as shown in §1, has the form, provided that $J_\varphi(0) = 0$:

$$J_\varphi = \frac{\sqrt{1 - \xi^2}}{\xi} \sum_{n=1}^{\infty} g_{2n-1} P'_{2n-1}(\xi). \tag{16}$$

Let now

$$\eta_1 = \sqrt{1 - \xi^2} \sum_{k=1}^{\infty} v_{2k-1} P'_{2k-1}(\xi),$$

then

$$\int \eta_1 \hat{K}_3 \eta_2 \xi \, d\xi = \int_0^1 \left(\sqrt{1 - \xi^2} \sum_{k=1}^{\infty} v_{2k-1} P'_{2k-1} \right)$$

$$\times \left(\sqrt{1 - \xi^2} \sum_{n=1}^{\infty} g_{2n+1} \beta_{2n-1} P'_{2n-1} \right) H_0 \, d\xi$$

$$= -\sum_{n=1}^{\infty} \beta_{2n-1}^{-1} \sum_{m=1}^{\infty} \chi_{2m-1} \int_0^1 (1 - \xi^2) P'_{2m-1} P'_{2n-1} H_0 \, d\xi$$

$$\times \sum_{k=1}^{\infty} v_{2k-1} \int_0^1 (1 - \xi^2) P'_{2k-1} P'_{2n-1} H_0(\xi) \, d\xi$$

$$= \int \eta_2 \hat{K}_3 \eta_1 \xi \, d\xi. \tag{17}$$

The self-conjugateness of the operator $\hat{K}_3$ [and consequently, the validity of the variational principle in (10)] is proved.

From (8), we have

$$\alpha_n^2 = \frac{\int \eta_n \hat{K} \eta_n \, dS}{\int s \eta_n^2 \, dS}, \tag{18}$$

where η_n is some solution of Eq. (8) with a fixed frequency α_n.

Let us make some general remarks concerning the variational principle.

1. The variational principle in the case of rotating systems exists only for radial perturbations (which conserve symmetry of the initial state). The formal possibility of the derivation of the variational principle for such perturbations is connected with the conservation of the angular momentum of the particle which allows one to exclude from equations of motion the azimuthal velocity component.

It is generally not necessary to derive the variational principle from equations of motion, as done above; one may directly vary the total energy of the system taking into account the available additional laws of conservation: of the magnetic flux for ideal frozen-in state, of the angular momentum of a mass unit, etc. Such an approach, in application to the cold disk system with a frozen magnetic field considered in this subsection, is adopted in [27].

2. From the positiveness of the second variation of the total energy for some particular perturbation

$$\delta^2 \varepsilon = \tfrac{1}{2} \int \eta \hat{K} \eta \, dS > 0,$$

the inference about its stability should be made with some care in order to avoid the characteristic mistake. For example, the variational principle can be used for the investigation of stability of perturbations of the type of uniform extensions–compressions; $\eta = \alpha r$. The positiveness of the second variation of the energy for these perturbations is readily proved. But the positiveness of $\delta^2 \varepsilon$ does not yet imply stability because these perturbations are generally speaking lacking a definite frequency. Indeed, let

$$\eta = \alpha r = \sum_n \alpha_n \eta_n,$$

where η_n is the solution with a definite frequency λ_n. We calculate the second variation of energy $\delta^2 \varepsilon$ by using the fact that the solutions of the self-conjugate equation in (8) with different frequencies are mutually orthogonal with a weight σ_0:

$$\delta^2 \varepsilon = \tfrac{1}{2} \sum_n \beta_n \lambda_n^2, \qquad \beta_n = |\alpha_n|^2 \int |\eta_n|^2 \sigma_0 \, dS.$$

The positiveness of $\delta^2 \varepsilon$ evidently does not mean that all squares of the frequencies λ_n^2 entering the expression for $\delta^2 \varepsilon$ are positive. At the same time, for instability, it is known that, it is enough that at least one of λ_n^2 be negative.

As the simplest example, one may consider the gravitating disk without magnetic field with a surface density $\sigma_0 = \alpha(1 - r^2/R^2)^{1/2} + \beta(1 - r^2/R^2)^{3/2}$.

In this case, one can easily make sure directly that the perturbation with $\eta = \alpha r$ is not the solution with a definite frequency, and, consequently, it is unstable since the spectrum of the model in question contains only one stable harmonic [228]. In reality the largest-scale modes having no potential and density nodes on the disk are stable, but for them $\eta \neq \alpha r$.

3.3 Short-Wave Approximation

The local dispersion equation in this case is derived in the same way as the Toomre dispersion equation (cf. Section 2.2). Assume, for the sake of simplicity, that perturbations are axially symmetrical. Then from (6) and (7) one can have

$$\Delta B_z = \frac{4\pi}{cr} \frac{d}{dr} (rJ_\varphi)\delta(z). \tag{19}$$

This equation is accurate so far. It has the same form as the Poisson equation

$$\Delta \psi = 4\pi G \sigma \delta(z) \tag{20}$$

and is utilized in a similar way. In the short-wave approximation, from (19), as in Section 2.2 for (20), we obtain

$$B_{1z} = i \frac{2\pi}{c} J_{\varphi_1} e^{-k|z|}, \tag{21}$$

where k is the radial wave number, $kr \gg 1$; the perturbed values are $e^{-i\omega t + ikr}$. Using (21) as well as a similar connection between Φ_1 and σ_1 (cf. Section 2.2), one can derive the sought-for dispersion equation[24]

$$\lambda^2 = \varkappa_H^2 - \left(s - \frac{\pi H_0^2}{s}\right)\pi^{-1}|k|, \qquad \omega^2 = \varkappa_H^2 - k\left(2\pi G\sigma_0 - \frac{B_{0z}^2}{2\pi\sigma_0}\right). \tag{22}$$

From (22) it is seen that at $\pi H_0^2 > s^2$, short waves are stabilized by the magnetic field. As we shall show in 3.3, this result remains valid also for long waves under condition $\varkappa_H^2 > 0$.

An interesting situation arises when the condition $\pi H_0^2 > s^2$ is valid only in a part of the disk. Call the region of the disk in which $s^2 < \pi H_0^2$, a "locally stable" region (LSR), and the region in which $s^2 > \pi H_0^2$, a "locally unstable" region (LUR). Note that even if LSR is a part of the disk comparable with the full mass, this does not mean stability of the main mass of the disk since long waves of the order of LUR size, which are not located in LUR but occupy the entire disk, may be unstable. Unfortunately, in self-consistent models of the disk with a magnetic field suggested in [27], the LUR size is of the order of the disk radius, and their instability is obvious.

---

[24] Taking into account the thermal dispersion, the right-hand side of (22) has an additional summand $k^2 v^2$, where v^2 defines the velocity dispersion. The term $\varkappa_H^2$ in (22) takes into account rotation and inhomogeneity of H_0 and σ_0 [cf. (8) and (9)].

Near the point $\xi = \xi_0$, separating LUR and LSR, local analysis is wrong and there is a need to investigate in more detail the initial equations. (As mentioned in §1, this situation always arises in consideration of the disk with a self-consistent magnetic field.) Such an investigation was performed in [95] (cf. Problem 1). It uses the technique of expansion of all perturbed values in series in Legendre polynomials similar to that applied earlier by Hunter in the problem of oscillations of the disk without the magnetic field (cf. Sections 2.1 and 2.2 and Section 3.4 below).

The component H_{z0} of the self-consistent magnetic field necessarily changes sign on the disk. And since the surface density diminishes to zero on the disk edge, then there is always at least one LUR.

One may easily make sure that, in order to create a significant magnitude of the magnetic field (i.e., such that the value of the magnetic pressure is of the order of density of the centrifugal energy), such small currents are sufficient so that the current velocity is many orders less than the azimuthal rotational velocity. The question of the existence of such currents finally refers to the problem of magnetic dynamo in flat galaxies; this problem is dealt with in a large number of works, and the description of their results lies beyond the scope of our book.

The model of the gravitating rotating disk with a self-consistent magnetic field can be applied as a model of a flat protogalaxy. Recently, this model has been discussed as one of the possible models of the quasar. In this case, the problem of stability plays a particularly important role. Indeed, the unstable model of a flat galaxy with a self-consistent magnetic field is good in that in its framework one can describe the origin of spirals, due to instability. On the other hand, it is obvious that the fact of instability is disastrous for the model of a quasar with a long lifetime.

To calculate the oscillation frequencies, we must equate the phase of the short-wave solution $\exp(i \int_0^R |k| \, dr)$ to the integer (N) of π. Then we shall obtain the following expression for the shortwave part of the spectrum of frequencies λ_N of the system:

$$\int_0^R \frac{\lambda_N^2 - \varkappa_H^2}{-s + \pi H_0^2/s} \, dr \sim N \qquad (N \gg 1), \qquad (23)$$

or

$$\lambda_N^2 = \frac{I_2}{I_1} + \frac{N}{I_1}, \qquad (24)$$

where

$$I_1 = \int_0^1 \frac{\xi \, d\xi}{(1 - \xi^2)^{1/2}(-s + \pi H_0^2/s)}, \qquad I_2 = \int_0^1 \frac{\varkappa_H^2 \xi \, d\xi}{(1 - \xi^2)^{1/2}(-s + \pi H_0^2/s)}.$$

The integrals lose their meaning if somewhere on the disk $s^2 = \pi H_0^2$. Near this point, the phase takes on arbitrarily large values. As shown in [228, 230], such a situation generally corresponds to the appearance of a continuous spectrum. A strict analysis (cf. Problem 1) confirms this expectation.

Remark 1. Dispersion equation (22), and consequently also the corresponding stability criterion, may be obtained from the dispersion equation for the gravitating flat uniform layer with a finite thickness in a uniform vertical magnetic field, which is derived by Aggarswal and Talwar [158]:

$$-\frac{\omega^2}{4\pi\lambda\rho}\left[\cosh ql - \frac{q^2}{k^2}\cosh X \frac{\cosh ql + (q/k)\sinh ql}{\cosh X + \sinh X}\right]$$

$$= \left(\frac{\cosh X}{\cosh X + \sinh X} - X\right)\left(\frac{q}{k}\sinh ql - \frac{q^2}{k^2}\sinh X \frac{\cosh ql + (q/k)\sinh ql}{\cosh X + \sinh X}\right),$$

where l is the half-thickness of the layer, k is the wave vector, $q = \gamma\sqrt{4\pi\rho}/H$, γ is the growth rate, $X = kl$. In case of short-wave perturbations, the terms associated with the rotation are nonessential, and therefore the stability criteria for the rotating disk and an infinitely thin layer coincide.

2. Instead of considering normal modes, one can (and in this case, this is even more natural, cf. §2) solve the problem of the initial perturbation evolution. We do not consider this point in detail here since the solution is carried out in a quite similar way as was done in §2 (for oscillations belonging to a continuous spectrum of class I). In particular, it is possible to show that the perturbations imposed in LSR near the circumference $\xi = \xi_0$ must drift toward this circumference.

3.4 Numerical Analysis of a Specific Model

In this subsection, we shall deal with the appearance of a continuous spectrum in the example of a specific model with an external magnetic field:

$$s = 3\xi, \qquad B_0 = h = \text{const}, \qquad J_{\varphi 0} = 0, \qquad \varkappa_H^2 = 3. \qquad (25)$$

The "inconsistency" of the model is not essential for the clarification of the problem of the appearance of a continuous spectrum; however, it allows one to reveal the manner of the approaching of the roots of the discrete spectrum as h decreases from $h > h_{cr}$ to $h < h_{cr}$.

For the model in (25), the condition

$$s^2 - \pi H_0^2 = 9\xi^2 - \pi h^2 = 0 \qquad (0 \le \xi \le 1)$$

shows that the continuous spectrum must be manifested at $h = h_{cr} = 3/\sqrt{\pi}$.

Further advantage provides that here we have a possibility of transforming from the system of two difference equations to one equation. In this subsection, we shall make use of another expansion for H and J, not as in subsection 3.2:

$$H = \sum_{n=1}^{\infty} g_{2n} P_{2n}(\xi), \qquad J_{\varphi} = \frac{\sqrt{1 - \xi^2}}{\pi} \sum_{n=1}^{\infty} \frac{g_{2n}}{\gamma'_{2n}} P'_{2n}(\xi). \qquad (26)$$

Substitute (26) into (1)–(4):

$$-(\lambda^2 - 3) \sum_{n=1}^{\infty} \frac{\Phi_{2n} p'_{2n}}{\gamma'_{2n}} = 3 \sum_{n=1}^{\infty} \Phi_{2n} p'_{2n} - \frac{h}{\pi} \sum_{n=1}^{\infty} \frac{g_{2n}}{\gamma'_{2n}} p'_{2n}, \tag{27}$$

$$9\xi^2 \sum_{n=1}^{\infty} g_{2n} p_{2n} - 3\xi h \sum_{n=1}^{\infty} \frac{\Phi_{2n}}{\gamma_{2n}} p_{2n} = 3h(1 - \xi^2) \sum_{n=1}^{\infty} \frac{\Phi_{2n}}{\gamma'_{2n}} p'_{2n}. \tag{28}$$

Eliminate g_{2n} from (27):

$$g_{2n} = \frac{\pi}{n} [\lambda^2 - 3(1 - \gamma'_{2n})] \Phi_{2n}$$

and substitute into (28). Setting equal in (28) the coefficients at p'_{2n}, we obtain the five-term difference equation [95].

The frequencies λ^2 will be defined by setting equal to zero the infinite determinant corresponding to this equation. In case $h > h_{cr}$, as shown in Problem 1, Φ_{2n} decreases exponentially with increasing n. This provides a possibility of restricting oneself to a certain (large enough) finite number of equations.

In calculations [95], it was assumed that $N = 40$. Figure 70 shows the process of approaching of discrete modes of oscillations as $h \to h_{cr}$. It is seen that, as $h \to h_{cr}$, the distance between the roots vanishes.

We emphasize that $\lambda^2 = \varkappa_H^2 = 3$ is not the point of accumulation of a discrete spectrum for $h = h_{cr}$, but the boundary of a continuous spectrum. The model with $h = h_{cr}(\xi_0 = 1)$ must be investigated separately (the case of multiple roots). It may be shown that in this model there is a continuous spectrum in the region $\lambda^2 > \varkappa_H^2 = 3$, and near $\xi = \xi_0$

$$\Sigma \sim |\xi - \xi_0|^{-3/2} \exp\left(i \frac{\sqrt{\lambda^2 - 3}}{\sqrt{|\xi - \xi_0|}} \right),$$

and the integral defining the perturbed mass has the form $\sim \int e^{ikx}\, dx$ as in models considered in [228]. For $h < h_{cr}$, there arises also a continuous

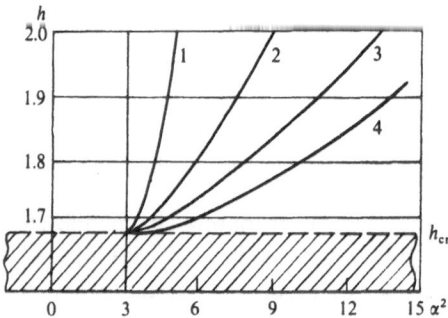

Figure 70. Trajectories of oscillation frequencies in models of solid-body rotating disk with an external magnetic field. The frequencies become closer without bound as $h \to h_{cr} = 3/\sqrt{\pi}$. For $h = h_{cr}$ the oscillation spectrum is continuous (the shaded area); 1, 2, 3, ... are the mode numbers.

spectrum in the region $\lambda^2 < \varkappa_H^2 = 3$. Instability is due, of course, to the presence of nodes in the perturbed quantities in LUR.

In [95] the modified model (25) is also numerically investigated, taking into account small pressure, which was assumed in the form $P_0 = \varepsilon \sigma_0^3$ for several values of the parameter ε. The frequency spectrum, as was to be expected, turned out to be discrete, and the distance between the frequencies diminished together with ε. The trajectories of frequencies in Fig. 70 did not merge at the point $\lambda^2 = 3$ but continued without intersecting into the region $\lambda^2 < 0$. With decreasing h, the first unstable modes become the most long-wave ones. This is, of course, explained by the fact that the modes with a large wavelength have a larger amplitude in LUR.

From these results it follows that the cold gravitating disk with a proper magnetic field is always unstable. To attain stable models, one may, for instance, introduce the gas pressure. Note also that if the magnetic field of the Galaxy is dealt with then, at least at the present epoch, it has, as we are aware, rather toroidal configuration. The model of a rotating disk with a frozen horizontal field is more difficult for the investigation than that with a vertical field.

As far as this issue is concerned, there is a work by Lynden-Bell [285] dealing with the stability of an infinitely thin gaseous disk (the field is assumed to be uniform). The magnetic field may produce a destabilizing effect on long waves. The instability growth rate found in [285] is proportional to B_0, in weak fields.

§ 4 Stability of a "Hot" Rotating Disk

4.1 Oscillations in the Plane of the Disk

4.1.1. Collisionless Disk. The Main Order of the WKB Approximation.
In Section 2.2, we have shown that the *cold* rotating disk is unstable with respect to many types of perturbations.[25] The influence of chaotic movements of the particles (of finite "temperature") in a collisionless disk on its stability was first taken into account by Toomre in the short-wave approximation [333]. He restricted himself to the consideration of radial perturbations. Later, Lin and Shu [270, 271] generalized the dispersion equation of Toomre for the case of arbitrary perturbations. Below, we shall mainly investigate this problem, followed by [271] which contains a detailed derivation and analysis of this more general dispersion equation.

---

[25] In this Section (as in 4.4), we shall restrict ourselves to the consideration of perturbations in the disk plane (similarly to Section 2.2 for a cold disk).

In the main order of the WKB approximation, the Poisson equation, as we are aware, yields the following coupling between the perturbed surface density σ_1 and gravitational potential Φ_1:

$$\Phi_1 = -\frac{2\pi G \sigma_1}{k}. \tag{1}$$

Therefore, our task is to calculate *the response* of the collisionless disk to the spiral gravitational field, which may be assumed to be given. Together with (1), this will lead to the sought-for dispersion equation.

In §5, Chapter II, we have faced a very similar problem, and below we use some formulae derived therein.

We denote by v_r and v_φ the components of chaotic velocities of the particles (in a cylindrical frame of reference r, φ, z, $v_z = 0$) and restrict ourselves to the consideration of systems with orbits which on the average do not strongly differ from circular ones. Then, v_r is coincident with the complete radial velocity, since there is no radial flux in the stationary state, and v_φ differs from the complete azimuthal velocity by the value $r\Omega_0$, where $r\Omega_0^2 = \partial\Phi_0/\partial r$ (Φ_0 is the axially symmetrical potential of the disk in the stationary state).

The distribution function $f(r, \varphi, v_r, v_\varphi, t)$ satisfies the equation

$$\frac{\partial f}{\partial t} + v_r \frac{\partial f}{\partial r} + \left(\Omega_0 + \frac{v_\varphi}{r}\right)\frac{\partial f}{\partial \varphi} + \left(-\frac{\partial \Phi}{\partial r} + \Omega_0^2 r + 2\Omega_0 v_\varphi + \frac{v_\varphi^2}{r}\right)\frac{\partial f}{\partial v_r}$$

$$- \left(\frac{\partial \Phi}{r\partial \varphi} + \frac{\varkappa^2}{2\Omega_0}v_r + \frac{v_r v_\varphi}{r}\right)\frac{\partial f}{\partial v_\varphi} = 0. \tag{2}$$

It is convenient to introduce the dimensionless velocities η and ξ by the formulae

$$v_r = \xi V_1 = \xi\left(\frac{2\Omega_0}{\varkappa}\right)(r\Omega_0), \qquad v_\varphi = \eta V_2 = \eta(r\Omega_0),$$

with the scale factors V_1 and V_2 according to epicycle theory (cf., e.g., §1). We linearize Eq. (2) by substituting

$$\Phi = \Phi_0(r) + \Phi_1, \qquad f = f_0(1 + \psi) \equiv e^{-Q_0}(1 + \psi),$$
$$\Phi_1 \ll \Phi_0, \qquad \psi \ll 1, \tag{3}$$

and then

$$\Phi_1, \psi \sim e^{-i\omega t + im\varphi + ikr} \qquad (kr \gg m). \tag{4}$$

Such a form of functions of (4) corresponds to perturbations of a spiral shape, and the equation of spiral is

$$kr + m\varphi = \text{const.} \tag{5}$$

Here, the case $k/m < 0$ corresponds to "trailing" spirals, while the case $k/m > 0$ corresponds to "leading" spirals (cf. Fig. 67).

In the main order in the small parameter $\varepsilon = 1/kr$ we obtain

$$\eta \frac{\partial \psi}{\partial \xi} - \xi \frac{\partial \psi}{\partial \eta} + i(v + \alpha \xi)\psi = -\frac{i\alpha\Phi_1}{V_1^2} \frac{\partial Q_0}{\partial \xi},$$ (6)

where

$$\alpha = (kr) \frac{2\Omega_0^2}{\varkappa^2},$$ (7)

$$v = \frac{\omega - m\Omega_0}{\varkappa}.$$ (8)

Further it is assumed that the stationary distribution function f_0 depends on ξ and η in the combination $\xi^2 + \eta^2$:

$$Q_0 = Q_0(\xi^2 + \eta^2, r).$$ (9)

This is always the case when only stars with small peculiar velocities are dealt with (§1). Just such a case is evidently of most interest from the point of view of theory applications (§3, Chapter XI). The contribution of a small part of particles possessing high peculiar velocities is negligible, particularly if one takes into account that the spiral gravitational field almost does not act on them (see below, the "reduction factor," by use of which this effect is taken into account).

The Schwarzschild distribution function, which is used most of all for specific calculations, also has the form of (9) and

$$Q_0 = \frac{1}{2}\left(\frac{v_r^2}{c_r^2} + \frac{v_\varphi^2}{c_\varphi^2}\right) + Q_{00}(r),$$ (10)

where c_r and c_φ are the velocity dispersions.

Introducing the polar coordinates τ and s in the space of dimensionless velocities ξ and η,

$$\xi = \tau \cos s, \qquad \eta = \tau \sin s,$$

we reduce Eq. (6) to the following:

$$-\frac{d\psi}{ds} + i(v + \varkappa\tau \cos s)\psi = -\frac{2i\alpha\Phi_1}{V_1^2} \frac{dQ_0}{d\tau^2} \tau \cos s.$$ (11)

This equation, within an accuracy of notations, coincides with the already familiar Eqs. (11), §5, Chapter II. The solution will be written as

$$\psi = -\frac{2\Phi_1}{V_1^2} \frac{dQ_0}{d\tau^2} [1 - q(\alpha\xi, \alpha\eta, v)],$$ (12)

where

$$q = \frac{v\pi}{2\pi \sin v\pi} \int_{-\pi}^{\pi} \exp\{i[vs - \alpha\xi \sin s + \alpha\eta(1 + \cos s)]\}\, ds.$$ (13)

To obtain the response of the surface density, one should integrate over velocity the value

$$f_1 = f_0 \psi = \frac{2\Phi_1}{V_1^2} \frac{df_0}{d\tau^2} (1 - q).$$ (14)

Lin and Shu [270, 271] do this for the case of the Schwarzschild distribution function

$$f_0 = \sigma_0(r) \frac{\Omega_0}{\pi \varkappa c_r^2} \exp\left[-\frac{V_1^2}{2c_r^2} (\xi^2 + \eta^2) \right],$$ (15)

where $\sigma_0(r)$ is the equilibrium surface density:

$$\sigma_0(r) = \iint f_0 \, dv_r \, dv_\varphi.$$

For the distribution function of (15), f_1 is expressed by the formula

$$f_1 = -\frac{\Phi_1}{c_r^2} f_0(1 - q).$$ (16)

The perturbation of the surface density corresponding to (16) is

$$\sigma_1 = -\sigma_0 \frac{\Phi_1}{c_r^2} \langle (1 - q) \rangle,$$ (17)

where $\langle \ \rangle$ means an average with the weight $\exp[-\mu_0(\xi^2 + \eta^2)/2]$, $\mu_0 = V_1^2/c_r^2$. To calculate (17), it is possible to make use of the familiar formula

$$\langle e^{i\lambda\xi} \rangle = e^{-\lambda^2/2\mu_0}.$$

Thus, one easily gets

$$\langle (1 - q) \rangle = 1 - \frac{v\pi}{\sin v\pi} G_v(x),$$ (18)

where

$$x = \frac{k^2 c_r^2}{\varkappa^2},$$ (19)

$$G_v(x) = \frac{1}{2\pi} \int_{-\pi}^{\pi} e^{-x(1 + \cos s)} \cos vs \, ds.$$ (20)

The value x has the order of magnitude

$$x \sim \left(\frac{\rho}{\lambda} \right)^2$$

where λ is the wavelength of perturbation and ρ is the mean size of the epicycle.

By using the relations (17)–(20), we represent the response of the surface density on the perturbed gravitational potential Φ_1 in the form

$$\frac{\sigma_1}{\sigma_0} = -\frac{k^2\Phi_1}{\varkappa^2}\frac{1}{(1-v^2)}\mathscr{F}_v(x), \tag{21}$$

where the "reduction factor"

$$\mathscr{F}_v(x) = \frac{1-v^2}{x}\left[1 - \frac{v\pi}{\sin v\pi}G_v(x)\right]. \tag{22}$$

is introduced. The dispersion equation ensues from the comparison of (21) with (1):

$$\frac{k_T}{|k|}\cdot(1-v^2) = \mathscr{F}_v(x), \tag{23}$$

where $k_T = \varkappa^2/2\pi G\sigma_0$ is the critical wave number of Toomre.

The alternative form of the dispersion equation also inferred by Lin and Shu [271] is the following:

$$\frac{k_T}{k} = \frac{1}{x}\left[1 - \sum_{n=1}^{\infty}\frac{(-1)^nI_n(x)e^{-x}}{(v/n)^2-1}\right], \tag{24}$$

where $I_n(x)$ is the Bessel function of imaginary argument. This form of writing is normally used in similar problems of plasma physics [86]. In customary notations, (23) implies

$$(\omega - m\Omega_0)^2 = \varkappa^2 - 2\pi G\sigma_0 k\mathscr{F}_v(x). \tag{25}$$

We recall that, for the cold disk, the dispersion equation is coincident with (25) for $\mathscr{F}_v(x) = \mathscr{F}_v(0) = 1$. Thus, the "reduction" factor $\mathscr{F}_v(x)$, which equals 1 at $x = 0$ and decreases with increasing x, i.e., with increasing mean velocity dispersion of the particles, quantitatively takes into account the fact that the gravitational field weakly affects the stars with high peculiar velocities (when the epicyclic radius is more than the wavelength: $\rho > \lambda$). The plots $\mathscr{F}_v(x)$ for some values of v are given in Fig. 71, which we took from the work of Lin

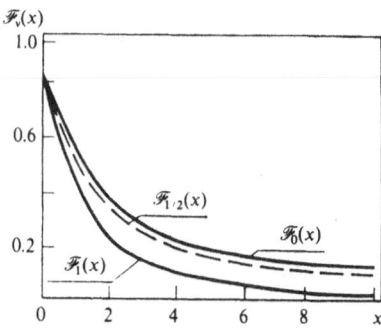

Figure 71. The dependence of the reduction factor $\mathscr{F}_v(x)$ on the radial velocity dispersion of stars [271].

and Shu [271].[26] The derivation of the reduction factor $\mathscr{F}_\nu(x)$ and of the corresponding dispersion equation (25) is the main result of their work.

For all interesting values of ν, the $\mathscr{F}_\nu(x)$ function is positively definite (at real ν^2). Therefore, according to (21), the maxima of the response of the surface density σ_1 correspond to the minima of the potential Φ_1 (as follows from the Poisson equation [cf. formula (1)]) only in the region where

$$\nu^2 < 1. \tag{26}$$

In other words, only in the range of (26), the density response agrees locally with that necessary for maintenance of the spiral gravitational field. The points r_1 and r_2 in which $\nu^2 = 1$ are called the points of Lindblad resonance; at these points $\omega/m = \Omega \pm \varkappa/m$. They play an important role in the spiral structure theory of Lin and Shu (cf. §3, Chapter XI), who believe, in accordance with the above, that the solution of spiral type rapidly oscillating in the radial direction lies between r_1 and r_2. Outside the resonances, the solution must have an essentially different form. To confirm this, it may be noted that, e.g., from the numerical calculation of radial oscillations of the entire disk performed by Toomre [333], it follows that at infinity ($r \to \infty$), the solutions decrease approximately exponentially.

Of course, in reality a special more accurate analysis of the solution near Lindblad resonances is required, even inside the range of (26) (cf. below).

In a particular case of radial oscillations ($m = 0$), the problem was solved by Toomre [333]. By restricting himself to radial perturbations independent of time, Toomre has obtained for them the following equation:

$$\frac{kc_r^2}{2\pi G \sigma_0} = 1 - \exp\left(-\frac{k^2 c_r^2}{\varkappa^2}\right) I_0\left(\frac{k^2 c_r^2}{\varkappa^2}\right). \tag{27}$$

From (27), at $c_r \to 0$ (k and $\varkappa \neq 0$) we obtain the critical wave number $k_{c_1} = \varkappa^2/2\pi G \sigma_0 = 2\pi/\lambda_T$. This result may also be obtained from the dispersion equation for a cold rotating disk. In the other limiting case, when $kc_r/\varkappa \to \infty$ (disk at rest), from (27) follows that $k_{c_2} \to 2\pi\sigma_0/c_r^2$. For arbitrary k, Eq. (27) defines the critical values of velocity dispersion of stars which stabilize the short-wave radial oscillations of the disk. Toomre represented (27) in the form of a curve (Fig. 72), the dependences of the value $x = k^2 c_r^2/\varkappa^2$ on the ratio λ/λ_T, where $\lambda_T = 4\pi^2 G \sigma_0/\varkappa^2$

$$\frac{1}{2\pi(\lambda/\lambda_T)} = \frac{1}{x}[1 - I_0(x)e^{-x}]. \tag{28}$$

The region above the curve corresponds to stability, while the region below the curve, to instability. Then it is seen (cf. Fig. 72) that the radial instabilities are completely suppressed if the velocity dispersion c_r at any

---

[26] [271] contains also the table of the values of $\mathscr{F}_\nu(x)$.

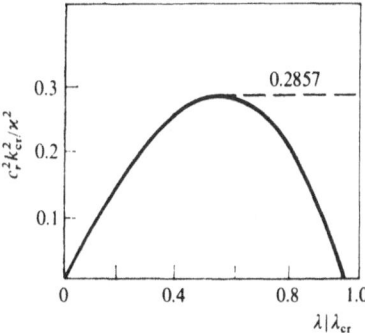

Figure 72. The curve of marginal stability of axisymmetric perturbations of a rotating stellar disk [333]. The region under the curve corresponds to instability.

point of the disk is more than the critical value of c_0:[27]

$$c_r > c_0, \qquad c_0 = (0.2857)^{1/2} \frac{\varkappa}{k_T} = 3.36 \frac{G\sigma_0}{\varkappa}. \qquad (28')$$

Strictly speaking, from Toomre's inference follows only that the curve (cf. Fig. 72) represents all admissible indifferently stable ($\omega^2 = 0$) perturbations. The conclusion drawn by Toomre about the fact that this curve separates stable and unstable solutions (respectively, above the curve and below it) requires, generally speaking, a more detailed analysis of the dispersion equation itself in (23) for $m = 0$. However, this conclusion is natural if, e.g., one bears in mind the above-mentioned limiting cases which can be considered elementarily and provide agreement with Toomre's prediction (cf. Fig. 72). It can also be substantiated on the basis of the theorem proved by Julian [188], according to which the distribution functions decreasing with energy must not lead to *oscillative instability* of the axially symmetric modes. From this theorem, evidently, it follows that the boundary between stable and unstable solutions passes on the line $\omega^2 = 0$, as suggested by Toomre.

---

[27] Hunter then obtained [68^{ad}] the stability criterion generalizing the Toomre criterion (28') for nonaxisymmetric disturbances taking into account nonuniformness of the rotation. Morozov [119^{ad}] took into account also the gradient of the surface density. The dispersion equation derived in [119^{ad}] is as follows:

$$\frac{\hat{k}c_r^2(1 + \hat{k}h)}{2\pi G\sigma_0} = 1 - \left\{ 1 - \frac{\omega_*}{v}\left(1 + \zeta + 2\eta z \frac{\partial}{\partial z} \right) \right\} \sum_{n=-\infty}^{\infty} \frac{v^2 I_n(z)e^{-z}}{v^2 - n^2\varkappa^2}. \qquad (23')$$

where

$$\hat{k} = (k^2 + m^2/r^2)^{1/2}, \ v = \omega - m\Omega, \ z = [k^2 + (2m\Omega/r\varkappa)^2](c_r^2/\varkappa^2),$$

$$\omega_* = (2m\Omega/r\varkappa^2)c_r^2 \, \partial \ln \sigma_0/\partial r, \ \xi = \partial \ln(2\Omega/\varkappa)/\partial \ln \sigma_0,$$

$$\eta = \partial \ln c_r/\partial \ln \sigma_0.$$

From (23') one can obtain such the stability criterion:

$$c_r \geq c_*; \qquad c_* = \frac{c_0(2\Omega/\varkappa)}{(1 + 0.98\varkappa^2 h/2\Omega c_r)}\left\{ 1 + 1.05 \left| \frac{\partial \ln \sigma_0}{k_T \partial r} (1 + \zeta - 1.1\eta) \right|^{2/3} \right\}. \qquad (28'')$$

For parameters of solar vicinity, the criterion (28") gives $c_* \simeq 48 \ ku/uv$.

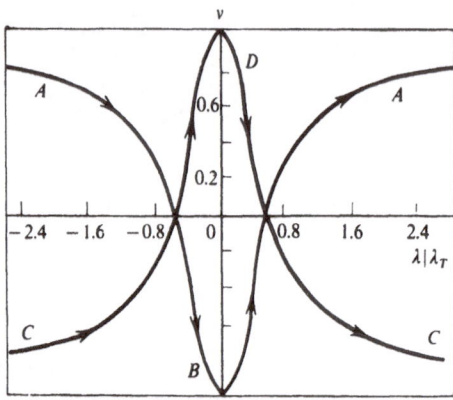

Figure 73. Dispersion relation (23) for the marginally stable stellar disk: (a) trailing; (b) leading.

The disks satisfying to the Toomre criterion (51) with equality sign are called marginally stable. Just such systems are usually considered in the Lin and Shu galactic spiral structure theory. For marginally stable disks, v^2 is nonnegative, and dispersion equation (23) provides a link between v and λ/λ_T ($\lambda = 2\pi/k$, $\lambda_T = 2\pi/k_T = 4\pi^2 G\sigma_0/\varkappa^2$), which is shown in Fig. 73 [325]. Normally, one chooses $\omega > 0, m > 0$; then the algebraically increasing values of v correspond to the increasing values of r. And the negative values of λ/λ_T correspond to trailing spirals, while the leading spirals correspond to the positive ones.

For each fixed ω, Eq. (23) defines, as is seen from Fig. 73, two functions $k(r)$ [325]. One of them at $v = -1$ (at the point of the inner Lindblad resonance) tends to infinity, while the other one vanishes. At the point of the outer Lindblad resonance $v = 1$, the situation is quite the reverse. Bearing in mind the behavior near the inner resonance which plays the principal role in the Lin and Shu theory, the former function is called the short-wave mode, and the latter the long-wave mode (Fig. 74).

As we have already mentioned, the spiral waves cannot propagate outside the region $-1 \leq v \leq 1$. For the case of marginally stable disks, this is the only restriction. However, when the disks are stable with some "reserve," i.e., the value

$$Q \equiv c_r/c_{r,\,\min} \qquad (29)$$

(where $c_{r,\min} \simeq (0.2857)^{1/2} \varkappa/k_T$ and c_r is the real velocity dispersion) is more than 1, there is another region forbidden for the waves. This is seen from Fig. 75, taken from the work of Toomre [334], who was the first to pay attention to this fact.

If one does not impose any boundary conditions, then dispersion equation (23) will yield the continuum of frequencies, ω. However, the boundary conditions for the spiral wave

$$\varphi - \left(\frac{\omega}{m}\right)t = -\frac{1}{m}\int^r k(r)\,dr + \varphi_0 \qquad (30)$$

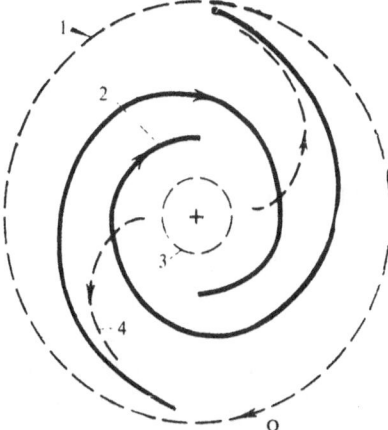

Figure 74. Long-wavelength and short-wavelength modes for the Schmidt model of the Galaxy [271]: (1) the corotation circle; (2) short-wavelength mode; (3) the inner Lindblad resonance; (4) long-wavelength mode; Ω = the direction of rotation of the Galaxy.

are very difficult to define without some arbitrariness. It can even be suggested [188] that in principle all the solutions of (30) are admissible. Lin and Shu at this point employ observations: by comparing the spirals of (30) corresponding to different ω with the observed spiral pattern of the Galaxy, they determine the frequency ω, at which there is better agreement (for more detail, cf. Chapter XI).

The continuous character of the oscillation frequency spectrum, as we shall see in §3, Chapter XI, presents the main problem of the primary theory of spiral structure developed by Lin and Shu, implying the drift of the spiral-wave packet toward the central part of the Galaxy.

4.1.2. The Dispersion Equation Taking into Account the Finite Thickness and the Gaseous Subsystem of the Disk. As we have already noted in the introduction to this chapter, for comparison with observations, dispersion equation (23) was slightly improved; first, it was necessary to introduce another reduction factor I in order to take into account the small finite

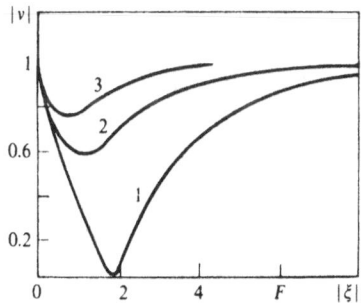

Figure 75. The dispersion relation due to Lin and Shu for $Q = 1$ (1); $Q = 1.4$ (2); $Q = 2$ (3); $\zeta = k/k_T$ [334].

thickness of the stellar disk, and, secondly, in (23), an additional term, the contribution of the gaseous component of the galaxy, was introduced. Thus, the full dispersion equation, the basis of the Lin and Shu spiral structure theory, is written as[28]

$$\frac{k_T}{|k|}(1 - v^2) = \mathscr{F}_v(x)I + \frac{\sigma_0}{\sigma_0^{(s)}}\mathscr{F}_v^{(g)}(x_g), \tag{31}$$

where $k_T = \varkappa^2/2\pi G\sigma_0^{(s)}$, $x_g = k^2D^2/\varkappa^2$, $\sigma_0^{(s)}$ is the surface density of stars, σ_0 is the full surface density, D is the speed of sound, or the mean velocity of turbulent movements of gas, and $\mathscr{F}_v^{(g)}(x_g)$ is the gas reduction factor. The last is determined by the formula similar to (21):

$$\frac{\sigma_1}{\sigma_0} = -\frac{k^2\Phi_1}{\varkappa^2}\frac{1}{(1 - v^2)}\mathscr{F}_v^{(g)}(x_g). \tag{32}$$

The $\mathscr{F}_v^{(g)}$ function is easily calculated following from the hydrodynamical equations

$$\mathscr{F}_v^{(g)}(x_g) = \frac{1}{1 + x_g/(1 - v^2)}. \tag{33}$$

For the reduction factor I providing the correction for thickness, several simple formulae have been suggested. Shu in [323] uses the following representation:

$$I = \frac{1}{1 + \frac{1}{2}|k|h}. \tag{34}$$

A similar formula is suggested already by Toomre [333], where it is justified by the following simple considerations. We calculate the decrease in the perturbed gravitational force, assuming for the sake of simplicity that the stars are confined between the planes $z = \pm h$ and the perturbed volume density is independent neither of the azimuthal angle φ nor of z:

$$\rho_1 \simeq \frac{a}{h}e^{ikr - i\omega t}, \qquad |z| < h.$$

Then from the equation

$$\Phi_1(r, z) = a\left(\frac{2\pi G}{k}\right)e^{ikr - k|z|}$$

it may be concluded that the radial component of the force at $z = 0$ from the layer $(z, z + dz)$ with such a density ρ_1 will be proportional to $\exp[-k|z|]\,dz$. Therefore, the complete perturbed force in the central plane $z = 0$ will be

$$\left.\left|\frac{\partial\Phi_1}{\partial r}\right|\right|_{z=0} = ia2\pi Ge^{ikr - i\omega t}I(k, h),$$

---

[28] The thickness of the gaseous disk may be neglected; therefore, the reduction factor I enters only into the stellar part of (31).

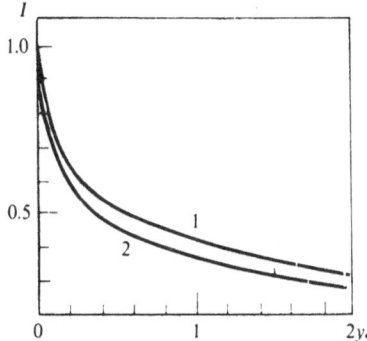

Figure 76. Reduction factors to take care of the finite thickness of a stellar disk according to Vandervoort [338] (1) and according to Shu [323] (2).

where the reduction factor is

$$I(k, h) = 2(1 - e^{-kh/2})/kh. \tag{35}$$

The most substantial investigation of the effect of the finite thickness or the stellar velocity dispersion in the z-direction on the disk stability is made by Vandervoort in [336–338] (also in the short-wave approximation). However, we shall not consider these papers in more detail, since the principal meaning lies in a more strict justification of the approximations adopted in the derivation of (34) or (35), which are quite obvious. We give only the plots constructed by Vandervoort for the reduction factor I (Fig. 76). In conclusion, we may note the dispersion equation (35), Section 3.1, Chapter IV, which takes into account the influence of the finite thickness of the disk not only on the short-wave, but also on the large-scale perturbations.

Applications of dispersion equation (31) for galactic spiral structure theory arc dealt with in §3, Chapter XI.

4.1.3. Accounting for a Resonance Damping. As we have already mentioned, from the Lin and Shu dispersion equation (23) it follows that the wave number k diverges in the case of the short-wave mode as it approaches the inner Lindblad resonance $v(r) = -1$, while for the long-wave mode, there takes place divergence of k at the outer Lindblad resonance $v(r) = +1$.

We have encountered similar divergences earlier in the investigation of oscillations of "cold" systems (cf. §2, 3, this chapter). However, in the problem of oscillations of "hot" disks which is considered here, these divergences are not natural: the thermal movement of the particles in resonant regions must, as is to be expected, smooth any singularities.

And, actually, as shown by Mark [290], the Lin and Shu dispersion equation (23) is valid only sufficiently far from resonant circumferences and must be substituted by another equation (see below) near them.

Mark has indicated also the specific mistake which makes dispersion equation (23) unsuitable in the vicinity of resonances. In the above derivation, this error is contained in a disguised form; it is easier to foresee it from a more systematic derivation of the dispersion equation (as well as, in parallel, also of other relations important to spiral structure theory) in [290, 325] which is based on a regular procedure of expansion of all values in powers of small parameters $1/kr$ and a/r (a is the epicyclic radius). It turns out that the non-correctness was in wrong expansions of singular denominators corresponding to resonance particles.

Mark [290] takes into account more exactly the role of resonant particles and derives the dispersion equation that replaces (23) for the region near the inner Lindblad resonance (of most important for the theory of spiral density waves in galaxies). In the simplest version, it has the following form:

$$\text{sgn}(\text{Re } k)\frac{k_T}{k} = \mathscr{F}_{nr} + \frac{i\sqrt{2}}{\varepsilon x}\left(\frac{L}{r}\right)\int_{-\infty}^{0} \exp(-\tau^2 + iz_L\tau - z)I_1(z)\,d\tau. \quad (36)$$

Here L is the constant of length dimension that enters the relation

$$v(r) + 1 = (r - r_L)L^{-1}, \quad (37)$$

which Mark assumes for the resonance region (L depends on Ω_p); practically, (37) proves to be valid for a sufficiently wide region of the disk. In addition, in (36), the notations

$$k_T(r) = \frac{\varkappa^2}{2\pi G\sigma_0}; \quad y = kr\varepsilon(r), \quad x = y^2, \quad \varepsilon = \frac{c_r(\varepsilon)}{\varkappa r}, \quad (38)$$

$$z = x + y\tau\sqrt{2}, \quad z_L = \frac{\sqrt{2(1 - r_L/r)}}{\varepsilon(r)}, \quad (39)$$

$$\mathscr{F}_{nr}(x, v) = \frac{1}{x}\left[1 - I_0(x)e^{-x} + \frac{2v - 1}{1 - v}I_1(x)e^{-x} + 2v^2\sum_{n=2}^{\infty}\frac{I_n(x)e^{-x}}{n^2 - v^2}\right] \quad (40)$$

are used.

The general behavior of k is more easily seen from the simplified version of Eq. (36) in which the term $\mathscr{F}_{nr}$ is omitted (as representing the contribution of the nonresonance particles). We also replace $I_1(z)$ by the asymptotic approximation for larger z. Then we obtain

$$\left[\frac{k(r)}{k_L}\right]^2 = iW\left(-\frac{z_L}{2}\right), \quad (41)$$

where z_L is defined by Eq. (39), $W(z)$ is the Kramp function, and

$$k_L^2 = \pi G\sigma_0(r_L)\varkappa^2(r_L)Lc^{-4}(r_L). \quad (42)$$

From (41), it is possible to infer that k is almost real when $(r - r_L) \gg \varepsilon r$. The positive and negative roots correspond to the leading and trailing spirals respectively. As $r \to r_L$, both modes have complex values of k; however, the real part of k remains larger in absolute value than the imaginary one. A more thorough investigation of Eq. (41) shows [290] that both the leading and trailing waves are damping. This follows from the fact that Im k does not change its sign as Im $\omega \to -\infty$ (Re $\omega = m\Omega_p$). For $r < r_L$, the asymptotic assumption used in the derivation of (41) from (36) is not true, and one should use Eq. (36).

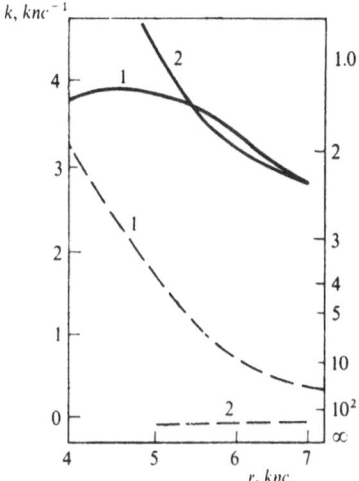

Figure 77. Dispersion curves $k = k(r)$. Solid lines $\mathrm{Re}(-k)$; Dashed lines $\mathrm{Im}(-k)$; (1) data given by Mark [290] for a disk; (2) data given by Shu [323].

The above dispersion equation was derived in the main order of perturbation theory respective to $1/kr \sim a/r$. Figure 77 shows the real and imaginary parts of $(-k)$ for a trailing spiral wave (in a stationary model of Galaxy by Schmidt [319]) as calculated by taking into account corrections of the subsequent order of smallness. It is seen first of all that k does not now diverge as $r \to r_L$. The imaginary part of k represents the resonant effect of damping omitted in Lin and Shu calculations. For comparison, there also are presented the real and imaginary parts of $-k$ taken from the paper of Shu [325], where the calculations are performed also within an accuracy of two first orders in the above small parameters (cf. §7, Supplement). In [325], an imaginary part $(-k)$ is negative and very small (in the main order, it is zero).

It is clear that Mark's calculations must correspond to Shu's results [325], when the resonance effects are small. The real parts practically coincide. The imaginary parts, however, differ since the solution of Shu has no effect of resonance damping, which happens to be essential (for the stationary model used) throughout the region covered by Fig. 77.

4.2 Bending Perturbations [28[ad], 37[ad]]

4.2.1. Derivation of the Equations of Membrane Oscillations of an Infinitely Thin Collisionless Disk. For a thin disk the disturbed potential may be assumed in the form ($\mathbf{r} = (x, y)$, $\mathbf{v} = (v_x, v_y)$)[29]

$$\Phi_1 = a(\mathbf{r}, t) \cdot z \tag{1}$$

---

[29] The membrane oscillations of the disk are treated by us as limiting, as the thickness c vanishes, for antisymmetrical oscillations with respect to z, of the rotating layer or ellipsoid with the wavelength $\lambda_\perp$, much greater than c (according to terminology, adopted in the elasticity theory [67], these are the so-called "weak" bendings). For $\lambda_\perp = \infty$, such perturbations degenerate simply to become a displacement of the system as a whole, parallel to the z-axis.

Let $Z(\mathbf{r}, \mathbf{v}, t)$ be the Lagrange displacements of stars of the middle plane $z = 0$. For the sake of simplicity assume that the unperturbed potential in the plane $z = 0$ is quadratic with respect to r: $\Phi_0 = \Omega_0^2 r^2/2$ [generalization for the case of arbitrary $\Phi_0(r)$ is simple]. Then for Z we have the equation

$$(\hat{D}^2 + \omega_0^2)Z = -\frac{\partial \Phi_1}{\partial z} = -a, \tag{2}$$

where ω_0 is the frequency of vertical oscillations of stars in the unperturbed state,

$$\hat{D} = \frac{\partial}{\partial t} + v\frac{\partial}{\partial r} - \Omega_0^2 r \frac{\partial}{\partial v}. \tag{3}$$

Due to small thickness of the disk $\omega_0^2 \gg \Omega_0^2$, the solution of (2) can also be written as

$$Z = -\frac{1}{\omega_0^2}\left(1 - \frac{\hat{D}^2}{\omega_0^2}\right)a. \tag{4}$$

For $\hat{D}^2 a$, it is easy to obtain

$$\hat{D}^2 a = \left[\left(\frac{\partial}{\partial t} + v\frac{\partial}{\partial r}\right)^2 - \Omega_0^2 r \frac{\partial}{\partial r}\right]a. \tag{5}$$

Averaging (4) with the distribution function of stars, $\bar{f}_0(\bar{r}, \bar{v})$ [which in turn is obtained by averaging the distribution function of the ellipsoid $f_0(r, v, z, v_z)$ over (z, v_z)], we find

$$\bar{Z} = \frac{a}{\omega_0^2} - \frac{1}{\omega_0^4}\left[\left(\frac{\partial}{\partial t} + \bar{v}\frac{\partial}{\partial r}\right)^2 + \frac{P_{ik}}{\sigma}\frac{\partial^2}{\partial x_i \partial x_k} - (\Omega_0^2 - \Omega^2)x_i\frac{\partial}{\partial x_i}\right]a, \tag{6}$$

where the "pressure" tensor is

$$P_{ik} = \int \bar{f}_0 v_i' v_k' \, dv', \tag{7}$$

$v = \bar{v}(r) + v'$; $\bar{v}(r)$ is the velocity of the centroid, $\bar{v}(r) = v_\varphi(r) = \Omega r$, v' are the peculiar velocities of stars; summing must be performed over the repeating indices; $i, k = 1, 2$; $x_1 = x, x_2 = y$.

From the equilibrium condition in the rotation plane we have

$$\frac{\partial P_{rr}}{\partial r} + \frac{P_{rr} - P_{\varphi\varphi}}{r} = -\sigma\left(\frac{\partial \Phi_0}{\partial r} - \Omega^2 r\right) = -\sigma r(\Omega_0^2 - \Omega^2). \tag{8}$$

Substituting $(\Omega_0^2 - \Omega^2)$ from (8) into (6), we obtain the final expression for $\bar{Z}$:

$$\bar{Z} = \frac{1}{\omega_0^2}\left\{a - \frac{1}{\omega_0^2}\left[\left(\frac{\partial}{\partial t} + \bar{v}\frac{\partial}{\partial r}\right)^2 a + \frac{1}{\sigma}\frac{\partial}{\partial x_i}P_{ik}\frac{\partial a}{\partial x_k}\right]\right\}. \tag{9}$$

We resolve (9) with respect to a:

$$a = \omega_0^2 \bar{Z} + \left(\frac{\partial}{\partial t} + \bar{v}\frac{\partial}{\partial r}\right)^2 Z + \frac{1}{\sigma}\frac{\partial}{\partial x_i}P_{ik}\frac{\partial \bar{Z}}{\partial x_k}. \tag{10}$$

On the other hand, the Poisson equation yields

$$a(r) = \omega_0^2 \overline{Z} + F_{gr}(\overline{Z}),$$ (11)

where

$$F_{gr}(\overline{Z}) = G \iint dr' \frac{\sigma(z')[\overline{Z}(r') - \overline{Z}(r)]}{|r - r'|^3}.$$ (12)

From (10) and (11) we find the sought-for equation for the vertical displacements of the disk $h \equiv Z$:

$$\left(\frac{\partial}{\partial t} + \bar{v}\nabla\right)^2 h = F_{gr}(h) - \frac{1}{\sigma}\frac{\partial}{\partial x_i}\left(P_{ik}\frac{\partial h}{\partial x_k}\right)$$ (13)

We note here that in such form Eq. (13) is valid for the arbitrarily rotating disks.

4.2.2. The Dispersion Equation is readily obtained for "Maclaurin" disks with the distribution function of (6) §1.4:

$$f_{0\gamma} = \frac{1}{2\pi\sqrt{1 - \gamma^2}}\left[(1 - \gamma^2)\left(1 - \frac{r^2}{R^2}\right) - v^2\right]^{-1/2}.$$ (14)

The "membrane" oscillation of such a disk obeys, as follows from (13), the equation

$$\left(\frac{\partial}{\partial t} + \gamma\frac{\partial}{\partial\varphi}\right)^2 h = F_1 + F_2 - \frac{1}{\sigma_0}\left(\frac{\partial P_0}{\partial r}\frac{\partial h}{\partial r} + P_0\Delta_\perp h\right),$$ (15)

where $\sigma_0 = \sqrt{1 - r^2} = \xi$ is the surface density, $P_0 = (1 - \gamma^2)\xi^3/3$ is the equilibrium "pressure," $\Phi_0 = r^2/2$ is the equilibrium potential, γ is the angular rate in units of circular velocity,

$$\Delta_\perp = \frac{\partial^2}{\partial r^2} + \frac{1}{r}\frac{\partial}{\partial r} + \frac{1}{r^2}\frac{\partial^2}{\partial\varphi^2}$$

is the two-dimensional Laplacian, while the forces F_1 and $F_2 = 2h$ have the same meaning as in Section 2.1 (where the bendings of a "cold" disk are considered). Substituting into Eq. (15) the displacement h in the form $\sim (1/\xi)P_{2n-1+m}^m(\xi)e^{im\varphi - i\omega t}$, it is easy to obtain [28^{ad}] the following dispersion equation:

$$(\omega - m\gamma)^2 = 4\Gamma_n^m - 2 - \tfrac{1}{3}(1 - \gamma^2)[(2n + m)(2n + m - 1) - m^2 - 2].$$ (16)

Calculations leading to this equation are quite similar to those given in Section 2.1; the expressions for the numbers Γ_n^m are also given therein.

The simplest nontrivial type of perturbation corresponding to $m = 0$, $n = 2$ (bell-shaped bending of the disk, the displacement is $h \sim P_3(\xi)/\xi \sim 2 - 5r^2$) is unstable, as follows from the dispersion equation obtained for $|\gamma| < \frac{1}{2}$. Asymptotically as $n \to \infty$ $\Gamma_n^m \to 2n/\pi$; at the same time the last term on the right-hand side of the dispersion equation (the contribution of the pressure) increases in absolute value $\sim n^2$, so that any "hot" disk ($|\gamma| \neq 1$) is unstable for perturbations of a sufficiently small scale.

Equation (16) admits the generalization for the case of superposition of the Maclaurin disks:

$$f_0 = \int_{-1}^{1} A(\gamma) f'_{0\gamma} \, d\gamma \tag{17}$$

(where $f'_{0\gamma}$ is the function $f_{0\gamma}$ written in the inertial system):

$$(\omega - m\bar{\gamma}\Omega_0)^2 = \{4\Gamma_n^m - 2 - \tfrac{1}{3}(1 - \bar{\gamma}^2)[(2n + m - 1)(2n + m) + m^2 - 2]$$
$$- m^2(\overline{\gamma^2} - \bar{\gamma}^2)\}\Omega_0^2, \tag{18}$$

where

$$\overline{\gamma^n} = \int_{-1}^{1} A(\gamma)\gamma^n \, d\gamma.$$

Generalization for the case of the disk with a "halo" is trivial:

$$(\omega - m\bar{\gamma}\Omega_0)^2 = \Omega_d^2(4\Gamma_n^m - 2) - \tfrac{1}{3}\Omega_0^2\{(1 - \bar{\gamma}^2)$$
$$\times [(2n + m - 1)(2n + m) - m^2 - 2]$$
$$- m^2(\overline{\gamma^2} - \bar{\gamma}^2)\} + \Omega_h^2, \tag{19}$$

where $\Omega_0^2 = \Omega_d^2 + \Omega_h^2$. Note by the way that practically all the disk systems under study in this chapter, must be unstable with respect to perturbations of the type under investigation (for wavelengths $\lambda < \beta R$, where R is the radius, while $\beta = $ the parameter characterizing the contribution of the "pressure" to the equilibrium of the disk, $v_{T\perp}^2 \simeq \beta G \sigma_0 R$). The development of the instability must evidently lead to a disk of finite thickness, for which the stability condition will be satisfied. The thickness of such a disk may still remain much smaller than the radius, so that in order to describe the perturbations in its "plane," one may again make use of the approximation of an infinitely-thin disk.

4.3 Methods of the Stability Investigation of General Collisionless Disk Systems [324, 249, 74[ad]]

4.3.1. Derivation of the Integral Equation. Derive, following Shu [324], the general integral equation for eigenmodes and eigenfrequencies of a self-gravitating collisionless disk. For that purpose, it is necessary first of all to calculate the *response* of such a disk to the perturbation of the gravitational field, omitting so far the question whether this field is self-consistent or not.

The kinetic equation describing the temporal variation of the distribution function of particles can be written in the form

$$\frac{\partial f}{\partial t} + [f, H] = 0, \tag{1}$$

where $[\cdots]$ is the Poisson bracket and H is the Hamiltonian. In the cylindrical coordinates (r, φ) [corresponding generalized impulses will be denoted via (p_r, p_φ)], the Hamiltonian H calculated for mass unit, is

$$H = \tfrac{1}{2}(p_r^2 + p_\varphi^2) + \Phi(r, \varphi, t) \tag{2}$$

[$\Phi(r, \varphi, t)$ is the gravitational potential in the disk plane], while the Poisson brackets are opened in the following way:

$$[f, H] = p_r \frac{\partial f}{\partial r} + \frac{p_\varphi}{r} \frac{\partial f}{\partial \varphi} + \left(\frac{p_\varphi^2}{r^3} - \frac{\partial \Phi}{\partial r} \right) \frac{\partial f}{\partial p_r} - \frac{\partial \Phi}{\partial \varphi} \frac{\partial f}{\partial p_\varphi}. \tag{3}$$

The unperturbed distribution function f_0 satisfies the equation

$$\frac{\partial f_0}{\partial t} + [f_0, H_0] = 0, \tag{4}$$

where H_0 has the form of (2), but the unperturbed potential is dependent only on the radius r: $\Phi_0 = \Phi_0(r)$.

The characteristics of Eq. (4) are of the form

$$dt = \frac{dr}{p_r} = \frac{d\varphi}{p_\varphi/r} = \frac{dp_r}{p_\varphi^2/r^3 - \partial \Phi_0/\partial r} = -\frac{dp_\varphi}{\partial \Phi_0/\partial \varphi = 0};$$

hence it follows that, in the most general case, f_0 may be an arbitrary function of the following four integrals of motion:

$$E_0 = \frac{1}{2} \left(p_r^2 + \frac{p_\varphi^2}{r^2} \right) + \Phi_0(r), \tag{5}$$

$$L = p_\varphi, \tag{6}$$

$$T = t - \int^r \left\{ 2[E_0 - \Phi_0(r')] - \frac{L^2}{(r')^2} \right\}^{-1/2} dr', \tag{7}$$

$$P = \varphi - \int^r \frac{L}{(r')^2} \left\{ 2[E_0 - \Phi_0(r')] - \frac{L^2}{(r')^2} \right\}^{-1/2} dr'. \tag{8}$$

But the phase integrals T and P are, as a rule, multivalued functions, so that it may be assumed that

$$f_0 = \begin{cases} F_0(E_0, L) & \text{for } E_0 < 0, L > 0, \\ 0 & \text{for } E_0 \geq 0 \text{ or } L < 0. \end{cases} \tag{9}$$

The condition $E_0 < 0$ is natural and means the presence only of stars bounded in the gravitational field of the disk. The second condition is already not so necessary; it implies that we restrict ourselves to the stars rotating in the same direction. It is clear, however, that the presence of a small number of stars rotating in the opposite direction cannot essentially alter the results.

Linearizing Eq. (1), we obtain

$$\frac{\partial f_1}{\partial t} + [f_1, H_0] = -[f_0, \Phi_1], \tag{10}$$

In accordance with the above, we shall assume the right-hand side of (10) to be given and find f_1 from this equation (by expressing it through Φ_1). The formal solution of (10) ensues by integration over unperturbed trajectories of stars [on the left-hand side of (10) is just the corresponding Stocks derivative d/dt]

$$f_1(t) - f_1(t_0) = \int_{t_0}^{t} [f_0, \Phi_1]_0 \, dt' = \int_{t_0}^{t} \left(\frac{\partial \Phi_1}{\partial r} \frac{\partial f_0}{\partial p_r} + \frac{\partial \Phi_1}{\partial \varphi} \frac{\partial f_0}{\partial p_\varphi} \right)_0 dt', \quad (11)$$

where the Poisson bracket $[\cdots]_0$ is calculated on the unperturbed trajectory It is convenient to switch to the variables $E_0, L : r, p_r, p_\varphi \to r, E_0, L$, i.e., in essence $p_r \to E_0$ because $p_\varphi = L$. Then the partial derivatives are evidently transformed thus:

$$\frac{\partial}{\partial r} \to \frac{\partial}{\partial r} + \left(-\frac{p_\varphi^2}{r^3} + \frac{\partial \Phi_0}{\partial r} \right) \frac{\partial}{\partial E_0}, \quad (12)$$

$$\frac{\partial}{\partial p_r} \to \Pi_0 \frac{\partial}{\partial E_0}, \quad (13)$$

$$\frac{\partial}{\partial p_\varphi} \to \frac{\partial}{\partial L} + \frac{L}{r^2} \frac{\partial}{\partial E_0}, \quad (14)$$

where Π_0 denotes the following function (r, E_0, L):

$$p_r = \Pi_0(r, E_0, L) = \left\{ 2[E_0 - \Phi_0(r)] - \frac{L^2}{r^2} \right\}^{1/2}. \quad (15)$$

In the new variables, Eq. (11) takes the form

$$f_1(t) - f_1(t_0) = \int_{t_0}^{t} \left\{ \left(\Pi_0 \frac{\partial \Phi_1}{\partial r} + \frac{L}{r^2} \frac{\partial \Phi_1}{\partial \varphi} \right) \frac{\partial F_0}{\partial E_0} + \frac{\partial \Phi_1}{\partial \varphi} \frac{\partial F_0}{\partial L} \right\} dt'. \quad (16)$$

The first term on the right-hand side of (16) can be rewritten in another form, by noting that along the unperturbed orbits

$$\frac{dr}{dt} = \Pi_0, \qquad \frac{d\varphi}{dt} = \frac{L}{r^2}, \quad (17)$$

so that

$$\Pi_0 \frac{\partial \Phi_1}{\partial r} + \frac{L}{r^2} \frac{\partial \Phi_1}{\partial \varphi} = \frac{d\Phi_1}{dt} - \frac{\partial \Phi_1}{\partial t}. \quad (18)$$

Now, (16) can be reduced to the form

$$f_1(t) = \frac{\partial F_0}{\partial E_0} \Phi_1(t) + f_1(t_0) - \frac{\partial F_0}{\partial E_0} \Phi_1(t_0)$$

$$+ \int_{t_0}^{t} \left(-\frac{\partial \Phi_1}{\partial t} \frac{\partial F_0}{\partial E_0} + \frac{\partial \Phi_1}{\partial \varphi} \frac{\partial F_0}{\partial L} \right)_0 dt'. \quad (19)$$

Let us seek the periodical solutions of the type of

$$\begin{aligned} \Phi_1(r, \varphi, t) &= \tilde{\Phi}_1(r)e^{i(\omega t - m\varphi)}, \\ f_1(r, \varphi, E_0, L, t) &= \tilde{f}_1(r, E_0, L)e^{i(\omega t - m\varphi)}. \end{aligned} \tag{20}$$

Substituting (20) into (19), we obtain

$$\begin{aligned} \tilde{f}_1(r(t), E_0, L) = {} & \frac{\partial F_0}{\partial E_0}\tilde{\Phi}_1(r(t)) + e^{-i[\omega t - m\varphi(t)]} \\ & \times \left[f_1(t_0) - \frac{\partial F_0}{\partial E_0}\Phi_1(t_0) - i\left(\omega\frac{\partial F_0}{\partial E_0} + m\frac{\partial F_0}{\partial L}\right) \right. \\ & \left. \times \int \Phi_1(r(t'))e^{i[\omega t' - m\varphi(t')]}\, dt' \right] \end{aligned} \tag{21}$$

Of course, only single-valued solutions are physically acceptable. We are now using this circumstance for the simplification of formula (21). The left-hand side of (21) is the function of only $(r(t), E_0, L)$ and is explicitly independent of $(\varphi(t), t)$. For the bounded stars, the $r(t)$ function is limited from above and from below at the points of rotation (r_1, r_2) and is the periodical function t with a period $T \equiv 2\tau_{1,2}(E_0, L)$. Since $\tilde{f}_1$ depends only on $r(t)$ and does not depend on $\varphi(t)$ and t, $\tilde{f}_1$ also must be the t function with the same period $2\tau_{1,2}$.

Assume that for one period of radial oscillations, the star rotates in the azimuth by an angle $2\varphi_{1,2}(E_0, L)$ (cf. Fig. 78). Let us now replace in Eq. (21) t by $t + 2\tau_{1,2}$ and make use of the fact that $\varphi(t + 2\tau_{1,2}) = \varphi(t) + 2\varphi_{1,2}$. Then, after some calculations, we obtain

$$\begin{aligned} \tilde{f}_1(t) = \tilde{f}_1(t + 2\tau_{1,2}) = {} & \frac{\partial F_0}{\partial E_0}\tilde{\Phi}_1(r(t)) + e^{-2i(\omega\tau_{12} - m\varphi_{12})} \\ & \times \left\{ \tilde{f}_1(t) - \frac{\partial F_0}{\partial E_0}\tilde{\Phi}_1(r(t)) - i\left(\omega\frac{\partial F_0}{\partial E_0} + m\frac{\partial F_0}{\partial L}\right) \right. \\ & \left. \times \int_t^{t+2\tau_{12}} \tilde{\Phi}_1(r(t'))\exp[i\omega(t' - t) - im(\varphi(t') - \varphi(t))]\, dt' \right\}. \end{aligned} \tag{22}$$

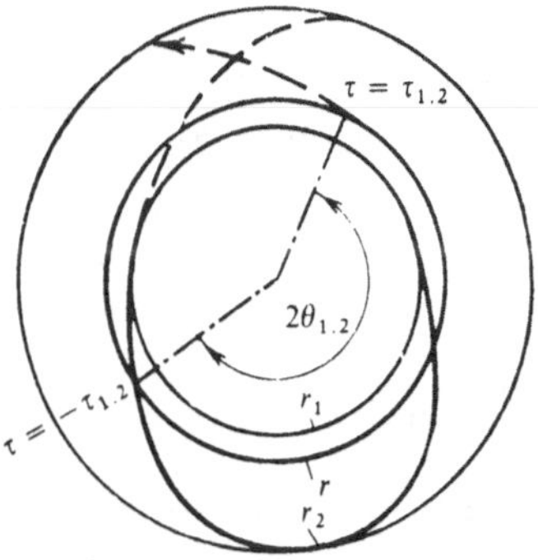

Figure 78. An unperturbed particle orbit in a gravitating disk [324] $(\theta_{1,2} = \varphi_{1,2})$.

Solving (22) with respect to $\tilde{f}_1(t)$, we find

$$\tilde{f}_1(t) = \frac{\partial F_0}{\partial E_0} \tilde{\Phi}_1(r(t)) - \left(\omega \frac{\partial F_0}{\partial E_0} + m \frac{\partial F_0}{\partial L} \right) \frac{e^{-i(\omega\tau_{12} - m\varphi_{12})}}{2 \sin(\omega\tau_{12} - m\varphi_{12})}$$

$$\times \int_t^{t + 2\tau_{12}} \tilde{\Phi}_1(r(t')) \exp\{i\omega(t' - t) - im[\varphi(t') - \varphi(t)]\} \, dt'. \quad (23)$$

Denote [for symmetrization of the integral in (23)]

$$r(t) \equiv r, \qquad \tau \equiv t' - t - \tau_{12}, \qquad \varphi_*(\tau) \equiv \varphi(t') - \varphi(t) - \varphi_{12},$$

$$r_*(\tau) = r(t'). \quad (24)$$

Then, from (23) we obtain the sought-for solution for $\tilde{f}_1$ in the form

$$\tilde{f}_1(r, E_0, L) = \frac{\partial F_0}{\partial E_0} \tilde{\Phi}_1(r) + \chi(r, E_0, L), \quad (25)$$

$$\chi(r, E_0, L) = - \frac{\omega \partial F_0/\partial E_0 + m \partial F_0/\partial L}{2 \sin(\omega\tau_{12} - m\varphi_{12})} \int_{-\tau_{12}}^{\tau_{12}} \tilde{\Phi}_1(r_*(\tau)) e^{i[\omega\tau - m\varphi_*(\tau)]} \, d\tau. \quad (26)$$

In (26) $r_*(\tau)$ and $\varphi_*(\tau)$ are the functions of (r, E_0, L, τ) uniquely defined from the following equations of motion (with periodical boundary conditions):

$$\frac{dr_*}{d\tau} = \Pi_0(r_*, E_0, L), \qquad r_* = r \quad \text{at } \tau = \pm\tau_{12}(E_0, L), \quad (27)$$

$$\frac{d\varphi_*}{d\tau} = \frac{L}{r_*^2(\tau)}, \qquad \varphi_* = \pm\varphi_{12}(E_0, L) \quad \text{at } \tau = \pm\tau_{12}(E_0, L). \quad (28)$$

For real ω, there is, as is evident from (26), a possibility of the *resonance* for stars, whose energies and angular moments are such that $\sin(\omega\tau_{1,2} - m\varphi_{1,2}) = 0$. It is clear that the cause of the resonance is the fact that in every cycle of the unperturbed movement, these stars fill the same phase of the perturbed gravitational field. The simplest form of resonance arises due to "corotation," $\omega\tau_{1,2} - m\varphi_{1,2} = 0$. But, for the spiral structure theory, the Lindblad resonances $\omega\tau_{1,2} - m\varphi_{1,2} = \pm\pi$ seem to be of more importance.

Calculate now the response of the stellar surface density σ_{*1} corresponding to the perturbed distribution function (25). If

$$\sigma_{*1}(r, \varphi, t) = \tilde{\sigma}_{*1}(r) e^{i(\omega t - m\varphi)}, \quad (29)$$

so it is evident that

$$r\tilde{\sigma}_{*1} = \iint \left\{ 2 \frac{\partial F_0}{\partial E_0} \tilde{\Phi}_1(r) + [\chi]_{\Pi_0 > 0} + [\chi]_{\Pi_0 < 0} \right\} \frac{dE_0 \, dL}{|\Pi_0|}, \quad (30)$$

where $1/|\Pi_0|$ is the Jacobian of the transform $(p_r, p_\varphi) \to (E_0, L)$. The need to distinguish explicitly the cases $\Pi_0 > 0$ and $\Pi_0 < 0$ is due to the fact that the transform $p_r \to E_0$ is not mutually single-valued: for a given

E_0, the value of Π_0 may be either positive or negative (the term $\sim \partial F_0/\partial E_0$ is even in $p_r = \Pi_0$). The Newtonian laws possess such properties at the time reversal that, for given E_0 and L, the solutions of Eqs. (27) and (28) have the following symmetry:

$$\left\{ \begin{matrix} r_*(\tau) \\ \varphi_*(\tau) \end{matrix} \right\}_{\Pi_0 > 0} = \left\{ \begin{matrix} r_*(-\tau) \\ -\varphi_*(-\tau) \end{matrix} \right\}_{\Pi_0 < 0}, \tag{31}$$

so that, finally, the perturbation of the surface density can be represented in the form

$$r\tilde{\sigma}_{*1}(r) = 2 \iint\limits_{\Pi_0 > 0} \left[\frac{\partial F_0(E_0, L)}{\partial E_0} \tilde{\Phi}_1(r) + \chi_*(r, E_0, L) \right] \frac{dE_0\, dL}{\Pi_0(r, E_0, L)}, \tag{32}$$

where the notation

$$\chi_* \equiv -\frac{\omega \partial F_0/\partial E_0 + m \partial F_0/\partial L}{2 \sin(\omega\tau_{12} - m\varphi_{12})} \int_{-\tau_{12}}^{\tau_{12}} \tilde{\Phi}_1(r_*(\tau)) \cos[\omega\tau - m\varphi_*(\tau)]\, d\tau \tag{33}$$

is introduced.

For real ω, the problem arises in the integration over $(E_0 L)$ in Eq. (32) if the system has resonance stars for which $(\omega\tau_{12} - m\varphi_{12})$ equals an integer π. Then the poles $[\sin(\omega\tau_{12} - m\varphi_{12})]^{-1}$ lie on the way of integration, and there arises an uncertainty customary in such cases: Is it necessary to take only the principal values of integrals or must the "imaginary contributions" from the poles be taken into account? A similar problem arises, for example, for electrostatic plasma oscillations. As is well known, it is solved (Landau, [65]) if one turns to the problem with initial conditions. In case the resonance stars are lacking (for example, in the analysis of unstable modes), there arise no uncertainties in the calculation of the integral (32).

We apply now the Poisson equation. The gravitational potential $\Phi_1 = \tilde{\Phi}_1(r)e^{i(\omega t - m\varphi)}$, linked with the surface density $\sigma_1 = \tilde{\sigma}_1(r)e^{i(\omega t - m\varphi)}$, can be written in the following form [324]:

$$\tilde{\Phi}_1(r) = -G \int_0^\infty \tilde{\sigma}_1(a)a\, da \oint \frac{e^{im\varphi}\, d\varphi}{\sqrt{r^2 + a^2 - 2ra \cos\varphi}}$$

$$= -2\pi G \int_0^\infty H_m(r, a)a\tilde{\sigma}_1(a)\, da, \tag{34}$$

where

$$H_m(r, a) \equiv H_m(a, r) = \frac{1}{r + a} h_m(\zeta), \tag{35}$$

$$h_m(\zeta) = \frac{2}{\pi} \int_0^{\pi/2} \frac{\cos 2mx}{\sqrt{1 - \zeta \cos^2 x}}\, dx, \qquad \zeta \equiv \frac{4ra}{(r + a)^2} \le 1. \tag{36}$$

The asymptotics of $h_m(\zeta)$ are

$$
h_m = \begin{cases}
\dfrac{(2m)!}{2^{4m}(m!)^2}\,\zeta^m & \text{for } \zeta \ll 1 \quad (a \ll r \text{ or } a \gg r), \\[3mm]
-\tfrac{1}{2}\ln(1 - \zeta^2) & \text{for } \zeta \simeq 1 \quad (\text{i.e., } a \simeq r).
\end{cases}
\tag{37}
$$

Hence it is seen that the existence of the integral in (34) assumes regular boundary conditions imposed on $\tilde{\sigma}_1(r)$ at $r = 0$ and as $r \to \infty$. For self-consistent oscillations, the density $\tilde{\sigma}_1(r)$ must be equal to the response of the surface density $\tilde{\sigma}_{1*}(r)$. This condition, together with Eqs. (32) and (33), leads to the sought-for integral equation for $\tilde{\sigma}_1(r)$:

$$
r\tilde{\sigma}_1(r) = \int_0^\infty K_{m\omega}(r, a)\,a\tilde{\sigma}_1(a)\,da.
\tag{38}
$$

The kernel of (38) is expressed by the formula

$$
K_{m\omega}(r, a) = -4\pi G \iint \frac{dE_0\, dL}{\Pi_0(r, E_0, L)} \left\{ \frac{\partial F_0}{\partial E_0} H_m(r, a) \right.
$$

$$
\left. - \frac{\omega \partial F_0/\partial E_0 + m\partial F_0/\partial L}{2\sin(\omega\tau_{12} - m\varphi_{12})} \int_{-\tau_{12}}^{\tau_{12}} H_m(r_*(\tau), a)\cos[\omega\tau - m\varphi_*(\tau)]\,d\tau \right\}.
\tag{39}
$$

4.3.2. Some General Properties of the Integral Equation (38).

Equation (38) is a uniform singular[30] integral equation, which must be solved for the definition of the characteristic values ω and eigenfunctions $r\tilde{\sigma}_1(r)$. It yields a full formulation of the problem of normal modes in any stellar disk with a velocity dispersion. A similar equation, however, in a epicyclic approximation for stellar orbits, was first found by Kalnajs [249].

From Eq. (37), it follows that $K_m(r, a)$ has logarithmic singularity in $r = a$. In spite of the integrability, it stresses the gravitational influence of the stars in a local vicinity and assumes a possibility of short-wave perturbations, which have already been dealt with.

This integral equation resembles the homogeneous Fredholm equation of the second kind, with the exception that the eigennumbers (frequencies ω) enter into (38) nonlinearly. The kernel of the integral equation (38), $K_m(r, \omega)$, is a very complicated function ω, r, r'.

The solutions of Eq. (38) contain all proper values ω and eigenfunctions $r\sigma_1(r)$ of the system. It is, however, easy to see that analytically this equation in the general form could not be solved due to the fact that the equation was very complex. Kalnajs [250] has solved this equation numerically for the model of the M31 galaxy (with the kernel Km_i in the epicyclic approximation). (cf. Section 4.1, Chapter XI).

Kalnajs (cf. [188]) proves that unstable modes are isolated. However, for small wavelengths, integral equation (38) can yield also more closely located

---

[30] Due to the logarithmic singularity in the kernel $K_{m\omega}$ at $a \sim r$ [cf. (37)].

modes ω. If ω is real (neutral oscillations), the spectrum of eigenvalues ω may form a continuum.

The accurate formulation of the problem of eigenvalues given above has so far provided a comparatively small number of particular results regarding stability of disk systems. Note here the Kalnajs theorem [251] referring to stability of the axially symmetrical modes that is the generalization of the respective result achieved by Toomre for the case of arbitrary (and not only short-wave) perturbations. The application of the generalized stability criterion leads, however, to conclusions which differ little from those drawn from the original criterion of Toomre.

In §7, Appendix, integral equation (38) will be employed to derive frequencies and amplitudes of oscillations within an accuracy of second order in the parameters $1/kr$ and a/r (which will be assumed to be small; cf. Section 4.1).

4.3.3. Method of the Matrix Equation. Kalnajs suggested [74^{ad}] also the alternative method of investigation of the small disturbances and of the stability of general disklike stellar systems. Namely, he derived the matrix equation which gives in principle all the eigenfrequencies and eigenfunctions of the system under consideration. This matrix equation has the following form:

$$\sum_{j=0}^{\infty} [M_{ij}(\omega) - \delta_{ij}]a_j = 0, \qquad (40)$$

where a_j are the coefficients of the expansions of the disturbed potential Φ_1 and of the disturbed surface density σ_1 over some biorthogonal system $\{\Phi_j, \sigma_j\}$:

$$\Phi_1 = \sum_{j=0}^{\infty} a_j\Phi_j, \qquad \sigma_1 = \sum_{j=0}^{\infty} a_j\sigma_j, \qquad -\frac{1}{2\pi G} \iint \Phi_i^* \sigma_j r \, dr \, d\varphi = \delta_{ij}, \quad (41)$$

and the matrix elements

$$M_{ij}(\omega) = \frac{1}{8\pi^3 G} \sum_l \iint F(I_1, I_2) \left\{ \left(l\frac{\partial}{\partial I_1} + m\frac{\partial}{\partial I_2} \right) \right.$$

$$\left. \times \left[\frac{(\Phi_i)_{lm}^*(\Phi_j)_{lm}}{l\Omega_1 + m\Omega_2 - \omega} \right] \right\} dI_1 \, dI_2. \qquad (42)$$

$F(I_1, I_2)$ is the equilibrium distribution function, I_1, I_2 are the action variables, Ω_1, Ω_2 are the oscillation frequencies of the particles at the equilibrium disk, the $(\Phi_i)_{lm}$ are the Fourier coefficients in the expansion of the function Φ_i,

$$(\Phi_i)_{lm}(I_1, I_2) = \int_0^{2\pi} \int_0^{2\pi} \Phi_i(I_i, w_i) \exp[-i(lw_1 + mw_2)], \qquad (43)$$

and w_1, w_2 are the angular variables conjugate with I_1, I_2. Eigenfrequencies ω must obviously be defined from the condition that the determinant of the set (40) be equal to zero. Derivation of Eq. (40) may be found in §5 of the Appendix. Recall also that the matrix equation for the general spherically symmetric systems, analogous to Eq. (40), was considered in §6, Chapter III (and in §4 of the Appendix). Equation (40) is well adopted for the numerical stability investigation of stellar disks.

Apart from the derivation of the matrix equation (40) itself, Kalnajs [74^{ad}] performed also some preliminary work for the further detailed analysis. In particular, a few distribution functions which are convenient for the numerical investigation, and potential-surface density of some biorthogonal systems are found there (we do not give them here in view of their cumbersome form). However, computations of eigenfrequencies and eigenfunctions have already taken very long and for the present are not finished. Toomre remarks [88^{ad}] two difficulties which are probably simultaneously the reasons of such delay in the investigations. First, it is very likely that stable self-gravitating disks have not many discrete eigenmodes; the oscillatory spectra have mainly to be continuous. Consequently, these modes must be singular near some resonance radii (see §7, Chapter II, or §2 of this chapter). Second, even if strictly oscillatory discrete modes are present, then it is not clear, by the expression of Toomre [88^{ad}], whether they include "any of spiral form." However, on the other hand, the investigation of unstable modes of a spiral form for which the difficulties marked above do not occur has very much interest. In the nonlinear regime modes may there transform into the quasistationary spirals with a finite amplitude (on nonlinear effects see Chapter VII).

4.4 Exact Spectra of Small Perturbations [93, 111, 113, 252]

4.4.1. Derivation of the Dispersion Equation for Maclaurin Disks. The exact spectra of small perturbations of "hot" systems are found in [93, 111, 113, 252] only for the uniformly rotating disks. One of the important stimuli for the investigation of such models was the fact (mentioned by Lin, Yuan, and Shu in [271]) that the suggestion about small eccentricities ceases to agree with the requirement of stability with respect to radial perturbations (by the Toomre criterion [333]) as one approaches the galactic center. In order that the central region might be made stable (in the local meaning of Toomre), there must be a fairly large velocity dispersion, so that the epicyclic approximation no longer fits the analysis. For the perturbations of the "barlike" type ($m = 2$), this occurs even earlier. Investigation of exact models allows one to better understand these defects of epicyclic modes. Another reason for the interest in accurate models is due to the numerical experiments of Miller and co-authors [294] and Hohl [215, 220], who showed that the disks originally maintained mainly by rotation evolve to states where the main role is played by "pressure" (velocity dispersion of the particles).

To begin with, we consider the stability of a disk system described by the distribution function of (6), Section 1.4:

$$f_0(r, v_r, v_\varphi) = \frac{c_0 \theta(\varkappa^2 - v_r^2 - v_\varphi^2)}{\sqrt{\varkappa^2 - v_r^2 - v_\varphi^2}}, \tag{1}$$

where

$$\Omega_0 = R = 1, \qquad c_0 = \frac{\sigma_0(0)}{2\pi\sqrt{1 - \gamma^2}}, \qquad \varkappa^2 = (1 - \gamma^2)(1 - r^2).$$

It has a root singularity on the boundary of the phase region occupied by the system: $v^2 = v_r^2 + v_\varphi^2 \leq (1 - \gamma^2)(1 - r^2)$, in this meaning is similar, e.g., to the distribution function of the uniform flat layer (cf. §1, Chapter I). Stability of the system (1) can be investigated by one of the methods described in §6, Chapter I.

In particular, one may write

$$f = \frac{\sigma_0(0)}{2\pi\sqrt{1 - \gamma^2}} \frac{\theta[(1 - \gamma^2)(1 - r^2) - v_r^2 - v_\varphi^2 + \varepsilon\chi]}{\sqrt{(1 - \gamma^2)(1 - r^2) - v_r^2 - v_\varphi^2 + \varepsilon\chi}},$$

where $\chi(r, \varphi, v_r, v_\varphi)$ is perturbation, $\varepsilon \ll 1$ is the expansion parameter of perturbation theory. Then the kinetic equation will yield the following equation for χ:

$$\frac{\partial \chi}{\partial t} + v_r \frac{\partial \chi}{\partial r} + \frac{v_\varphi}{r} \frac{\partial \chi}{\partial \varphi} - (1 - \gamma^2)r \frac{\partial \chi}{\partial v_r} + 2\gamma\left(v_r \frac{\partial \chi}{\partial v_\varphi} - v_\varphi \frac{\partial \chi}{\partial v_r} \right)$$

$$+ \frac{v_\varphi}{r}\left(v_\varphi \frac{\partial \chi}{\partial v_r} - v_r \frac{\partial \chi}{\partial v_\varphi} \right) = -2v_r \frac{\partial \Phi_1}{\partial r} - 2v_\varphi \frac{\partial \Phi_1}{r \partial \varphi}, \tag{2}$$

while the perturbation of the surface density will be calculated by the formulae

$$\sigma_1 = \sigma_1^{(1)} + \sigma_1^{(2)},$$

$$\sigma_1^{(1)} = \frac{\sigma_0(0)}{2\pi\sqrt{1 - \gamma^2}} \int_0^{2\pi} d\alpha \int_0^{\sqrt{(1 - \gamma^2)(1 - r^2)}} dv \left. \frac{\partial \chi^{(1)}}{\partial v} \right/ \sqrt{(1 - \gamma^2)(1 - r^2) - v^2},$$

$$\sigma_1^{(2)} = \sigma_0(0)\left(\sqrt{1 - r^2 - \frac{\chi^{(2)}}{1 - \gamma^2}} - \sqrt{1 - r^2} \right), \tag{3}$$

$$\chi^{(2)} \equiv \chi(v = 0), \qquad \chi^{(1)} = \chi - \chi^{(2)}, \qquad v_r = v \cos \alpha, \qquad v_\varphi = v \sin \alpha.$$

The kinetic equation for the function χ is most readily solved by the method of "integration over trajectories." For that purpose, note that (2) can be presented in the form

$$\frac{d\chi}{dt} = -2\left(\frac{d\Phi_1}{dt} - \frac{\partial \Phi_1}{\partial t} \right), \tag{4}$$

where d/dt denotes the "full" (Lagrange) time derivative along the unperturbed trajectory of a particle. The latter can be written in the following manner (cf. §4, Chapter II):

$$r'^2 = v^2 \sin^2 t + \gamma^2 r^2 + r^2(1 - \gamma^2) \cos^2 t + r v_r \sin 2t$$
$$+ 2\gamma r v_\varphi \sin^2 t,$$
$$r' e^{i(\varphi' - \varphi)} = \tfrac{1}{2}\{e^{-i(1 + \gamma)t}[r(1 - \gamma) - (v_\varphi - ir)]$$
$$+ e^{i(1 - \gamma)t}[r(1 + \gamma) + (v_\varphi - ir)]\},$$

where $r, \varphi, v_r, v_\varphi, v$ are the coordinates and velocities of particles at the time $t = 0$.

It may be shown (see below) that the eigenfunctions also in this case will be as (61) and (62), Section 2.2 (i.e., same as in a cold disk). By the way, to obtain the oscillation frequencies corresponding to the mode with the indices (n, m) $[\Phi_1^{n, m} \sim e^{im\varphi}(r^{n+m} + \cdots)$, $n > 0$, n even], it is sufficient to maintain in the calculations only the term with the higher degree r. If the perturbed density σ_1 is presented in the form

$$\sigma_1 = \sigma_1^{(1)} + \sigma_1^{(2)}, \qquad \sigma_1^{(1)} = A^{(1)} e^{im\varphi}(r^{n+m-2} + \cdots), \tag{5}$$

$$\sigma_1^{(2)} = \sigma_0(0)[\sqrt{1 - r^2} - 2A^{(2)} e^{im\varphi}(r^{n+m} + \cdots) - \sqrt{1 - r^2}], \tag{6}$$

then the dispersion equation can be written in the following manner [111, 113]:

$$4\gamma_{n+m}^m(A^{(1)} + A^{(2)}) = 1, \tag{7}$$

where

$$\gamma_{n+m}^m = \frac{(n + 2m)! \, n!}{2^{2(m+n)+1}\{[(n + 2m)/2]!\}^2 [(n/2)!]^2}.$$

The calculation of $A^{(1)}$ and $A^{(2)}$ is easy; this is done in [111, 113, 252].[31] In the most compact form, dispersion equation (7) is given in [252]:

$$1 = \frac{2}{\pi} \frac{\Gamma((n + 1)/2)\Gamma(m + 1)}{\Gamma(n/2 + m + 1)2^m} \sum_{c=0}^{n/2} \sum_{d=-c}^{c} \sum_{l=-d}^{m-d} \frac{\Gamma(n/2 + m + c + \tfrac{1}{2})}{\Gamma(c - d + 1)}$$

$$\times \frac{(-1)^{c+d}(1 - \gamma)^{c+d+l-1}(1 + \gamma)^{m+c+d-l-1}[2l - m(1 - \gamma)]}{\Gamma(c + d + 1)\Gamma(n/2 - c + 1)\Gamma(l + d + 1)\Gamma(m - l - d + 1)(\omega - m + 2l)}$$

$$\tag{8}$$

---

[31] The dispersion equation may be derived also by the method similar to that used in the investigation of stability of the cylinder in §4, Chapter II (Antonov [14]). The technique of calculation remains mainly the same. X and Y retain the same meaning as earlier, but instead of the earlier law of averaging, we have a new one:

$$\overline{X} = \frac{1}{2\pi} \int_0^1 \frac{s \, ds}{\sqrt{1 - s^2}} \int_0^{2\pi} X(x, y, -\gamma y + s\varkappa \cos \varphi, \gamma x + s\varkappa \sin \varphi) \, d\varphi.$$

Here $s\varkappa$ is the current value of the modulus of the peculiar velocity.

4.4.2. Investigation of the Dispersion Equation (8). Note first some limiting cases. If in (8) one switches over to the hydrodynamical limit $\gamma \to 1$, we shall obtain Eq. (57), Section 2.2, that has been studied earlier by Hunter [226].

In the case $A^{(1)} = 0$, from (7) (in the frame of reference rotating with an angular velocity γ) we obtain the following dispersion equation:

$$1 = \lambda_m A_{m\gamma}, \tag{9}$$

where

$$A_{m\gamma} = \frac{1}{1 - \gamma^2} - \frac{1}{2^m} \sum_{\sigma=0}^{m} C_m^\sigma (1 - \gamma)^{m-\sigma-1}(1 + \gamma)^{\sigma-1} \frac{\omega}{\omega + m(1 + \gamma) - 2\sigma}.$$

For oscillations with $m = 2$ (the largest-scale modes) this yields the following equation for $\omega' = \omega + 2\gamma$ [93, 252]:

$$2(\omega')^3 - 5\omega' + 6\gamma = 0. \tag{10}$$

In the perturbation in question, the original circular disk converts to the elliptical one. The critical value γ_c from (10) separating the stable and unstable solutions is $\gamma_c \approx 0.507$. As we shall see below, the stabilization conditions of this mode are more difficult in comparison with axially symmetrical perturbations. For the next type of oscillations with $m = 3$ (less large-scale ones) the equation similar to (10) is the following ($\omega' = \omega + 3\gamma$):

$$[(\omega')^2 - 9][(\omega')^2 - 1] + \tfrac{5}{8}[3(\omega')^2 + 12\gamma\omega' - 9(1 - 2\gamma^2)] = 0. \tag{11}$$

The critical value of γ in this case turns out to be larger than that for the previous mode: $\gamma_c > 0.507$. It may be shown that the stabilization conditions (of the modes in question) "improve" continually with decreasing scale of perturbation, i.e., with increasing number of the mode m.

Figure 79 shows the plots of the dependence of the growth rate of instability, Im ω, on the number of the mode m for different values of γ: $\gamma = 0.7, 0.8, 0.9, 0.95, 1$. It is seen how the range of unstable solutions is shifted with decreasing γ, i.e., with increasing value of the velocity dispersion towards the largest-scale perturbations. However, for the disks with $\gamma \approx 1$, the maximal growth rates of instability lie, as was to be expected, in the

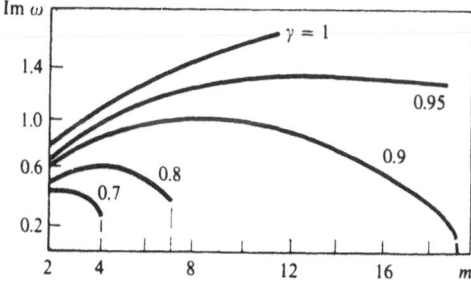

Figure 79. The dependence of the instability growth rate Im ω on the mode number m for various disks (6) in Section 1.4.

range of large values of m. Note that the eigenfunctions in this case, $\Phi_1 \sim r^m \cos m\varphi$, are essentially nonzero only in the immediate vicinity of the disk boundary $r = 1$.

A similar situation of course takes place also in the general case $A^{(1)} \neq 0$ until we deal with the stabilization of the hydrodynamic branch of oscillations of the unstable cold disk. Most difficult is the stabilization of the modes $(m = 1, n = 2)$ and $(m = 2, n = 2)$. The equation for the oscillation frequencies of the former is the following $(\omega' = \omega + \gamma)$:

$$[(\omega')^2 - 1][(\omega')^2 - 9] + \tfrac{3}{8}[11(\omega')^2 + 20\gamma\omega' - 30\gamma^2 - 9] = 0. \quad (12)$$

For the case $(m = 2, n = 2)$, we obtain $(\omega' = \omega + 2\gamma)$

$$\omega'[(\omega')^2 - 4][(\omega')^2 - 16] + \tfrac{5}{4}[4(\omega')^3 + 15\gamma(\omega')^2 - 22\omega' + 12\gamma - 84\gamma^3] = 0. \quad (13)$$

Figure 80 gives the plots of the value of instability growth rate $\operatorname{Im}\omega$ of these modes versus the parameter γ. The parts of the plots near $\gamma = 1$ have a normal increasing form denoting the stabilization of the hydrodynamical instability of the cold disk by the stellar velocity dispersion. Further variation of the curves, however, is anomalous, which evidently is due to the fact that the stationary distribution functions considered are increasing functions of the particle energy (in a rotating frame of reference).

In [93, 252] it is proved that the zones of different instabilities completely cover the segment $0 \leq \gamma \leq 1$, i.e., all the models of (1) are unstable (with respect to excitation of some mode). This is seen, for example, from Fig. 81, taken from the work of Kalnajs [252]. In this figure, the solid lines denote instability regions. Any vertical line traced in this figure [and consequently, corresponding to some definite model from the number of (1)] will necessarily intersect at least one of the solid lines.

It is important to note that the nonaxially symmetrical modes are unstable in the region, where the radial oscillations are deliberately stable. The stability boundary of the latter lies in the region $\gamma = 0.85$ (the $m = 0, n = 4$ mode), which coincides rather well with the criterion of marginal stability

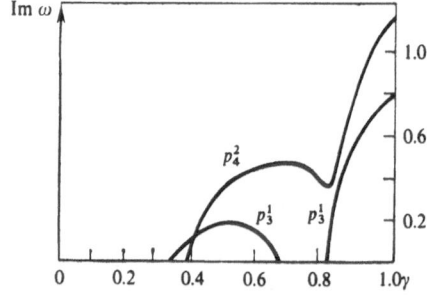

Figure 80. The dependence of the instability growth rate $\operatorname{Im}\omega$ on γ for the modes P_3^1 and P_4^2.

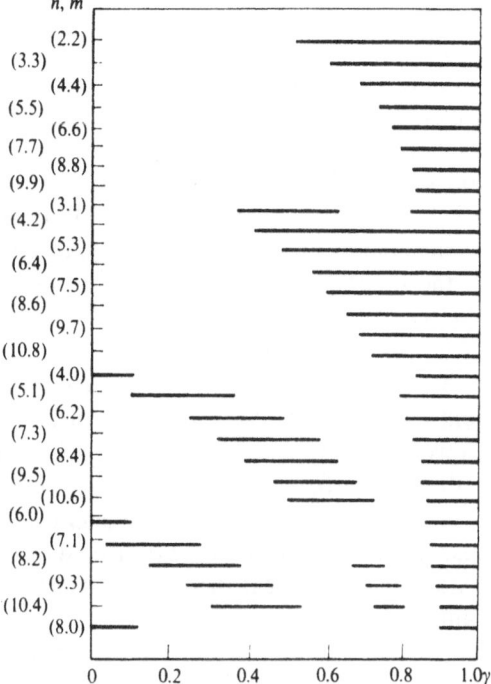

Figure 81. Instability of all models of the solid-body rotating disks (6) in Section 1.4 [252]. Solid lines are boundaries of instability regions for various modes. The mode numbers are indicated on the ordinate axis.

obtained from the results of local analysis in the extrapolation of them on the region of the largest-scale modes (they are, of course, the most unstable). It must be noted, however, that the local analysis and such extrapolation provide wrong information regarding the stability of large-scale nonaxially symmetrical modes which still remain unstable (cf. Problem 4).

The significance of the investigation of oscillation spectra of disk models in (1) lies mainly in the fact that they are used to automatically establish the functions (1) per se are of certain practical interest, although at first sight it seems that they cannot provide any information regarding the behavior of real stellar systems: These are characterized, e.g., by a decreasing distribution of stars in velocities.

Take first the distribution function of a cold disk with circular orbits of the particles, $f_0 \sim \delta(v_r)\delta(v_\varphi)$, in a rotating system. It is investigated in detail for stability by Hunter [226]. The next task was to take into account the finite velocity dispersion of stars, which in all cases of interest is much smaller than the regular velocity (of rotation). Consequently, it should be that the δ functions are in some way "smeared." But it happens that the method of "smearing" is indifferent if one restricts oneself to the analysis of systems having small velocity dispersion.

The point is that they are lacking kinetic instabilities. There is always only one unstable root corresponding to the "Hunter" root in the limit of circular orbits. This is the hydrodynamic branch of oscillations. The be-eigenfrequencies of any "composite" model. However, the distribution havior of this only oscillation branch of interest is, of course, independent of the details of the distribution function but is defined by its first moments only. In this respect, it does not matter which distribution function should be considered: (1) or, say, the Maxwellian one (of course, properly "cut off" at higher velocities). The results must be similar. With large velocity dispersions, such a coincidence is no longer necessary. In particular, as shown by Julian [244], the distribution functions decreasing with energy must not lead to oscillative instability (overstability) of the axially symmetrical modes. At the same time, distribution function (1) leads at $\gamma \approx 0$ to just such instability. At small dispersion of residual velocities of the particles, taking into account its finite value leads to expansions of the corresponding dispersion equations in the parameter $\delta = c/\lambda\Omega = \rho/\lambda$, where c is dispersion, λ is the wavelength of perturbation, Ω is the angular velocity of rotation, ρ is the average size of the epicycle. When this parameter is small, the correction terms for the dispersion equation of Toomre (of a cold disk) coincide for any distribution functions with identical dispersions (the second moments of the distribution function).

Thus, even the simplest models of disks in (1) are quite suitable for a quantitative investigation of the behavior of stabilization of unstable oscillations of the cold disk with a small (in comparison with the regular velocity) velocity dispersion of stars. Apparently, in all real spiral galaxies, the velocity dispersion is indeed small in this implication.

From the expression for the parameter δ, it is seen that the accuracy of this correspondence of the results of investigation of model stability in (1) and (6), Section 1.4, and of real stellar distributions (for instance, Maxwellian) must improve with increasing scale of perturbation. The behavior of large-scale modes is thoroughly defined by a few first moments. We can make sure directly as well by obtaining equations for characteristic frequencies of large-scale modes of arbitrary uniformly rotating disks in (6), Section 1.4 {cf. [93, 252] and below, (16)–(21)}. Indeed, it turns out that these frequencies depend only on the mean values of lower degrees of γ: $\bar{\gamma}, \overline{\gamma^2}, \overline{\gamma^3}$ [with the weight of $A(\gamma)$]. To investigate the stability of these modes, consequently, one does not have to specify in any way the distribution function of the system at all: it is quite enough to specify these means (which may be equal for many specific systems).

The advantage of this exact method of investigation is first of all just the possibility of investigating not only small-scale but also large-scale perturbations. Thus, in particular, one is able to prove that just the large-scale nonaxially symmetrical modes become most "dangerous," and therefore of most interest in real systems with finite velocity dispersion of stars.

Hohl in [219] investigated numerically the evolution of a collisionless disk with distribution functions in (1) at the values of the parameter $\gamma = 0.8, 0.6, 0.4, 0$. The first of these values roughly corresponds, as we have seen above, to the stability boundary with respect to radial perturbations.

Numerical experiment shows, however, in full agreement with the theory above presented, explicit instability of such a disk ($\gamma = 0.8$) with respect to nonradial perturbations. For approximately two rotations of the disk about its axis, it develops an apparent ellipticity, and then a position arises that resembles barred spiral.

In the remaining three cases in [219], there is only weak instability (which may be due to oscillative instability–overstability, or to the action of the collisions).

4.4.3. Oscillation Spectra of Composite Models.

Consider now the "composite" models of (8), §1. As we have already mentioned, the oscillation spectra of all these disks can be obtained automatically from the spectrum of the initial distribution functions $f_{0\gamma}$, as given above. One has only to replace, in the equations for characteristic frequencies of the models in (1), degrees of the parameter γ by corresponding mean values with a weight $A(\gamma)$.

If dispersion equation (7) is written symbolically in the form

$$D_\gamma^{(m,\,n)}(\omega) = 0, \tag{14}$$

then the sought-for dispersion equation defining the oscillation frequencies of the system described by the function in (8), §1, for the same spatial mode, will be [93, 252]

$$\int_{-1}^{1} d\gamma\, A(\gamma) D_\gamma^{(n,\,m)}(\omega + m\gamma) = 0. \tag{15}$$

Here it is important, of course, that the eigenfunctions are the same for all γ [93, 111, 252]. The frequency shift in (15) $\omega \to \omega + m\gamma$ corresponds to the Doppler shift: the rule of derivation of Eq. (15), using the familiar equation in (7), means simply that we must sum the contributions of perturbations of all γ-constituents of a particular composite model with frequencies calculated in an inertial frame of reference.

The importance of investigation of the "composite" distribution functions is defined, above all, by the fact that despite the proved instability of each of the constituents, the whole composite system may nevertheless be stable (cf. below).

The dispersion equations for distribution function (8) of Section 1.4, for large-scale modes, is easily obtained from similar dispersion equations for the function (1) by replacing the parameter γ by the corresponding average values:

$$\gamma^p \to \overline{\gamma^p} \equiv \int_{-1}^{1} A(\gamma)\gamma^p\, d\gamma, \qquad \int_{-1}^{1} A(\gamma)\, d\gamma = 1. \tag{16}$$

We give some simple examples. The dispersion equations for largest-scale modes:

(1) the ($m = 2, n = 0$) mode

$$2\omega^3 - 5\omega + 6\bar\gamma = 0; \tag{17}$$

(2) the $(m = 0, n = 4)$ mode

$$(\omega^2 - 4)(\omega^2 - 16) - \tfrac{45}{8}(-\omega^2 + 2 + 14\overline{\gamma^2}) = 0; \qquad (18)$$

(3) the $(m = 1, n = 2)$ mode

$$(\omega^2 - 9)(\omega^2 - 1) + \tfrac{3}{8}(11\omega^2 + 20\overline{\gamma}\omega - 30\overline{\gamma^2} - 9) = 0; \qquad (19)$$

(4) the $(m = 3, n = 0)$ mode

$$(\omega^2 - 9)(\omega^2 - 1) + \tfrac{5}{8}[3\omega^2 + 12\overline{\gamma}\omega - 9(1 - 2\overline{\gamma^2})] = 0; \qquad (20)$$

(5) the $(m = 2, n = 2)$ mode

$$\omega(\omega^2 - 4)(\omega^2 - 16) + \tfrac{5}{4}(4\omega^3 + 15\overline{\gamma}\omega^2 - 22\omega + 12\overline{\gamma} - 84\overline{\gamma^3}) = 0. \qquad (21)$$

An outstanding fact easily seen from (17)–(21) implies that the stability of large-scale modes depends only on mean values of few lower degrees of the parameter γ. The last have a clear physical meaning: $\overline{\gamma}$ equals the angular velocity of the system, $\overline{\gamma^2}$ is linked with the velocity dispersions of particles averaged in volume of the system, etc.

The stability of the "barlike" mode $(m = 2, n = 0)$ converting the initial circular disk to the elliptical one, is defined, in accordance with (17), only by the mean angular velocity of the disk as a whole, so that the stability condition is: $\overline{\gamma} \lesssim 0.507$. This condition implies that, for stability, the part of the contribution of chaotic movements to equilibrium of the system must be sufficiently large.

According to (18), the stability of the simplest axially symmetrical mode $(m = 0, n = 4)$ is defined only by the value $\overline{\gamma^2}$. Consequently [93, 252], instability will take place if either $\overline{\gamma^2} \gtrsim 0.65$ or $\overline{\gamma^2} \lesssim 0.01$.

Take now the largest-scale nonaxially symmetrical modes $(m = 1, n = 2)$ and $(m = 2, n = 2)$ the anomalous behavior of which, at sufficiently small values of the parameter γ in models (1), was mentioned by us above (cf. Fig. 80). From Eqs. (19) and (21) it follows that the behavior of small perturbations of arbitrary composite models of (8), §1, is defined here by the pairs of values: $(\overline{\gamma}, \overline{\gamma^2})$ in the case of the $(m = 1, n = 2)$ mode and $(\overline{\gamma}, \overline{\gamma^3})$ for the $(m = 2, n = 2)$ mode. The quantities inside each of these pairs are generally independent [unlike the models in (1)] and are defined by setting a specific distribution function, i.e., the $A(\gamma)$ function.

The composite models are investigated in detail in [93, 252]. Let us give a table (Table IX) taken from a paper of Kalnajs [252], which lists some characteristic data on equilibrium states and on stability of the four composite models.[32]

It is assumed that all "γ-disks" of the form of (1) entering in these composite systems rotate in the same direction: $\gamma > 0$.

Kalnajs investigated the stability of these models by calculating the oscillation frequencies for all modes with $n, m \leq 20$. This happens to be

---

[32] In [93] the stability conditions of a large number of models with the weight function of the form $A(\gamma) \sim \gamma^{n_1}(1 - \gamma^2)^{n_2/2}$ (where n_1, n_2 are integers) are considered.

Table IX Parameters of some composite models [252]

Parameter	Model A	Model B_0	Model B_1	Model B_3
$A(\gamma)$	1	$4(1-\gamma^2)^{1/2}\pi$	$3\gamma(1-\gamma^2)^{1/2}$	$15\gamma^3(1-\gamma^2)^{1/2}/2$
$\bar{\gamma}$	$\frac{1}{2}$	$\frac{4}{3}$	$3\pi/16$	$15\pi/64$
Energy of chaotic motion vs. energy of rotation	2	3	$\frac{3}{2}$	$\frac{3}{4}$
Number of unstable modes with $n, m \leq 20$	0	0	3	13

enough, in order that the asymptotic behavior of frequencies of different families of modes might be manifested, which are classified according to the number of radial nodes of the eigenfunctions $(n - m)/2$ (see below). All instabilities occur among the modes with the least (n, m) and the asymptotic behavior clearly indicates that the remaining modes are stable.

The models A and B_0 are stable. The model B_0 is somewhat "hotter," but otherwise it very much resembles A. The model B_0 is included because its distribution function (as for the models B_1 and B_3) is the elementary function E and L, while the distribution function of A contains elliptical integrals.

The model B_3 has 13 unstable modes. The $(2, 2), (3, 3), \ldots, (7, 7)$ modes rotating most rapidly in the direct direction form one group of unstable modes, while the $(4, 2), (5, 3), \ldots, (10, 8)$ modes (also the most rapid) form another. The $(3, 3)$ mode of the former group has the largest growth rate (Im $\omega = 0.5547$), and the $(7, 7)$ mode, the least (Im $\omega = 0.204$), whereas, in the latter group, the maximum growth rate (Im $\omega = 0.469$) refers to the $(6, 4)$ mode, and the minimum one (Im $\omega = 0.223$), to the $(4, 2)$ mode.

The model B_1 has three unstable modes. Out of them, the barlike directly rotating $(2, 2)$ mode has the largest growth rate.

The models B_1 and B_3 show once again that the disks, which are hot enough to suppress radial instabilities, may still be very unstable with respect to nonaxially symmetrical modes.

Let us now clarify the problem of the form of eigenfunctions, i.e., the spatial dependence of the perturbed density and potential in the exact models of disk systems considered.

In the hydrodynamical limit $\gamma = 1$ in (1), the eigenfunctions for the arbitrary mode characterized by a pair of indices (m, n) were found by Hunter. They have (inside the disk) the form of (61), Section 2.2.

In [93, 111, 252], it was shown that also in the general case $\gamma \neq 1$, i.e., for arbitrary disks, the eigenfunctions have the same form. This can be directly tested for the first modes if one calculates the terms of expansion of the radial part of the perturbed potential $\Phi_1 = r^{n+m} + Cr^{n+m-2} + \cdots$ followed after r^{n+m}. We give several examples of such calculations in Problem 2. Kalnajs [252] gives the general proof for this statement valid for all modes at once.

Since the eigenfunctions are alike for any values of the parameter γ, they will remain the same also for arbitrary distribution functions of uniformly rotating disks of the form of (8), §1, which are the linear combinations of the disks in (1) with different γ.

We reproduce briefly (mainly, the scheme) of the general proof of Kalnajs.

By restricting ourselves to perturbations with a finite potential energy, let us make the functional space be Hilbertian by using (minus) potential energy of interaction of two perturbations as a scalar product.

The potential energy of interaction of two perturbations with the indices (n, m) and (n', m') is

$$W = \tfrac{1}{2} \int \rho \Phi \, dV = \tfrac{1}{2} \int \sigma \Phi \, dS, \tag{22}$$

for

$$\Phi = \psi_n^m \equiv 4\gamma_n^m P_n^m(\xi)e^{im\varphi}, \qquad \sigma = \sigma_n^{m'} \equiv P_{n'}^{m'}(\xi)e^{im'\varphi}/\xi,$$

$$(\sigma_{n'}^{m'}, \psi_n^m) = -4\gamma_n^m \int_0^{2\pi} d\theta$$

$$\times \int_0^1 \frac{r \, dr}{\sqrt{1 - r^2}} e^{i(m - m')\theta} P_n^m[(1 - r^2)^{1/2}] P_{n'}^{m'}[(1 - r^2)^{1/2}]. \tag{23}$$

It vanishes at $m \neq m'$. If, however, $m = m'$, then instead of (23) we have

$$-2\pi 4\gamma_n^m \int_0^1 P_n^m(\xi)P_{n'}^{m'}(\xi) \, d\xi \qquad (\xi \equiv \sqrt{1 - r^2}). \tag{24}$$

This integral vanishes when $(n - n')/2$ is a nonzero integer.

We determine now the operator of the response $\tilde{R}$: $\tilde{R}\psi_n^m$ is the potential produced by the density response on ψ_n^m, calculated by solving the linearized kinetic equation. ψ_n^m will be the eigenfunction of the operator $\tilde{R}$, if

$$\tilde{R}\psi_n^m = \lambda\psi_n^m. \tag{25}$$

If, moreover, $\lambda = 1$, then ψ_n^m is the normal mode.

Another way of proving that ψ_n^m is the eigenfunction is to show the orthogonality of $\tilde{R}\psi_n^m$ with respect to all the remaining potentials with the indices different from (n, m):

$$(\sigma_{n'}^{m'}, \tilde{R}\psi_n^m) = 0, \qquad \text{if } (n, m) \neq (n', m'). \tag{26}$$

The expression for the scalar product in terms of the potential was derived by Kalnajs [251, 252]. We do not give it here because it is somewhat complicated. It is essential to us only that if ω is real and is not equal to the linear combination of the radial and azimuthal frequencies $(l\Omega_1 + m\Omega_2)$ of any particles, then the response operator is self-conjugate,[33] so that (26) at the

---

[33] One can find the explicit expression for R and make sure that there is self-conjugateness of this operator if one makes use of the formulae derived in the previous section in the derivation of integral equation (38).

same time means that also

$$(\sigma_n^m, \tilde{R}\psi_{n'}^{m'})^* = 0, \qquad (n, m) \neq (n', m'). \tag{27}$$

Since in the uniformly rotating disk $\Omega_1 = 2$, $\Omega_2 = 1$, then $\tilde{R}$ will be self-conjugate when ω is real and not integer. But since, on the left-hand sides of (26) and (27), there are the analytical functions ω (with their poles at integer ω), then if it is shown that ψ_n^m are eigenfunctions for real ω, their analytical continuations will be eigenfunctions also for complex ω.

The orthogonality condition in (23) may be simplified: it is sufficient to show that the disk response to the potential of the form

$$\Phi_s = \exp(im\varphi - i\omega t)r^m r^{2s} \qquad \left(s \leq \frac{n - m}{2}\right) \tag{28}$$

also is a polynomial in r of the same degree. In other words,

$$\tilde{R}\Phi_s = a_0\Phi_s + a_1\Phi_{s-1} + \cdots + a_s. \tag{29}$$

Since ψ_n^m is the linear combination Φ_s, then the response $\tilde{R}\psi_n^m$ is the linear combination ψ_{m+2l}^m with $0 \leq l \leq (n - m)/2$ and therefore is orthogonal to all ψ_{n+2j}^m with $j = 1, 2, \ldots$. But from the self-conjugateness of $\tilde{R}$ it follows that $\tilde{R}\psi_n^m$ must be orthogonal to all $\psi_{n-2j}^m, j = 1, 2, \ldots, (n - m)/2$.

Thus, to prove that ψ_n^m are eigenfunctions, one has to demonstrate that the response to (28) is (29). This is proved by a direct calculation (by its self-consistency).

Dispersion equation (7) can be presented in the form

$$\sum_{k=-n}^{n} \frac{a_k}{\omega - k} = 1, \tag{30}$$

and for $m = 0$, $a_0 = 0$. Hence it follows that for each ψ_n^m in a hot disk, there are $(n + 1)$ for $m \neq 0$ and n for $m = 0$ (due to preservation of angular momentum) degrees of freedom [according to the number of different roots of Eq. (30)]. Thus, to investigate the "evolution" of ψ_n^m for $m \neq 0$, one has to determine $(n + 1)$ independent equations for the moments of the distribution function. Recall that in a "cold" disk we have had respectively three (for $m \neq 0$) and two ($m = 0$) degrees of freedom.

Stability of the model (9), §1. A strict proof for stability of each composite model of the type of (8), §1, with respect to *arbitrary* perturbations is a rather difficult task. It is relatively simple for the model (9), §1 [14].

The technique of calculation changes little in comparison with the case of the cylinder (cf. §4, Chapter II). We introduce again the horizontal deviations X and Y satisfying similar equations. They are represented by polynomials of degree N. However, the averaging formula is substituted by a somewhat different one. We establish the manner of averaging in velocities. It may be approached by using the formulae

$$v_x = \frac{xv_r - yv_\varphi}{\sqrt{x^2 + y^2}}, \qquad v_y = \frac{yv_r + xv_\varphi}{\sqrt{x^2 + y^2}} \tag{31}$$

and by taking all means from the expression of the type $v_\varphi^\alpha v_r^\beta$. If α or β is odd, then according to symmetry we have zero. If both indices are even, we introduce temporarily the polar coordinates in the velocity diagram using the formulae

$$v_r = r\sqrt{a^2 - x^2 - y^2}\ \sin\varphi, \qquad v_\varphi = ra\cos\varphi$$

(a is the radius of the disk), and then we have ($\Omega_0 = 1$)

$$\overline{v_\varphi^\alpha v_r^\beta} = \frac{a^\alpha}{\pi}(a^2 - x^2 - y^2)^{\beta/2} \int_0^1 r^{\alpha+\beta+1}\ dr \int_0^{2\pi} \cos^\alpha\varphi \sin^\beta\varphi\ d\varphi. \quad (32)$$

If $\alpha \geq 2$, then the result of averaging has, with respect to $R = \sqrt{x^2 + y^2}$, a degree less than $\alpha+\beta$, and does not play any role in separating the higher terms (which is formally performed by dividing by R^N with a subsequent transition to $R \to \infty$). One should take into account only the products of v_r^β and their averages

$$\overline{v_r^\beta} = \frac{2(\beta - 1)!!}{(\beta + 2)!!}(a^2 - x^2 - y^2)^{\beta/2} = \frac{2(\beta - 1)!!}{(\beta + 2)!!}(iR)^\beta + \cdots. \quad (33)$$

By comparing this with (31), we obtain the sought-for rule of averaging

$$\overline{X} = \frac{1}{\pi}\int_0^{2\pi} X(x, y, ix\sin\varphi, iy\sin\varphi)\cos^2\varphi\ d\varphi. \quad (34)$$

We write the continuity equation

$$\sigma_1 = -\frac{\partial}{\partial x}(X\sigma) - \frac{\partial}{\partial y}(Y\sigma). \quad (35)$$

The density perturbation equals some polynomial of degree $(N + 1)$ divided by $\sqrt{a^2 - x^2 - y^2}$. The values of the azimuthal wave number $m = -N - 1, -N + 1, \ldots, N + 1$ are possible.

Since σ vanishes on the boundary, the boundary displacement makes a contribution of the higher order of smallness to the potential. Therefore, there are no special boundary modes.

Taking Φ_1 in the form

$$\Phi_1 \sim (x + iy)^n (x - iy)^{n_1} \equiv R^{N+1} e^{im\varphi} \quad (36)$$

and following everywhere the older term of polynomials, in the same way we obtain the dispersion equation

$$\frac{n_1(n + \tfrac{1}{2}) + n(n_1 + \tfrac{1}{2})}{\pi} \int_0^\infty e^{-\omega\tau} \sinh\tau\ d\tau$$

$$\times \int_0^{2\pi} (\cosh\tau - \sin\varphi \sinh\tau)^N \cos^2\varphi\ d\varphi$$

$$= -\frac{(2n)!!(2n_1)!!}{4(2n - 1)!!(2n_1 - 1)!!} \qquad (N = n + n_1 - 1). \quad (37)$$

Separating in the integrand the terms with e^{τ} and $e^{-\tau}$, we find

$$\int_0^{2\pi} (\cosh \tau - \sin \varphi \sinh \tau)^N \cos^2 \varphi \, d\varphi = 2^{-N} \sum_{m=0}^{N} \frac{N! \, e^{(2m-N)\tau}}{m!(N-m)!}$$

$$\times \int_0^{2\pi} (1 - \sin \varphi)^{m+1}(1 + \sin \varphi)^{N-m+1} \, d\varphi$$

$$= 2\pi \sum_{m=0}^{N} \frac{(2m+1)!!(2N-2m+1)!! \, e^{(2m-N)\tau}}{(2m)!!(2N-2m)!!(N+1)(N+2)}. \tag{38}$$

Performing integration over τ, we reduce the dispersion equation to the form

$$\sum_{m=0}^{N} \frac{(2m+1)!!(2N-2m+1)!!}{(2m)!!(2N-2m)!![(\omega+N-2m)^2-1]} = -\delta(n, n_1), \tag{39}$$

where

$$\delta(n, n_1) = \frac{(2n)!!(2n_1)!!(n+n_1)(n+n_1+1)}{4(2n-1)!!(2n_1-1)!!(4nn_1+n+n_1)}. \tag{40}$$

As earlier, the equation can be represented also in the multiplicative form, namely

$$\frac{\omega^2(\omega^2-4)\cdots(\omega^2-N^2)}{(\omega^2-1)(\omega^2-9)\cdots[\omega^2-(N+1)^2]} - 1 + \delta(n, n_1) = 0 \qquad (N \text{ even}), \tag{41}$$

or

$$\frac{(\omega^2-1)(\omega^2-9)\cdots(\omega^2-N^2)}{(\omega^2-4)(\omega^2-16)\cdots[\omega^2-(N+1)^2]} - 1 + \delta(n, n_1) = 0 \qquad (N \text{ odd}). \tag{42}$$

The model proves to be stable. To prove this, establish first of all the inequality

$$\delta(n, n_1) \geq 1. \tag{43}$$

Since the factor $(2n)!!/(2n-1)!!$ increases at $n = 1, 2, 3, \cdots$ and the factor

$$\frac{(n+n_1)(n+n_1+1)}{4nn_1+n+n_1},$$

as an analysis by usual methods shows, increases also over n at $n > n_1 > 0$, then the problem reduces to the consideration of the pairs (n, n_1) of the form of $(1, 0)$ and (n, n); for the remaining pairs, inequality (43) is further strengthened. But $\delta(1, 0) = 1$ (connected to the trivial case of the system displacement as a whole), and

$$\delta(n, n) = \frac{1}{4}\left[\frac{(2n)!!}{(2n-1)!!}\right]^2 \geq 1 \qquad (n \geq 1).$$

Taking into account (43), all the roots are separated. They each lie within the interval $(0, 1), (2, 3), \ldots, (N, N + 1)$ at even N or $(1, 2), (3, 4), \ldots, (N, N + 1)$ at odd N.

Bar-like perturbations of uniformly rotating disks are considered in detail in a special paper of Kalnajs and Athanasoula [254]. Indeed, these perturbations evidently deserve a separate investigation. We have already mentioned the main cause of such a preference in Chapter IV (cf. §3). It implies that it is the barlike mode that dominates in the evolution of flattened systems stable with respect to axially symmetric perturbations. This was convincingly demonstrated in numerical experiments of Ostriker and Peebles [301], and earlier by Hohl [220] and Miller and co-authors [294] (for systems having no significant "halo").

Practically, it is important, of course, to get an answer to the question how the oscillation modes of the models considered above are modified in the transition to differentially rotating systems. It is difficult to answer this question with certainty, but it is possible to state the following natural suggestion [254].

"Weakly collective" modes [i.e., modes linked to small "residues" of a_i in (30) and whose existence is due to a large time interval of coherency in the absence of differential rotation[34]] will hardly "survive" in the presence of the latter. Least of all is the elliptical (barlike) deformation $(2, 2)$ subjected to differential rotation; in addition, to excite it, there is needed the smallest transfer of energy and angular momentum. Therefore, it is possible to expect the existence of analogs of the $(2, 2)$ mode also in the differentially rotating disks, which is just supported by the above numerical experiments.

In the previous chapter (§3), we have described the Peebles and Ostriker hypothesis, which argues that stability (or instability) with respect to the barlike mode is not sensible to the detailed structure of the disk and in all events is defined by the value of the ratio t of the rotational energy to (minus) potential energy, where the critical value $t_{cr} = 0.14 \pm 0.03$. The investigation of uniformly rotating disks serves, in particular, as a test of the Peebles and Ostriker hypothesis. Above we have seen that stability of the barlike modes for a particular family of uniformly rotating disks (8), §1, is defined only by a mean velocity of the disk rotation. This stability criterion is easily expressed through the parameter t. It turns out that $t_{cr} = 0.1286$, which does not contradict the stability boundary indicated by Peebles and Ostriker.

In [254], it is shown that the $(2, 2)$ modes of the rotating disks in (8), §1, take place in a wider class of equilibria and that their stability is defined by the same criterion.

The perturbed distributions of the density and potential corresponding to the barlike mode $(R = \Omega_0 = \sigma_0(0) = 1)$

$$\sigma_1(r, \varphi, t) = \varepsilon \, \text{Re}\{r^2(1 - r^2)^{-1/2} \exp[i(2\varphi - \omega t)]\},$$

$$\Phi_1(r, \varphi, t) = -\tfrac{3}{4} \, \text{Re}\{r^2 \exp[i(2\varphi - \omega t)]\}.$$

(44)

---

[34] The frequencies of these modes are close to resonance frequencies.

The interpretation of $\sigma_1(r, \varphi, 0)$ is natural: this is the change in density if the equilibrium is deformed by giving to the stars originally at the point $\mathbf{r}$, a small mean displacement $\delta\mathbf{r}$:

$$\delta x = \varepsilon x, \qquad \delta y = -\varepsilon y.$$

Such deformation can be considered as a result of the extension of the disk along the x-axis by $(1 + \varepsilon)$ times and compression along the y-axis by $(1 - \varepsilon)$ times. Unperturbed motions of the stars are harmonical oscillations along x and y at a frequency $\gamma = 1$. If the perturbed field (44) is slowly introduced, the stellar orbits will deviate from the unperturbed ones: $\mathbf{r} \rightarrow \mathbf{r} + \Delta\mathbf{r}$. We express the deviations of $\Delta\mathbf{r}$ in the function of the unperturbed location, stellar velocity, and time t. At a small ε, $\Delta\mathbf{r}$ is also small, so that (within an accuracy of the first order of magnitude in ε) $\Delta\mathbf{r}$ can be calculated by substituting the perturbed field along the unperturbed orbit. Corresponding linear equations for forced harmonical oscillations are readily integrated, and one thus gets

$$\Delta x - i\Delta y = \tfrac{3}{4}\varepsilon\{(x_0 + iy_0)A - i(u_0 + iv_0)B\}, \qquad (45)$$

where $A \equiv 2/(4 - \omega^2)$, $B \equiv 4/(4\omega - \omega^3)$, and (x_0, y_0) and (u_0, v_0) are the unperturbed values of coordinates and stellar velocities (at the "moment" $t = -\infty$). Hence, the mean value of $\langle\Delta\mathbf{r}\rangle$ for all the velocities is

$$\langle\Delta x\rangle - i\langle\Delta y\rangle = \tfrac{3}{4}\varepsilon\{(x_0 + iy_0)A - i(\langle u_0\rangle + i\langle v_0\rangle)B\}. \qquad (46)$$

Finally, in order that these deviations might lead to variations of the density which will reproduce the shape of the assumed field Φ_1, it is necessary that the mean displacement be proportional to $\delta\mathbf{r}$, i.e., that there be

$$\langle u_0\rangle = -\gamma y_0, \qquad \langle v_0\rangle = \gamma x_0. \qquad (47)$$

It is thus seen that the mean velocity field of the equilibrium disk must be a field of uniform rotation with an angular velocity which may be taken from the interval $0 < \gamma \leq 1$. This limitation and the requirement that the surface density be equal to $\sigma_0(0)\sqrt{1 - r^2/R}$ is satisfied by a much wider class of equilibria than those considered earlier in (8), §1.

The self-consistency of oscillations yields

$$\tfrac{3}{4}(A + \gamma B) = -\frac{3(1 + \gamma)/8}{\omega - 2} + \frac{3\gamma/4}{\omega} + \frac{3(1 - \gamma)/8}{\omega + 2} = 1. \qquad (48)$$

This equation is identical to Eq. (17) for the (2, 2) mode of the disks from (8), Section 1.4. There are three possible oscillation frequencies; their dependence on γ is presented in Fig. 82. Instability starts at $\gamma > 5/9\sqrt{5/6} = 0.5072$, which corresponds to the ratio of the rotational kinetic energy to the (minus) potential energy $t = 125/972 = 0.1286\ldots$. The velocities of unstable modes in the azimuth are enclosed within a rather narrow interval, between $(5/24)^{1/2}$ and $1/2$ of the angular velocity of a star in circular orbit.

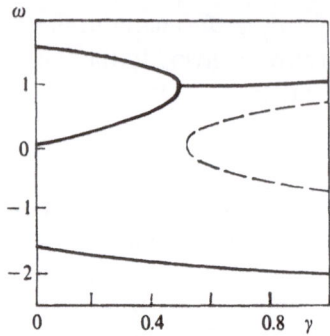

Figure 82. The dependence of ω on γ [254]. The imaginary part (dashed line) belongs to direct rotating (Re $\omega > 0$) modes; solid lines are Re ω.

4.5 Global Instabilities of Gaseous Disks. Comparison of Stability Properties of Gaseous and Stellar Disks [95, 56^{ad}]

The linear stability theory of a rotating gas polytropic layer with an adiabatic exponent $\gamma = 1, 2, \infty$ relative to localized disturbances (with scales λ small as compared with the angular velocity variation scale a) was constructed by Goldreich and Lynden-Bell[35] [209], and then was generalized to the case of an arbitrary γ in paper [31^{ad}]. Investigating in §1, Chapter VII (Section 1.2) *nonlinear* evolution of such disturbances, we shall briefly describe also corresponding results of linear theory.

In the present section we consider global instabilities of rotating gas disks having some finite radii R ($\lambda \sim R$). For the present we do not have the complete solution to the problem. For example, there is no definitive answer to the question of just how much halo is necessary to stabilize completely a gas disk close to the local stability boundary ($Q \simeq 1$). Data, obtained for the present, show that a disk which does not have a massive hot "spheroidal component" in its center requires for the stability a halo of perhaps three times the mass of the disk, or more [56^{ad}]. For certain conclusions of such a kind it is necessary to consider more variety of models, first of all those models which have essentially higher mass concentration to the disk center than occurs in the models considered so far. Nevertheless, some interesting results in this field are obtained. Below we try to give an account of these results.

4.5.1. The Case of Uniform Rotation. As the simplest example of a "hot" *hydrodynamical* system, let us consider [95] a uniformly rotating disk with an angular velocity $\gamma\Omega_0$, whose equilibrium is ensured by the corresponding

---

[35] Another work [210] by the same authors is devoted to an evolution of disturbances in a strongly nonuniformly rotating disk ($\lambda \sim a$).

pressure gradient which is taken into account in the model hydrodynamical equations

$$P_z = 0, \qquad P_\perp = \frac{\Omega_0^2 \sigma_0(0) R^2}{3} (1 - \gamma^2) \left(\frac{R^2 - r^2}{R^2} \right)^{3/2} \sim \sigma^3. \qquad (1)$$

The gravitational potential $\Phi_0 = (\Omega_0^2 r^2)/2$, while the surface density $\sigma_0(r) = \sigma_0(0) \sqrt{1 - r^2/R^2}$.

The equilibrium equation is written in the following way:

$$-\frac{1}{\sigma_0} \frac{\partial P_0}{\partial r} + \Omega_0^2 \gamma^2 r = \frac{\partial \Phi_0}{\partial r}, \qquad (2)$$

or, for the model in question,

$$(1 - \gamma^2) \Omega^2 r + \Omega^2 \gamma^2 r = \Omega^2 r. \qquad (3)$$

The oscillation spectrum of such a model can be obtained by a method quite similar to those used by Hunter for the case of a cold disk (Section 2.2). It turns out that the eigenfrequencies must satisfy the following characteristic equation:

$$1 - \frac{4\Omega^2}{\lambda^2 - 4\gamma^2 \Omega^2} \left[\tfrac{1}{4}(1 - \gamma^2) - \gamma_l^m \right] \left[l(l + 1) - m^2 - \frac{2m\gamma\Omega}{\lambda} \right] = 0. \qquad (4)$$

Here $\lambda \equiv \omega - m\gamma\Omega$, ω is the frequency, and

$$\gamma_l^m = \frac{(l + m)!(l - m)!}{2^{2l+1} \{[(l + m)/2]!\}^2 \{[(l - m)/2]!\}^2}. \qquad (5)$$

The eigenfunctions (potential), as usual for such systems, are expressed through the associated Legendre functions

$$\Phi_{l,m} \sim P_l^m \left(\sqrt{\frac{R^2 - r^2}{R^2}} \right) e^{im\varphi}. \qquad (6)$$

In particular, for radial oscillations from (4) we just get expressions for the square of the frequency

$$\omega^2 = 4\gamma^2 \Omega^2 + 4\Omega^2 l(l + 1) \left\{ \frac{1}{4}(1 - \gamma^2) - \frac{(l!)^2}{2^{2l+1}} \left[\left(\frac{l}{2} \right)! \right]^{-4} \right\}. \qquad (7)$$

Hence it follows that radial oscillations are stable for $\gamma^2 < 0.9$. More difficult is the stabilization of nonaxially symmetrical modes, where the most unstable happens to be the mode with $l = 2$, $m = 2$ corresponding to deformation of the circular disk to an elliptical one. For its stabilization, it is necessary that there be $\gamma^2 < 0.5$, i.e., the contribution to equilibrium made by pressure must exceed the contribution made by the centrifugal force.

If the disk with mass M_D and radius $R = 1$ is immersed into the homogeneous spherical halo with mass M_H (within R), which is not subjected to disturbances, instead of (4) we have the following characteristic equation:

$$\frac{1}{\eta} - \frac{4\Omega^2}{\lambda^2 - 4\gamma^2\Omega^2}\left[\frac{1}{4}(1 - \gamma^2) - \gamma_l^m\right]\left[l(l + 1) - m^2 - \frac{2m\gamma\Omega}{\lambda}\right] = 0,$$

where

$$\eta \equiv \frac{1}{1 + 4\mu/3\pi}, \qquad \mu \equiv \frac{M_H}{M_D}.$$

The halo stabilizes most effectively the very large-scale models. Let us consider, for example, modes with $l = m$. From the dispersion relation one can easily obtain in this case that

$$\frac{\lambda}{\Omega} = -\gamma \pm \sqrt{\gamma^2 - m\eta(\gamma^2 + 4\gamma_m^m - 1)},$$

whence it follows that the given mode is stabilized when

$$\eta = \frac{\gamma^2}{m(\gamma^2 + 4\gamma_m^m - 1)}.$$

In particular, for the cold case ($\gamma = 1$)

$$\eta_c = \frac{1}{4m\gamma_m^m},$$

so that

$$\mu_c = \frac{3\pi}{4}(4m\gamma_m^m - 1).$$

The last formula gives: for $m = 2$, $\mu_c = 3\pi/8 \simeq 1.19$; for $m = 3$, $\mu_c = 21\pi/32 \simeq 2.06$; for $m = 4$, $\mu_c = 57\pi/64 \simeq 2.75$; for $m = 5$, $\mu_c = 561\pi/512 \simeq 3.44$; and so on.

4.5.2. More General Models were considered by Bardeen [56[ad]]. These models are non-uniformly-rotating disks with a moderate mass concentration to the center.

It is clear that stability criteria relative to large-scale modes, formulated in suitable terms, must not essentially change when compared to the case of the uniform rotation. This was confirmed by results of computations in [56[ad]] (see below). However, if we take into account nonuniformness of a rotation, the form of the eigenfunctions changes. Namely, lines of equi-density for the most unstable modes transform into the trailing spirals. Then, it is to be expected that the appearance of a spiral pattern depends considerably on which region of a disk [(1) central or (2) middle] is more unstable in local meaning. In the first case, a spiral pattern over the whole disk is induced by the large-amplitude barlike disturbance in the center, a rather weakly tightened (open) spiral in a middle part of a disk and in its periphery is then in essence the density response in a nonuniformly rotating gas onto this bar disturbance. In the second case when the amplitude of the barlike central disturbance must be small, a spiral pattern for the most unstable mode is defined by dispersion characteristics of the medium in a middle part of a disk. If a halo with a sufficiently large mass is present, then a typical (characteristic) radial scale of disturbances decreases so that one can expect the spiral modes with a rather high degree of tightness to appear.

All that was said above is really confirmed by the results of concrete computations in [56[ad]]. As in the previous subsection 4.5.1, we assume the pressure to be flat: $P_z = 0$, $P_\perp \neq 0$. Consequently, the dynamical equations governing motions in the plane of the disk can be written

$$\frac{dv_\perp}{dt} = -\nabla_\perp(\Phi + \psi(\sigma)), \tag{8}$$

where Φ is the gravitational potential in the plane of the disk and the potential of the pressure

$$\psi(\sigma) = \int_0^\sigma \sigma^{-1} \frac{dp}{d\sigma} \, d\sigma. \tag{9}$$

Below, all quantities will be written in units such that the radius of the disk $R = 1$, the mass of the disk $M_D = 1$, and the gravitational constant $G = 1$. The models considered in detail by Bardeen have a $\psi(\sigma)$ of the form

$$\psi(\sigma) = \beta(\pi\sigma)^\alpha, \tag{10}$$

where the exponent α is connected with the adiabatic exponent γ by the relation $\alpha = \gamma - 1$; the value β characterizes the relative contribution of the pressure into the disk's equilibrium.

The stability investigation, for the given equilibrium model, is performed by the standard method (see, for example, §2, present chapter). For the angular dependence corresponding to the azimuthal number $m = 2$, the radial dependence of all the disturbances are represented by expansions over associated Legendre functions $P_l^m(\sqrt{1 - r^2})$. The linearized dynamical equations then reduce to an infinite set of equations for the coefficients of the Legendre functions. The Legendre expansions were truncated at about 20–30 terms, and then the corresponding truncated characteristic equation for the eigenfrequencies was solved.

The computations showed that the stability boundary corresponded always to the value of the parameter $t = T_{\text{rot}}/|W|$ of order of $t_l \simeq 0.26$ (as to the uniformly rotating disk).

What is more interesting are the results concerning the influence of a halo on the stability and on the form of the unstable modes. In Fig. 83

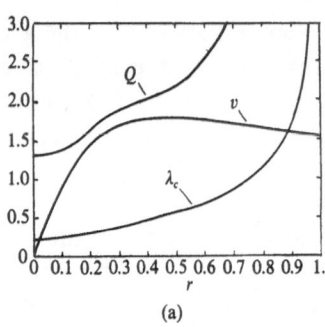
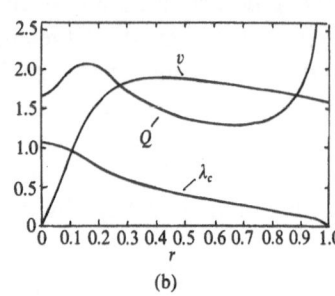

(a)   (b)

Figure 83. Functions $Q(r)$, $v(r)$, $\lambda_c(r)$ for two different models ((a) and (b)) of disk-like galaxies [56^{ad}].

(borrowed from [56^{ad}]) the distributions of the most essential equilibrium parameters are represented (for two extreme cases) by

$$v(r) = \Omega(r) \cdot r,$$

$$Q(r) = \frac{\varkappa^2}{\pi\sigma} \frac{d\psi}{d(\pi\sigma)} \qquad [\varkappa^2 = 4\Omega^2 + r(\Omega^2)'], \tag{11}$$

$$\lambda_c(r) = 2\frac{d\psi}{d\sigma}.$$

The parameter $Q(r)$ is the analog of Toomre's local stability parameter for stellar disks ("the stability reserve"), and $\lambda_c(r)$ is the wavelength corresponding to the minimum of the local dispersion curve; the latter can be written as

$$(\omega - m\Omega)^2 = \varkappa^2 - 2\pi Gk + \sigma\frac{d\psi}{d\sigma}k^2 \tag{12}$$

The minimum is less than zero if $Q < 1$. Figure 83(a) corresponds to the disk with more unstable center ($\beta = 0.2605$, $\alpha = 2/3$, $t = 0.2121$[36]), and Fig. 83(b) corresponds to the disk with more unstable middle part ($\beta = 0.050$, $\alpha = 4/3$, $t = 0.2225$, $Q_{min} = 1.289$). The halo mass was the same in both cases: $M_H = 1.433$. Lines of the maximum surface density for these two models are represented in Fig. 84(a) and 84(b) respectively. Some characteristics of the spiral disturbance in the first case are $\Omega_p \equiv \text{Re }\omega/m = 2.75$, $\text{Im }\omega = 0.09$, the corotation radius $r_{corot} = 0.63$, and the amplitude of the disturbance has the maximum within the central bar; in the second case: $\Omega_p = 2.5$, $\text{Im }\omega = 0.328$, $r_{corot} = 0.66$, and the amplitude of the disturbance has here the broad maximum in the vicinity of the corotation radius but is very small in the central part of the disk. Remarkably, the eigenmodes in the both cases have an explicit spiral form in spite of the rather small increment of the instability ($\text{Im }\omega/\text{Re }\omega \ll 1$).

---

[36] Note that W in the definition of the parameter $t = T_{rot}/|W|$ includes in that case (apart from the potential energy of the disk and the energy of interaction between the disk and the halo) also the potential energy of the halo itself. Then, t is here defined not for the infinitely thin disk but for the finite-thickness disk corresponding to the given pressure of a gas (this results in some increase in the value of t).

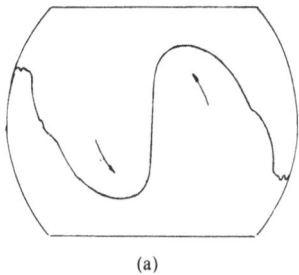
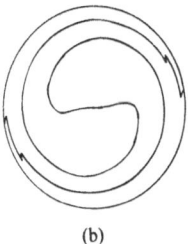

(a)   (b)

Figure 84. The most unstable two-armed spiral modes for the disk models corresponding to Figs. 82^a (a) and 82^b (b) [56^{ad}].

Besides that, from the consideration of different distributions of the density it turns out to be possible to state the dependence of the growth rate on the relative degree of the mass concentration to the center in the halo and in the disk (speaking more exactly, the trends of this dependence since only the rather weakly nonuniformly rotating disks and the halo's models with sufficiently smooth variation of the density in the radius are considered). It turned out that at least for the models investigated in [56^{ad}] the central condensation of the halo decreases somewhat the stabilizing effect for the bar-like mode. The reason, according to Bardeen [56^{ad}] is the following. The central condensation does help damp the central bar, but simultaneously the angular velocity of the spiral pattern is increased so that the outer Lindblad resonance moves inwards the disk and has a stronger destabilizing effect.

4.5.3. Comparison with Stellar Disks. In concluding this section, we say a few words about the relevance of the stability investigations for gas and stellar disks. It is necessary to do so because, for example, if we speak about the possible applications of the results of the linear large-scale stability investigations of gas disks to the galactic spirals (the way borne in mind by Bardeen [56^{ad}]), then such applications are possible only if the likeness in the behaviors of the corresponding disturbances for gas and stellar disks is present. Indeed, the largest part of the mass of spiral galaxies is contained in the stars. So the gas disk as the model of the whole galaxy is first of all the simplified model of just the stellar disk of this galaxy. As to the disturbances in the proper gas component of the real galaxy, they are, according to the observations, essentially nonlinear (in particular, in the spiral arms the large-scale shock waves are formed).

As we know, in gas (or fluid) systems secular instability may also develop apart from the dynamical instability investigated above. The former become possible in the case when on some sequence of the equilibrium states the bifurcation point with another sequence of equilibria, having more small total energy under the fixed total angular momentum and mass, occurs. The growth rates of the secular instability are proportional to the viscosity ν of a gas. One can show [56^{ad}] that the secular instability of uniformly rotating gas disks (without a halo) occurs at $t = 0.125$, which exactly coincides with the point of the stability loss for the collisionless disk. The latter may obviously have only dynamical instability. A similar coincidence occurs also for more general models. Thus, the instability (dynamical) of a stellar disk is analogous to the secular instability of a gas disk. On increasing the halo mass, the difference between the dynamical and secular instabilities decreases.

Above, we saw that gas disks with a moderate degree of mass concentration to the center are more stable (dynamically) than corresponding stellar disks. However, in contrast, strongly centrally condensed stellar disks having an inner Lindblad resonance may be more stable compared with gas disks considered in this section. Note in this connection the work [91^{ad}] of Zang, devoted to the stability investigation of the disk models with the distribution functions $f_0 = $ const $\cdot L_z^q e^{-E/\sigma_r^2}$ [a partial case of the models (10) §1]. In these disks the velocity of rotation is not dependent on the radius, $V_0(r) = V_0 = $ const; the surface density is singular, $\sigma_0(r) = V_0^2/2\pi Gr$, so that the mass increases linearly with r. The two last circumstances lead to the necessity of introducing some suitable truncations for small and for large r. It turns out that the axisymmetric ($m = 0$) modes are globally stabilized when the velocity dispersion $\sigma_r \geq 0.378V_0$, in good agreement with the local criterion of Toomre. It is, however, interesting that already for such a critical value of the velocity dispersion (i.e., for $Q = 1$) Zang was unable to find any two- or multiarmed (i.e., $m \geq 2$) instabilities despite much systematic searching. Instead of this, Zang in this case found strongly unstable one-armed ($m = 1$) modes. It is, however, possible that some unexpected results (in particular those connected with two-armed modes) are due simply to the peculiarities of the computer algorithm assumed in [91^{ad}], and principally to the manner of cutoff for smaller r.

Problems

1. Investigate the conditions of the existence of a continuous frequency spectrum of oscillations of "cold" disks with a poloidal magnetic field (cf. §3).

Solution. Rewrite the system (1)–(4) (§3) in the form of two equations:

$$s^2(\lambda^2 - \chi_H^2) \int_1^\xi \xi \sum d\xi = \frac{s^3(1 - \xi^2)}{\xi} \frac{\partial \psi}{\partial \xi} - s^2\sqrt{1 - \xi^2} H_0 J,$$

$$\xi s^2 H - sH_0 \sum \xi = (H_0's - H_0s') \int_1^\xi \xi \sum d\xi. \tag{1}$$

To analyze the behavior of eigenfrequencies λ, we make use of the "matrix" method [95, 229]. We seek solutions for the potential, density, current, and magnetic field in the form of the following expansions:

$$\psi(\xi) = \sum_{n=1}^\infty \Phi_{2n} P_{2n}(\xi), \qquad \xi \sum = \sum_{n=1}^\infty \frac{\Phi_{2n}}{\gamma_{2n}} P_{2n}(\xi),$$

$$J = \frac{\sqrt{1 - \xi^2}}{\xi} \sum_{n=1}^\infty g_{2n-1} P_{2n-1}'(\xi), \qquad H = \frac{1}{\xi} \sum_{n=1}^\infty g_{2n-1} \beta_{2n-1} P_{2n-1}(\xi), \tag{2}$$

where Φ_{2n} and g_{2n-1} are the unknown coefficients. The relations $\psi \leftrightarrow \sum$ and $J \leftrightarrow H$ are easily established [95, 228] from the conditions of matching on the disk surface of vacuum solutions for ψ and H,

$$\gamma_{2n} = \tfrac{1}{2}\left[\frac{(2n-1)!!}{(2n)!!}\right]^2.$$

By using the relation

$$\int_1^\xi \xi \sum d\xi = \int_1^\xi \sum_{n=1}^\infty \frac{\Phi_{2n}}{\gamma_{2n}} p_{2n}\, d\xi = -(1-\xi^2)\sum_{n=1}^\infty \frac{\Phi_{2n}}{\gamma_{2n}} p'_{2n},$$

we obtain

$$-s^2(\lambda^2 - \chi_H^2)\sum_{n=1}^\infty \frac{\Phi_{2n}}{\gamma'_{2n}} p'_{2n} = \frac{s^3}{3}\sum_{n=1}^\infty \Phi_{2n} p'_{2n} - \frac{s^2 H_0}{\xi}\sum_{n=1}^\infty g_{2n-1} p'_{2n-1},$$

$$\xi s^2 H - s H_0 \sum_{n=1}^\infty \frac{\Phi_{2n}}{\gamma_{2n}} p_{2n} = -(H'_0 s - H_0 s')(1-\xi^2)\sum_{n=1}^\infty \frac{\Phi_{2n}}{\gamma_{2n}} p'_{2n}. \tag{3}$$

Let S and H_0 be represented in the form of series in degrees ξ:

$$s^2(\lambda^2 - \chi_H^2) = \sum_{j=0}^N A_j \xi^{2j}, \qquad \frac{s^3}{3} = \sum_{j=0}^N B_j \xi^{2j+2},$$

$$s^2 H_0 = \sum_{j=0}^N c_j \xi^{2j+2}, \qquad \xi s^2 = \xi \sum_{j=0}^N D_j \xi^{2j+2}, \tag{4}$$

$$s H_0 = \xi \sum_{j=0}^N F_j \xi^{2j}, \qquad (1-\xi^2)(H'_0 s - H_0 s') = \sum_{j=0}^N G_j \xi^{2j}.$$

Any model with an arbitrary s and H_0 may with desired accuracy be approximated by choosing a sufficiently large N.

The further procedure is as follows: substitute (2) and (4) into (1) and go in the two equations to p'_{2n}; here it is necessary to make use of recurrent relations (for normalized Legendre polynomials):

$$\xi^2 p'_{2n} = a_n p'_{2n+2} + b_n p'_{2n} + c_n p'_{2n-2},$$

$$a_n = \frac{2n(2n+1)}{x_{n+1}}, \qquad b_n = \frac{8n^2 + 4n - 3}{(4n+3)(4n-1)}, \qquad c_n = \frac{2n(n+1)}{x_n},$$

$$x_n = (4n-1)[(4n+1)(4n-3)]^{1/2},$$

$$p_{2n-1} = e_n p'_{2n} - d_n p'_{2n-2},$$

$$e_n = [(4n-1)(4n+1)]^{-1/2}, \qquad d_n = [(4n-1)(4n-3)]^{-1/2},$$

$$\xi p_{2n} = u_n p'_{2n+2} + v_n p'_{2n} + z_n p'_{2n-2},$$

$$u_n = \frac{(2n+1)}{x_{n+1}}, \qquad v_n = [(4n-1)(4n+3)]^{-1}, \qquad z_n = -\frac{2n}{x_n},$$

$$\xi p'_{2n-1} = f_n p'_{2n} + l_n p'_{2n-2},$$

$$f_n = \frac{2n-1}{[(4n-1)(4n+1)]^{1/2}}, \qquad l_n = \frac{2n}{[(4n-1)(4n-3)]^{1/2}}.$$

By equalling the coefficients for equal p'_{2n}, we obtain the system of two difference equations with respect to g_{2n-1} and Φ'_{2n}. Further, we will be interested in the asymptotic form of Φ_{2n} and g_{2n-1} at $n \to \infty$; therefore, we write the equations by retaining in the coefficients at Φ_{2n} and g_{2n-1} only the terms of higher orders in $1/n$:

$$\sum_j B_j 4^{-(1+j)}(E + 2 + E^{-1})^{j+1}\Phi_{2n} - \tfrac{1}{2}\sum_j C_j 4^{-j}(E + 2 + E^{-1})^j(g_{2n-1} + g_{2n+1})$$

$$+ \frac{1}{n}\left[\frac{\pi}{2}\sum_j A_j 4^{-j}(E + 2 + E^{-1})^j\Phi_{2n} + \tfrac{1}{4}\sum_j C_j 4^{-j}(E + 2 + E^{-1})^j(g_{2n-1} - g_{2n+1})\right.$$

$$- \tfrac{1}{2}\sum_j C_j(4E)^{-j}j(E + 1)^{2j-1}(E - 1)(g_{2n-1} + g_{2n+1})$$

$$\left. + \sum_j B_j(4E)^{-(j+1)}(E + 1)^{2j+1}(E - 1)(j + 1)\Phi_{2n}\right] + O\!\left(\frac{1}{n^2}\right) = 0, \tag{5}$$

$$\sum_j D_j(E + 2 + E^{-1})^{j+1}4^{-(1+j)}(g_{2n-1} - g_{2n+1}) - \frac{\pi}{2}\sum_j F_j(E + 2 + E^{-1})^j 4^j$$

$$\times (\Phi_{2n-2} - \Phi_{2n+2}) + \frac{1}{n}\left[\sum_j D_j 4^{-(1+j)}(E + 2 + E^{-1})^{j+1}\left(-\frac{g_{2n+1}}{4} - \frac{g_{2n-1}}{4}\right)\right.$$

$$+ \sum_j D_j(4E)^{-1-j}(j + 1)(E - 1)(E + 1)^{2j+1}(g_{2n-1} - g_{2n+1})$$

$$+ \frac{\pi}{2}\sum_j F_j(4E)^{-j}(E + 1)^{2j-1}j(E - 1)(\Phi_{2n+2} - \Phi_{2n-2})$$

$$+ \frac{\pi}{2}\sum_j F_j 4^{-j}(E + 2 + E^{-1})^j(\tfrac{3}{4}\Phi_{2n-2} - \tfrac{1}{2}\Phi_{2n} + \tfrac{3}{4}\Phi_{2n+2})$$

$$\left. + \pi\sum_j G_j 4^{-j}(E + 2 + E^{-1})^j\Phi_{2n}\right] + O\!\left(\frac{1}{n^2}\right) = 0. \tag{6}$$

Here, as in [228], the operator E is introduced:

$$E\Phi_{2n} = \Phi_{2n+2}, \qquad Eg_{2n-1} = g_{2n+1}.$$

The system (5), (6) must be solved with the boundary condition $\Phi_{2n}, g_{2n-1} \to 0$ as $n \to \infty$.

In the higher order of magnitude in $1/n$, we have the system of difference equations with constant coefficients. Following the general theory, let us seek the solution of these equations in the form

$$\Phi_{2n} = \Phi t^{2n}, \qquad g_{2n-1} = g t^{2n-1}. \tag{7}$$

Substituting (7) into (5) and (6), we obtain

$$\Phi\sum_j B_j \zeta^{2j+2} - g\sum_j C_j \zeta^{2j+1} = 0, \tag{8}$$

$$g\sum_j D_j \zeta^{2j+2} - \Phi\pi\sum_j F_j \zeta^{2j+1} = 0, \tag{9}$$

where

$$\zeta = \frac{t^2 + 1}{2t}. \tag{10}$$

Comparing with (4), one can write

$$s(\zeta)g - H_0(\zeta)\Phi = 0, \qquad \pi H_0(\zeta)g - s(\zeta)\Phi = 0. \tag{11}$$

The condition of nontriviality of the solution yields

$$s^2(\zeta) - \pi H_0^2(\zeta) = 0. \tag{12}$$

The value of t is easily found from (10). Let $\zeta = \zeta_0$ be the root of Eq. (12). Then from (10), we have the following characteristic equation:

$$t^2 - 2t\zeta_0 + 1 = 0, \qquad t_{1,2} = \zeta_0 \pm \sqrt{\zeta_0^2 - 1}. \tag{13}$$

Each root of ζ_0 has a pair of values of $t: t_1$ and t_2. There are two cases possible.

(1) If all $\zeta_0 > 1$, then we have two series of real solutions for t, and for one series $t_1 < 1$, while for the other $t_2 > 1$. The values $t = t_1$ give decreasing solutions for g_{2n-1} and Φ_{2n} and, consequently, satisfy the boundary conditions. The exponential decreasing of the solutions corresponds to the discrete set of eigenvalues [228].

(2) There is a root $\zeta_0 < 1$, which just corresponds to the vanishing of the value $s^2 - \pi H_0^2$ on the disk. As we have already noted, the presence of the root $\zeta_0 < 1$ always takes place for models with the self-consistent magnetic field. In this case $|t_1| = |t_2| = 1$, and, to clarify the asymptotic behavior of Φ_{2n} and g_{2n-1}, one needs to take into account the terms of order n^{-1} in Eqs. (5) and (6) (cf. [228]).

Let us seek the solution in the form

$$\Phi_{2n} = \Phi(2n)^\alpha e^{\pm 2in\beta}, \qquad g_{2n-1} = g(2n-1)^\alpha e^{\pm(2n-1)i\beta}, \tag{14}$$

with the unknown α and β. The substitution of (14) into (5) and (6) and setting the determinant of the system, similar to (11), equal to zero yields

$$\cos\beta = \xi_0, \qquad \alpha = -\tfrac{1}{2} \mp i\pi \frac{s(\lambda^2 - \chi_H^2)}{\sqrt{1-\xi^2}} \frac{\zeta_0}{2ss' - 2\pi H_0 H_0'}. \tag{15}$$

In (15), the values of s and H_0 are taken at the point $\xi = \xi_0 \equiv \zeta_0$, in which $s^2(\xi) = \pi H_0^2(\xi)$.

We have obtained the negative real part of λ at any values of λ^2. This means that the solution of (14) satisfies the needed boundary conditions $\Phi_{2n} \to 0, g_{2n-1} \to 0$ as $n \to \infty$ for any λ^2 ($-\infty < \lambda^2 < \infty$). Consequently, the decreasing solution exists throughout the range of real values of λ^2, and we have a continuous spectrum. Instability of the disk with a self-consistent magnetic field is proved for any wavelength.

Being aware of the solution of (14) for Φ_{2n} and g_{2n-1}, we can easily sum the series in (2) for larger n. As a result, we obtain

$$\psi_{asymp} \sim \frac{1}{\sqrt{|\xi - \xi_0|}} \exp\left[\pm i \frac{(\lambda^2 - \chi_H^2)\pi\xi_0 s(\xi_0)}{\sqrt{1-\xi_0^2}} \frac{\ln|\xi - \xi_0|}{2(ss' - \pi H_0 H_0')|_{\xi=\xi_0}}\right], \tag{16}$$

where, for $\lambda^2 < \chi_H^2(\xi_0)$, ψ_{asymp} is defined by expression (16) in LUR and is zero in LSR. At $\lambda^2 > \chi_H^2(\xi_0)$, we face the opposite case.[37]

Formula (16) for ψ does not take into account harmonics with lower n; it just leads to localization of perturbations. From physical considerations, it is clear that perturbations with a mean wavelength (averaged over small-scale oscillations near $\xi = \xi_0$) of the order

---

[37] Note that expression (16) for ψ_{asymp} ensues also from the WKB analysis, taking into account the pre-exponential terms.

of the size of the unstable region will excite the entire disk even if the size of this unstable region is much less than the disk size [note also that, for $\chi_H^2(\xi_0) < 0$, the LSR turns out to be really unstable].

In conclusion, it is necessary to make a further observation. As is seen from formula (16), in the case of a continuous spectrum, there is uncertainty of the mass in the vicinity of the point $\xi = \xi_0$ at which $s^2 = \pi H_0^2$. This circumstance is associated with the selection of an idealized model in the form of an infinitely thin cold disk. If, for example, one introduces into the equations a term with pressure in the form $P \sim \Sigma\sigma^\gamma$ (assuming Σ to be as small as desired), then this uncertainty is eliminated. Here, the shortest waves prove to be stabilized, but they do not present danger for the disk stability due to their localization. The continuous spectrum becomes discrete with very close values of frequencies. Indeed, the presence of the continuous spectrum is due to divergence of the integral of the type of (24), §3 (defining the phase of perturbations) which vanishes as an arbitrary small pressure is introduced.

Considerations similar to those in Section 2.1 allow one to assume that the development of perturbations in time in the real problem with initial conditions must in this case cause production of a ring with increased density near $\xi = \xi_0$.

2. Calculate the eigenfunction of the problem of small perturbations of the disks in (6), §1, for any simplest large-scale mode.

Solution. Consider the simplest non-axial-symmetrical mode ($m = 1, n = 2$): the parameter γ may be arbitrary. Prove that the eigenfunction in this case is proportional to

$$\Phi \sim P_3^1(\eta) \sim re^{i\varphi}(r^2 - \tfrac{4}{5}). \tag{1}$$

We seek the solution in the form

$$\Phi = re^{i\varphi}(r^2 + \alpha) \qquad (\alpha = \text{const}). \tag{2}$$

For the χ function, by solving the kinetic equation, we obtain the following expression:

$$\chi = -2(1 - \gamma^2)A^{(2)} + A^{(1)}rv^2 - 2\,dr, \tag{3}$$

where

$$A^{(1)} = \frac{4}{5}\frac{-10(\omega')^2 - 20\gamma\omega' + 30\gamma^2}{[(\omega')^2 - 1][(\omega')^2 - 9]}, \tag{4}$$

$$A^{(2)} = -\frac{3(\omega')^2 + 4\gamma\omega' - 6\gamma^2 - 9}{[(\omega')^2 - 1][(\omega')^2 - 9]}, \tag{5}$$

$$d = -\frac{1 - \gamma^2}{(\omega')^2 - 1} \qquad (\omega' \equiv \omega + \gamma). \tag{6}$$

Now we calculate $\sigma^{(1)}$ and $\sigma^{(2)}$ by formulae (3) of Section 4.3:

$$\sigma^{(1)} = \frac{\sigma_0 A^{(1)}r(1 - r^2)}{\sqrt{1 - r^2}}, \tag{7}$$

$$\sigma^{(2)} = \sigma_0\sqrt{1 - r^2} - \left[2A^{(2)}r^3 + \frac{2d\alpha r}{1 - \gamma^2}\right]e^{i\varphi} - \sigma_0\sqrt{1 - r^2}, \tag{8}$$

$$\sigma = \sigma^{(1)} + \sigma^{(2)} = \sigma_0\left\{\eta + \frac{e^{i\varphi}}{\eta}\left[r^2\left(A^{(2)} + \frac{d\alpha}{1 - \gamma^2}\right) - (A^{(1)} + A^{(2)})r^3\right.\right.$$

$$\left.\left. + \left(A^{(1)} - \frac{d\alpha}{1 - \gamma^2}\right)r\right]\right\},$$

where

$$\eta = \sqrt{1 - \frac{r^2}{R^2}}, \qquad R^2 = 1 - \left[2A^{(2)} + \frac{2d\alpha}{1 - \gamma^2}\right]e^{i\varphi}. \tag{9}$$

This, first of all, yields the dispersion equation

$$4\lambda_3^1(A^{(1)} + A^{(2)}) = 1, \qquad 4\lambda_3^1 = \tfrac{3}{8}. \tag{10}$$

As may be proved, it is reduced to Eq. (12), Section 4.4. We should verify [cf. formula (1)] that

$$4(A^{(1)} + A^{(2)}) = 5\left(A^{(1)} - \frac{d\alpha}{1 - \gamma^2}\right), \tag{11}$$

$$(A^{(1)} + A^{(2)}) = 5\left(A^{(2)} + \frac{d\alpha}{1 - \gamma^2}\right). \tag{12}$$

By a direct test, one can make sure that relations (11) and (12) are indeed satisfied.

3. Same as for the anisotropic disk in (9), §1.

Solution. We construct the perturbation theory using a standard sample. Assume the perturbation distribution function to be of the form

$$f = \frac{\sigma_0}{\pi}\,\theta[(1 - r^2)(1 - v_\varphi^2) - v_r^2 + \varepsilon\chi]. \tag{1}$$

The linearized equation for the function χ is the following:

$$\frac{\partial\chi}{\partial t} = -2v_r\,\frac{\partial\Phi}{\partial r} - 2(1 - r^2)v_\varphi\,\frac{\partial\Phi}{r\,\partial\varphi}. \tag{2}$$

Consider, as an example, one of the simplest oscillation modes: the radial perturbation of the form

$$\Phi - r^4 + ar^2 + b. \tag{3}$$

The solution of Eq. (2) is found by integrating over unperturbed trajectories:

$$\chi = 4r^4\left(\frac{1}{\omega^2 - 16} + \frac{1}{\omega^2 - 4}\right) + 4ar^2\,\frac{1}{\omega^2 - 4} + 4\left(\frac{1}{\omega^2 - 16} - \frac{1}{\omega^2 - 4}\right)$$

$$\times (v_r^4 + v_\varphi^4 + 2v_r^2 v_\varphi^2) - 16\,\frac{1}{\omega^2 - 16}\,(\tfrac{3}{2}r^2 v_r^2 + \tfrac{1}{2}r^2 v_\varphi^2) - \frac{4a}{\omega^2 - 4}\,(v_r^2 + v_\varphi^2). \tag{4}$$

The perturbed surface density corresponding to (4) is, as usual, composed of two parts:

$$\sigma_1^{(2)} = \sigma_0\sqrt{1 - r^2 + \varepsilon\chi(0, 0)} - \sigma_0\sqrt{1 - r^2}, \tag{5}$$

$$\sigma_1^{(1)} = \frac{\sigma_0}{2\pi}\iint \theta[(1 - r^2)(1 - v_\varphi^2) - v_r^2]$$

$$\times \left\{\frac{\chi(0, 0) + (\partial/\partial v_\varphi)[\chi(0, v_\varphi)/v_\varphi]}{1 - r^2} + \frac{\partial}{\partial v_r}\left(\frac{\chi_1}{v_r}\right)\right\}dv_r\,dv_\varphi, \tag{6}$$

where

$$\chi(0, 0) = \chi(v_r = 0, v_\varphi = 0),$$

$$\chi(0, 0) = 4r^4\left(\frac{1}{\omega^2 - 16} + \frac{1}{\omega^2 - 4}\right) + 4ar^2 \frac{1}{\omega^2 - 4} \equiv pr^4 + qr^2,$$

$$\chi(0, v_\varphi) = 4\left(\frac{1}{\omega^2 - 16} - \frac{1}{\omega^2 - 4}\right)v_\varphi^4 - \frac{16}{\omega^2 - 16}\frac{r^2v_\varphi^2}{2} - \frac{4a}{\omega^2 - 4}v_\varphi^2,$$

$$\chi_1 = \chi - \chi(0, 0) - \chi(0, v_\varphi).$$

Calculating $\sigma_1^{(1)}$, according to (6), we obtain

$$\sigma_1^{(1)} = \frac{\sigma_0}{2}\left(\frac{1}{\omega^2 - 4} - \frac{1}{\omega^2 - 16}\right)\frac{1}{\sqrt{1 - r^2}} + \frac{\sigma_0}{2}\sqrt{1 - r^2}\left[\left(\frac{9}{\omega^2 - 16} - \frac{9 + 8a}{\omega^2 - 4}\right)\right.$$

$$\left. - r^2\left(\frac{31}{\omega^2 - 16} + \frac{1}{\omega^2 - 4}\right)\right] \equiv \frac{\alpha}{\sqrt{1 - r^2}} + \beta\sqrt{1 - r^2} + \gamma r^2\sqrt{1 - r^2}. \quad (7)$$

The potential perturbation corresponding to σ_1 will be calculated by the formula

$$\Phi = -4G\int_0^r \frac{ds}{\sqrt{r^2 - s^2}} \int_s^a \frac{r'\sigma_1(r')\, dr'}{\sqrt{r'^2 - s^2}}, \quad (8)$$

where the disk radius a is defined from the condition

$$1 - a^2 + \varepsilon\chi(0, 0)|_{r=1} = 0; \quad (9)$$

therefore,

$$a \approx 1 + \tfrac{1}{2}\varepsilon\chi(0, 0)|_{r=1}. \quad (10)$$

We calculate first of all the contribution to Φ by $\sigma_1^{(1)}$:

$$\Phi^{(1)} = -4G\pi^2\left(\frac{1}{4}\alpha + \frac{1}{8}\beta - \frac{1}{16}\beta r^2 + \frac{1}{32}\gamma + \frac{1}{32}\gamma r^2 - \frac{9}{8\cdot 32}\gamma r^4\right). \quad (11)$$

We calculate the contribution to Φ of $\sigma_1^{(2)}$ by the formula

$$\Phi^{(2)} = -4G\sigma_0\frac{\varepsilon}{2}\int_0^r \frac{ds}{\sqrt{r^2 - s^2}} \int_s^1 \frac{r'\chi_0(r')\, dr'}{\sqrt{(1 - r'^2)(r'^2 - s^2)}}$$

$$= -4G\sigma_0\frac{\varepsilon}{4}\pi^2\left[\left(\frac{3}{16}p + \frac{1}{4}q\right) + \left(\frac{1}{16}p + \frac{1}{8}q\right)r^2 + \frac{9}{8\cdot 16}pr^4\right]. \quad (12)$$

The full potential perturbation is $\Phi = \Phi^{(1)} + \Phi^{(2)}$. Comparing this expression for Φ with the initial equation (3), we obtain the dispersion equation for a given mode,

$$\frac{45}{64}\left(\frac{7}{\omega^2 - 16} + \frac{1}{\omega^2 - 4}\right) = -1, \quad (13)$$

as well as the following equations for the coefficients a and b:

$$a = -2(\tfrac{1}{8}\gamma - \tfrac{1}{4}\beta + \tfrac{1}{16}p + \tfrac{1}{8}q), \quad (14)$$

$$b = -2(\alpha + \tfrac{1}{2}\beta + \tfrac{1}{8}\gamma + \tfrac{3}{16}p + \tfrac{1}{4}q). \quad (15)$$

Equation (13) is reduced to the biquadratic one

$$16\omega^4 - 230\omega^2 + 529 = 0. \tag{16}$$

Hence $\omega_{1,2}^2 = (115 \pm 69)/16$; $\omega_1^2 = 23/2$ and $\omega_2^2 = 23/8$, so that the mode in question is stable. From Eq. (14), we obtain the expression for a:

$$a = \frac{3}{2} \frac{2\omega^2 + 13}{(\omega^2 - 16)(\omega^2 - 1)}, \tag{17}$$

which, by taking into account Eq. (13), yields $a = -8/7$. In a similar way, from (15), we have $b = 8/35$. Thus, as was to be expected, also in this case we have the already familiar eigenfunctions for a given mode:

$$\Phi \sim r^4 - \tfrac{8}{7}r^2 + \tfrac{8}{35} \sim P_4^0(\sqrt{1 - r^2}).$$

For the mode $\Phi \sim P_2(\sqrt{1 - r^2})$ it is easy to obtain (from expressions written above if one retains in them only the terms containing r^2) the respective frequency of oscillations $\omega^2 = 1$.

4. Derive the dispersion equation for short-wave oscillations of a sufficiently cold $(1 - \gamma^2 \ll 1)$ disk (6), §1.

Solution. The derivation is performed in a quite similar way to the one made by Toomre [333] or Lin, Yuan, and Shu [271] (cf. Section 4.1), but, only owing to the singular behavior of function (6), Section 1.4, is the linearization performed, for example, according to the scheme adopted in Section 4.4.

In such an analysis, it is assumed (apart from the natural $\lambda \ll R$) that the movement of the particles is sufficiently close to the circular one $c_{r,\varphi} \ll \Omega r$; however, it is not required that the parameter $kc_{r,\varphi}/\Omega$ be small (for the distribution functions $1 - \gamma^2 \ll 1$ under consideration).

The general equation ensuing for the χ function becomes essentially simplified if one is interested in short-wave oscillations $(\Phi_1 \sim e^{ikr}\Phi; \; kr \gg m)$ of a sufficiently cold disk $(1 - \gamma^2 \ll 1)$. It is reduced then to the well-known equation

$$-i(v - kv_r)\chi + \left(v_r \frac{\partial\chi}{\partial v_\varphi} - v_\varphi \frac{\partial\chi}{\partial v_r}\right) = 2ik'v_r\Phi, \tag{1}$$

where $v \equiv (\omega - m)/2$, $k' \equiv k/2$. This equation is readily solved by the method of "integration over angle." The solution is easily written in the form used by Lin, Yuan, and Shu [271] (Section 4.1):

$$\chi = 2\Phi Q, \tag{2}$$

where

$$Q = 1 - \frac{v\pi}{\sin v\pi} \cdot \frac{1}{2\pi} \int_{-\pi}^{\pi} \exp\{i[vs + k'v_r \cdot \sin s - k'v_\varphi(1 + \cos s)]\}\, ds. \tag{3}$$

Now one needs to calculate the perturbation of the surface density σ_1, by using, for example, formulae (5) and (6) of Section 4.4. Since in this case $\chi_1 = \chi$, so we get

$$\sigma_1 = \frac{c_0 \Phi}{2\pi} \frac{v\pi}{\sin v\pi} \int_0^{2\pi} \int_0^\infty d\alpha\, dv \int_{-\pi}^{\pi} ds \exp\left\{i\left[vs - 2vk'\cos\frac{s}{2}\cos\left(\frac{s}{2} + \alpha\right)\right]\right\}$$

$$\times i\left(-2k'\cos\frac{s}{2}\right)\cos\left(\frac{s}{2} + \alpha\right). \tag{4}$$

In (4), one can perform integration over velocities. For that purpose, rewrite the integrals in (4) in the form (changing the order of integration)

$$-\int_{-\pi}^{\pi} e^{ivs}\, ds \int_0^{\varkappa} dv\, \frac{iB}{\sqrt{\varkappa^2 - v^2}} \int_0^{2\pi} d\alpha\, e^{iBv\cos(s/2+\alpha)}\cos\left(\frac{s}{2}+\alpha\right),\qquad(5)$$

where $B \equiv -2k'\cos s/2$. The internal integral is

$$\int_0^{2\pi} d\alpha\, e^{iBv\cos\alpha}\cos\alpha = \int_0^{2\pi} d\alpha \cos\alpha\left[1 + 2\sum_{k=1}^{\infty} i^k J_k(Bv)\cos k\alpha\right] = 2\pi i J_1(Bv).\qquad(6)$$

The integral over v is also known:

$$\int_0^{\varkappa} dv\, \frac{J_1(Bv)}{\sqrt{\varkappa^2 - v^2}} = \frac{\pi}{2}J_{1/2}^2\left(\frac{B\varkappa}{2}\right) = \frac{2}{B\varkappa}\sin^2\left(\frac{B\varkappa}{2}\right),\qquad(7)$$

so that the perturbed density is expressed by Φ in the following manner:

$$\sigma_1 = \Phi c_0\, \frac{v\pi}{\sin v\pi}\frac{2}{\varkappa}\int_{-\pi}^{\pi} ds\, e^{ivs}\sin^2\left(\varkappa k'\cos\frac{s}{2}\right).\qquad(8)$$

Within the short-wave limit, instead of the integral, in the general case, the relation between Φ and σ_1, the local relation

$$\sigma_1 = -\frac{k}{2\pi G}\Phi\qquad(9)$$

takes place.

Now, from (8) and (9), we finally obtain the following simple dispersion equation:

$$\frac{k_1\sqrt{1-\gamma^2}}{G\sigma_0(0)} = \frac{v\pi}{\sin v\pi}\int_{-\pi}^{\pi} ds\, e^{ivs}\sin^2\left(k_1\cos\frac{s}{2}\right),\qquad(10)$$

where

$$k_1 \equiv k'\varkappa = \frac{k}{2}\sqrt{(1-\gamma^2)(1-r^2)}.$$

In the particular case of a cold disk $|\gamma| \to 1$ this yields the Toomre dispersion equation (in the usual notation).

$$(\omega - m\Omega_0)^2 = 4\Omega_0^2 - 2\pi G\sigma_0(r)k_0.\qquad(11)$$

Find now, similarly to [333] (§4.1), the marginal curve of dispersion equation (10). It results from the assumption in (10) that $v^2 = 0$:

$$\frac{\sqrt{1-\gamma^2}}{2\pi\sigma_0 G} = \frac{\pi}{4}\sqrt{1-\gamma^2} = \frac{[1 - J_0(2k_1)]}{2k_1} \equiv F(2k_1),\qquad(12)$$

where $J_0(x)$ is the Bessel function of the zero order of the first kind. The plot $F(2k_1)$ is given in Fig. 85. The points of the curve $F(2k_1) = \frac{1}{4}\pi\sqrt{1-\gamma^2}$ give the pairs of the values of $(2k_1, \gamma)$ corresponding to indifferently equilibrium perturbations. One may, as in [333] assume (Section 4.1) that the curve in Fig. 85 separates the regions of stability (upwards from the curve) and instability (below the curve). From this figure it is evident that for γ, satisfying the inequality $F_{max} < \frac{1}{4}\pi\sqrt{1-\gamma^2}$, the disk is unstable. Since $F_{max} \approx 0.41$ $(2k_c \approx 2.75)$, then the critical value of $\gamma_c \approx 0.85$. Thus, it is seen that only the disks with trajectories of particles very close to circular are unstable with respect to

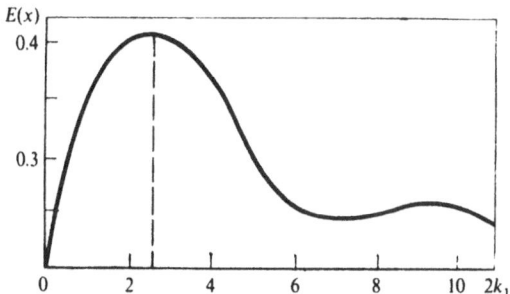

Figure 85. The marginal stability curve for the disks (6), Section 1.4 calculated in the short wave approximation.

shortwave perturbations. We have, however, seen that the most dangerous for the disks with $\gamma \neq 1$ are just the large-scale (nonaxially symmetrical) perturbations leading to instability of the disk models in question at any value of the parameter γ. Relation (12) [as the dispersion equation itself in (10)] resembles very much that obtained by Toomre for marginal stability of the "Schwarzschild" disk (cf. Section 4.1).

5. Derive the equation for oscillation frequencies of a uniformly rotating disk being inside the "halo" of uniform density which makes a contribution to the equilibrium of the disk but is not subjected to any perturbations [252].

Solution. For the sake of simplicity, assume that the halo has a spherical shape and the same radius as the disk. Elementary calculation shows that the part of the equilibrium force which is produced by the disk itself is

$$\eta = \frac{M_D}{M_D + (4/3\pi)M_H},\tag{1}$$

where M_D and M_H are the mass of the disk and the halo, respectively. If, as in [252], the dispersion equation for one disk in (6), §1, is represented in the form of

$$\lambda(n, m, \gamma, \omega) = 1,\tag{2}$$

then in this case we have

$$\eta\lambda(n, m, \gamma, \omega) = 1.\tag{3}$$

In order that the model B_1 might be stabilized, it is enough that $\eta = 0.9221$, and, in order that the model B_3 might be stabilized, we need $\eta = 0.7793$. These η correspond to 17% and 40% of the total mass in the halo of the system. The presence of the halo allows one to construct stable disks colder than the model A without a halo. Since the parameter of Toomre Q changes $\sim 1/\eta$, then the new $\check{Q}$ for B_1 is 1.446 and for B_3 is 1.460, while, for the model A, Q is 1.333.

References

[1]   T. A. Agekyan, *Stars, Galaxies, Metagalaxy*, Nauka, Moscow, 1966 (in Russian).

[2]   T. A. Agekyan, Vestn. Leningr. Univ. **1**, 152 (1962) (in Russian).

[3]   C. W. Allen, *Astrophysical Quantities*, 3d ed., Athlone Press, London.

[4]   V. A. Antonov, Astron. Zh. **37**, 918 (1960) [Sov. Astron. **4**, 859 (1961)].

[5]   V. A. Antonov, Vestn. Leningr. Univ. **13**, 157 (1961) (in Russian).

[6]   V. A. Antonov, Vestn. Leningr. Univ. **19**, 96 (1962) (in Russian).

[7]   V. A. Antonov, in *Itogi Nauki, Ser. Astron.: Kinematika i Dinamika Zvezdnykh Sistem (Scientific Findings, Astron. Ser.: Kinematics and Dynamics of Stellar Systems)*, VINITI, Moscow, 1968 (in Russian).

[8]   V. A. Antonov, Uch. Zap. Leningr. Univ. No. 359, 64 (1971) (in Russian).

[9]   V. A. Antonov, Dokl. Akad. Nauk SSSR **209**(3), 584 (1973) [Sov. Phys.— Dokl. **18** (3), 159 (1973)].

[10] V. A. Antonov and E. M. Nezhinskii, Uch. Zap. Leningr. Univ. **363**, 122 (1973) (in Russian).

[11] V. A. Antonov, in *Dinamika Galaktik i Zvezdnykh Skoplenii (Dynamics of Galaxies and Star Clusters)*, Nauka, Alma-Ata, 1973 (in Russian).

[12] V. A. Antonov and S. N. Nuritdinov, Vestn. Leningr. Univ. **7**, 133 (1975) (in Russian).

[12a] V. A. Antonov and S. N. Nuritdinov, Astron. Zh. **54**, 745 (1977).

[13] V. A. Antonov, in *Itogi Nauki, Ser. Astron., T. 10: Ravnovesije i Ustoichivost' Gravitirujushchikh Sistem (Scientific Findings, Astron. Ser.): Equilibrium and Stability of Gravitating Systems)*, Nauka, Moscow, 1975 (in Russian).

445

[13a] V. A. Antonov, in *Dinamika i Evolutsija Zvezdnykh Sistem* (*Dynamics and Evolution of Stellar Systems*), VAGO, Moscow and Leningrad, 1975, p. 269 (in Russian).

[14] V. A. Antonov, Uch. Zap. Leningr. Univ., Tr. Astron. Obs. **24**, 98 (1968) (in Russian).

[15] P. Appell, *Figures d'equilibre d'une masse liquide homogene en rotation*, Russian translation, ONTI, Leningrad and Moscow, 1936.

[16] H. Arp, in *Astrofizika* (*Astrophysics*), Nauka, Moscow, 1961 (in Russian).

[17] W. Baade, *Evolution of Stars and Galaxies*, Cecilia Payne-Gaposhkin, Ed., Harvard University Press, Cambridge, Mass. (Russian translation: Mir, Moscow, 1966.)

[18] R. Bellman, *Stability Theory of Differential Equations*, McGraw-Hill, New York, 1953 (Russian translation: IL, Moscow, 1954).

[19] G. S. Bisnovatyi-Kogan, Ya. B. Zel'dovich, and A. M. Fridman, Dokl. Akad. Nauk SSSR **182**, 794 (1968) [Sov. Phys.—Dokl. **13**, 960 (1969)].

[20] G. S. Bisnovatyi-Kogan, Ya. B. Zel'dovich, R. Z. Sagdeev, and A. M. Fridman, Zh. Prikl. Mekh. Tekh. Fiz. **3**, 3 (1969) (in Russian).

[21] G. S. Bisnovatyi-Kogan and Ya. B. Zel'dovich, Astrofizika **5** (3), 425 (1969) [Astrophysics **5**(3), 198–200 (1969)].

[22] G. S. Bisnovatyi-Kogan and Ya. B. Zel'dovich, Astrofizika **5** (2), 223 (1969) [Astrophysics **5**(3), 105–109 (1969).

[23] G. S. Bisnovatyi-Kogan and Ya. B. Zel'dovich, Astrofizika **6** (3), 387 (1970) [Astrophysics **6**(3), 207–212 (1973)].

[24] G. S. Bisnovatyi-Kogan and Ya. B. Zel'dovich, Astron, Zh. **47**, 942 (1970) [Sov. Astron. **14**, 758 (1971)].

[25] G. S. Bisnovatyi-Kogan, Astrofizika **7**, 121 (1971) [Astrophysics **7**, 70 (1971)].

[26] G. S. Bisnovatyi-Kogan, Astron. Zh. **49**, 1238 (1972) [Sov. Astron. **16**, 997 (1973)].

[27] G. S. Bisnovatyi-Kogan and S. I. Blinnikov, preprint, Inst. Prikl. Mat. Akad. Nauk SSSR **34**, Moscow, 1972 (in Russian).

[28] G. S. Bisnovatyi-Kogan and A. B. Mikhailovskii, Astron. Zh. **50**, 312 (1973) [Sov. Astron. **17**, 205 (1973)].

[29] G. S. Bisnovatyi-Kogan, Pis'ma Astron. Zh. **1**(9), 3 (1975) [Sov. Astron. Lett. **1**(5), 177 (1975)].

[30] M. S. Bobrov, *Kol'tsa Saturna* (*Saturn's Rings*), Nauka, Moscow, 1970 (in Russian).

[31] A. A. Vedenov, E. P. Velikhov, and R. Z. Sagdeev, Usp. Fiz. Nauk **73**, 701 (1961) [Sov. Phys.—Usp. **4**, 332 (1961)].

[32] Yu.-I. K. Veltmann, in *Itogi Nauki, Ser. Astron.: Kinematika i Dinamika Zvezdnykh Sistem* (*Scientific Findings, Astron. Ser.: Kinematics and Dynamics of Stellar Systems*), VINITI, Moscow, 1968 (in Russian).

[33] N. Ya. Vilenkin, *Spetsial'nyje Funktsii i Teorija Group* (*Special Functions and Group Theory*), Nauka, Moscow, 1965 (in Russian).

[34] M. A. Vlasov, *Pis'ma Zh. Eksp. Teor. Fiz.* **2**, 274 (1965) [JETP Lett. **2**, 174 (1965)].

[35] G. de Vaucouleurs, in *Strojenije Zvezdnykh Sistem* (*The Structure of Stellar Systems*), IL, Moscow, 1962 (in Russian).

[36] B. A. Vorontsov-Velyaminov, *Atlas i Katalog Vzaimodeistvuyushchikh Galaktik* (*Atlas and Catalogue of Interconnected Galaxies*), Gos. Astron. Inst. imeni P. K. Shternberg, Moscow, 1959 (in Russian).

[37] B. A. Vorontsov-Velyaminov, *Vnegalakticheskaya Astronomia* (*Extragalactic Astronomy*), Nauka, Moscow, 1972 (in Russian).

[38] S. K. Vsekhsvjatskii, in *Problemy Sovremennoi Kosmogonii* (*Problems of Modern Cosmogony*), V. A. Ambartsumjan, Ed., Nauka, Moscow, 1969 (in Russian).

[39] A. A. Galeev and R. Z. Sagdeev, Vopr. Teor. Plazmy (Plasma Theory Problems) **7**, 3 (1973) (in Russian).

[40] I. F. Ginzburg, V. L. Polyachenko, and A. M. Fridman, Astron. Zh. **48**, 815 (1971) [Sov. Astron. **15**, 643 (1972)].

[41] I. M. Glazman, *Direct Methods of Qualitative Spectral Analysis*, IPST, Jerusalem, 1965 (Engl. translation).

[42] I. S. Gradshtein and I. M. Ryznik, *Table of Integrals, Series, and Products*, Academic Press, New York, 1965 (Engl. translation).

[43] L. E. Gurevich, Vopr. Kosmog. (The Problems of Cosmogony), **2**, 150 (1954) (in Russian).

[44] L. E. Gurevich, Astron. Zh. **46**, 304 (1969) [Sov. Astron, **13**, 241 (1969)].

[45] A. G. Doroshkevich and Ya. B. Zel'dovich, Astron. Zh. **40**, 807 (1963) [Sov. Astron. **7**, 615 (1964)].

[46] B. M. Dzyuba and V. B. Yakubov, Astron. Zh. **47**, 3 (1970) [Sov. Astron. **14**, 1 (1970)].

[47] Ya. B. Zel'dovich and M. A. Podurets, Astron. Zh. **42**, 963 (1965) [Sov. Astron. **9**, 742 (1966)].

[48] Ya. B. Zel'dovich and I. D. Novikov, *Relativistic Astrophysics*, translated by David Arnett, University of Chicago Press, Chicago, 1971 (Engl. translation).

[48a] Ya. B. Zel'dovich and I. D. Novikov, *Strojenije i Evolutsija Vselennoi* (*The Structure and Evolution of the Universe*), Nauka, Moscow, 1975 (in Russian).

[49] Ya. B. Zel'dovich and I. D. Novikov, Preprint, Inst. Prikl. Mat. Akad. Nauk SSSR **23**, Moscow, 1970 (in Russian).

[50] Ya. B. Zel'dovich, V. L. Polyachenko, A. M. Fridman, and I. G. Shukhman, preprint, Inst. Zemn. Magn. Ionsf. Rasprostr. Radiovoln Sibir. Otd. Akad. Nauk SSSR, No. 7–72, Irkutsk, 1972 (in Russian).

[51] G. M. Idlis, Astron. Zh. **29**, 694 (1952) (in Russian).

[52] G. M. Idlis, in *Itogi Nauki, Ser. Astron., Kinematika i Dinamika Zvezdnykh Sistem* (*Scientific Findings, Astron. Ser.: Kinematics and Dynamics of Stellar Systems*), VINITI, Moscow, 1968 (in Russian).

[53] B. B. Kadomtsev, Vopr. Teor. Plazmy (The Problems of Plasma Theory) **2**, 132 (1963) (in Russian).

[54] B. B. Kadomtsev, A. B. Mikhailovskii, and A. V. Timofeev, Zh. Eksp. Teor. Fiz. **47**, 2266 (1964) Sov. Phys.—[JETP **20**, 1517 (1965)].

[55] E. Kamke, *Differentialgleichungen Reeller Funktionen*, Chelsea Publishing Co., New York, 1947 (Russian translation: Nauka, Moscow, 1965).

[55a] V. I. Karpman, *Nelinejnyje Volny v Dispergirujushchikh Sredakh* (*Nonlinear Waves in Dispersing Media*), Nauka, Moscow, 1973 (in Russian).

[56] S. V. Kovalevskaya, in *S. V. Kovalevskaya, Nauchnyje Raboty* (*Scientific Reports*), Izd. Akad. Nauk SSSR, Moscow, 1948 (in Russian).

[57] V. I. Korchagin and L. S. Marochnik, Astron. Zh. **52**, 15 (1975) [Sov. Astron. **19**, 8 (1975)].

[58] G. G. Kuz'min, Publ. Tartus. Astron. Obs. **32**, 211 (1952) (in Russian).

[59] G. G. Kuz'min, Publ. Tartus. Astron. Obs. **35**, 285 (1956) (in Russian).

[60] G. G. Kuz'min, Astron. Zh. **33**, 27 (1956) (in Russian).

[61] G. G. Kuz'min, Izv. Akad. Nauk Eston. SSR **5**, 91 (1956) (in Russian).

[62] G. G. Kuz'min and Yu.-I. K. Veltmann, Publ. Tartus. Astron. Obs. **36**, 5 (1967) (in Russian).

[63] M. A. Lavrentjev and B. V. Shabat, *Metody Teorii Funktsii Kompleksnogo Peremennogo* (*Methods of the Theory of Functions of a Complex Variable*), Fizmatgiz, Moscow, 1958 (in Russian).

[64] H. Lamb, *Hydrodynamics*, Cambridge University Press, Cambridge, 1932 (Russian translation: Gostekhizdat, Moscow, 1947).

[65] L. D. Landau, Zh. Eksp. Teor. Fiz. **16**, 574 (1946) (in Russian).

[66] L. D. Landau, Izv. Akad. Nauk SSSR, Ser. Fiz. **17**, 51 (1953) (in Russian).

[67] L. D. Landau and E. M. Lifshitz, *Fluid Mechanics*, Pergamon Press, London, and Addison-Wesley Publishing Co., Reading, Mass., 1959 (Engl. translation).

[68] L. D. Landau and E. M. Lifshitz, *Quantum Mechanics*, Pergamon Press, Oxford and New York, and Addison-Wesley Publishing Co., Reading, Mass., 1965 (Engl. translation).

[69] L. D. Landau and E. M. Lifshitz, *Mechanics*, Pergamon Press, Oxford and New York, 1976 (Engl. translation).

[70] L. D. Landau and E. M. Lifshitz, *The Classical Theory of Fields*, Pergamon Press, Oxford and New York, 1971 (Engl. translation).

[71] N. N. Lebedev, *Problems in Mathematical Physics*, Pergamon Press, Oxford and New York, 1966 (Engl. translation).

[72] V. I. Lebedev, M. N. Maksumov, and L. S. Marochnik, Astron. Zh. **42**, 709 (1965) [Sov. Astron. **9**, 549 (1966)].

[73] B. Lindblad, in *Strojenije Zvezdnykh Sistem* (*The Structure of Stellar Systems*), IL, Moscow, 1962 (Russian translation).

[74] C. C. Lin, *The Theory of Hydrodynamic Stability*, Cambridge University Press, Cambridge, 1966 (Russian translation: IL, Moscow, 1958).

[75] E. M. Lifshitz, Zh. Eksp. Teor. Fiz. **16**, 587 (1946) (in Russian).

[76] E. M. Lifshitz and I. M. Khalatnikov, Usp. Fiz. Nauk **30**, 391 (1963) [Sov. Phys.—Usp. **6**, 495 (1964)].

[77] L. Lichtenshtein, *Figury Ravnovesija Vrashchayushchejsja Zhidkosti* (*Equilibrium Configurations of Rotating Fluid*), Nauka, Moscow, 1965 (in Russian).

[78]   A. M. Lyapunov, *Selected Works*, Izd. Akad. Nauk SSSR, Moscow, 1954–1965 (in Russian).

[78a]   R. K. Mazitov, Prikl. Mat. Tekh. Fiz. **1**, 27 (1965) (in Russian).

[79]   V. A. Mazur, A. B. Mikhailovskii, A. L. Frenkel, and I. G. Shukhman, preprint, Instituta Atomnoi Energii, No. 2693, 1976 (in Russian).

[80]   M. N. Maksumov, Dokl. Akad. Nauk Tadzh. SSR **13**, 15 (1970) (in Russian).

[81]   M. N. Maksumov, Bul. Inst. Astrofiz. Akad. Nauk Tadzh. SSR **64**, 3 (1974) (in Russian).

[82]   M. N. Maksumov, Bul. Inst. Astrofiz. Akad. Nauk Tadzh. SSR **64**, 22 (1974) (in Russian).

[83]   M. N. Maksumov and Yu. I. Mishurov, Bul. Inst. Astrofiz. Akad. Nauk Tadzh. SSR **64**, 16 (1974) (in Russian).

[84]   L. S. Marochnik and A. A. Suchkov, Usp. Fiz. Nauk **112**(2), 275 (1974) [Sov. Phys.—Usp. **17**(1), 85 (1974)].

[85]   A. B. Mikhailovskii, A. L. Frenkel', and A. M. Fridman, Zh. Eksp. Teor. Fiz. **73**, 20 (1977) [Sov. Phys.—JETP **46**, 9 (1977)].

[86]   A. B. Mikhailovskii, *Theory of Plasma Instabilities*, Consultants Bureau, New York, 1974, Vol. I (Engl. translation).

[87]   A. B. Mikhailovskii, A. M. Fridman, and Ya. G. Epel'baum, Zh. Eksp. Teor. Fiz. **59**, 1608 (1970) [Sov. Phys.—JETP **32**, 878 (1971)].

[88]   A. B. Mikhailovskii and A. M. Fridman, Zh. Eksp. Teor. Fiz. **61**, 457 (1971) [Sov. Phys.—JETP **34**, 243 (1972)].

[89]   A. B. Mikhailovskii, *Theory of Plasma Instabilities, Vol. 2, Instabilities of an Inhomogeneous Plasma*, Consultants Bureau, New York, 1974 (Engl. translation).

[89a]   A. B. Mikhailovskii, V. I. Petviashvili, and A. M. Fridman, Astron. Zh. **56**, 279 (1979).

[90]   A. B. Mikhailovskii and A. M. Fridman, Astron. Zh. **50**, 88 (1973) [Sov. Astron. **17**, 57 (1973)].

[90a]   A. B. Mikhailovskii, V. I. Petviashvili, and A. M. Fridman, Pis'ma Zh. Eksp. Teor. Fiz. **26**, 129 (1977) [JETP Lett. **24**(2), 43 (1976)].

[91]   A. G. Morozov, V. L. Polyachenko, and I. G. Shukhman, preprint, Inst. Zemn. Magn. Ionosf. Rasprostr. Radiovoln Sibir. Otd. Akad. Nauk SSSR, No. 3–72, Irkutsk, 1972, (in Russian).

[92]   A. G. Morozov, V. L. Polyachenko, and I. G. Shukhman, preprint, Inst. Zemn. Magn. Ionosf. Rasprostr. Radiovoln Sibir. Otd. Akad. Nauk SSSR, No. 6–72, Irkutsk, 1972, (in Russian).

[93]   A. G. Morozov, V. L. Polyachenko, and I. G. Shukhman, preprint, Inst. Zemn. Magn. Ionosf. Rasprostr. Radiovoln Sibir. Otd. Akad. Nauk SSSR, No. 1–73, Irkutsk, 1973, (in Russian).

[94]   A. G. Morozov and A. M. Fridman, Astron. Zh. **50**, 1028 (1973) [Sov. Astron. **17**, 651 (1974)].

[95]   A. G. Morozov, V. L. Polyachenko, and I. G. Shukhman, preprint, Inst. Zemn. Magn. Ionosf. Rasprostr. Radiovoln Sibir. Otd. Akad. Nauk SSSR, No. 5–74, Irkutsk, 1974, (in Russian).

[96] A. G. Morozov, V. L. Polyachenko, and I. G. Shukhman, Astron. Zh. **51**, 75 (1974) [Sov. Astron. **18**, 44 (1974)].

[97] A. G. Morozov, V. L. Polyachenko, A. M. Fridman, and I. G. Shukhman, in *Dinamika i Evolutsija Zvezdnykh Sistem* (*Dynamics and Evolution of Stellar Systems*), VAGO, Akad. Nauk SSSR, Moscow, 1975 (in Russian).

[98] A. G. Morozov, V. L. Polyachenko, and I. G. Shukhman, preprint, Inst. Zemn. Magn. Ionosf. Rasprostr. Radiovoln Sibir. Otd. Akad. Nauk SSSR, No. 3-75, Irkutsk, 1975, (in Russian).

[98a] A. G. Morozov and A. M. Fridman, in *Dinamika i Evolutsija Zvezdnykh Sistem* (*Dynamics and Evolution of Stellar Systems*), VAGO, Akad. Nauk SSSR, Moscow, 1975, p. 238 (in Russian).

[99] A. G. Morozov, V. L. Polyachenko, V. G. Fainshtein, and A. M. Fridman, Astron. Zh. **53**, 946 (1976) [Sov. Astron. **20**, 535 (1976)].

[99a] A. G. Morozov, V. L. Fainshtein, and A. M. Fridman, Dokl. Akad. Nauk SSSR **231**, 588 (1976) [Sov. Phys.—Dokl. **21**(11), 661 (1976)].

[100] A. G. Morozov, V. G. Fainshtein, and A. M. Fridman, Zh. Eksp. Teor. Fiz. **71**, 1249 (1976) [Sov. Phys.—JETP **44**, 653 (1976)].

[101] K. F. Ogorodnikov, *Dynamics of Stellar Systems*, Pergamon, Oxford, 1965 (Engl. translation).

[102] L. M. Ozernoi and A. D. Chernin, Astron. Zh. **44**, 321 (1967) [Sov. Astron. **11**, 907 (1968)].

[103] L. M. Ozernoi and A. D. Chernin, Astron. Zh. **45**, 1137 (1968) [Sov. Astron. **12**, 901 (1969)].

[104] J. H. Oort, In *Strojenije Zvezdnykh Sistem* (*The Structure of Stellar Systems*), IL, Moscow, 1962 (Russian translation).

[105] M. Ya. Pal'chik, A. Z. Patashinskii, V. K. Pienus, and Ya. G. Epel'baum, preprint, Instituta Yadernoi Fiziki Sibir. Otd. Akad. Nauk SSSR, 99-100, Novosibirsk, 1970 (in Russian).

[106] M. Ya. Pal'chik, A. Z. Patashinskii, and V. K. Pienus, preprint, Instituta Yadernoi Fiziki Sibir. Otd. Akad. Nauk SSSR, 100, Novosibirsk, 1970 (in Russian).

[107] A. G. Pakhol'chik, Astron. Zh. **39**, 953 (1962) [Sov. Astron. **6**, 741 (1963)].

[108] S. B. Pikel'ner, *Osnovy Kosmicheskoi Elektrodinamiki* (*Principles of Cosmical Electrodynamics*), Fizmatgiz, Moscow, 1961 (in Russian).

[108a] V. L. Polyachenko, V. S. Synakh, and A. M. Fridman, Astron. Zh. **48**, 1174 (1971) [Sov. Astron. **15**, 934 (1972)].

[109] V. L. Polyachenko and A. M. Fridman, Astron. Zh. **48**, 505 (1971) [Sov. Astron. **15**, 396 (1971)].

[110] V. L. Polyachenko and A. M. Fridman, Astron. Zh. **49**, 157 (1972) [Sov. Astron. **16**, 123 (1972)]:

[111] V. L. Polyachenko and I. G. Shukhman, preprint, Inst. Zemn. Magn. Ionosf. Rasprostr. Radiovoln Sibir. Otd. Akad. Nauk SSSR, 1-72, Irkutsk, 1972 (in Russian).

[112] V. L. Polyachenko and I. G. Shukhman, preprint, Inst. Zemn. Magn. Ionosf. Rasprostr. Radiovoln Sibir. Otd. Akad. Nauk SSSR, 2-72, Irkutsk, 1972, (in Russian).

[113]   V. L. Polyachenko and I. G. Shukhman, Astron. Zh. **50**, 97 (1973) [Sov. Astron. **17**, 62 (1973)].

[114]   V. L. Polyachenko and I. G. Shukhman, Astron. Zh. **50**, 649 (1973) [Sov. Astron. **17**, 413 (1973).

[115]   V. L. Polyachenko and I. G. Shukhman, Astron. Zh. **50**, 721 (1973) [Sov. Astron. **17**, 460 (1974)].

[116]   V. L. Polyachenko, kandidatskaja dissertatsija (doctoral dissertation), Leningrad, 1973 (in Russian).

[117]   V. L. Polyachenko, Dokl. Akad. Nauk SSSR **229**, 1335 (1976) [Sov. Phys.—Dokl. **21**(8), 417 (1976)].

[118]   V. L. Polyachenko and I. G. Shukhman, Pis'ma Astron. Zh. **3**, 199 (1977) [Sov. Astron. Lett. **3**, 105 (1977)].

[119]   V. S. Safronov, *Evolutsija Doplanetnogo Oblaka i Obrazovanije Zemli i Planet* (*The Evolution of a Protoplanetary Cloud and Formation of the Earth and Planets*), Nauka, Moscow, 1969 (in Russian).

[120]   V. P. Silin and A. A. Rukhadze, *Elektromagnitnyje Svoistva Plazmy i Plazmopodobnykh Sred* (*The Electromagnetic Properties of Plasma and Plasmalike Media*), Gosatomoizdat, Moscow, 1961 (in Russian).

[121]   L. J. Slater, *Confluent Hypergeometric Functions*, Cambridge University Press, Cambridge, 1960 (Russian translation: Comp. Centr. Akad. Nauk SSSR, Moscow, 1966).

[122]   H. B. Sawyer Hogg, in *Strojenije Zvezdnykh Sistem* (*The Structure of Stellar Systems*), IL, Moscow, 1962 (Russian translation).

[123]   Th. H. Stix, *The Theory of Plasma Waves*, McGraw-Hill, New York, 1962 (Russian translation: Atomoizdat, Moscow, 1966).

[124]   V. S. Synakh, A. M. Fridman, and I. G. Shukhman, Dokl. Akad. Nauk SSSR **201**(4), 827 (1971) [Sov. Phys.—Dokl. **16**(12), 1062 (1972)].

[125]   V. S. Synakh, A. M. Fridman, and I. G. Shukhman, Astrofizika **8**(4), 577 (1972) [Astrophysics **8**(4), 338 (1972)].

[126]   S. I. Syrovatskii, Tr. Fiz. Inst. Akad. Nauk **8**, 13 (1956) (in Russian).

[127]   M. F. Subbotin, *Kours Nebesnoi Mekhaniki* (*Celestial Mechanics*), GTTI, Moscow–Leningrad, 1949, Vol. 3 (in Russian).

[128]   A. V. Timofeev, preprint, Inst. Atomnoi Energii im. Kurchatova, Moscow, 1968 (in Russian).

[129]   A. V. Timofeev, Usp. Fiz. Nauk **102**, 185 (1970) [Sov. Phys.—Usp. **13**(5), 632 (1971)].

[130]   A. N. Tikhonov and A. A. Samarskii, *Equations in Mathematical Physics*, Macmillan, New York, 1963 (Engl. translation).

[131]   B. A. Trubnikov, Vopr. Teor. Plazmy (The Problems of Plasma Theory) **1**, 98 (1963) (in Russian).

[132]   E. T. Whittaker and G. N. Watson, *A Course of Modern Analysis*, The University Press, New York, 1947, Vol. 1 (Russian translation: GIFML, Moscow, 1959).

[133]   E. T. Whittaker and G. N. Watson, *A Course of Modern Analysis*, The University Press, New York, 1947, Vol. 2 (Russian translation: GIFML, Moscow, 1963).

[134] V. N. Fadeeva and N. M. Terentjev, *Tablitsy Znachenii Integrala Verojatnosti ot Kompleksnogo Argumenta* (*Tables of Integral Values of Probability of a Complex Argument*), GITTL, Moscow, 1954 (in Russian).

[135] Ya. B. Fainberg, At. Energ. (At. Energy) **11**, 391 (1961) (in Russian).

[136] E. L. Feinberg, Usp. Fiz. Nauk **104**, 539 (1971) [Sov. Phys.—Usp. **14**, 455 (1972)].

[137] V. G. Fesenkov, Astron. Zh. **28**, 492 (1951) (in Russian).

[138] D. A. Frank-Kamenetskii, *Lektsii po Fizike Plazmy* (*Lectures on Plasma Physics*), Atomizdat, Moscow, 1968 (in Russian).

[139] A. M. Fridman, in *Itogi Nauki, Ser. Astron., T. 10: Ravnovesije i Ustoichivost' Gravitirujushchikh Sistem* (*Scientific Findings, Astron. Ser.: Equilibrium and Stability of Gravitating Systems*), Moscow, 1975 (in Russian).

[140] A. M. Fridman, Astron. Zh. **43**, 327 (1966) [Sov. Astron. **10**, 261 (1966)].

[141] A. M. Fridman, Astron. Zh. **48**, 910 (1971) [Sov. Astron. **15**, 720 (1972)].

[142] A. M. Fridman, Astron. Zh. **48**, 320 (1971) [Sov. Astron. **15**, 250 (1971)].

[143] A. M. Fridman and I. G. Shukhman, Dokl. Akad. Nauk SSSR, **202**, 67 (1972) [Sov. Phys.—Dokl. **17**, 44 (1972)].

[144] A. M. Fridman, doktorskaja dissertatsija (doctoral thesis), Moscow, 1972 (in Russian).

[145] L. G. Khazin and E. E. Shnol', Dokl. Akad. Nauk SSSR, **185**, 1018 (1969) [Sov. Phys.—Dokl. **14**, 332 (1969)].

[146] F. Zwicky, in *Strojenije Zvezdnykh Sistem* (*The Structure of Stellar Systems*), IL, Moscow, 1962 (Russian translation).

[147] S. Chandrasekhar, *Principles of Stellar Dynamics*, Chicago University Press, Chicago, 1942 (Reprinted: Dover, New York, 1960) (Russian translation: IL, Moscow, 1948).

[148] S. Chandrasekhar, *Ellipsoidal Figures of Equilibrium*, Yale University Press, New Haven, 1969 (Russian translation: Mir, Moscow, 1973).

[149] V. D. Shafranov, Vopr. Teor. Plazmy (The Problems of Plasma Theory) **3**, 3 (1963) (in Russian).

[150] M. Schwarzschild, *Structure and Evolution of the Stars*, Princeton University Press, Princeton, N.J., 1958 (Reprinted: Dover, New York, 1965) (Russian translation: IL, Moscow, 1961).

[151] O. Yu. Schmidt, *Chetyre Lektsii o Teorii Proiskhozhdenija Zemli* (Four Lectures on the Theory of Origin of the Earth), Izd. Akad. Nauk SSSR, Moscow, 1950 (in Russian).

[152] E. E. Shnol', Astron. Zh. **46**, 970 (1969) [Sov. Astron. **13**, 762 (1970)].

[153] I. G. Shukhman, kanadidatskaja dissertatsija (doctoral dissertation), Leningrad, 1973 (in Russian).

[154] I. G. Shukhman, Astron. Zh. **50**, 651 (1973) [Sov. Astron. **17**, 415 (1973)].

[155] A. Einstein, *Sobr. Sochin., T. 2* (*Selected Works, Vol. 2*), Nauka, Moscow, 1967 (in Russian).

[156] L. E. El'sgoltz, *Differential Equations and the Calculus of Variations*, Mir, Moscow, 1970 (in English).

[157]   E. Yanke, F. Emde, and F. Loesh, *Spetsialnyje Funktsii. Formuly, Grafiki, Tablitsy* (*Special Functions. Formulas, Graphs, and Tables*), Nauka, Moscow, 1964 (in Russian).

[158]   M. Aggarswal and S. P. Talwar, Monthly Notices Roy. Astron. Soc. **146**, 187 (1969).

[159]   E. S. Avner and I. R. King, Astron. J. **72**, 650 (1967).

[160]   B. Barbanis and K. H. Prendergast, Astron. J., **72**(2), 215 (1967).

[161]   J. M. Bardeen and R. V. Wagoner, Astrophys. J. **158**(2), 65 (1969).

[162]   J. M. Bardeen and R. V. Wagoner, Astrophys. J., **167**(3), 359 (1971).

[163]   L. Bel, Astrophys. J. **155**, 83 (1969).

[164]   H. P. Berlage, Proc. K. Ned. Akad. Wet. Amsterdam **51**, 965 (1948).

[165]   H. P. Berlage, Proc. K. Ned. Akad. Wet. Amsterdam **53**, 796 (1948).

[166]   A. B. Bernstein, F. A. Frieman, H. D. Kruskal, and R. M. Kulsrud, Proc. Roy. Soc. London **17**, 244 (1958).

[167]   P. Bodenheimer and J. P. Ostriker, Astrophys. J. **180**, 159 (1973).

[168]   W. B. Bonnor, Appl. Math. **8**, 263 (1967).

[169]   W. H. Bostick, Rev. Mod. Phys. **30**, 1090 (1958).

[170]   J. C. Brandt, Astrophys. J. **131**, 293 (1960).

[171]   J. C. Brandt, Monthly Notices Roy. Astron. Soc. **129**, 309 (1965).

[172]   G. H. Bryan, Philos. Trans. **180**, 187 (1888).

[173]   E. M. Burbidge, G. R. Burbidge, and K. H. Prendergast, Astrophys. J. **130**, 739 (1959).

[174]   E. M. Burbidge, G. R. Burbidge, and K. H. Prendergast, Astrophys. J. **137**, 376 (1963).

[175]   E. M. Burbidge, G. R. Burbidge, and K. H. Prendergast, Astrophys. J. **140**, 80, 1620 (1964).

[176]   E. M. Burbidge and G. R. Burbidge, Astrophys. J. **140**, 1445 (1964).

[177]   B. F. Burke, Astron. J. **62**, 90 (1957).

[178]   W. B. Burton, Bull. Astron. Netherl. **18**, 247 (1966).

[179]   G. L. Camm, Monthly Notices Roy. Astron. Soc. **101**, 195 (1941).

[180]   G. L. Camm, Monthly Notices Roy. Astron. Soc. **112**(2), 155 (1952).

[181]   G. Carranza, G. Courtes, Y. Georgellin, and G. Monnet, C. R. Acad. Sci. Paris **264**, 191 (1967).

[182]   G. Carranza, G. Courtes, Y. Georgellin, G. Monnet, and A. Pourcelot, Ann. Astrophys. **31**, 63 (1968).

[183]   G. Carranza, R. Crillon, and G. Monnet, Astron. Astrophys. **1**, 479 (1969).

[184]   K. M. Case, Phys. Fluids **3**, 149 (1960).

[185]   S. Chandrasekhar and E. Fermi, Astrophys. J. **118**, 113 (1953).

[186]   S. Chandrasekhar, *Hydrodynamics and Hydromagnetic Stability*, Clarendon Press, Oxford, 1961.

[187]   G. Contopoulos, Astrophys. J. **163**, 181 (1971).

[188]   G. Contopoulos, Astrophys. Space Sci. **13**(2), 377 (1971).

[189]   G. Contopoulos, Astrophys. J. **160**, 113 (1970).

[190]   G. Courtes and R. Dubout-Crillon, Astron. Astrophys. **11**(3), 468 (1971).

[191]   M. Crezé and M. O. Mennessier, Astron. Astrophys. **27**(2), 281 (1973).

[192]   J. M. A. Danby, Astron. J. **70**, 501 (1965).

[193]   G. Danver, Ann. Obs. Lund. **10**, 134 (1942).

[194]   M. E. Dixon, Astrophys. J. **164**, 411 (1971).

[194a] J. P. Doremus and M. R. Feix, Astron. Astrophys. **29**(3), 401 (1973).

[195]   O. J. Eggen, D. Lynden-Bell, and A. R. Sandage, Astrophys. J. **136**, 748 (1962).

[196]   A. S. Eddington, Monthly Notices Roy. Astron. Soc. **75**(5), 366 (1915).

[197]   A. S. Eddington, Monthly Notices Roy. Astron. Soc. **76**(7), 572 (1916).

[198]   G. Elwert and D. Z. Hablick, Astrophys. J., **61**, 273 (1965).

[199]   S. I. Feldman, and C. C. Lin. Stud. Appl. Math. **52**, 1 (1973).

[200]   E. Fermi, Progr. Theor. Phys. **5**, 570 (1950).

[201]   K. C. Freeman, Monthly Notices Roy. Astron. Soc. **130**, 183 (1965).

[202]   K. C. Freeman, Monthly Notices Roy. Astron. Soc. **133**(1), 47 (1966).

[203]   K. C. Freeman, Monthly Notices Roy. Astron. Soc. **134**(1), 1 (1966).

[204]   K. C. Freeman, Monthly Notices Roy. Astron. Soc. **134**(1), 15 (1966).

[205]   K. C. Freeman, Astrophys. J. **160**(3), 811 (1970).

[206]   K. C. Freeman and G. de Vaucouleurs, Astron. J. **71**(9), 855 (1966).

[207]   M. Fujimoto, Publ. Astron. Soc. Jpn. **15**, 107 (1963).

[208]   M. Fujimoto, *IAU Symposium No. 29*, D. Reidel, Dordrecht, 1966.

[209]   P. Goldreich and D. Lynden-Bell, Monthly Notices Roy. Astron. Soc. **130**, (2–3), 97 (1965).

[210]   P. Goldreich and D. Lynden-Bell, Monthly Notices Roy. Astron. Soc. **130**, (2–3), 125 (1965).

[211]   J. Guibert, Astron. Astrophys. **30**(3), 353 (1974).

[212]   D. ter Haar, Rev. Mod. Phys. **22**, 119 (1950).

[213]   A. P. Henderson, Ph.D. thesis, University of Maryland, 1967.

[214]   M. Henon, Ann. Astrophys. **29**(2), 126 (1959).

[215]   F. Hohl, Astron. J. **73**(5), 98, 611 (1968).

[216]   M. Henon, Bull Astron, **3**, 241 (1968).

[217]   M. Henon, Astron. Astrophys. **24**(2), 229 (1973).

[218]   R. W. Hockney and D. R. K. Brownrigg, Monthly Notices Roy. Astron. Soc. **167**(2), 351 (1974).

[219]   F. Hohl, J. Comput. Phys. **9**, 10 (1972).

[220]   F. Hohl, Astrophys. J. **168**, 343 (1971).

[221]   R. J. Hosking, Austr. J. Phys. **22**(4), 505 (1969).

[222]   F. Hoyle and M. Schwarzschild, Astrophys. J., Suppl. **2**(13), (1955).

[223]   F. Hoyle, *Frontiers of Astronomy*, New York, 1960.

[224]   F. Hoyle and W. A. Fowler, Nature **213**, 373 (1967).

[225]   E. Hubble, *The Realm of the Nebulae*, Yale University Press, New Haven, 1937.

[226]   C. Hunter, Monthly Notices Roy. Astron. Soc. **126**(4), 299 (1963).

[227]   C. Hunter, Monthly Notices Roy. Astron. Soc. **129**(3–4), 321 (1965).

[228]   C. Hunter, Stud. Appl. Math. **48**(1), 55 (1969).

[229]   C. Hunter, Astrophys. J. **157**(1), 183 (1969).

[230]   C. Hunter and A. Toomre, Astrophys. J. **155**(3), 747 (1969).

[231]   C. Hunter, Astrophys. J. **162**(1), 97 (1970).

[232]   C. Hunter, in *Dynamics of Stellar Systems*, Hayli, ed., D. Reidel, Dordrecht and Boston, 1970.

[233]   C. Hunter, Ann. Rev. Fluid Mech., **4**, 219 (1972).

[234]   C. Hunter, Monthly Notices Roy. Astron. Soc. **166**, 633 (1974).

[235]   C. Hunter, Astron. J. **80**(10), 783 (1975).

[236]   G. M. Idlis, Astron. Zn. **3**, 860 (1959).

[237]   K. A. Innanen, J. Roy. Astron. Soc. Can. **63**(5), 260 (1969).

[238]   J. R. Ipser and K. S. Thorne, preprint, OAP-121 California Inst. Technol., Pasadena, 1968.

[239]   J. R. Ipser and K. S. Thorne, Astrophys. J., **154**(1), 251 (1968).

[240]   J. D. Jackson, Plasma Phys. **1**, 171 (1960).

[241]   J. H. Jeans, Monthly Notices Roy. Astron. Soc. **76**(7), 767 (1916).

[242]   J. Jeans, *Astronomy and Cosmology*, Cambridge University Press, Cambridge, 1929.

[243]   H. M. Johnson. Astrophys. J. **115**, 124 (1952).

[244]   W. H. Julian, Astrophys. J. **155**(1), 117 (1969).

[245]   W. H. Julian and A. Toomre, Astrophys. J. **146**(3), 810 (1966).

[246]   B. B. Kadomtzev and O. P. Pogutze, Phys. Rev. Lett. **25**(17), 1155 (1970).

[247]   F. D. Kahn and L. Woltjer, Astrophys. J. **130**, 705 (1959).

[248]   F. D. Kahn and J. E. Dyson, Ann. Rev. Astron. Astrophys **3**, 47 (1965).

[249]   A. J. Kalnajs, Ph.D. thesis, Harvard University, 1965.

[250]   A. J. Kalnajs, in *IAU Symposium No. 38*, D. Reidel, Dordrecht, 1970.

[251]   A. J. Kalnajs, Astrophys. J. **166**(2), 275 (1971).

[252]   A. J. Kalnajs, Astrophys. J. **175**(1), 63 (1972).

[253]   A. J. Kalnajs, Astrophys. J. **180**, 1023 (1973).

[254]   A. J. Kalnajs and G. E. Athanassoula, Monthly Notices Roy. Astron. Soc. **168**, 287 (1974).

[255]   S. Kato, Publ. Astron. Soc. Jpn. **23**, 467 (1971).

[256]   S. Kato, Publ. Astron. Soc. Jpn. **25**, 231 (1973).

[257]   F. J. Kerr, Monthly Notices Roy. Astron. Soc. **123**, 327 (1962).

[258]   F. J. Kerr and G. Westerhout, in *Stars and Stellar Systems*, Chicago University Press, Chicago and London, 1965.

[259]   F. J. Kerr, Austr. J. Phys. Astrophys. Suppl., No. 9, (1969).

[260]   I. R. King, Astron. J. **70**(5), 376 (1965).

[261]   N. Krall and M. Rosenbluth, Phys. Fluids **6**, 254 (1963).

[262]   G. P. Kuiper, *Astrophysics*, J. A. Hynek, ed., New York, 1951.

[263] R. M. Kulsrud, J. W.-K. Mark, and A. Caruso, Astrophys. Space Sci. **14**(1), 52 (1971).

[264] R. M. Kulsrud and J. W.-K. Mark, Astrophys. J. **160**, 471 (1970).

[265] P. S. Laplace, Mem. Acad. Sci. (Mécanique Celeste, k. 3, p. VI), 1789 (1787).

[265a] M. J. Lighthill, J. Inst. Math. Appl. **1**, 269 (1965).

[266] E. P. Lee, Astrophys. J. **148**, 185 (1967).

[267] C. C. Lin and F. H. Shu, Astrophys. J. **140**(2), 646 (1964).

[268] C. C. Lin, L. Mestel, and F. Shu, Astrophys. J. **142**(4), 1431 (1965).

[269] C. C. Lin, SIAM J. Appl. Math. **14**(4), 876 (1966).

[270] C. C. Lin and F. H. Shu, Proc. Nat. Acad. Sci. USA **55**(2), 229 (1966).

[271] C. C. Lin, C. Yuan, and F. H. Shu, Astrophys. J. **155**(3), 721 (1969).

[272] C. C. Lin, in *IAU Symposium, No.. 38*, D. Reidel, Dordrecht, 1970.

[273] B. Lindblad, Stockholm Obs. Ann. **20**(6), (1958).

[274] B. Lindblad, Stockholm. Obs. Ann. **22**, 3 (1963).

[275] P. O. Lindblad, Popular Arstok Tidschr. **41**, 132 (1960).

[276] P. O. Lindblad, Stockholm Obs. Ann. **21**, 3 (1960).

[277] P. O. Lindblad, in *Interstellar Matter in Galaxies*, L. Woltjer, ed., New York, 1962.

[278] C. Lundquist, Phys. Rev. **83**, 307 (1951).

[279] D. Lynden-Bell, Monthly Notices Roy. Astron. Soc. **120**(3), 204 (1960).

[280] D. Lynden-Bell, Monthly Notices Roy. Astron. Soc. **123**, 447 (1962).

[281] D. Lynden-Bell, Monthly Notices Roy-Astron. Soc. **124**, 279 (1962).

[282] D. Lynden-Bell, Astrophys. J. **139**, 1195 (1964).

[283] D. Lynden-Bell, Monthly Notices Roy. Astron. Soc. **129**, 299 (1965).

[284] D. Lynden-Bell, *The Theory of Orbits in a Solar System and in Stellar Systems*, 1966.

[285] D. Lynden-Bell, Lect. Appl. Math. **9**, 131 (1967).

[286] D. Lynden-Bell, Monthly Notices Roy. Astron. Soc. **136**, 101 (1967).

[287] D. Lynden-Bell and J. P. Ostriker, Monthly Notices Roy. Astron. Soc. **136**(3), 293 (1967).

[288] D. Lynden-Bell and N. Sanitt, Monthly Notices Roy. Astron. Soc. **143**(2), 176 (1969).

[289] D. Lynden-Bell and A. J. Kalnajs, Monthly Notices Roy. Astron. Soc. **157**, 1 (1972).

[289a] J. W.-K. Mark, Astrophys. J. **169**, 455 (1971).

[290] J. W.-K. Mark, Proc. Nat. Acad. Sci. USA **68**(9), 2095 (1971).

[290a] J. W.-K. Mark, Astrophys. J. **193**, 539 (1974).

[291] J. C. Maxwell, *The Scientific Papers*, Cambridge University Press, Cambridge, 1859, Vol. 1, p. 287.

[292] L. Mestel, Monthly Notices Roy. Astron. Soc. **126**(5–6), 553 (1963).

[293] R. W. Michie, Monthly Notices Roy. Astron. Soc. **125**(2), 127 (1963).

[294] R. W. Miller, K. H. Prendergast, and W. J. Quirk, Astrophys. J. **161**(3), 903 (1970).

[295] G. Münch, Publ. Astron. Soc. Pacific, **71**, 101 (1959).

[295a] T. O'Neil, Phys. Fluids **8**, 2255 (1965).

[296] J. H. Oort, Bull. Astron. Netherl., **6**, 249 (1932).

[297] J. H. Oort, Scientific Am. **195**, 101 (1956).

[298] J. H. Oort, F. J. Kerr, and G. Westerhout, Monthly Notices Roy. Astron. Soc. **118**, 319 (1958).

[299] J. H. Oort, in *Interstellar Matter in Galaxies*, L. Woltjer, ed., W. A. Benjamin, New York, 1962.

[300] J. P. Ostriker and P. Bodenhiemer, Astrophys. J. **180**, 171 (1973).

[301] J. P. Ostriker and P. J. E. Peebles, Astrophys. J. **186**(2), 467 (1973).

[302] P. J. Peebles and R. H. Dicke, Astrophys. J. **154**, 898 (1968).

[303] J. H. Piddington, Monthly Notices Roy. Astron. Soc. **162**, 73 (1973).

[304] J. H. Piddington, Astrophys. J. **179**, 755 (1973).

[305] H. C. Plummer, Monthly Notices Roy. Astron. Soc. **71**, 460 (1911).

[306] K. H. Prendergast, Astron. J. **69**, 147 (1964).

[307] K. H. Prendergast and E. Tomer, Astron. J. **75**, 674 (1970).

[308] W. J. Quirk, Astrophys. J. **167**(1), 7 (1971).

[309] R.-G. Rohm, Ph.D. thesis, MIT, Cambridge, Mass., 1965.

[310] P. H. Roberts and K. Stewartson, Astrophys. J. **137**(3), 777 (1963).

[311] W. W. Roberts, Astrophys. J. **158**, 123 (1969).

[312] W. W. Roberts, M. S. Roberts, and F. H. Shu, Astrophys. J. **196**, 381 (1975).

[313] M. N. Rosenbluth, N. Krall, and N. Rostocker., Nucl. Fusion, Suppl. **2**, 143 (1962).

[314] G. W. Rougoor, Bull. Astron. Inst. Netherl. **17**, 318 (1964).

[315] V. C. Rubin and W. K. Ford, Astrophys. J. **159**(2), 379 (1970).

[316] H. N. Russel, *Astronomy, Part 1*, 1926.

[317] A. Sandage, *The Hubble Atlas of Galaxies*, Carnegie Inst., Washington, 1961.

[318] A. Sandage, K. C. Freeman, and N. R. Stokes, Astrophys. J. **160**, 831 (1970).

[319] M. Schmidt, in *Galactic Structure*, A. Blaauw and M. Schmidt, eds., University of Chicago Press, Chicago, 1965.

[320] W. W. Shane and G. P. Bieger-Smith, Bull. Astron. Netherl. **18**, 263 (1966).

[321] H. Shapley and H. B. Sawyer, Harv. Obs. Bull. No. 852 (1927).

[322] F. H. Shu, Astron. J. **73**(10), 201 (1968).

[323] F. H. Shu, Ph.D. thesis, Harvard University Press, Cambridge, Mass., 1968.

[324] F. H. Shu, Astrophys. J. **160**(1), 89 (1970).

[325] F. H. Shu, Astrophys. J. **160**(1), 99 (1970).

[326] F. H. Shu, R. V. Stachnic, and J. C. Yost, Astrophys. J. **166**(3), 465 (1971).

[327] E. A. Spiegel, *Symp. Origine Syst. Solaire, Nice, 1972*, Paris, 1972.

[328] P. Strömgren, in *Proc. IAU Symp. No. 31*, Noordwick, 1966.

[329] P. Strömgren, in IAU Symp. *No. 31*, D. Reidel, Dordrecht, 1967.

[330] P. Sweet, Monthly Notices Roy. Astron. Soc. **125**, 285 (1963).

[331] A. Toomre, Lectures in Geophysical Fluid Dynamics at the Woods Hole Oceanographic Institution, 1966.

[332] A. Toomre, Astrophys. J. **138**, 385 (1963).

[333] A. Toomre, Astrophys. J. **139**(4), 1217 (1964).

[334] A. Toomre, Astrophys. J. **158**, 899 (1969).

[335] S. D. Tremaine, preprint, California Inst. Technol., Pasadena, 1976.

[336] P. O. Vandervoort, Astrophys. J. **147**(1), 91 (1967).

[337] P. O. Vandervoort, Mem. Soc. Roy. Sci. Liege **15**, 209 (1967).

[338] P. O. Vandervoort, Astrophys. J. **161**, 67, 87 (1970).

[339] G. de Vaucouleurs, Mem. Mt. Stromlo Obs. **111**(3), (1956).

[340] G. de Vaucouleurs, Astrophys. J. Suppl. **8**(76), 31 (1963).

[341] G. de Vaucouleurs, Rev. Popular Astron. **57**(520), 6 (1963).

[342] G. de Vaucouleurs, Astrophys. J. Suppl. **8**(74), 31 (1964).

[343] G. de Vaucouleurs, A. de Vaucouleurs, and K. C. Freeman, Monthly Notices Roy. Astron. Soc. **139**(4), 425 (1968).

[344] G. de Vaucouleurs and K. C. Freeman, Vistas Astron. **14**, 163 (1973).

[345] L. Volders, Bull. Astron. Netherl. **14**, 323 (1959).

[346] H. Weaver, in *IAU Symp. No. 38*, D. Reidel, Dordrecht, 1970.

[347] C. F. Von Weizsäcker, Z. Astrophys. **22**, 319 (1944).

[348] C. F. Von Weizsäcker, Naturwiss. **33**, 8 (1946).

[349] G. Westerhout, Bull. Astron. Inst. Netherl. **14**, 215 (1958).

[350] R. Wielen, Astron. Rechen-Inst., Heidelberg Mitt. Ser. A, No. 47, (1971).

[351] C. P. Wilson, Astron. J. **80**, 175 (1975).

[352] R. van der Wooley, Monthly Notices Roy. Astron. Soc. **116**(3), 296 (1956).

[353] R. van der Wooley, Observatory, **81**(924), 161 (1961).

[354] C.-S. Wu, Phys. Fluids **11**(3), 545 (1968).

[355] A. B. Wyse and N. U. Mayall, Astrophys. J. **95**, 24 (1942).

[356] S. Yabushita, Monthly Notices Roy. Astron. Soc. **143**(3), (1969).

[357] S. Yabushita, Monthly Notices Roy. Astron. Soc. **133**(3), 247 (1966).

[358] S. Yabushita, Monthly Notices Roy. Astron. Soc. **142**(2), 201 (1969).

[359] C. Yuan, Astrophys. J. **158**(3), 871 (1969).

[360] C. Yuan, Astrophys. J. **158**(3), 889 (1969).

Additional References

[1] L. M. Al'tshul', Dep. No. 50295, VINITI, 1972.

[2] N. N. Bogolyubov and Yu. A. Mitropol'skiy, *Asymptotic Methods in the Theory of Nonlinear Oscillations*, M., Nauka, Moscow, 1974.

[3] N. P. Buslenko, *Statistical Test Method (the Monte-Carlo Method)*, SMB, M., Fizmatgiz, Moscow 1962.

[4] Yu.-I. K. Veltmann, Trudy Astrofiz. Inst. AN Kaz. SSR **5**, 57 (1965).

[5] Yu.-I. K. Veltmann, Publications of the Tartusk. Astr. Observ. **34**, 101 (1964); **35**, 344, 356 (1966).

[6] B. I. Davydov, Dokl. AN SSSR **69**, 165 (1949).

[7] B. P. Demidovich, I. A. Maron and E. Z. Shuvalova, *Numerical Analysis Methods*. GIFML, M., 1963.

[8] V. I. Dokuchayev and L. M. Ozernoy, Preprint FIAN im. P. N. Lebedev, No. 133, S.; ZhETF, **73**, 1587 (1977); Letters to Astron. Zh. **3**, 391 (1977).

[9] S. M. Yermakov and G. A. Mikhaylov, *Course of Statistical Modeling*, M., Nauka, Moscow 1976.

[10] V. Ye. Zakharov, ZhETF **60**, 1713 (1971).

[11] V. Ye. Zakharov, Izvestiya vyzov. Radiofizika **17**, 431.

[12] V. Ye. Zakharov, PMTF No. 2, 86 (1968).

[13] V. Ye. Zakharov, ZhETF **62**, 1945 (1972).

[14] G. M. Idlis, Astron. Zh. **33** (1), 53 (1956).

[15] B. B. Kadomtsev, *Collective Phenomena in Plasma*, M., Nauka, Moscow, 1976.

[16] B. P. Kondrat'yev and L. M. Ozernoy, Letters to Astron. Zh. **5**, 67 (1979).

[17] V. I. Korchagin and L. S. Marochnik, Astron. Zh. **52**(4), 700 (1975).

[18] G. G. Kuzmin and Yu.-I. K. Veltmann, Publ. Tartusk. Astr. Observ. **36**, 3, 470 (1968).

[19] G. G. Kuzmin and Yu.-I. K. Veltmann, Coll: "*Dynamics of Galaxies and Stellar Clusters*," 1973, Alma-Ata, Nauka, Moscow, p. 82.

[20] A. B. Mikhaylovskiy, V. I. Petviashvili and A. M. Fridman, Letters to ZhETF **26**, 341 (1977).

[21] L. S. Marochnik, Astrofizika **5**, 487 (1969).

[22] A. G. Morozov and I. G. Shukhman, Letters to Astron. Zh. **6**, 87 (1980).

[23] A. G. Morozov, Letters to Astron. Zh. **3**, 195 (1977).

[24] A. G. Morozov and A. M. Fridman, *Report at the All-Union Conference "Latent Mass in the Universe*," Tallin, January, 1975.

[25] A. G. Morozov, Astron. Zh **56**, 498 (1979).

[26] S. N. Nuritdinov, Author's abstract of Thesis, Leningrad, 1975, Astrofizika **11**, 135 (1975).

[27] L. N. Osipkov, Letters to Astron. Zh. **5**, 77 (1979).

[28] V. L. Polyachenko, Letters to Astron. Zh. **3**, 99 (1977).

[29] V. L. Polyachenko and A. M. Fridman, Letters to Astron. Zh. **7**, 136 (1981).

[30] V. L. Polyachenko and I. G. Shukhman, Letters to Astron. Zh. **3**, 199 (1977).

[31] V. L. Polyachenko, S. M. Churilov and I. G. Shukhman, Preprint SibIZMIR SO AN SSSR, No. 1–79, Irkutsk, 1979; Astron. Zh. **57**, 197 (1980).

[32] V. L. Polyachenko and I. G. Shukhman, Astron. Zh. **56**(5), 957 (1979).

[33] V. L. Polyachenko and I. G. Shukhman, Letters to Astron. Zh. **3**(6), 254 (1977).

[34] V. L. Polyachenko and I. G. Shukhman, Preprint SibIZMIR SO AN SSSR, No. 31–78, Irkutsk, 1978.

[35] V. L. Polyachenko and I. G. Shukhman, Astron. Zh. **57**(2), 268 (1980).

[36] V. L. Polyachenko, Letters to Astron. Zh. (1983), to appear.

[37] V. L. Polyachenko and I. G. Shukhman, Astron. Zh. **56**(4), 724 (1979).

[38] V. L. Polyachenko, Astron. Zh. **56**, 1158 (1979).

[39] V. L. Polyachenko and A. M. Fridman, Astron. Zh., (1983), to appear.

[40] V. L. Polyachenko and I. G. Shukhman, Preprint SibIZMIR SO AN SSSR, No. 1–78, Irkutsk, 1978.

[41] V. L. Polyachenko, Letters to Astron. Zh. (1983), to appear.

[42] V. L. Polyachenko, Letters to Astron. Zh. (1983), to appear.

[43] V. L. Polyachenko, Letters to Astron. Zh. **7**(3), 142 (1981).

[44] V. L. Polyachenko and I. G. Shukhman, Astron. Zh. **58**, 933 (1981).

[45] Yu. M. Rozenraukh, Thesis, IGU, Irkutsk, 1977.

[46] R. Z. Sagdeev, Vopr. Teor. Plazmy (Plasma Theory Problems) Ed. M. A. Leontovich, No. 4, M., Atomizdat, 1963.

[47] M. A. Smirnov and B. V. Komberg, Letters to Astron. Zh. **4**, 245 (1978).

[48] I. M. Sobol', *Numerical Monte-Carlo Methods*, M., Nauka, Moscow, 1973.

[49]   A. M. Fridman, Uspekhi fiz. nauk **125**, 352 (1978).

[50]   A. M. Fridman, Letters to Astron. Zh. **4**, 243 (1978).

[51]   A. M. Fridman, Letters to Astron. Zh. **4**, 207 (1978).

[52]   A. M. Fridman, Letters to Astron. Zh. **5**, 325 (1979).

[53]   S. M. Churilov and I. G. Shukhman, Astron. Zh. **58**, 260 (1981); **59**, 1093 (1982).

[54]   V. D. Shapiro and V. I. Shevchenko, ZhETF **45**, 1612 (1963).

[55]   J. N. Bahcall, Astrophys. J. **209**, 214 (1976).

[56]   J. M. Bardeen, in *IAU Symposium No. 69*, D. Reidel, Dordrecht, 1975.

[57]   F. Bertola and M. Capaccioli, Astrophys. J. **219**, 404 (1978).

[58]   J. Binney, Monthly Notices Roy. Astron. Soc. **177**, 19 (1976).

[59]   M. Clutton-Brock, Astrophys. Space Sci. **16**, 101 (1972).

[60]   G. Contopoulos, Astron. Astrophys. **64**, 323 (1978).

[61]   M. J. Dunkan and J. C. Wheeler, preprint, Astrophys. J. Lett. (1980), Dept.
       Astron., Univ. Texas, Austin, 1979.

[62]   J. Frank and M. J. Rees, Monthly Notices Roy. Astron. Soc. **176**, 633 (1976).

[63]   A. M. Fridman, preprint, Inst. Zemn, Magn. Ionosf. Rasprostr. Radiovoln
       Sibir. Otd. Akad. Nauk SSSR, 6–78, Irkutsk, 1978.

[64]   A. M. Fridman, Y. Palous and I. I. Pasha, Monthly Notices Roy. Astron. Soc.
       194, 705 (1981).

[64a] A. M. Fridman and V. L. Polyachenko, Zh ETP **81**, 13 (1981).

[65]   P. B. Globa-Mikhailenko, J. Math. (7 serie) **II**, 1 (1916).

[66]   P. Goldreigh and S. Tremaine, Astrophys. J. **222**, 850 (1978).

[67]   F. Hohl, in *IAU Symposium No. 69*, D. Reidel, Dordrecht, 1975.

[68]   C. Hunter, Astrophys. J. **181**, 685 (1973).

[69]   C. Hunter, Astron. J. **82**, 271 (1977).

[70]   S. Ikeuchi, Progr. Theor. Phys. **57**, 1239 (1977).

[71]   G. Illingworth, Astrophys. J. Lett. **218**, L43 (1977).

[72]   A. J. Kalnajs, Proc. Astron. Soc. Austr. **2**, 174 (1973).

[73]   A. J. Kalnajs, Astrophys. J. **205**, 745, 751 (1976).

[74]   A. J. Kalnajs, Astrophys. J. **212**, 637 (1977).

[75]   J. Katz, Monthly Notices Roy. Astron. Soc. **183**, 765 (1978).

[76]   I. R. King, Astron. J. **71**, 64 (1966).

[77]   I. R. King, in *IAU Symposium No. 69*, D. Reidel, Dordrecht, 1975.

[78]   A. P. Lightman and S. L. Shapiro, Astrophys. J. **211**, 244 (1977).

[79]   D. Lynden-Bell, *C.N.R.S. International Colloquium No. 241*, Centre Nat. de la
       Rech. Sci., Paris, 1975.

[80]   R. H. Miller, J. Comput. Phys. **21**, 400 (1976).

[81]   R. H. Miller, Astrophys. J. **223**, 122 (1978).

[82]   L. M. Ozernoy and B. P. Kondrat'ev, Astron. Astrophys. **79**, 35 (1979).

[83]   P. J. E. Peebles, Astron. J. **75**, 13 (1970).

[84]   P. J. E. Peebles, Gen. Relativity Gravity, **3**, 63 (1972).

[85]   C. J. Peterson, Astrophys. J. **222**, 84 (1978).

[86]   W. L. W. Sargent, P. J. Young, A. Boksenberg, K. Shortrigge, C. R. Lynds, and F. D. A. Hartwick, Astrophys. J. **221**, 731 (1978).

[87]   L. Schipper and I. R. King, Astrophys. J. **220**, 798 (1978).

[88]   A. Toomre, Annu. Rev. Astron. Astrophys. **15**, 437 (1977).

[89]   P. J. Young, W. L. W. Sargent, A. Boksenberg, C. R. Lynds, and F. D. A. Hartwick, Astrophys. J. **222**, 450 (1978).

[90]   P. J. Young, J. A. Westphal, J. Kristian, C. P. Wilson, and F. P. Landauer, Astrophys. J. **221**, 721 (1978).

[91]   T. A. Zang, Ph.D. thesis, M.I.T., Cambridge, Mass., 1976.

[92]   M. Nishida and T. Ishizawa, 1976, preprint, Kyoto Univ.

[93]   M. Abramowitz and I. A. Stegun, eds., *Handbook of Mathematical Function*, 1964.

[94]   P. L. Schechter and J. E. Gunn, Astrophys. J. **229**, 472 (1979).

[95]   J. R. Cott, Astrophys. J. **201**, 2961 (1975).

[96]   R. B. Larson, Monthly Notices Roy. Astron. Soc. **173**, 671 (1975).

[97]   G. Illingworth, Astrophys. J. **43**, 218 (1977).

[98]   J. J. Binney, Monthly Notices Roy. Astron. Soc. **183**, 501 (1978).

[99]   J. J. Binney, Monthly Notices Roy. Astron. Soc. **190**, 421 (1980).

[100] T. C. Chamberlin, Astrophys. J. **14**, 17 (1901).

[101] E. Holmberg, Astrophys. J. **94**, 385 (1941).

[102] J. Pfleiderer and H. Siedentopf, Z. Astrophys. **51**, 201 (1961).

[103] J. Pfleiderer, Z. Astrophys. **58**, 12 (1963).

[104] N. Tashpulatov, Astron. Zh. **46**, 1236 (1969).

[105] A. Toomre, in *IAU Symposium No. 79*, D. Reidel, Dordrecht, 1977.

[106] S. Yabushita, Monthly Notices Roy. Astron. Soc. **153**, 97 (1971).

[107] F. Zwicky, Naturwissenschaften, **26**, 334 (1956).

[108] N. N. Kozlov, T. M. Eneev, and R. A. Syunyaev, Dokl. Akad. Nauk SSSR **204**, 579 (1972).

[109] T. M. Eneev, N. N. Kozlov, and R. A. Syunyaev, Astron. Astrophys. **22**, 41 (1973).

[110] A. Toomre, *IAU Symposium No. 38*, D. Reidel, Dordrecht, 1970.

[111] A. M. Fridman, *IAU Symposium No. 79*, D. Reidel, Dordrecht, 1977.

[112] S. M. Churilov and I. G. Shukhman, Astron. Tsirk., No. 1157 (1981).

[113] L. W. Esposito, J. P. Dilley, and J. W. Fountain, J. Geophys. Res. **85** (A11), 5948 (1980).

[114] P. Goldreich and S. D. Tremaine, Icarus **34**, 227 (1978).

[115] S. S. Kumar, Publ. Astron. Soc. Japan **12**, 552 (1960).

[116] R. H. Sanders and G. T. Wrikon, Astron. Astrophys., **26**, 365 (1973).

[117] G. Chew, M. Goldberger and F. Low, Proc. Roy. Soc. A, **236**, 112 (1956).

[118] A. G. Doroshkevich, A. A. Klypin, preprint IPM, No. 2 (1980).

[119]   A. G. Morozov, Astron. Zh. **57**, 681 (1980).

[120]   A. Lane *et al.*, Science **215**, No. 4532 (1982).

[121]   N. C. Lin and P. Bodenheimer, Astrophys. J. Letters **248**, L83 (1981).

[122]   G. W. Null, E. L. Lau, and E. D. Biller, Astron. J. **86**, 456 (1981).

[123]   V. L. Polyachenko, and A. M. Fridman, Astron. Tsirk., No. 1204 (1981).

[124]   Voyager Bulletin, Mission status report, No. 57, Nov. 7, NASA (1980).

[125]   W. R. Ward, Geophys. Res. Lett. **8**, No. 6 (1981).

[126]   B. B. Kadomtsev, Letters to ZhETP **33**(7), 361 (1981).

[127]   I. V. Igumenshev, Thesis, ChGU, Chelyabinsk (1982).

[128]   V. A. Ambartsumian, The Scientific Papers, v. 1, Erevan, 1960.

[129]   V. A. Ambartsumian, Astron. Zh. **14**, 207 (1937).

[130]   M. G. Abrahamian, Astrophysics **14**, 579 (1978).

Index

The page numbers which appear in upright figures refer to Volume I and those which appear in italic refer to Volume II.

Action density *196*
Anisotropic distribution functions 142
Antispiral theorem *189*

Barlike mode
  linear 284, *415*
  nonlinear *125, 128*
  with a halo *128*
Barlike structure *207*
Beam instability 20
  in a cylinder model
    kinetic 5, *35*
    hydrodynamical *8*
  in a layer model *34*
  in multi-component systems with
        homogeneous flows *2*
Biorthonormalized sets *286–290*
Black hole 29, *152, 153–157*
Boltzmann kinetic equation 4

Camm distribution functions 144–145
  nonlinear evolution 240
  stability investigation of 186, 219

Cauchi–Bunyakovsky inequality 155
Characteristic frequencies. *See*
        Eigenfrequencies
Characteristics of equation 7
Cherenkov resonance 7, *33*, 66
Cold medium 6
Collapse of nonlinear waves *124*
Collisionless
  damping. *See* Landau damping
  kinetic equation 4
  shocks *62*
  systems 2
  universe *242*
Collisions *143*
Compressible medium 2
Computer simulation. *See* Statistical
        simulation
Consistency condition 50
Continuous spectra. *See* Spectra
Corotation resonance. *See* Resonance
Cylinders
  beam instability *5, 8, 35*
  composite models 80
  firehose instability 83, 97
  Jeans instability 83, 96
  nonlinear stability *131, 132*

465

Debye radius   6, 13
Decay instability   *93*
Decrement of damping   10
Derivative along the trajectory   4
Discrete spectra. *See* Spectra
Disks
  composite models   340
  continuous spectra   350, 354, 374
  Lindblad relation between velocity
     dispersions   339
Dispersion equation (relation),
     linear   11, 393
Dispersion relation, nonlinear. *See*
     Nonlinear dispersion relation
Distribution function
  Camm   144–145
  Eddington   149
  Freeman   146, 190
  Kuz'min–Veltmann   147, 231
  Osipkov–Idlis   146, 209
  "playing" of   201
  Schwarzschild   391
  trapped particles of   *69, 78*
  untrapped (transit) particles of   *70, 78*
Drift of a spiral wave packet   *190*
Duffing's equation   *61*

Eddington's distribution function   149
Eigenfrequencies   10
  of a cylinder   106
  of a disk
    for bending perturbations   403
    for plane perturbations
      in composite models   419
      in Maclaurin's model   414
      with a halo   443
  matrix formulation of a problem of
    for disks   *291*
    for spheres   *282*
Eigenfunctions   10
  for Camm's spheres   187
  for Freeman's spheres   192
  for homogeneous cylinders   107
  for homogeneous layer   71
  for Maclaurin's disks   422
Ellipsoids
  inhomogeneous   260
  oblate   248
  prolate   313
  superposition of Freeman's
     models   284
Elliptical cylinder   132, 299, 319
Elliptical disks   298
Envelope solitons   *62*

Epicycle
  radius of   82
Epicyclic
  approximation   81
  frequency   82
Equation
  Euler   3
  of a spiral wave   396
Equations for perturbations   9
Expansions in δ-series   150
Explosive instability   *55*
Exponential instability   10

Firehose instability
  dispersion relation
    for a cylinder   99
    for a disk   403
    for a layer   40
  physics of   37, 42
Flute instability
  for a cylinder   *24*
  for a layer   *20*
  in the Galaxy   204
Freeman's sphere   146, 190

Galaxies   *216–221*
Galaxy
  parameters of   *213*
  rotation curve   *219*
  Schmidt's model   *213*
  spiral structure
    Lin–Shu theory of   389–401, *213*

Hubble's "tuning-fork" diagram   *139*
Hydrodynamics   2

Incompressible
  cylinder   117
  ellipsoids   285
  fluid   1
  layer   34
Increment of instability   10
Infinitely-thin layer   32
Instability
  beam. *See* Beam instability
  beam-gradient   *9*
  decay   *93*
  explosive   *55*
  firehose. *See* Firehose instability
  flute-like. *See* Flute instability
  gradient   *11*

hydrodynamical   *16*
Jeans   11
Kelvin–Helmholtz   *16*
modulational   *123*
temperature-gradient   *11*
Instability of circular orbits   83
Integration over trajectories   44, 166

Jacobi ellipsoids   246, 285
Jeans
   frequency   11
   instability   11
   suppression of   20
   wavelength   12

Kinetic equation
   of Boltzmann   4
   collisionless, in various
         coordinates   *271*
   for waves   97, *318*
Kramp function   33
Kuz'min–Veltmann's distribution
         function   147, 231

Landau bypass rule   32
Landau damping (or growth)
   linear   91, *180*
   nonlinear   *69, 78*
Langmuir oscillations   13
Layer
   eigenfunctions of   71
   long wavelength perturbations   69
   nonlinear evolution   74
   short wavelength perturbations   69
Leading spirals   370
Lindblad relation between velocity
         dispersions   339
Lindblad resonances   394, *171*
Liouville's equation   5
Local perturbations   122

Mach number   *19*
Maclaurin's collisionless disks
   distribution function   340
   eigenfrequencies   414
   eigenfunctions   422
Maclaurin's ellipsoids
      (spheroids)   246, 285
Marginal stability   394
Material spiral arms   *167*

Method
   of characteristics. *See* Integration
         over trajectories   44
   of Lagrangian shifts   53
   of phase volume variation   58
Migration of stars   *215*
Modulational instability   *123*

Negative energy waves   *173*
Nonhomogeneous layer   30
Non-Jeans instabilities   *2*
Nonlinear corrections for a critical
            wavelength
   in incompressible layer   *116*
   in isothermal layer   *111, 113*
Nonlinear dispersion relation
   for a disk   *40, 42*
   for a gaseous disk   *48*
   for a stellar disk   *52*
Nonlinear evolution
   of a distribution function
      in a cylinder   *69*
      in a disk near the corotation   75
   of a monochromatic wave   *70*
Nonlinear Jeans instability   *55*
Nonlinear parabolic equation   *100*
Nonlinear waves. *See* Solitons and
         nonlinear waves
Nonrotating liquid cylinder   117

Osipkov–Idlis distribution function   146
   stability   209

Permutational modes   68
Phase fluid   4
Plasma and gravitating medium
         (comparison)   13–14
Plateau formation in distribution
         function   75–78
Playing of distribution functions   200
   method of inversion   *279*
   Neumann's method   *279*
Poisson bracket   5, *274*
Poisson equation   3, *271*
Power instability   10, 65, 130
Prolate spheroid   313

Quadrupole momenta tensor   206

Reduction factor
  of Lin–Shu ($\mathscr{F}_\nu$)   393
  for a thickness (I)   398
Resonance
  Cherenkov   7, *33, 66*
  corotation   *180*
  Lindblad   394, *171*
  rotational   7
Resonance conditions   *93*
Response of a system   390
Rotational resonance   7

Schwarzschild's distribution
        function   391
Self-consistent field   4
Separation of angular variables   *275*
Shuster's sphere   147, 231
Singular solutions   123
Solitary waves. *See* Solitons
Solitons and nonlinear waves
  in a disk   *40, 42*
  in a gaseous disk   *48*
  in a stellar disk   *42*
Spectra. *See also* Dispersion equation.
        Eigenfrequencies
  continuous   350, 374
Spheres
  circular orbits with   137, 143, 164, 238
  elliptical orbits with   138, 146, 186
  general type of   138, 207
  isotropic distribution function
        with   136, 139, 152
  nearly-circular orbits
        with   137, 143, 181
  radial trajectories with   138, 147–148,
        193
  with a rotation   150
Stability. *See also* Instability
  of solitons   *54*
  of a system   10

Statistical simulation of stellar
        systems   199, 203
  See also Playing of distribution
        functions
Strongly elongated ellipsoids   263
Suppression of Jeans instability   20
Surface perturbations   121

Temperature-gradient instability   20, *11*
Thin prolate spheroid   313
Third integral   8
Time-independent perturbations   69
Toomre's
  stability criterion   395
  stability reserve   396
  wavelength   364
Trailing spirals   370, *206*
Transit particles. *See* Untrapped
        particles
Trapped particles   69, 78
Turbulence   *83*
Two-stream instability. *See* Beam
        instability

Uniformly rotating disk   346, 370
Unperturbed trajectory   9
Untrapped (transit) particles   *70, 78*

Variational method   377, 381
Vlasov equations   4, 6
  linearization approximation   9
  solution by method of characteristic.
        See Integration over trajectories

Waves of negative energy   *173*
Weak turbulence. *See* Turbulence
WKB-approximation   389, *300*